new mathematics: revision and pra

M. E. Wardle
Senior Lecturer (Mathematical Education), University of Warwick

C. J. Weeks
Lecturer, Rolle College of Education, Exmouth

Oxford University Press 1978

Oxford University Press
Walton Street, Oxford OX2 6DP

OXFORD LONDON GLASGOW
NEW YORK TORONTO MELBOURNE WELLINGTON
KUALA LUMPUR SINGAPORE JAKARTA HONG KONG TOKYO
DELHI BOMBAY CALCUTTA MADRAS KARACHI
IBADAN NAIROBI DAR ES SALAAM CAPE TOWN

ISBN 0 19 832617 3

First published 1978

Set by Hope Services, Abingdon, Oxon.
Printed and bound at William Clowes & Sons Limited
Beccles and London

Preface

The purpose of this book is to meet the needs of pupils preparing for a modern syllabus 'O' level GCE examination or for one of the CSE examinations with modern options. The scope of the book is such as to cover all topics likely to be met in any of these examinations. It is not expected that a pupil would need to work through all the material provided.

The book is subdivided into sections with easily recognizable headings, e.g. quadratic equations, number bases, etc. Any such division is arbitrary and is not intended to raise questions about the unity of the subject or to contravene the philosophy of approach to mathematics that is implied in modern courses for schools. However, the authors believe that practice of specific skills is needed in any course.

Each section begins with revision notes and/or worked examples. The practice exercises are for the most part confined to simple and direct applications of the mathematical ideas through which the pupil can develop accuracy and confidence.

May 1978

M.E.W.
C.J.W.

Contents

Part 1: Arithmetic and computation

1.1 Number bases

The number 1245 in base six can be converted to base ten by using the appropriate column headings as shown below.

$$1245_{six} = 1 \times (6^3) + 2 \times (6^2) + 4 \times (6^1) + 5 \times 1 = 317_{ten}$$

To convert 177 in base ten to base six you have to find the number of 36's, 6's, and 1's.

$$177_{ten} = 4 \times (36) + 5 \times (6) + 3 \times (1)$$
$$= 453_{six}$$

36 | 177
4 Rem 33

6 | 33
5 Rem 3

1 | 3
3

The base six addition 1245 + 132 is shown on the right.

Remember in base six the only numerals used are 0, 1, 2, 3, 4, and 5.

Column headings

	$6^3 = 216$	$6^2 = 36$	$6^1 = 6$	$6^0 = 1$
	1	2	4	5
+		1	3	2
	1	4	2	1
		1	1	

Exercise 1.1a

1. Write the following base six numbers in base ten:
   (a) 32 (b) 111 (c) 203 (d) 1001 (e) 2315
2. Write the following base four numbers in base ten:
   (a) 21 (b) 101 (c) 220 (d) 1103 (e) 3021
3. Write as a number in base ten:
   (a) 14_{six} (b) 231_{six} (c) 31_{four} (d) 321_{four}
4. Write as a number in base ten:
   (a) 23_{five} (b) 417_{eight} (c) 1111_{three} (d) 631_{nine}
5. Write the following base ten numbers in base six:
   (a) 25 (b) 43 (c) 72 (d) 135 (e) 461
6. Write the following base ten numbers in base four:
   (a) 14 (b) 25 (c) 65 (d) 118 (e) 258
7. Write the following base ten numbers in the base shown:
   (a) 41, base nine (b) 72, base five (c) 49, base seven
   (d) 123, base eight (e) 241, base three (f) 100, base two
8. Set out as shown in the example above, and hence find:
   (a) 123 + 42, in base six (b) 453 + 25, in base six
   (c) 2004 + 555, in base six (d) 1234 + 452, in base six
9. Find in base four:
   (a) 32 + 13 (b) 102 + 12 (c) 213 + 111 (d) 333 + 101
10. Find in base seven:
   (a) 25 + 34 (b) 110 + 63 (c) 354 + 263 (d) 222 + 666

The calculations for 1011 + 101, 1011 − 101 and 1011 × 101 in base two are set out below.

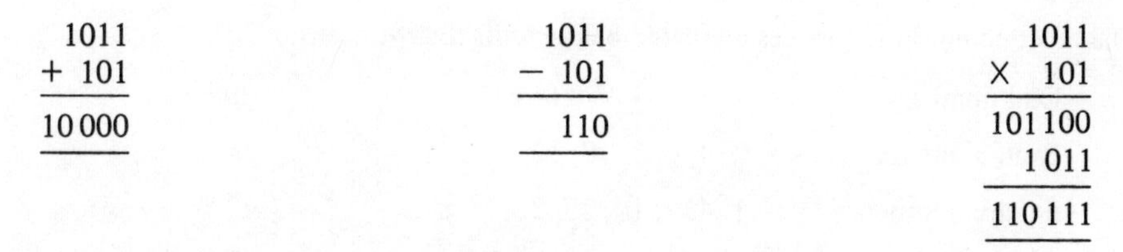

Remember the column headings in base two are . . . 32, 16, 8, 4, 2, 1.

Exercise 1.1b

1. Write the following base two numbers in base ten:
   (a) 101 (b) 1101 (c) 1001 (d) 10101 (e) 11011
2. Write the following base ten numbers in base two:
   (a) 6 (b) 13 (c) 25 (d) 33 (e) 75
3. Find in base two:
   (a) 101 + 11 (b) 1101 + 110 (c) 1111 + 101 (d) 10111 + 1011
4. Find in base two:
   (a) 101 − 11 (b) 1101 − 110 (c) 1111 − 101 (d) 10111 − 1011
5. Write each of the base two numbers in questions 3 and 4 in base ten and check your results.
6. Find in base two:
   (a) 1101 × 101 (b) 1110 × 11 (c) 1010 × 111 (d) 10101 × 110
7. Find in base six:
   (a) 235 − 44 (b) 431 − 52 (c) 505 − 341 (d) 1234 − 552
8. Find in base four:
   (a) 123 − 31 (b) 321 − 13 (c) 111 − 22 (d) 1201 − 312
9. Find in the base shown:
   (a) 246 − 57, in base nine (b) 413 − 24, in base seven
   (c) 1011 − 234, in base five (d) 6712 − 431, in base eight
10. Find in the base shown:
    (a) 23 × 2, in base four (b) 53 × 4, in base six
    (c) 123 × 7, in base eight (d) 214 × 12, in base five
11. Write down the next two terms of each sequence:
    (a) 1, 11, 101, 111, 1001, 1011, . . . , in base two.
    (b) 11, 22, 110, 121, 202, 220, . . . , in base three.
    (c) 1, 4, 14, 31, 100, 121, . . . , in base five.
    (d) 10 000, 2000, 1000, 200, 100, 20, . . . , in base four.
    (e) 200, 142, 124, 110, 52, 34, . . . , in base six.
12. Find:
    (a) 123 + 234 + 345, in base six, and write your answer in base four.
    (b) $243_{\text{five}} + 135_{\text{seven}}$, and write your answer in base nine.
    (c) $1421_{\text{six}} + 222_{\text{nine}} - 431_{\text{five}}$, and write your answer in base three.

1.2 Number patterns and sequences

The most common sequences are listed below with their nth term

Odd numbers	1, 3, 5, 7, 9, 11, ...	$(2n-1)$
Even numbers	2, 4, 6, 8, 10, 12, ...	$2n$
Square numbers	1, 4, 9, 16, 25, 36, ...	n^2
Cube numbers	1, 8, 27, 64, 125, 216, ...	n^3
Triangle numbers	1, 3, 6, 10, 15, 21, ...	$\frac{1}{2}n(n+1)$
Multiples of three	3, 6, 9, 12, 15, 18, ...	$3n$
Powers of two	1, 2, 4, 8, 16, 32, ...	2^{n-1}
Powers of three	1, 3, 9, 27, 81, 243, ...	3^{n-1}
Fibonacci numbers	1, 1, 2, 3, 5, 8, 13, 21, ...	–
Prime numbers	2, 3, 5, 7, 11, 13, 17, 19, ...	–

Many other sequences are formed by adding the same number to produce successive terms, e.g. 2, 5, 8, 11, 14, 17, 20, 23, ... $(3n-1)$.
These are called *arithmetic sequences* (progressions).

Exercise 1.2a

1. Write down the difference between each pair of terms in the sequence and describe the new sequence formed in this way.
(a) square numbers (b) triangle numbers (c) powers of two
(d) powers of three (e) Fibonacci numbers
2. By comparing each sequence with the first given in each group write down the nth term of the sequence:
(a) 4, 8, 12, 16, 20, 24, 28, ...
5, 9, 13, 17, 21, 25, 29, ...
7, 11, 15, 19, 23, 27, 31, ...
(b) 1, 2, 4, 8, 16, 32, ...
0, 1, 3, 7, 15, 31, ...
3, 4, 6, 10, 18, 34, ...
(c) 1, 4, 9, 16, 25, 36, ...
3, 6, 11, 18, 27, 38, ...
6, 9, 14, 21, 30, 41, ...
3. Write down the next two terms of each sequence and by comparing it with the known sequences find the nth term.
(a) 4, 7, 10, 13, 16, 19, ... (b) 2, 5, 10, 17, 26, 37, ...
(c) 2, 4, 7, 11, 16, 22, ... (d) 2, 6, 18, 54, 162, 486, ...
4. Write down the nth and $(n+1)$th square number. Show that their difference is always an odd number.
5. Write down the nth and $(n+1)$th triangle number. Show that their sum is always a square number.
6. Show, by taking any three consecutive numbers in the Fibonacci sequence, that the product of the outer pair is always within one of the square of the middle number.

Pascal's Triangle

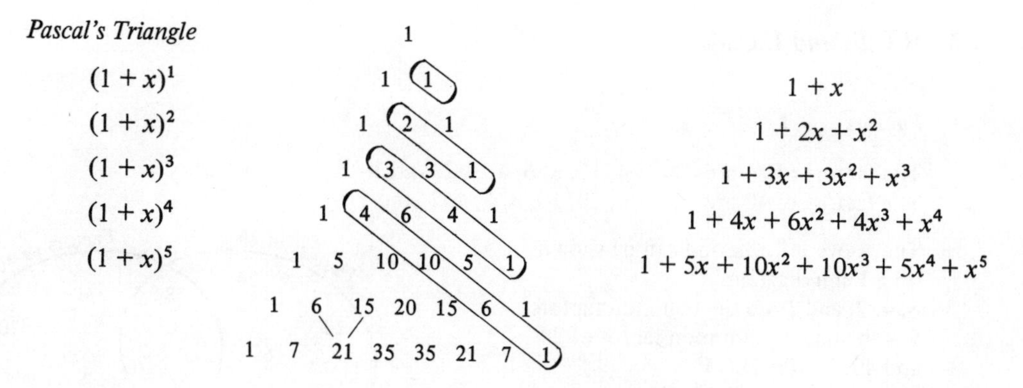

Note: each entry is obtained by adding the two adjacent entries in the previous row, e.g. $21 = 6 + 15$

Exercise 1.2b

1. Find the sum of each horizontal row in Pascal's Triangle. Is the result the same as each value of $(1+x)^n$ when $x = 1$?
2. Find the sum of each diagonal in Pascal's Triangle. Four have been marked with loops in the diagram at the top of the page. What is the sequence of numbers formed in this way?
3. Use Pascal's Triangle to write down the terms of $(1+x)^8$.

Arithmetic progressions $a + (a+d) + (a+2d) + (a+3d) + \dots + (a+(n-1)d)$

e.g. $2 + 5 + 8 + 11 + \dots$

$a = 2, d = 3$ nth term $2 + 3(n-1) = 3n - 1$

The sum of n terms is $\frac{1}{2}n(2a + (n-1)d)$

Geometric progressions $a + ar + ar^2 + ar^3 + \dots + ar^{n-1}$

e.g. $2 + 6 + 18 + 54 + \dots$

$a = 2, r = 3$ nth term 2.3^{n-1}

The sum of n terms is $a\dfrac{(r^n - 1)}{(r-1)}$ or $a\dfrac{(1-r^n)}{(1-r)}$

Exercise 1.2c

1. Say whether the sequence is an arithmetic or geometric progression.
   (a) 3, 7, 11, 15, 19, 23, 27, ...
   (b) 4, 12, 36, 108, 324, 972, ...
   (c) 1, 0·1, 0·01, 0·001, 0·0001, 0·000 01, ...
   (d) 71, 65, 59, 53, 47, 41, 35, ...
2. Write down the nth term of each sequence in question 1.
3. Use the corresponding formula above to find the sum of 20 terms of
   (a) $3 + 7 + 11 + 15 + 19 \dots$
   (b) $11 + 13 + 15 + 17 + 19 + \dots$
   (c) $4 + 12 + 36 + 108 + 324 \dots$
   (d) $4 + 2 + 1 + 0{\cdot}5 + 0{\cdot}25 + 0{\cdot}125 \dots$
4. Write down the first six terms of an arithmetic progression where $a = 7$ and $d = 11$.
5. Write down the first six terms of a geometric progression where $a = 36$ and $r = \frac{1}{3}$.
6. Find the sum of the first ten terms in each of questions 4 and 5.

1.3 H.C.F. and L.C.M.

Factors

The factors of 24 are 24, 12, 8, 6, 4, 3, 2, and 1
The factors of 40 are 40, 20, 10, 8, 5, 4, 2, and 1

The above information can be shown on a Venn diagram.
8, 4, 2, and 1 are the common factors.
8 is the highest common factor of 24 and 40, i.e. the H.C.F.

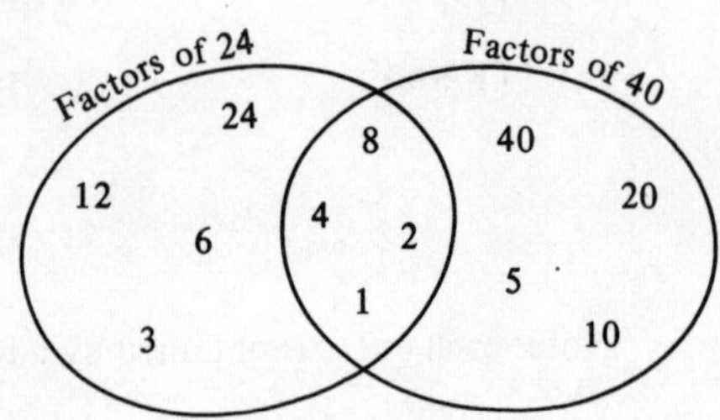

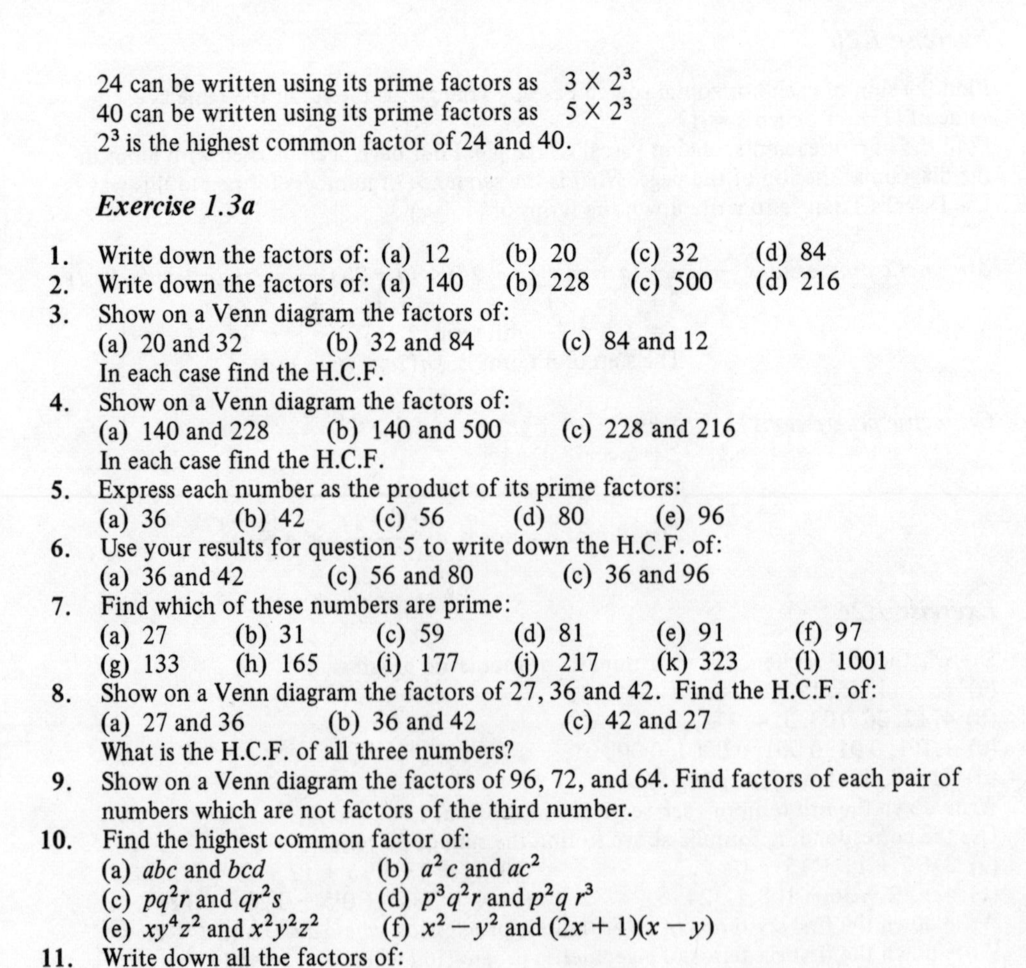

24 can be written using its prime factors as 3×2^3
40 can be written using its prime factors as 5×2^3
2^3 is the highest common factor of 24 and 40.

Exercise 1.3a

1. Write down the factors of: (a) 12 (b) 20 (c) 32 (d) 84
2. Write down the factors of: (a) 140 (b) 228 (c) 500 (d) 216
3. Show on a Venn diagram the factors of:
   (a) 20 and 32 (b) 32 and 84 (c) 84 and 12
   In each case find the H.C.F.
4. Show on a Venn diagram the factors of:
   (a) 140 and 228 (b) 140 and 500 (c) 228 and 216
   In each case find the H.C.F.
5. Express each number as the product of its prime factors:
   (a) 36 (b) 42 (c) 56 (d) 80 (e) 96
6. Use your results for question 5 to write down the H.C.F. of:
   (a) 36 and 42 (c) 56 and 80 (c) 36 and 96
7. Find which of these numbers are prime:
   (a) 27 (b) 31 (c) 59 (d) 81 (e) 91 (f) 97
   (g) 133 (h) 165 (i) 177 (j) 217 (k) 323 (l) 1001
8. Show on a Venn diagram the factors of 27, 36 and 42. Find the H.C.F. of:
   (a) 27 and 36 (b) 36 and 42 (c) 42 and 27
   What is the H.C.F. of all three numbers?
9. Show on a Venn diagram the factors of 96, 72, and 64. Find factors of each pair of numbers which are not factors of the third number.
10. Find the highest common factor of:
    (a) abc and bcd (b) a^2c and ac^2
    (c) pq^2r and qr^2s (d) p^3q^2r and p^2qr^3
    (e) xy^4z^2 and $x^2y^2z^2$ (f) $x^2 - y^2$ and $(2x + 1)(x - y)$
11. Write down all the factors of:
    (a) pqr (b) x^2y (c) u^3v^2 (d) pq^2r^3

Multiples

The multiples of 3 are 3, 6, 9, 12, 15, 18, 21, 24, 27, 30, 33, 36, 39, . . .
The multiples of 4 are 4, 8, 12, 16, 20, 24, 28, 32, 36, 40, . . .

The above information can be shown on a Venn diagram: 12, 24, and 36 are the common multiples.
12 is the lowest common multiple of 3 and 4, i.e. the L.C.M.

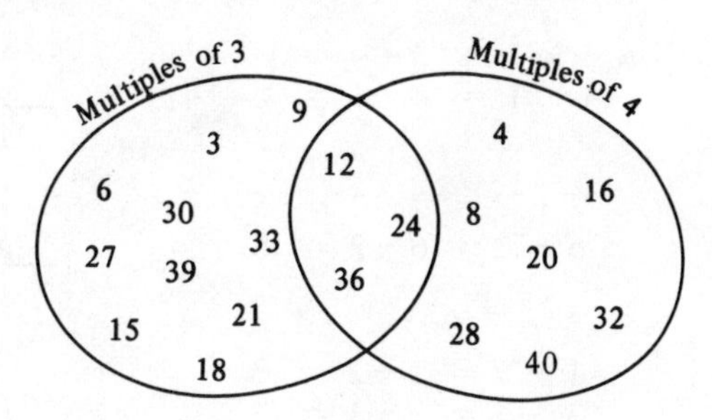

42 can be written using its prime factors as $2 \times 3 \times 7$
60 can be written using its prime factors as $2^2 \times 3 \times 5$
$2^2 \times 3 \times 5 \times 7$ is the lowest common multiple of 42 and 60.

Exercise 1.3b

1. Write down the first six multiples of: (a) 7 (b) 15 (c) 6
2. Write down the first six multiples of: (a) 24 (b) 16 (c) 30
3. Show on a Venn diagram the first six multiples of:
   (a) 7 and 6 (b) 6 and 15 (c) 12 and 8
   In each case find the L.C.M.
4. Show on a Venn diagram the first six multiples of:
   (a) 24 and 16 (b) 16 and 30 (c) 24 and 30
   In each case find the L.C.M.
5. Express each number as the product of its prime factors:
   (a) 24 (b) 30 (c) 45 (d) 64 (e) 84 (f) 96
6. Use your results for question 5 to write down the L.C.M. of:
   (a) 24 and 30 (b) 30 and 45 (c) 45 and 96
   (d) 24 and 84 (e) 64 and 96 (f) 45 and 64
7. Find which of these numbers are multiples of 24:
   (a) 48 (b) 60 (c) 96 (d) 144 (e) 264 (f) 456
8. Express each number as the product of its prime factors and find whether the larger is a multiple of the smaller.
   (a) 4928 and 616 (b) 840 and 36 (c) 2730 and 210
9. Show on a Venn diagram the first ten multiples of 24, 27, and 36. Find the L.C.M. of:
   (a) 24 and 27 (b) 27 and 36 (c) 36 and 24
10. Three grandchildren call on their grandmother every 12 days, 16 days and 18 days respectively. Find how often:
    (a) two call on the same day (b) all three call on the same day
11. Find the lowest common multiple of:
    (a) pqr and qrs (b) p^2q and qr
    (c) x^2yz and xy^2z (d) uv^2w^3 and u^2vw^2
    (e) $24ab$ and $36bc$ (f) $x^2 - y^2$ and $(2x + 1)(x - y)$

1.4 Negative numbers

Addition and subtraction

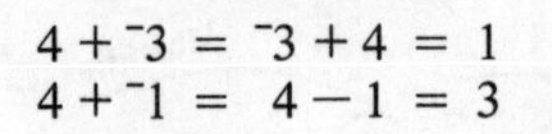

$4 + {}^{-}3 = {}^{-}3 + 4 = 1$
$4 + {}^{-}1 = 4 - 1 = 3$

$^{-}2 + {}^{-}3 = {}^{-}2 - 3 = {}^{-}5$
$^{-}6 + {}^{-}2 = {}^{-}6 - 2 = {}^{-}8$

$1 - {}^{-}2 = 1 + 2 = 3$
$^{-}3 - {}^{-}1 = {}^{-}3 + 1 = {}^{-}2$

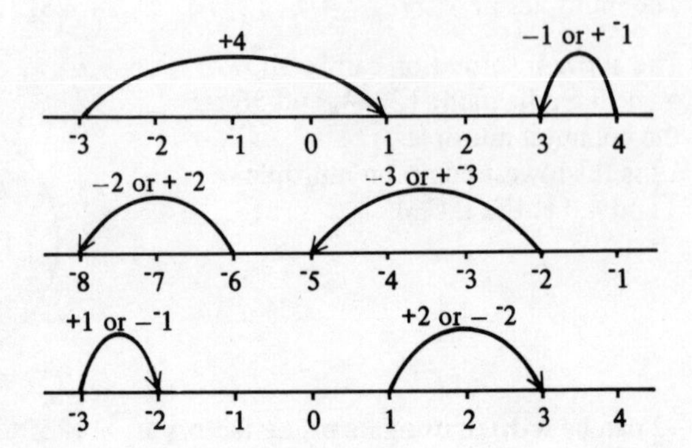

Remember: adding a negative number is the same as subtracting the corresponding positive number. i.e. $+({}^{-}n) = -n$

subtracting a negative number is the same as adding the corresponding positive number. i.e. $-({}^{-}m) = +m$

Exercise 1.4a

1. Find:
(a) $4 + 7$ (b) $4 - 7$ (c) $4 + {}^{-}7$ (d) $^{-}4 + 7$ (e) $^{-}4 + {}^{-}7$
2. Find:
(a) $12 - 5$ (b) $12 + {}^{-}5$ (c) $^{-}12 + 5$ (d) $5 - 12$ (e) $^{-}5 + {}^{-}12$
3. Find:
(a) $6 - 2$ (b) $^{-}6 - 2$ (c) $^{-}2 + 6$ (d) $^{-}2 - 6$ (e) $2 - 6$
4. Find:
(a) $13 - 7$ (b) $^{-}13 + 7$ (c) $7 - 13$ (d) $^{-}7 + 13$ (e) $^{-}13 - 7$
5. Find:
(a) $3 - 4$ (b) $4 - {}^{-}3$ (c) $3 - {}^{-}4$ (d) $^{-}3 - {}^{-}4$ (e) $^{-}4 - {}^{-}3$
6. Find:
(a) $15 - 9$ (b) $^{-}9 - 15$ (c) $15 - {}^{-}9$ (d) $9 - {}^{-}15$ (e) $^{-}15 - {}^{-}9$
7. Find:
(a) $4 + 9 + {}^{-}3$ (b) $5 + {}^{-}4 + {}^{-}7$ (c) $^{-}11 + 3 + {}^{-}5$ (d) $^{-}2 + {}^{-}3 + {}^{-}4$
8. Find:
(a) $14 + {}^{-}5 - 6$ (b) $^{-}7 - 4 + 13$ (c) $^{-}12 - 2 + {}^{-}6$ (d) $5 - 8 + {}^{-}12$
9. Find:
(a) $5 - {}^{-}3 + 2$ (b) $^{-}4 + 3 - {}^{-}5$ (c) $^{-}7 + {}^{-}2 - {}^{-}9$ (d) $6 - {}^{-}4 - {}^{-}5$
10. Find the sum of the first five:
(a) negative numbers (b) multiples of three (c) negative odd numbers
11. Find the sum of the numbers between:
(a) $^{-}3$ and 5 inclusive (b) $^{-}4$ and 1 inclusive (c) $^{-}9$ and $^{-}3$ inclusive
12. What number in the box will make the statement true?
(a) $5 + \square = 8$ (b) $3 + \square = 1$ (c) $5 - \square = {}^{-}2$ (d) $6 - \square = {}^{-}1$
(e) $^{-}2 + \square = 5$ (f) $^{-}4 + \square = {}^{-}7$ (g) $^{-}3 - \square = 2$ (h) $^{-}5 - \square = {}^{-}3$
13. At 6 a.m. the temperature was $^{-}5$°C. By 10 a.m. it had risen to 7°C. Find the difference in temperatures, and the average increase per hour.
14. Find the distance between the points:
(a) $(2, {}^{-}3)$ and $(2, 4)$ (b) $(5, {}^{-}2)$ and $({}^{-}1, {}^{-}2)$ (c) $(6, {}^{-}4)$ and $(6, {}^{-}8)$

Multiplication and division

The results for multiplying and dividing negative numbers can be obtained from patterns as below.

$3 \times 2 = 6$	$^{-}3 \times 3 = {}^{-}9$	$12 \div 3 = 4$	$^{-}16 \div 4 = {}^{-}4$
$3 \times 1 = 3$	$^{-}3 \times 2 = {}^{-}6$	$12 \div 2 = 6$	$^{-}16 \div 2 = {}^{-}8$
$3 \times 0 = 0$	$^{-}3 \times 1 = {}^{-}3$	$12 \div 1 = 12$	$^{-}16 \div 1 = {}^{-}16$
$3 \times {}^{-}1 = {}^{-}3$	$^{-}3 \times 0 = 0$	$12 \div {}^{-}1 = {}^{-}12$	$^{-}16 \div {}^{-}1 = 16$
$3 \times {}^{-}2 = {}^{-}6$	$^{-}3 \times {}^{-}1 = 3$	$12 \div {}^{-}2 = {}^{-}6$	$^{-}16 \div {}^{-}2 = 8$
$3 \times {}^{-}3 = {}^{-}9$	$^{-}3 \times {}^{-}2 = 6$	$12 \div {}^{-}3 = {}^{-}4$	$^{-}16 \div {}^{-}4 = 4$

Remember: When a positive number and a negative number are multiplied, the result will be a negative number i.e. $n \times ({}^{-}m) = {}^{-}nm$.

When two negative numbers are multiplied or divided, the result will be a positive number i.e. $({}^{-}p) \times ({}^{-}q) = pq$.

Example: $5 - 2(3-6) + {}^{-}4 - (7-11)$
$= 5 - 6 + 12 - 4 - 7 + 11 = 11$

Exercise 1.4b

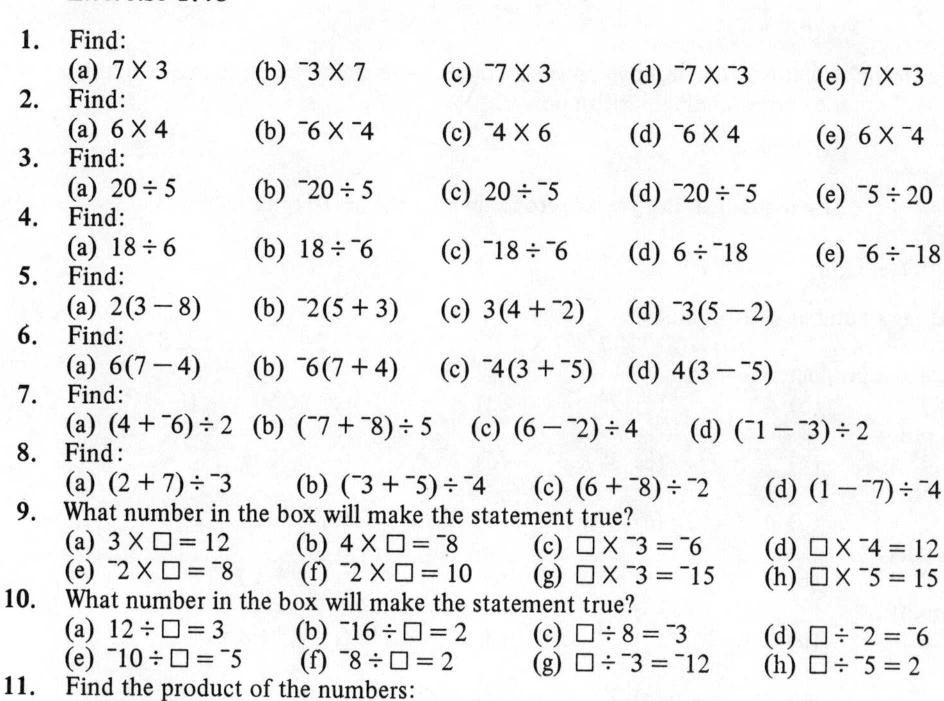

1. Find:
(a) 7×3 (b) $^{-}3 \times 7$ (c) $^{-}7 \times 3$ (d) $^{-}7 \times {}^{-}3$ (e) $7 \times {}^{-}3$
2. Find:
(a) 6×4 (b) $^{-}6 \times {}^{-}4$ (c) $^{-}4 \times 6$ (d) $^{-}6 \times 4$ (e) $6 \times {}^{-}4$
3. Find:
(a) $20 \div 5$ (b) $^{-}20 \div 5$ (c) $20 \div {}^{-}5$ (d) $^{-}20 \div {}^{-}5$ (e) $^{-}5 \div 20$
4. Find:
(a) $18 \div 6$ (b) $18 \div {}^{-}6$ (c) $^{-}18 \div {}^{-}6$ (d) $6 \div {}^{-}18$ (e) $^{-}6 \div {}^{-}18$
5. Find:
(a) $2(3-8)$ (b) $^{-}2(5+3)$ (c) $3(4+{}^{-}2)$ (d) $^{-}3(5-2)$
6. Find:
(a) $6(7-4)$ (b) $^{-}6(7+4)$ (c) $^{-}4(3+{}^{-}5)$ (d) $4(3-{}^{-}5)$
7. Find:
(a) $(4+{}^{-}6) \div 2$ (b) $({}^{-}7+{}^{-}8) \div 5$ (c) $(6-{}^{-}2) \div 4$ (d) $({}^{-}1-{}^{-}3) \div 2$
8. Find:
(a) $(2+7) \div {}^{-}3$ (b) $({}^{-}3+{}^{-}5) \div {}^{-}4$ (c) $(6+{}^{-}8) \div {}^{-}2$ (d) $(1-{}^{-}7) \div {}^{-}4$
9. What number in the box will make the statement true?
(a) $3 \times \square = 12$ (b) $4 \times \square = {}^{-}8$ (c) $\square \times {}^{-}3 = {}^{-}6$ (d) $\square \times {}^{-}4 = 12$
(e) $^{-}2 \times \square = {}^{-}8$ (f) $^{-}2 \times \square = 10$ (g) $\square \times {}^{-}3 = {}^{-}15$ (h) $\square \times {}^{-}5 = 15$
10. What number in the box will make the statement true?
(a) $12 \div \square = 3$ (b) $^{-}16 \div \square = 2$ (c) $\square \div 8 = {}^{-}3$ (d) $\square \div {}^{-}2 = {}^{-}6$
(e) $^{-}10 \div \square = {}^{-}5$ (f) $^{-}8 \div \square = 2$ (g) $\square \div {}^{-}3 = {}^{-}12$ (h) $\square \div {}^{-}5 = 2$
11. Find the product of the numbers:
(a) $6, {}^{-}2, {}^{-}3$ (b) $5, {}^{-}4, 2$ (c) $^{-}6, {}^{-}5, {}^{-}4$ (d) $^{-}4, {}^{-}11, 2, {}^{-}3$
12. Evaluate $2p - 3q - r^2$ where:
(a) $p = {}^{-}2, q = 4$, and $r = 5$ (b) $p = {}^{-}4, q = {}^{-}3$ and $r = 4$
(c) $p = 5, q = {}^{-}7$, and $r = {}^{-}3$ (d) $p = {}^{-}1, q = 6$ and $r = {}^{-}2$

1.5 Powers and indices

$3 \times 3 \times 3 \times 3 \times 3 = 3^5$ 3^5 is read as 'three to the power five'; the ^{5} is called the index (plural indices).

$3^5 \times 3^2 = (3 \times 3 \times 3 \times 3 \times 3) \times (3 \times 3) = 3^7$

$3^5 \div 3^2 = (3 \times 3 \times 3 \times 3 \times 3) \div (3 \times 3) = 3^3$

Remember: When multiplying two numbers with different indices, you must add the indices, i.e. $3^a \times 3^b = 3^{(a+b)}$

When dividing two numbers with different indices, you must subtract the indices, i.e. $3^a \div 3^b = 3^{(a-b)}$

$$3^2 \div 3^5 = \frac{3 \times 3}{3 \times 3 \times 3 \times 3 \times 3} = 3^{(2-5)} = 3^{-3} = \frac{1}{3 \times 3 \times 3} \text{ or } \frac{1}{3^3}$$

$$2^3 \div 2^4 = \frac{2 \times 2 \times 2}{2 \times 2 \times 2 \times 2} = 2^{(3-4)} = 2^{-1} = \frac{1}{2}$$

$$4^3 \div 4^3 = \frac{4 \times 4 \times 4}{4 \times 4 \times 4} = 4^{(3-3)} = 4^0 = 1$$

Remember: A number to a negative power is equal to the reciprocal of the same number to the corresponding positive power.

i.e. $3^{-a} = \frac{1}{3^a}$

Any number to the power zero is always one, i.e. $n^0 = 1$.

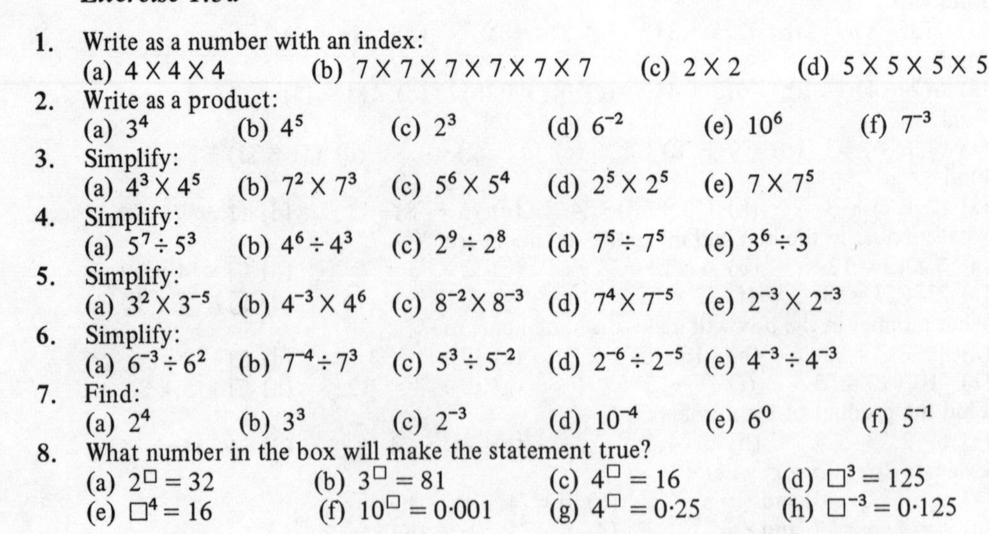

Exercise 1.5a

1. Write as a number with an index:
(a) $4 \times 4 \times 4$ (b) $7 \times 7 \times 7 \times 7 \times 7 \times 7$ (c) 2×2 (d) $5 \times 5 \times 5 \times 5$
2. Write as a product:
(a) 3^4 (b) 4^5 (c) 2^3 (d) 6^{-2} (e) 10^6 (f) 7^{-3}
3. Simplify:
(a) $4^3 \times 4^5$ (b) $7^2 \times 7^3$ (c) $5^6 \times 5^4$ (d) $2^5 \times 2^5$ (e) 7×7^5
4. Simplify:
(a) $5^7 \div 5^3$ (b) $4^6 \div 4^3$ (c) $2^9 \div 2^8$ (d) $7^5 \div 7^5$ (e) $3^6 \div 3$
5. Simplify:
(a) $3^2 \times 3^{-5}$ (b) $4^{-3} \times 4^6$ (c) $8^{-2} \times 8^{-3}$ (d) $7^4 \times 7^{-5}$ (e) $2^{-3} \times 2^{-3}$
6. Simplify:
(a) $6^{-3} \div 6^2$ (b) $7^{-4} \div 7^3$ (c) $5^3 \div 5^{-2}$ (d) $2^{-6} \div 2^{-5}$ (e) $4^{-3} \div 4^{-3}$
7. Find:
(a) 2^4 (b) 3^3 (c) 2^{-3} (d) 10^{-4} (e) 6^0 (f) 5^{-1}
8. What number in the box will make the statement true?
(a) $2^{\square} = 32$ (b) $3^{\square} = 81$ (c) $4^{\square} = 16$ (d) $\square^3 = 125$
(e) $\square^4 = 16$ (f) $10^{\square} = 0{\cdot}001$ (g) $4^{\square} = 0{\cdot}25$ (h) $\square^{-3} = 0{\cdot}125$

Laws of indices $a^m \times a^n = a^{(m+n)}, \quad a^m \div a^n = a^{(m-n)}$

$$(a^m)^n = a^{mn}, \quad a^{-n} = \frac{1}{a^n}, \quad a^0 = 1$$

Fractional indices $a^{\frac{1}{2}} \times a^{\frac{1}{2}} = a^1, \quad a^{\frac{1}{2}} = \sqrt{a}$

$$b^{\frac{1}{3}} \times b^{\frac{1}{3}} \times b^{\frac{1}{3}} = b^1, \quad b^{\frac{1}{3}} = \sqrt[3]{b}$$

Example $\left(\frac{27}{125}\right)^{-\frac{1}{3}} = \left(\frac{125}{27}\right)^{\frac{1}{3}} = \frac{(125)^{\frac{1}{3}}}{(27)^{\frac{1}{3}}} = \frac{5}{3}$

Exercise 1.5b

1. Simplify:

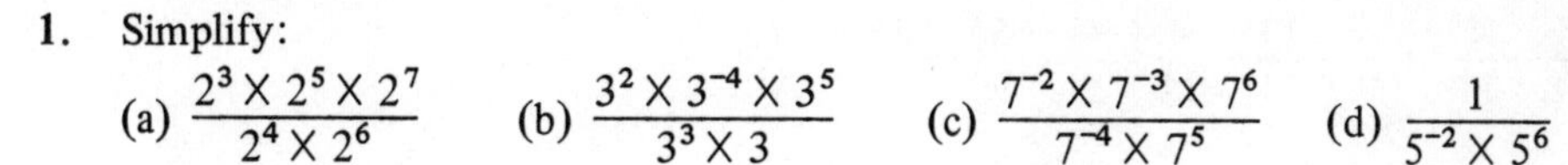

(a) $\frac{2^3 \times 2^5 \times 2^7}{2^4 \times 2^6}$ (b) $\frac{3^2 \times 3^{-4} \times 3^5}{3^3 \times 3}$ (c) $\frac{7^{-2} \times 7^{-3} \times 7^6}{7^{-4} \times 7^5}$ (d) $\frac{1}{5^{-2} \times 5^6}$

2. Simplify:
(a) $4^{\frac{1}{3}} \times 4^{\frac{1}{3}}$ (b) $2^{\frac{1}{2}} \times 2^2 \times 2^{\frac{3}{2}}$ (c) $5^{\frac{1}{4}} \times 5^{\frac{3}{4}} \times 5^{\frac{1}{2}}$ (d) $(6^{\frac{1}{3}} \times 6^{\frac{4}{3}}) \div 6^{\frac{2}{3}}$

3. Find:
(a) $(2^3)^2$ (b) $(3^2)^3$ (c) $4^{\frac{1}{2}}$ (d) $27^{\frac{1}{3}}$ (e) $(8^{\frac{1}{3}})^2$ (f) $(10^2)^{-1}$

4. Simplify:
(a) $\frac{3^2 \times 2^4 \times 3^5}{2^3 \times 3^4}$ (b) $\frac{5^2 \times 7^3 \times 7^4}{7^2 \times 5^4}$ (c) $\frac{6^{-2} \times 5^3 \times 6^4}{5^{-1} \times 6^3}$ (d) $\frac{2^6 \times 4^5 \times 3^2}{4^2 \times 3^5}$

5. Simplify:
(a) $p^2 \times p^3 \times p^7$ (b) $x^7 \times x^6 \times x^{-5}$ (c) $w^{-2} \times w^6 \times w^{-7}$ (d) $l^{\frac{1}{2}} \times l^{\frac{1}{2}} \times l^{\frac{1}{2}}$

6. Simplify:
(a) $(y^3)^4$ (b) $(x^2)^p$ (c) $(w^n)^n$ (d) $(z^8)^{\frac{1}{2}}$ (e) $(l^{-4})^{\frac{1}{4}}$ (f) $(p^{2n})^{\frac{1}{n}}$

7. What number in the box will make the statement true?
(a) $3^5 \times 3^{\square} = 3^9$ (b) $4^{\square} \times 4^2 = 4^{-5}$ (c) $7^{\square} \times 7^{-3} = 7^{-6}$

8. What number in the box will make the statement true?
(a) $(3^2)^{\square} = 3^{12}$ (b) $(5^3)^{\square} = 5^{18}$ (c) $(4^{\square})^3 = 4^9$ (d) $(7^{\square})^{\frac{1}{2}} = 7^2$

9. Simplify:
(a) $(3 \times 10^2) \times (2 \times 10^3)$ (b) $(12 \times 10^5) \div (2 \times 10^3)$ (c) $(8 \times 10^{-5}) \div (4 \times 10^2)$

10. Simplify and then write your answer as a number between 1 and 10 multiplied by a power of 10, (i.e. in standard form).
(a) $(4 \times 10^3) \times (3 \times 10^2)$ (b) $(25 \times 10^6) \times (2 \times 10^2)$ (c) $(128 \times 10^5) \div (32 \times 10^7)$

11. Find:
(a) $(\frac{2}{3})^2$ (b) $(\frac{3}{4})^3$ (c) $(\frac{2}{5})^{-1}$ (d) $(\frac{1}{7})^{-2}$ (e) $(1\frac{1}{2})^3$ (f) $(\frac{3}{2})^{-3}$

12. Find:
(a) $\sqrt{16}$ (b) $\sqrt{\frac{4}{9}}$ (c) $(25)^{\frac{1}{2}}$ (d) $(\frac{16}{25})^{\frac{1}{2}}$ (e) $\sqrt[3]{27}$ (f) $(\frac{8}{27})^{\frac{1}{3}}$

13. Find:
(a) $(16)^{-\frac{1}{2}}$ (b) $(\frac{49}{4})^{-\frac{1}{2}}$ (c) $(125)^{-\frac{1}{3}}$ (d) $(\frac{64}{125})^{-\frac{1}{3}}$ (e) $(100)^{\frac{3}{2}}$ (f) $(125)^{\frac{2}{3}}$

14. Find:
(a) $2^{\frac{1}{3}} \times 2^{\frac{1}{2}} \times 2^{\frac{1}{6}}$ (b) $(16)^{\frac{3}{4}} \times (16)^{-\frac{1}{2}}$ (c) $5^{\frac{3}{4}} \times 5^{-\frac{1}{4}} \times 5^{-\frac{1}{2}}$ (d) $(27)^{\frac{2}{3}} \div (27)^{-\frac{2}{3}}$

1.6 Fractions

Addition and subtraction

Fractions may be added if their denominators are the same: $\frac{11}{15}+\frac{2}{15}=\frac{13}{15}$

In each case below the fractions are replaced by equivalent fractions. Since they then have the same denominators, they may be combined.

$$\frac{4}{5}+\frac{2}{15}=\frac{4\times 3}{5\times 3}+\frac{2}{15}=\frac{12}{15}+\frac{2}{15}=\frac{14}{15}$$

$$\frac{2}{3}+\frac{5}{7}=\frac{2\times 7}{3\times 7}+\frac{5\times 3}{7\times 3}=\frac{14}{21}+\frac{15}{21}=\frac{29}{21}$$

$$\frac{5}{9}-\frac{4}{11}=\frac{5\times 11}{9\times 11}-\frac{4\times 9}{11\times 9}=\frac{55}{99}-\frac{36}{99}=\frac{19}{99}$$

Exercise 1.6a

1. Find:
(a) $\frac{1}{5}+\frac{3}{5}$ (b) $\frac{5}{7}-\frac{2}{7}$ (c) $\frac{5}{11}+\frac{9}{11}$ (d) $\frac{37}{53}-\frac{16}{53}$ (e) $\frac{17}{30}+\frac{7}{30}$
2. Find:
(a) $\frac{1}{15}+\frac{3}{5}$ (b) $\frac{5}{7}-\frac{2}{21}$ (c) $\frac{5}{33}+\frac{9}{11}$ (d) $\frac{37}{106}-\frac{16}{53}$ (e) $\frac{17}{30}+\frac{7}{90}$
3. Find:
(a) $\frac{1}{5}+\frac{2}{3}$ (b) $\frac{5}{7}-\frac{3}{4}$ (c) $\frac{5}{11}+\frac{3}{7}$ (d) $\frac{37}{53}-\frac{1}{2}$ (e) $\frac{5}{24}+\frac{3}{16}$

Multiplication and Division

$6\times\frac{1}{2}=3$	$\frac{3}{7}\times 2=\frac{6}{7}$	$6\div\frac{1}{2}=12$	$\frac{3}{7}\div 2=\frac{3}{14}$
$3\times\frac{1}{2}=\frac{3}{2}$	$\frac{3}{7}\times 1=\frac{3}{7}$	$3\div\frac{1}{2}=6$	$\frac{3}{7}\div 1=\frac{3}{7}$
$\frac{1}{3}\times\frac{1}{2}=\frac{1}{6}$	$\frac{3}{7}\times\frac{1}{2}=\frac{3}{14}$	$\frac{1}{3}\div\frac{1}{2}=\frac{2}{3}$	$\frac{3}{7}\div\frac{1}{2}=\frac{6}{7}$

In general $\frac{3}{7}\times\frac{2}{5}=\frac{3\times 2}{7\times 5}=\frac{6}{35}$ $\quad\frac{3}{7}\div\frac{2}{5}=\frac{3}{7}\times\frac{5}{2}=\frac{15}{14}$

i.e. $\frac{a}{b}\times\frac{c}{d}=\frac{ac}{bd}$ $\quad$ i.e. $\frac{a}{b}\div\frac{c}{d}=\frac{a}{b}\times\frac{d}{c}=\frac{ad}{bc}$

Exercise 1.6b

1. Find:
(a) $8\times\frac{1}{2}$ (b) $15\times\frac{1}{3}$ (c) $8\times\frac{3}{2}$ (d) $15\times\frac{4}{3}$ (e) $11\times\frac{3}{4}$
2. Find:
(a) $\frac{3}{5}\times\frac{1}{2}$ (b) $\frac{4}{7}\times\frac{1}{3}$ (c) $\frac{3}{10}\times\frac{3}{4}$ (d) $\frac{7}{8}\times\frac{4}{5}$ (e) $\frac{6}{35}\times\frac{15}{16}$
3. Find:
(a) $8\div\frac{1}{2}$ (b) $15\div\frac{1}{3}$ (c) $8\div\frac{3}{2}$ (d) $15\div\frac{4}{3}$ (e) $16\div\frac{4}{9}$
4. Find:
(a) $\frac{3}{5}\div\frac{1}{2}$ (b) $\frac{4}{7}\div\frac{1}{3}$ (c) $\frac{3}{10}\div\frac{7}{11}$ (d) $\frac{7}{8}\div\frac{4}{5}$ (e) $\frac{18}{35}\div\frac{16}{25}$

Decimal equivalents

$\frac{1}{2} = 0{\cdot}5$ $\frac{1}{4} = 0{\cdot}25$ $\frac{1}{8} = 0{\cdot}125$ $\frac{1}{5} = 0{\cdot}2$ $\frac{1}{10} = 0{\cdot}1$

$\frac{1}{3} = 0{\cdot}3$ $\frac{1}{6} = 0{\cdot}16$ $\frac{1}{9} = 0{\cdot}1$ $\frac{1}{7} = 0{\cdot}142857$

To find the decimal equivalent of $\frac{3}{7}$

either use $\frac{1}{7}$ and multiply by 3 i.e. $0{\cdot}142857 \times 3 = 0{\cdot}428571\ldots$
or divide 3 by 7 i.e. $7\,\underline{)3{\cdot}000000}$
$0{\cdot}428571\ldots$

Exercise 1.6c

1. Write down the decimal equivalent of:
(a) $\frac{3}{4}$ (b) $\frac{3}{8}$ (c) $\frac{5}{8}$ (d) $\frac{3}{5}$ (e) $\frac{5}{9}$ (f) $\frac{5}{6}$
2. Find the decimal equivalent of:
(a) $\frac{5}{7}$ (b) $\frac{4}{11}$ (c) $\frac{5}{12}$ (d) $\frac{5}{24}$ (e) $\frac{3}{16}$ (f) $\frac{4}{25}$
3. Write as a fraction in its lowest form:
(a) 0·7 (b) 0·6 (c) 0·11 (d) 0·75 (e) 0·45 (f) 0·08

Exercise 1.6d

1. Find the distance between the points:
(a) $(2, \frac{4}{5}), (2, \frac{1}{5})$ (b) $(\frac{1}{3}, 3), (\frac{2}{9}, 3)$ (c) $(4, \frac{2}{5}), (4, \frac{3}{4})$
2. Find the vector equivalent to:
(a) $\begin{pmatrix} 2 \\ \frac{1}{2} \end{pmatrix} + \begin{pmatrix} 3 \\ \frac{1}{4} \end{pmatrix}$ (b) $\begin{pmatrix} \frac{2}{3} \\ 1 \end{pmatrix} - \begin{pmatrix} \frac{2}{5} \\ 3 \end{pmatrix}$ (c) $\begin{pmatrix} \frac{1}{4} \\ \frac{2}{7} \end{pmatrix} + \begin{pmatrix} \frac{5}{9} \\ \frac{1}{3} \end{pmatrix}$
3. Find the average (mean) of:
(a) $\frac{1}{2}, \frac{5}{2}$ (b) $\frac{2}{7}, \frac{3}{7}$ (c) $\frac{1}{2}, \frac{3}{4}$ (d) $\frac{1}{2}, \frac{1}{3}, \frac{1}{4}$ (e) $\frac{2}{7}, \frac{3}{4}, \frac{1}{28}$
4. Find:
(a) $\frac{1}{2}$ of $3\frac{1}{3}$ kg (b) $\frac{3}{4}$ of $2\frac{1}{2}$ km (c) $\frac{2}{5}$ of $1\frac{1}{4}$ litres
5. An enlargement of scale factor $\frac{3}{5}$ is followed by another of scale factor $\frac{4}{9}$. Find the scale factor of the single enlargement equivalent to these two.
6. A boy spends $\frac{1}{4}$ of his pocket money on sweets, $\frac{1}{3}$ on comics, and $\frac{2}{7}$ on stamps. What fraction of his pocket money is left?
7. A firm uses $\frac{11}{16}$ of its oil supply during the four winter months. $\frac{5}{7}$ of this is used for heating. What fraction of the year's supply is left for other purposes during the winter?
8. Find:
(a) $(\frac{5}{6} + \frac{3}{4}) \div \frac{2}{7}$ (b) $(2\frac{1}{2} - 1\frac{1}{3}) \times \frac{3}{4}$ (c) $(1\frac{3}{4} + 2\frac{7}{8}) \times 3\frac{1}{2}$

1.7 Decimals

Addition and subtraction

When setting out it is important to keep the decimal points in line.

$$\begin{array}{r} 13{\cdot}26 \\ 1{\cdot}59 \\ +\ 4{\cdot}23 \\ \hline 19{\cdot}08 \\ \hline \end{array} \qquad \begin{array}{r} 13{\cdot}26 \\ -\ 6{\cdot}59 \\ \hline 6{\cdot}67 \\ \hline \end{array}$$

Exercise 1.7a

1. Set out your working as above and find:
(a) 1·7 + 2·1 (b) 2·9 + 3·3 (c) 17·5 + 27·6 (d) 4·2 + 99·9
(e) 2·8 − 1·6 (f) 7·5 − 4·5 (g) 16·3 − 5·8 (h) 49·6 − 29·8
2. Set out your working as above and find:
(a) 2·36 + 3·23 (b) 5·97 + 1·03 (c) 8·53 + 7·68 (d) 23·16 + 7·09
(e) 9·75 − 4·63 (f) 2·66 − 1·04 (g) 7·82 − 1·67 (h) 29·43 − 9·67
3. Find:
(a) 2·1 + 0·21 + 3·07 (b) 6·13 + 7·5 + 0·68 (c) 7·04 + 8·13 − 4·69
(d) 0·612 + 0·735 + 0·416 (e) 17·064 − 8·192 (f) 0·007 + 0·011 + 0·123

Multiplication

2 × 6 = 12·0	0·6 × 0·1 = 0·06	To find 6·<u>12</u> × 3·<u>4</u>
0·2 × 6 = 1·2	0·6 × 0·01 = 0·006	first find
0·02 × 6 = 0·12	0·6 × 0·001 = 0·0006	612 × 34 = 20808
0·002 × 6 = 0·012	0·6 × 0·0001 = 0·00006	So result is 20·<u>808</u>

Remember that the number of decimal places in the result will be the sum of the number of decimal places in the two numbers being multiplied.

Exercise 1.7b

1. Find:
(a) 0·3 × 7 (b) 1·2 × 4 (c) 2·7 × 5 (d) 6·3 × 8 (e) 12·4 × 9
2. Find:
(a) 0·21 × 3 (b) 1·34 × 4 (c) 2·76 × 8 (d) 6·93 × 5 (e) 17·26 × 9
3. Find:
(a) 7 × 0·2 (b) 14 × 0·6 (c) 29 × 0·5 (d) 65 × 0·7 (e) 153 × 0·9
(f) 8 × 0·02 (g) 14 × 0·03 (h) 29 × 0·04 (i) 65 × 0·06 (j) 153 × 0·02
4. Find:
(a) 1·26 × 10 (b) 2·53 × 12 (c) 6·81 × 25 (d) 5·34 × 62 (e) 0·71 × 98
5. Find:
(a) 0·2 × 0·4 (b) 0·6 × 0·7 (c) 1·9 × 0·5 (d) 2·3 × 0·7 (e) 7·8 × 0·9
(f) 0·02 × 0·3 (g) 0·06 × 0·8 (h) 0·12 × 0·7 (i) 3·29 × 0·6 (j) 4·36 × 0·5
6. Find:
(a) 1·6 × 4·7 (b) 2·8 × 3·5 (c) 2·41 × 1·2 (d) 2·07 × 6·8 (e) 6·51 × 2·48

Division

$12 \div 6 = 2$ $\quad$ $6 \div 0{\cdot}1 = 60$

$1{\cdot}2 \div 6 = 0{\cdot}2$ $\quad$ $6 \div 0{\cdot}01 = 600$

$0{\cdot}12 \div 6 = 0{\cdot}02$ $\quad$ $6 \div 0{\cdot}001 = 6000$

$0{\cdot}012 \div 6 = 0{\cdot}002$ $\quad$ $6 \div 0{\cdot}0001 = 60000$

$$\frac{6{\cdot}88}{4{\cdot}3} = \frac{6{\cdot}88 \times 10}{4{\cdot}3 \times 10} = \frac{68{\cdot}8}{43} = 1{\cdot}6$$

The example on the right above shows how the division of decimals can be simplified by first making the denominator a whole number.

Exercise 1.7c

1. Find:
   (a) $0{\cdot}8 \div 4$ (b) $1{\cdot}6 \div 2$ (c) $2{\cdot}8 \div 7$ (d) $7{\cdot}5 \div 5$ (e) $13{\cdot}2 \div 11$
2. Find:
   (a) $0{\cdot}64 \div 4$ (b) $1{\cdot}44 \div 8$ (c) $2{\cdot}82 \div 3$ (d) $0{\cdot}054 \div 6$ (e) $2{\cdot}184 \div 7$
3. Find:
   (a) $7 \div 0{\cdot}1$ (b) $14 \div 0{\cdot}1$ (c) $8 \div 0{\cdot}01$ (d) $9 \div 0{\cdot}001$ (e) $23 \div 0{\cdot}01$
   (f) $8 \div 0{\cdot}2$ (g) $15 \div 0{\cdot}3$ (h) $6 \div 0{\cdot}03$ (i) $24 \div 0{\cdot}04$ (j) $52 \div 0{\cdot}002$
4. Find:
   (a) $1{\cdot}26 \div 10$ (b) $2{\cdot}52 \div 12$ (c) $6{\cdot}75 \div 25$ (d) $0{\cdot}06 \div 15$ (e) $0{\cdot}24 \div 48$
5. Find:
   (a) $0{\cdot}4 \div 0{\cdot}1$ (b) $0{\cdot}8 \div 0{\cdot}4$ (c) $6{\cdot}4 \div 0{\cdot}8$ (d) $7{\cdot}2 \div 0{\cdot}9$ (e) $8{\cdot}4 \div 0{\cdot}7$
   (f) $0{\cdot}4 \div 0{\cdot}01$ (g) $0{\cdot}9 \div 0{\cdot}03$ (h) $0{\cdot}08 \div 0{\cdot}2$ (i) $5{\cdot}6 \div 0{\cdot}08$ (j) $8{\cdot}1 \div 0{\cdot}03$
6. Find:
   (a) $14{\cdot}4 \div 3{\cdot}6$ (b) $24{\cdot}3 \div 2{\cdot}7$ (c) $9{\cdot}92 \div 3{\cdot}1$ (d) $0{\cdot}76 \div 1{\cdot}9$ (e) $5{\cdot}525 \div 1{\cdot}3$

Exercise 1.7d

1. Find 16×6 and hence write down the result for:
   (a) $1{\cdot}6 \times 6$ (b) $16 \times 0{\cdot}6$ (c) $1{\cdot}6 \times 0{\cdot}6$ (d) $0{\cdot}16 \times 0{\cdot}6$ (e) $0{\cdot}16 \times 0{\cdot}06$
2. Find 23×12 and hence write down the result for:
   (a) $2{\cdot}3 \times 12$ (b) $230 \times 1{\cdot}2$ (c) $2{\cdot}3 \times 1{\cdot}2$ (d) $23 \times 0{\cdot}12$ (e) $2{\cdot}3 \times 0{\cdot}012$
3. Find $51 \div 3$ and hence write down the result for:
   (a) $5{\cdot}1 \div 3$ (b) $51 \div 0{\cdot}3$ (c) $5{\cdot}1 \div 0{\cdot}3$ (d) $51 \div 0{\cdot}03$ (e) $5{\cdot}1 \div 30$
4. Find $102 \div 17$ and hence write down the result for:
   (a) $10{\cdot}2 \div 17$ (b) $1020 \div 1{\cdot}7$ (c) $1{\cdot}02 \div 1{\cdot}7$ (d) $10{\cdot}2 \div 0{\cdot}17$ (e) $1{\cdot}02 \div 0{\cdot}017$
5. What number in the box will make the statement true?
   (a) $0{\cdot}3 \times \square = 0{\cdot}6$ (b) $0{\cdot}4 \times \square = 0{\cdot}08$ (c) $0{\cdot}7 \times \square = 0{\cdot}21$
   (d) $\square \times 1{\cdot}3 = 0{\cdot}26$ (e) $\square \times 2{\cdot}1 = 0{\cdot}84$ (f) $\square \times 0{\cdot}12 = 0{\cdot}096$
6. What number in the box will make the statement true?
   (a) $0{\cdot}9 \div \square = 0{\cdot}3$ (b) $0{\cdot}6 \div \square = 2$ (c) $0{\cdot}8 \div \square = 80$
   (d) $\square \div 0{\cdot}1 = 21$ (e) $\square \div 0{\cdot}5 = 0{\cdot}25$ (f) $\square \div 0{\cdot}02 = 0{\cdot}16$
7. Find the area of a rectangle $3{\cdot}21$ cm long and $1{\cdot}7$ cm wide.
8. The area of a rectangle is $29{\cdot}14\ cm^2$. If the length is $6{\cdot}2$ cm find the width of the rectangle.

1.8 Approximations and standard form

Results of calculations are often given in an approximate form, either to a fixed number of decimal places, or to a fixed number of significant figures. The first method is useful if one knows in advance the likely size of the result; in the second method it is the number of actual figures which is more important.

Example

15 741·231	is 15 741·23	to 2 decimal places
	or 15 700	to 3 significant figures
0·013 926	is 0·01	to 2 decimal places
	or 0·013 9	to 3 significant figures

Remember when the next figure after those in question is a 5 or more, the previous figure is rounded up to the next number.

i.e.	71·2573	is 71·26	to 2 decimal places	71·25(7)
		or 71·3	to 3 significant figures	71·2(5)

Exercise 1.8a

1. Write down 13 714·261 to the number of significant figures stated:
(a) 3 (b) 7 (c) 2 (d) 6 (e) 1

2. Write down 0·013 426 91 to the number of significant figures stated:
(a) 3 (b) 6 (c) 4 (d) 5 (e) 1

3. Write down 1·672 351 8 to the number of decimal places stated:
(a) 2 (b) 5 (c) 4 (d) 6 (e) 1

4. Write down 0·032 618 19 to the number of decimal places stated:
(a) 2 (b) 4 (c) 7 (d) 3 (e) 1

5. Write down the number correct to 2 decimal places:
(a) 12·6341 (b) 0·1207 (c) 0·017 53 (d) 135·625 41

6. Write down the number correct to 3 decimal places:
(a) 9·763 45 (b) 0·483 126 (c) 14·235 512 (d) 0·001 782

7. Write down the number correct to 3 significant figures
(a) 12·6317 (b) 56 421·251 (c) 0·001 237 81 (d) 1395·21

8. Write down the number correct to 2 significant figures:
(a) 56 180·342 (b) 0·036 158 (c) 0·003 156 (d) 49·7614

9. State how many decimal places are given in each number:
(a) 36·25 (b) 431·2 (c) 0·007 61 (d) 2·0001 (e) 460

10. State how many significant figures are given in each number:
(a) 43·7 (b) 128 (c) 0·0013 (d) 47 000 (e) 0·1362

11. Write down the result to 2 decimal places:
(a) 4·2 × 0·6 (b) 0·07 × 1·6 (c) 0·5 × 0·09 (d) 1·63 × 1·2

12. Write down the result to 3 significant figures:
(a) 14·3 × 8 (b) 263 × 4 (c) 7·52 × 3 (d) 0·467 × 6

Estimation

When using a calculator or a slide rule for calculations, it is helpful to have an idea of what result you expect before starting. This will act as a check.

$$\frac{362 \times 4{\cdot}975}{72 \times 9{\cdot}86} \approx \frac{400 \times 5}{70 \times 10} = \frac{20}{7} \approx 3$$

$$\frac{0{\cdot}265 \times 0{\cdot}00713}{10{\cdot}26 \times 0{\cdot}097} \approx \frac{0{\cdot}3 \times 0{\cdot}007}{10 \times 0{\cdot}1} = \frac{0{\cdot}021}{10} \approx 0{\cdot}002$$

An approximate answer is easily found (as shown above) by correcting each of the numbers involved to one significant figure.
The actual results are 2·54 and 0·001 90 correct to 3 sig. figs.

Standard form

Numbers are more easily compared by writing them as a number between one and ten multiplied by a power of ten. This is called *standard form.*

i.e. $362 = 3{\cdot}62 \times 10^2$; $0{\cdot}007\,16 = 7{\cdot}16 \times 10^{-3}$

This is also a useful way of estimating results.

$$\frac{576 \times 43{\cdot}2 \times 0{\cdot}0071}{0{\cdot}0628 \times 4623} = \frac{(5{\cdot}76 \times 10^2) \times (4{\cdot}32 \times 10^1) \times (7{\cdot}1 \times 10^{-3})}{(6{\cdot}28 \times 10^{-2}) \times (4{\cdot}623 \times 10^3)} \approx \frac{6 \times 4 \times 7}{6 \times 5 \times 10} \approx 6 \times 10^{-1}$$

Exercise 1.8b

1. Write the number in standard form:
(a) 462 (b) 51 060 (c) 0·125 (d) 0·002 61 (e) 96·5
2. Write the number in standard form:
(a) 56×10^2 (b) 5172×10^{-1} (c) $0{\cdot}0067 \times 10^{-3}$ (d) $0{\cdot}024 \times 10^6$
(e) $127 \div 10$ (f) $1786 \div 10^2$ (g) $0{\cdot}0712 \div 10^3$ (h) $12{\cdot}5 \div 10^{-3}$
3. Find the result and then write it in standard form:
(a) 761×4 (b) $0{\cdot}006\,15 \times 7$ (c) $1{\cdot}751 \times 12$ (d) $0{\cdot}0061 \times 0{\cdot}02$
4. Find the result and then write it in standard form:
(a) $(3 \times 10^2) \times (4 \times 10^3)$ (b) $(17 \times 10^{-2}) \times (6 \times 10^{-3})$ (c) $(28 \times 10^5) \times (5 \times 10^{-2})$
5. Find the result and then write it in standard form:
(a) $2{\cdot}6 \times 10^2 + 3{\cdot}7 \times 10^2$ (b) $19{\cdot}2 \times 10^{-3} + 165 \times 10^{-3}$ (c) $62{\cdot}1 \times 10^3 + 1{\cdot}9 \times 10^2$
(d) $634 \times 10^5 - 124 \times 10^4$ (e) $2{\cdot}4 \times 10^4 \times 5 \times 10^{-2}$ (f) $(504 \times 10^{-3}) \div (12 \times 10^5)$
6. Write each number correct to one significant figure and hence estimate:
(a) $2759 \times 81{\cdot}3$ (b) $0{\cdot}0069 \times 31{\cdot}6$ (c) $(593)^2 \times 0{\cdot}083$
(d) $\dfrac{42{\cdot}3 \times 123{\cdot}2}{81{\cdot}7 \times 5{\cdot}9}$ (e) $\dfrac{0{\cdot}0072 \times 659{\cdot}3}{51{\cdot}26 \times 0{\cdot}117}$ (f) $\dfrac{32\,610 \times 43{\cdot}57}{215{\cdot}3 \times 0{\cdot}081}$
7. Write each number in standard form and hence estimate:
(a) $493 \times 81{\cdot}2$ (b) $0{\cdot}078 \times 0{\cdot}51$ (c) $2361 \times 0{\cdot}00896$
(d) $\dfrac{26 \times 5431}{814 \times 2{\cdot}16}$ (e) $\dfrac{1916 \times 0{\cdot}217}{43{\cdot}2 \times 0{\cdot}012}$ (f) $\dfrac{(72{\cdot}3)^2 \times 0{\cdot}000\,21}{5931 \times 38{\cdot}26}$

1.9 Percentages and interest

A percentage is a fraction where the denominator is 100.

i.e. 17% is the same as $\frac{17}{100}$.

To express a fraction as a percentage, it is necessary to find an equivalent fraction whose denominator is 100.

i.e. $\frac{3}{4} \times \frac{100}{1} = 75$ so $\frac{3}{4} = \frac{75}{100} = 75\%$.

In the same way any decimal may be expressed as a percentage;

i.e. $0{\cdot}125 = \frac{125}{1000} = \frac{12{\cdot}5}{100} = 12{\cdot}5\%$.

Example 1

Bill earns £50 per week and Jim earns £55 per week. Bill receives a rise of 15% and Jim a rise of 5%. Find who now earns more.

Bill Rise is 15% of £50 $= \frac{15}{100} \times 50 =$ £7·50. New pay is £57·50.

Jim Rise is 5% of £55 $= \frac{5}{100} \times 55 =$ £2·75. New pay is £57·75.

Remember: a percentage is always a percentage *of* some quantity.

Exercise 1.9a

1. Express as a fraction in its lowest form:
(a) 50% (b) 80% (c) 15% (d) 64% (e) 5% (f) 120%

2. Express as a percentage:
(a) $\frac{3}{4}$ (b) $\frac{2}{5}$ (c) $\frac{3}{25}$ (d) $\frac{7}{20}$ (e) $\frac{1}{8}$ (f) $\frac{5}{4}$

3. Find:
(a) 10% of 80 (b) 6% of 50 (c) 20% of 60 (d) 75% of 88
(e) 12% of 350 (f) 7·5% of 40 (g) 1% of 126 (h) 0·5% of 400

4. Find:
(a) 40% of 3 kg (b) 15% of 150 m (c) 6·5% of £1500 (d) 83% of £20 000

5. Find to the nearest penny:
(a) $3\frac{1}{2}$% of £12 (b) $8\frac{1}{2}$% of £35·40 (c) $5\frac{1}{4}$% of £1421

6. Increase:
(a) 70 by 10% (b) 125 by 40% (c) £1560 by 8% (d) 76 by 0·1%

7. Decrease:
(a) 40 by 5% (b) 264 by 25% (c) £2100 by 7·5% (d) 27 by 0·2%

8. Express:
(a) 7 as a percentage of 28 (b) 8 as a percentage of 24 (c) 7·5 as a percentage of 60

9. Express as examination percentages:
(a) 32 marks out of 40 (b) 60 marks out of 96 (c) 215 marks out of 250

10. Express the difference between the two numbers as a percentage of the first:
(a) 100, 116 (b) 32, 40 (c) 60, 80 (d) 75, 60 (e) 512, 448

11. A man who earns £3200 per annum has his salary increased by 5% plus £4 per week. Find his new salary and the total percentage increase.

12. A dairy herd of 1200 cows is increased by 5% in the first year and by 10% in the second year. Find the new size of the herd and the overall percentage increase.

Interest, profit and discount

Example 2

The advertised price of a car is £1640. A discount of 5% is allowed for a cash sale, whereas the interest charges would be 15% if the car is bought on credit terms. Find the difference between the two methods.

5% of 1640 is 82, so the cash price is £1558
15% of 1640 is 246, so the credit price is £1886
It costs £328 more to buy the car on credit terms.

Example 3

By selling a T.V. for £294 a shop makes a $22\frac{1}{2}\%$ profit. Find the original cost of the T.V.

Let £C be the original cost of the T.V.

$$C + 22\tfrac{1}{2}\% \text{ of } C = 294 \quad \text{so} \quad C + \frac{22\frac{1}{2}}{100}C = 294$$

$$\text{so} \quad C = \frac{294 \times 100}{122{\cdot}5} \quad \text{i.e. the cost was £240.}$$

Exercise 1.9b

1. A man pays 12% interest on a loan of £2500. How much is the interest and how much more would it have cost him if the interest were 15%?
2. A cash discount of 7% is allowed on an article costing £36. Find the saving involved and the additional saving if the discount were 8%.
3. A shop buys a T.V. for £240. Find the sale price for a profit of:
(a) 5% (b) 10% (c) $12\frac{1}{2}\%$ (d) 15% (e) $33\frac{1}{3}\%$
4. A man buys a car for £1600. Express his loss as a percentage if he sells it for:
(a) £1200 (b) £1520 (c) £1408
5. A shop buys a suite for £500. Express the profit as a percentage if they sell it for:
(a) £600 (b) £540 (c) £562·50
6. A savings account gains interest at 4% per annum. Find the value of an investment of £1000 after each of the first four years assuming that the interest is re-invested.
7. The annual premium for a car insurance is £200. Discounts of 10% and 60% are allowed for one driver only, and 4 years of no claims bonus respectively. Find the premium required if the 10% is deducted first followed by the 60%. What difference would it make if the 60% had been deducted first? Find the overall percentage discount in either case.
8. A house is sold for £16 500 and a 10% profit is made. Find the original cost of the house.
9. A caravan is sold for £736, having depreciated 8% in value. Find the original cost of the caravan.
10. A dealer sells a painting to one man at a 10% profit. It is then resold to a second man at a 5% loss for £1254. Find the original price and the overall percentage increase in value.

1.10 Squares and square roots

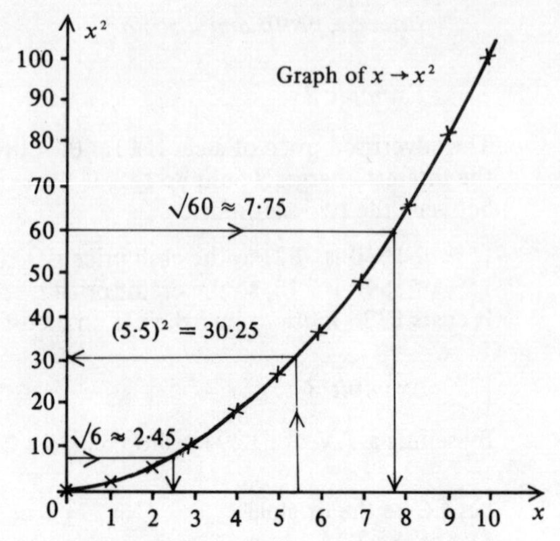

The graph of $x \to x^2$ is shown on the right. It can be used to find the squares of numbers between 1 and 10.
For example: $(5{\cdot}5)^2 = 30{\cdot}25$.

By using the inverse we can find the square root of any number between 1 and 100.
For example:
$\sqrt{6} \approx 2{\cdot}45$ and $\sqrt{60} \approx 7{\cdot}75$.

The square root of any number can always be found from the square roots of numbers between 1 and 100.

For example: $\sqrt{600} = \sqrt{6 \times 100} = \sqrt{6} \times \sqrt{100} \approx 2{\cdot}45 \times 10 = 24{\cdot}5$
and $\sqrt{0{\cdot}006} \quad \sqrt{60 \times 10^{-4}} = \sqrt{60} \times \sqrt{10^{-4}} \approx 7{\cdot}75 \times 10^{-2} = 0{\cdot}0775$

Exercise 1.10a

1. Use the graph above to find an approximate value for:
(a) $(3{\cdot}5)^2$ (b) $(4{\cdot}5)^2$ (c) $(6{\cdot}5)^2$ (d) $(8{\cdot}5)^2$ (e) $(9{\cdot}5)^2$
2. Use the graph above to find an approximate value for:
(a) $\sqrt{20}$ (b) $\sqrt{40}$ (c) $\sqrt{50}$ (d) $\sqrt{70}$ (e) $\sqrt{80}$
3. Find:
(a) $(300)^2$ (b) $(30)^2$ (c) $(0{\cdot}3)^2$ (d) $(0{\cdot}03)^2$ (e) $(0{\cdot}003)^2$
4. Using your answers for question 3 and the fact that $\sqrt{90} \approx 9{\cdot}5$ find:
(a) $\sqrt{900}$ (b) $\sqrt{9000}$ (c) $\sqrt{0{\cdot}09}$ (d) $\sqrt{0{\cdot}9}$ (e) $\sqrt{0{\cdot}009}$
5. Write the number as 4×10^n and hence find the square root of:
(a) 400 (b) 40 000 (c) 0·04 (d) 0·0004
Check your results by squaring them.
6. Write the number as 9×10^n and hence find the square root of:
(a) 90 000 (b) 0·09 (c) 900 (d) 0·0009
7. Write down the square root of:
(a) 16×10^2 (b) 25×10^4 (c) 36×10^{-4} (d) 49×10^{-2} (e) 81×10^6
8. Write down the square root of:
(a) 9×10^4 (b) 4×10^{-6} (c) 40×10^3 (d) 90×10^{-3} (e) $8{\cdot}1 \times 10^5$
9. Given that $\sqrt{5} \approx 2{\cdot}236$ and $\sqrt{50} \approx 7{\cdot}071$ find:
(a) $\sqrt{500}$ (b) $\sqrt{5000}$ (c) $\sqrt{0{\cdot}5}$ (d) $\sqrt{0{\cdot}05}$ (e) $\sqrt{0{\cdot}005}$
10. Given that $\sqrt{3} \approx 1{\cdot}732$ and $\sqrt{30} \approx 5{\cdot}477$ find:
(a) $\sqrt{3000}$ (b) $\sqrt{300}$ (c) $\sqrt{300\,000}$ (d) $\sqrt{0{\cdot}03}$ (e) $\sqrt{0{\cdot}0003}$
11. Given that $\sqrt{12} \approx 3{\cdot}464$ and $\sqrt{1{\cdot}2} \approx 1{\cdot}095$ find:
(a) $\sqrt{120}$ (b) $\sqrt{1200}$ (c) $\sqrt{12\,000}$ (d) $\sqrt{0{\cdot}12}$ (e) $\sqrt{0{\cdot}012}$
12. Given that $\sqrt{7{\cdot}1} \approx 2{\cdot}665$ and $\sqrt{71} \approx 8{\cdot}426$ find:
(a) $\sqrt{710}$ (b) $\sqrt{0{\cdot}71}$ (c) $\sqrt{7100}$ (d) $\sqrt{0{\cdot}071}$ (e) $\sqrt{71\,000}$

Using square root tables (three-figure tables)

There are two sets of square root tables. One is for numbers between 1 and 10, the other for numbers between 10 and 100.

	0	1	2	3
2·0	1·41	1·42	1·42	1·42
2·1	1·45	(1·45)	1·46	1·46
2·2	1·48	1·49	1·49	1·49

Using a copy of part of these tables as shown on the right, we find that:
$\sqrt{2{\cdot}11} = 1{\cdot}45$ and $\sqrt{21{\cdot}1} = 4{\cdot}59$

	0	1	2	3
20	4·47	4·48	4·49	4·51
21	4·58	(4·59)	4·60	4·62
22	4·69	4·70	4·71	4·72

To find the square root of a number not between 1 and 100 you have to find which of the two tables is appropriate. A pairing method is used. The first pair indicates which table to use. Remember there will be one figure for each pair on either side of the decimal point.

Example Find $\sqrt{2110}$ and $\sqrt{0{\cdot}000211}$

Using the 10 to 100 table: $\sqrt{21|10|{\cdot}00} = 4\ 5\ {\cdot}9$

Using the 1 to 10 table: $\sqrt{0{\cdot}|00|02|11|00} = 0{\cdot}\ 0\ 1\ 4\ 5$

Exercise 1.10b

1. Use the tables above to write down:
(a) $\sqrt{2{\cdot}13}$ (b) $\sqrt{2{\cdot}01}$ (c) $\sqrt{20{\cdot}3}$ (d) $\sqrt{22{\cdot}2}$ (e) $\sqrt{20{\cdot}0}$
2. Use the tables above to write down:
(a) $\sqrt{213}$ (b) $\sqrt{201}$ (c) $\sqrt{2030}$ (d) $\sqrt{2220}$ (e) $\sqrt{2000}$
(f) $\sqrt{2130}$ (g) $\sqrt{20100}$ (h) $\sqrt{0{\cdot}0203}$ (i) $\sqrt{0{\cdot}222}$ (j) $\sqrt{0{\cdot}002}$
3. Given that $\sqrt{5{\cdot}62} = 2{\cdot}37$ and $\sqrt{56{\cdot}2} = 7{\cdot}50$ find:
(a) $\sqrt{562}$ (b) $\sqrt{5620}$ (c) $\sqrt{0{\cdot}562}$ (d) $\sqrt{0{\cdot}0562}$ (e) $\sqrt{56\ 200}$
4. Given that $\sqrt{8{\cdot}71} = 2{\cdot}95$ and $\sqrt{87{\cdot}1} = 9{\cdot}33$ find:
(a) $\sqrt{8710}$ (b) $\sqrt{871}$ (c) $\sqrt{0{\cdot}871}$ (d) $\sqrt{0{\cdot}008\ 71}$ (e) $\sqrt{0{\cdot}0871}$
5. Find from tables $\sqrt{4{\cdot}16}$ and $\sqrt{41{\cdot}6}$, hence write down:
(a) $\sqrt{41\ 600}$ (b) $\sqrt{0{\cdot}416}$ (c) $\sqrt{416}$ (d) $\sqrt{0{\cdot}0416}$ (e) $\sqrt{0{\cdot}000\ 416}$
6. Find from tables $\sqrt{1{\cdot}77}$ and $\sqrt{17{\cdot}7}$, hence write down:
(a) $\sqrt{177}$ (b) $\sqrt{0{\cdot}0177}$ (c) $\sqrt{17\ 700}$ (d) $\sqrt{0{\cdot}177}$ (e) $\sqrt{0{\cdot}001\ 77}$
7. Write down from your tables:
(a) $\sqrt{35}$ (b) $\sqrt{7{\cdot}1}$ (c) $\sqrt{96}$ (d) $\sqrt{4{\cdot}5}$ (e) $\sqrt{1{\cdot}6}$
(f) $\sqrt{23{\cdot}6}$ (g) $\sqrt{81{\cdot}9}$ (h) $\sqrt{5{\cdot}68}$ (i) $\sqrt{1{\cdot}34}$ (j) $\sqrt{60{\cdot}2}$
8. Find using your tables:
(a) $\sqrt{350}$ (b) $\sqrt{0{\cdot}71}$ (c) $\sqrt{9600}$ (d) $\sqrt{0{\cdot}045}$ (e) $\sqrt{160}$
(f) $\sqrt{23\ 600}$ (g) $\sqrt{0{\cdot}0819}$ (h) $\sqrt{0{\cdot}568}$ (i) $\sqrt{1340}$ (j) $\sqrt{0{\cdot}006\ 02}$
9. Use the pairing method above to estimate to one significant figure:
(a) $\sqrt{5170}$ (b) $\sqrt{462{\cdot}1}$ (c) $\sqrt{0{\cdot}8061}$ (d) $\sqrt{0{\cdot}0914}$ (e) $\sqrt{0{\cdot}000\ 426}$

1.11 Logarithms

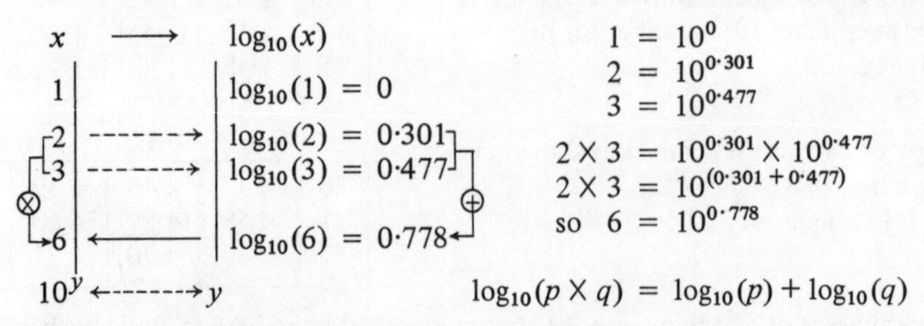

Any positive number x can be written as the power y of another number a. This power y is called the logarithm of x to the base a.

i.e. if $x = a^y$ then $y = \log_a(x)$

Remember: $\log_{10}(1000) = 3$ since $1000 = 10^3$
$\log_{10}(100) = 2$ since $100 = 10^2$
$\log_{10}(10) = 1$ since $10 = 10^1$
$\log_{10}(0{\cdot}01) = \bar{2}$ since $0{\cdot}01 = 10^{-2}$

Exercise 1.11a

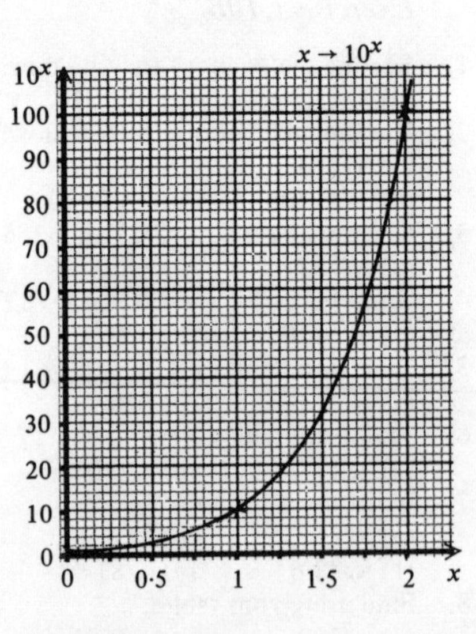

1. The graph of $x \to 10^x$ is shown on the right. The inverse of this graph is $x \to \log_{10}x$.
   Use the graph to write down:
   (a) $10^{0{\cdot}5}$ (b) $10^{1{\cdot}5}$
2. Explain why the result of multiplying your answers for (a) and (b) in question 1 should be 100.
3. Use the graph to write down:
   (a) $\log_{10}20$ (b) $\log_{10}30$
   (c) $\log_{10}40$ (d) $\log_{10}60$
4. Use your results from question 3 to write down:
   (a) $\log_{10}(20 \times 60)$ (b) $\log_{10}(30 \times 40)$
   (c) $\log_{10}(1200)$
   Do you agree that your results for (a), (b) and (c) should be the same?
5. $8 = 2^3$ and $243 = 3^5$; write down:
   (a) $\log_2(8)$ (b) $\log_3(243)$
6. $\log_4(64) = 3$ and $\log_5(25) = 2$. Re-write these two results using powers.
7. Write down:
   (a) $\log_{10}(10\,000)$ (b) $\log_{10}(1\,000\,000)$ (c) $\log_{10}(0{\cdot}001)$
8. Write down:
   (a) $\log_2(64)$ (b) $\log_3(81)$ (c) $\log_5(125)$

Logarithm tables (three-figure tables)

The diagram on the right shows part of a set of three-figure $\log_{10}$ tables.

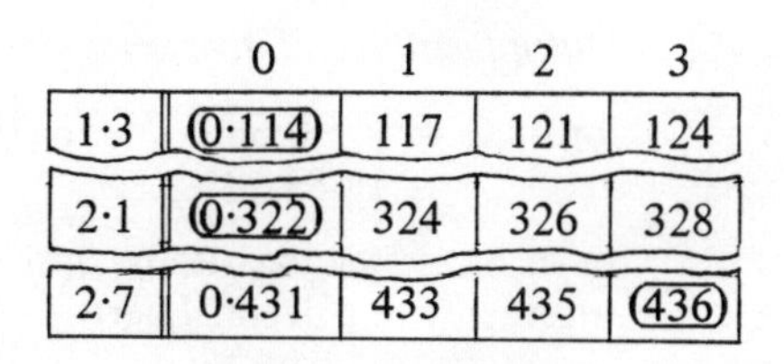

	0	1	2	3
1·3	(0·114)	117	121	124
2·1	(0·322)	324	326	328
2·7	0·431	433	435	(436)

To multiply 1·3 by 2·1 you find the logarithm of each number, add them together and then find which number has the result as its logarithm.

No.	Log.
1·3 →	0·114
2·1 →	0·322
2·73 ←	0·436

Exercise 1.11b (Assume all logs are logarithms to the base 10)

1. Using log (2) = 0·301 and log (3) = 0·477 find:
   (a) log (6) (b) log (12) (c) log (18) (d) log (4)
   (e) log (9) (f) log (8) (g) log (27) (h) log (72)
2. Using log (5) = 0·699 and log (7) = 0·845 find:
   (a) log (35) (b) log (25) (c) log (49) (d) log (125)
   (e) log (343) (f) log (175) (g) log (625) (h) log (245)
3. Find log (11) and log (13). Hence write down:
   (a) log (11 × 11) (b) log (13 × 13) (c) log (11 × 13) (d) log (11 × 11 × 11)
4. Find log (2·5) and log (3·1). Hence write down:
   (a) log $(2{\cdot}5)^2$ (b) log (3·1 × 2·5) (c) log $(3{\cdot}1)^3$ (d) log $(3{\cdot}1 \times 2{\cdot}5)^2$
5. If $\log(p) = x$ and $\log(q) = y$ write down:
   (a) $\log(pq)$ (b) $\log(p^2)$ (c) $\log(q^3)$ (d) $\log(p^3q^2)$

Exercise 1.11c

1. Write down the logarithm of:
   (a) 2·15 (b) 3·76 (c) 5·02 (d) 7·50 (e) 8·99 (f) 9·17
   (g) 21·5 (h) 376 (i) 5020 (j) 0·075 (k) 0·899 (l) 91 700
2. Find the number whose logarithm is:
   (a) 0·204 (b) 0·491 (c) 0·693 (d) 0·864 (e) 0·830 (f) 0·021
   (g) 1·204 (h) 3·491 (i) 2·693 (j) $\bar{1}{\cdot}864$ (k) $\bar{3}{\cdot}830$ (l) $\bar{4}{\cdot}021$
3. Use logarithms to find:
   (a) 1·7 × 2·6 (b) 4·3 × 1·2 (c) 5·3 × 6·7 (d) 9·7 × 8·5 (e) $(2{\cdot}1)^2$
4. Use logarithms to find:
   (a) 1·42 × 3·15 (b) 5·71 × 1·34 (c) 8·32 × 7·16 (d) 32·5 × 74·3 (e) $(1{\cdot}58)^2$
5. Use logarithms to find:
   (a) 25·3 × 0·47 (b) 171 × 0·39 (c) 2·97 × 0·0061 (d) 0·007 × 0·04
   (e) $(0{\cdot}165)^2$

Multiplication and division

$\log 48{\cdot}2 = 1 + 0{\cdot}683 = 1{\cdot}683$ $\qquad \log 0{\cdot}482 = \bar{1} + 0{\cdot}683 = \bar{1}{\cdot}683$ $\qquad \log 0{\cdot}00482 = \bar{3} + 0{\cdot}683 = \bar{3}{\cdot}683$

In each case the logarithm consists of a whole number part which may be positive or negative, together with a decimal part.

The whole number part is called the *characteristic.* The decimal part is called the *mantissa.* The mantissa is always positive.

Example 1 Find $26{\cdot}2 \times 0{\cdot}0071$

Note: $1 + \bar{3} + 1 = \bar{1}$

No.	Log.
26·2 →	$1{\cdot}418$
0·0071 →	$\bar{3}{\cdot}815$
0·171 ←	$\bar{1}{\cdot}233$

Example 2 Find $417 \div 0{\cdot}25$

Note: $2 - \bar{1} = 3$

No.	Log.
417	$2{\cdot}620$
0·25	$\bar{1}{\cdot}398$
1670	$3{\cdot}222$

Remember $\bar{1}$ is negative 1, so subtracting $\bar{1}$ is the same as adding 1.

Exercise 1.11d

1. Find:
(a) $2 + \bar{1} + 1$ (b) $\bar{1} + \bar{2} + 1$ (c) $1 - \bar{3} - 1$ (d) $\bar{2} - \bar{1} - 1$
2. Find:
(a) $2{\cdot}716 + \bar{3}{\cdot}429$ (b) $1{\cdot}517 - \bar{2}{\cdot}482$ (c) $\bar{1}{\cdot}715 - \bar{3}{\cdot}923$
3. Write down the numbers whose logarithms are the results for:
(a) question 1 (b) question 2
4. Use logarithms to find:
(a) $175 \times 0{\cdot}0261$ (b) $0{\cdot}178 \times 0{\cdot}253$ (c) $96{\cdot}3 \times 0{\cdot}00047$ (d) $(0{\cdot}023)^2$
5. Use logarithms to find:
(a) $\dfrac{175}{26}$ (b) $\dfrac{4570}{1{\cdot}73}$ (c) $\dfrac{43{\cdot}7}{0{\cdot}35}$ (d) $\dfrac{7{\cdot}61}{0{\cdot}0041}$ (e) $\dfrac{51{\cdot}8}{0{\cdot}0812}$
6. Use logarithms to find:
(a) $15{\cdot}2 \div 371$ (b) $2{\cdot}96 \div 0{\cdot}038$ (c) $0{\cdot}572 \div 41{\cdot}9$ (d) $0{\cdot}03 \div 0{\cdot}007$
7. Find the area of a rectangle 21·3 m long and 7·6 m wide.
8. Find the volume of a cuboid 76·3 mm long, 42·8 mm wide, and 12·7 mm high.
9. The area of a rectangle is 147 cm^2. If the length is 28·3 cm find the width.
10. Find the cost of 144 tins of coffee if each tin costs £3·75.
11. Write down an estimate first and then use logarithms to find:
(a) $\dfrac{17{\cdot}2 \times 43{\cdot}8}{1{\cdot}96}$ (b) $\dfrac{453 \times 7{\cdot}15}{28{\cdot}3}$ (c) $\dfrac{1860 \times 0{\cdot}732}{14{\cdot}2 \times 0{\cdot}021}$
12. Write down an estimate first and then use logarithms to find:
(a) $\dfrac{236 \times 718}{54 \times 17}$ (b) $\dfrac{4{\cdot}92 \times 0{\cdot}0637}{153 \times 0{\cdot}814}$ (c) $\dfrac{0{\cdot}00538 \times (3{\cdot}47)^2}{835 \times 0{\cdot}162}$

Calculating powers and roots

Example 3 Find $(14{\cdot}7)^3$

Note: $(14{\cdot}7)^3 = 14{\cdot}7 \times 14{\cdot}7 \times 14{\cdot}7$

No.	Log.
14·7 ⟶	1·167 × 3
3170 ⟵	3·501

Example 4 Find $(6{\cdot}37)^{\frac{1}{3}}$

Note: $(1{\cdot}85)^3 = 6{\cdot}37$

No.	Log.
6·37 ⟶	0·804 ÷ 3
1·85 ⟵	0·268

Example 5 Find $(0{\cdot}384)^{\frac{1}{4}}$

Note: $\dfrac{\bar{1}{\cdot}584}{4} = \dfrac{\bar{4} + 3{\cdot}584}{4} = \bar{1} + 0{\cdot}896$

No.	Log.
0·384 ⟶	$\bar{1}{\cdot}584$ ÷ 4
0·787 ⟵	$\bar{1}{\cdot}896$

Remember: to calculate the nth power or nth root of a number, you first find the logarithm of the number, then multiply or divide by n, and finally read back the result from the tables.

Exercise 1.11e

1. Use logarithms to find:
(a) $(6{\cdot}35)^2$ (b) $(3{\cdot}42)^5$ (c) $(11{\cdot}5)^3$ (d) $(0{\cdot}571)^2$ (e) $(0{\cdot}026)^3$
2. Use logarithms to find:
(a) $(4{\cdot}37)^{\frac{1}{2}}$ (b) $(263)^{\frac{1}{2}}$ (c) $(7320)^{\frac{1}{3}}$ (d) $(72{\cdot}5)^{\frac{1}{5}}$ (e) $(100)^{\frac{1}{4}}$
3. Use logarithms to find:
(a) $(0{\cdot}038)^{\frac{1}{2}}$ (b) $(0{\cdot}0093)^{\frac{1}{3}}$ (c) $(0{\cdot}000\,57)^{\frac{1}{4}}$ (d) $(0{\cdot}591)^{\frac{1}{2}}$ (e) $(0{\cdot}027)^{\frac{1}{3}}$
4. Work through the flow chart. Write down the result and the calculation you have just completed.

(a) 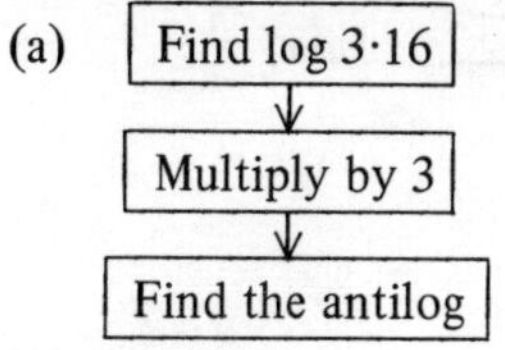

(b) 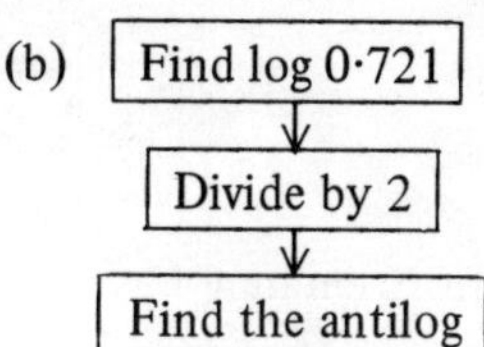

5. Solve the equation:
(a) $x^3 = 371$ (b) $x^5 = 1000$ (c) $x^{\frac{1}{2}} = 13{\cdot}1$ (d) $x^{\frac{1}{4}} = 2{\cdot}6$
6. Solve the equation:
(a) $10^x = 5{\cdot}82$ (b) $10^x = 31{\cdot}6$ (c) $10^x = 0{\cdot}073$ (d) $10^x = 0{\cdot}16$
7. Solve the equation:
(a) $\log x = 3{\cdot}759$ (b) $x \log 2 = \log 8$ (c) $\log(x^3) = 3{\cdot}603$
8. Solve the equation:
(a) $2^x = 5$ (b) $3^x = 7$ (c) $5^x = 137$ (d) $2^x = 0{\cdot}2$

Part 2: Mensuration

2.1 Pythagoras' Theorem

The Theorem of Pythagoras connects the squares of the sides of any right-angled triangle. In triangle ABC,

$$AC^2 = AB^2 + BC^2.$$

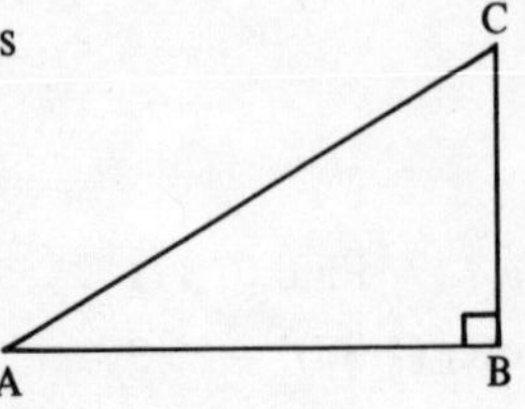

AC is the hypotenuse of the triangle.

Pythagoras' Theorem is mainly used to calculate lengths of sides and of diagonals in plane and solid shapes.

Example 1

Calculate the height of an isosceles triangle of base 8 cm whose equal sides are of length 10 cm.

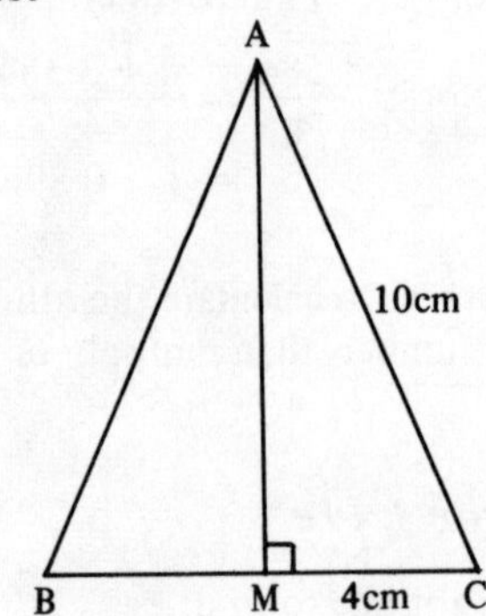

In the diagram, AMC is a right-angled triangle. Using Pythagoras' Theorem

$$\begin{aligned} AM^2 &= AC^2 - MC^2 \\ &= 10^2 - 4^2 \\ &= 100 - 16 \end{aligned}$$

so $AM = \sqrt{84} = 9{\cdot}17$ cm

Example 2

Find the length of the longest straight rod that will just fit inside a box of dimensions 60 cm × 30 cm × 20 cm.

To simplify the working, let the dimensions be 6 × 3 × 2 units.
In the diagram,

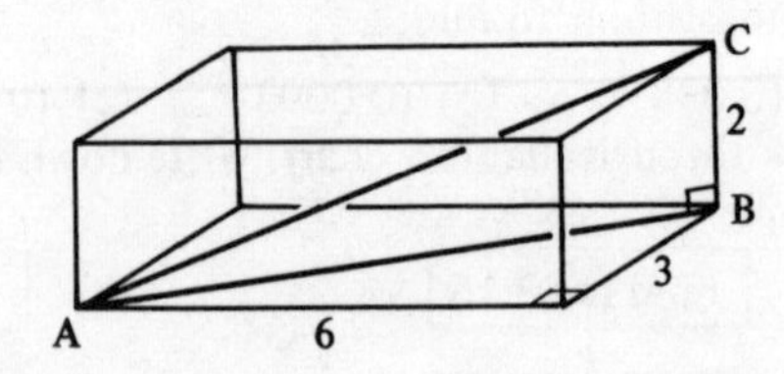

$$\begin{aligned} AB^2 &= 6^2 + 3^2 = 45 \\ AC^2 &= AB^2 + 2^2 = 49 \end{aligned}$$

so $AC = 7$ units

The longest straight rod is therefore 70 cm long.

Exercise 2.1

1. Calculate the length of the hypotenuse of a right-angled triangle whose other two sides in centimetres are:
   (a) 6, 8 (b) 10, 24 (c) 10, 7·5 (d) 7, 24
2. Calculate the length of the shortest side of a right-angled triangle whose other two sides in centimetres are:
   (a) 60, 61 (b) 4·0, 4·1 (c) 2·4, 2·6 (d) 2·9, 2·1
3. Calculate to three significant figures the length of the diagonal of a square of side:
   (a) 1 cm (b) 10 cm (c) 100 cm
4. Calculate to three significant figures the length of the diagonal of a rectangle of length 8 cm and width 7 cm.
5. One diagonal of a rhombus of side 11 cm is 5 cm long. Find the length of the other diagonal and hence the area of the rhombus.
6. The length of the sides of an isosceles triangle are 12 cm, 12 cm, and 10 cm. Find the height of the triangle and hence its area.
7. Find the height of an equilateral triangle of side 10 metres. How far up a wall will a 20 m ladder reach inclined at 60° to the horizontal?
8. The length of the diagonal of a square is 10 metres. Find the length of the side of the square. How far up a wall will a 20 m plank reach inclined at 45° to the horizontal?
9. The vertices A, B, C, D of a certain quadrilateral have co-ordinates (0, 2), (3, 1), (0, −4), (−2, 0). Find the length of each side of the quadrilateral and the length of the diagonal BD.
10. Show that the shape with vertices (0, 0), (4, 2), (6, −2), (2, −4) is a square by first calculating the lengths of its sides and then calculating the lengths of its diagonals.
11. The diagram shows a rectangular box of dimensions 10 cm × 7 cm × 6 cm.
    (a) Draw the triangle XTY and calculate the length length XY.
    (b) Draw the triangle XYZ and calculate the length XZ.
    (c) If W is the mid-point of XY, calculate the length WZ.

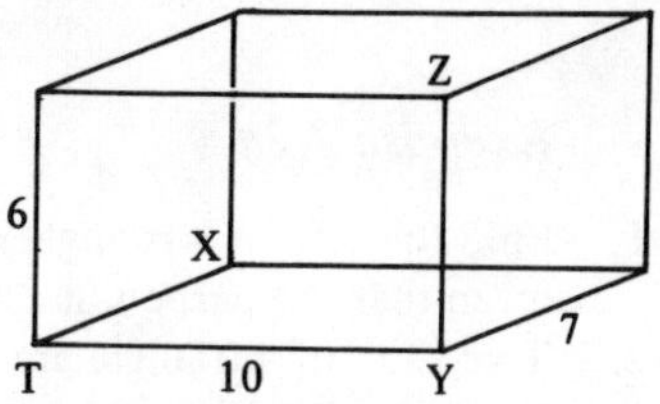

Z
6
X
7
T
10
Y

12. ABCDXYZW is a cube as shown in the diagram. P, Q, R, are the mid-points of DW, DB, and DZ respectively. The side of the cube is 2 units.
    Which of the following statements are true and which are false?
    (a) $DQ = \sqrt{2}$ (b) $DR = 1/\sqrt{2}$ (c) $PQ = \sqrt{3}$
    (d) $QR = 2$ (e) $BR = \sqrt{10}$

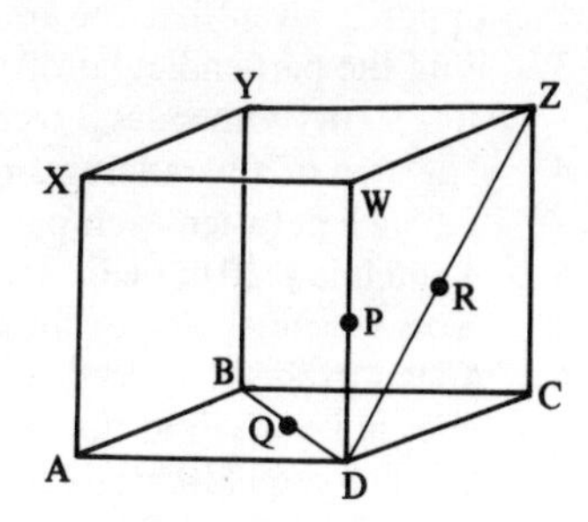

2.2 Areas and volumes

Triangle

The area of a triangle $= \frac{1}{2}$ base $\times$ height

$= \frac{1}{2}bh$

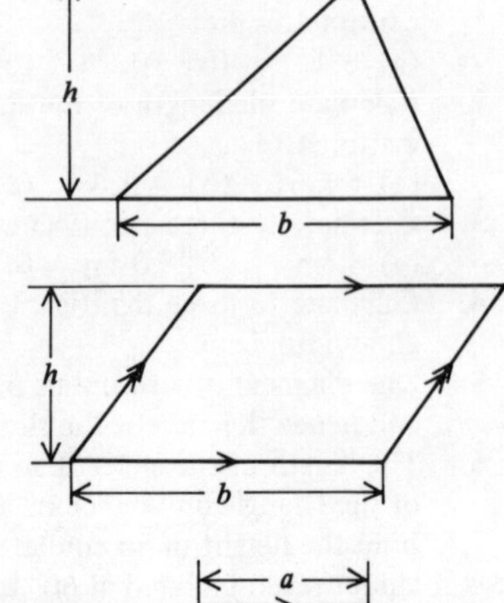

Parallelogram

The area of a parallelogram is (length of side) $\times$ (perpendicular distance from the side parallel to it).

Area $= bh$

Trapezium

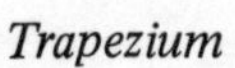

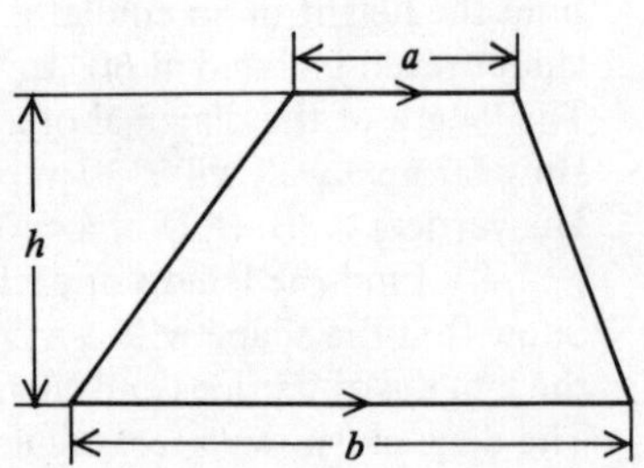

The area of a trapezium is (half the sum of the lengths of the parallel sides) $\times$ (distance between the parallel sides).

Area $= \frac{1}{2}(a + b)h$

Exercise 2.2a

1. Find the area of a triangle of base 6·2 cm, height 5 cm. Find also the length of another side given that the perpendicular distance from the vertex opposite this side is 4 cm.
2. Two sides of a triangle are 6 cm and 8 cm. Find the distance of each from the vertex opposite given that the area of the triangle is 15 cm^2.
3. Find the perpendicular distance between each pair of opposite sides of a parallelogram of area 9 cm^2 with sides 3·6 cm and 5·4 cm.
4. The area of a parallelogram is 84 cm^2. Its sides are 14 cm and 8 cm. Find the perpendicular distance between each pair of opposite sides.
5. A ditch is 1·20 m deep. Its base is level and 30 cm wide. The top is 75 cm wide. Calculate the cross-sectional area of the ditch.
6. A lean-to shed has vertical sides of 1·20 m and 1·50 m. The width of the shed between these sides is 1·0 m. Find the area of an end wall of the shed.
7. ABCD is a quadrilateral in which AC bisects the area of ABCD. The area of ABCD is 40 cm^2 and AC = 10 cm. What is the minimum possible length of BD?
8. The diagonals of a kite are 12 cm and 9 cm. Find its area.
9. The light from an open doorway throws a patch of light on level ground in the shape of a trapezium. The doorway is 72 cm wide and the furthest part of the patch of light is 2·40 m from the doorway. Find the greatest width of the patch of light given that its area is 2·16 m^2.

Prisms

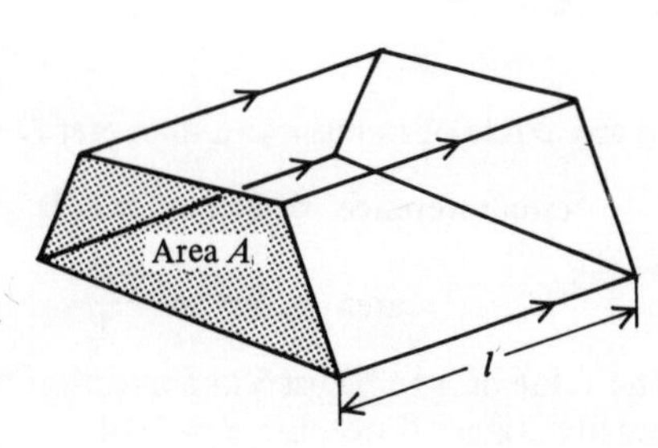

A prism is a solid with a pair of parallel identical end faces. The edges joining the end vertices are all parallel.

In a *right prism*, the end faces are perpendicular to the joining edges.

The volume of a right prism = (area of end face) × length

$$V = Al$$

Exercise 2.2b

1. The area of one end face of a rectangular box is 20 cm². Its length is 8 cm.
   (a) Find its volume.
   (b) How many such boxes would fill a crate of exactly five times the dimensions of the box?

2. The length of a rectangular prism is 9 cm and its volume is 270 cm³. Find
   (a) the area of an end face
   (b) the third edge length if one of the edges of the end face is 5 cm.

3. The diagram shows the dimensions of the end face of a tent 2·80 m long. Find:
   (a) the area of the end in m²
   (b) the volume of the tent in m³.

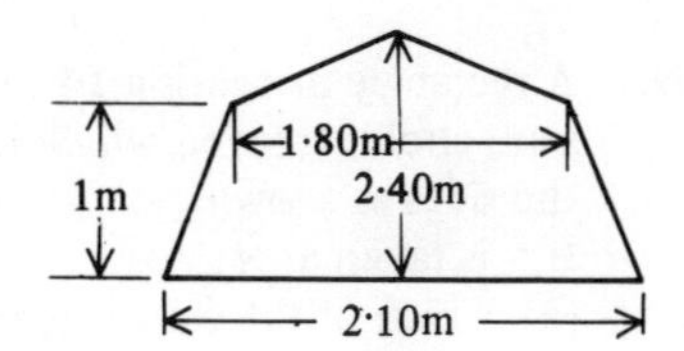

4. The end of a pencil is a regular hexagon with dimensions as shown. The pencil is 16 cm long. Calculate:
   (a) the area of the triangle shown
   (b) the area of the hexagon
   (c) the volume of the pencil to the nearest 0·1 cm³.

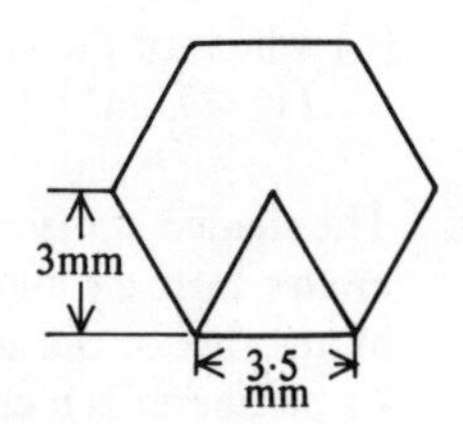

5. A drinking trough 70 cm deep and 2 m long has a trapezium cross section. The width at the top is 65 cm and at the bottom is 50 cm. Find in litres the maximum volume of water the trough will hold. (1 litre = 1000 cm³).

6. A wooden chicken house has an end section with the dimensions shown. The house is 2 m long. Find the volume of the house and the volume of space per hen if 12 hens are kept in the house.

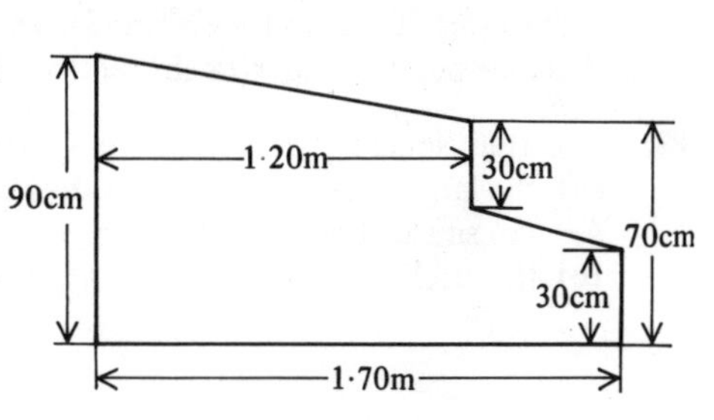

2.3 Circles

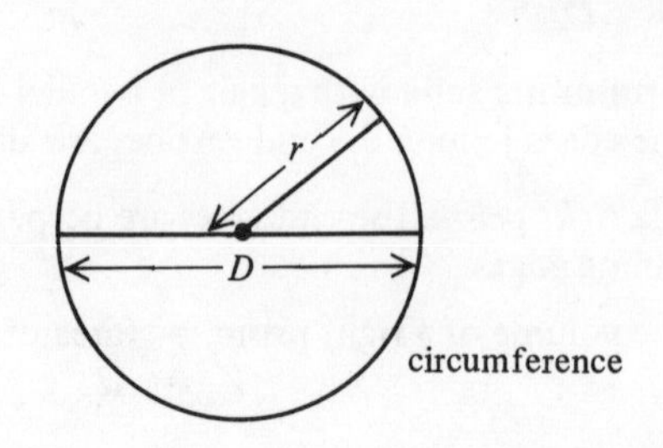

In any circle of radius r and diameter D

$$\text{circumference} = 2\pi r = \pi \times D$$

$$\text{area} = \pi r^2 = \frac{\pi D^2}{4}$$

The value of π to be used in a question is usually stated. If not, use $\pi = 3{\cdot}14$.

Exercise 2.3a

1. Calculate the area and circumference of a circle of radius 7 cm.
2. A piece of ribbon 150 cm long is wrapped round a circular cake of diameter 40 cm. Calculate the length of overlap.
3. Find the radius of a circle of circumference 942 cm.
4. Find the radius of a circle of circumference 45 cm.
5. Find the radius of a circle of area 50·24 cm².
6. Find the radius of a circle of area 200 cm².
7. Calculate the area of a circle of circumference 20 cm.
8. Calculate the circumference of a circle of area 100 cm².

9. A rectangle measuring 14 cm × 7 cm has a semi-circle removed which just touches the sides as shown.
   If π is taken as $\frac{22}{7}$,
   (a) which of the values below is the perimeter of the shaded shape?
   (1) 72 cm (2) 50 cm (3) 86 cm (4) 144 cm (5) 42 cm
   (b) which of the values below is the area of the shaded shape?
   (1) 77 cm² (2) 42 cm² (3) 21 cm² (4) 109 cm² (5) 10·5 cm²

10. The shaded star-shaped figure shown is formed by removing parts of circles from a square of side 10 cm. The centre of each circular arc is at a vertex of the square. If the area of the star shape is A cm² and its perimeter is p cm, which of the following statements are true?
    (a) $A < 100$ (b) $A > 50$ (c) $p < 40$ (d) $p > A$
    (e) $p = 5\pi$

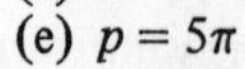

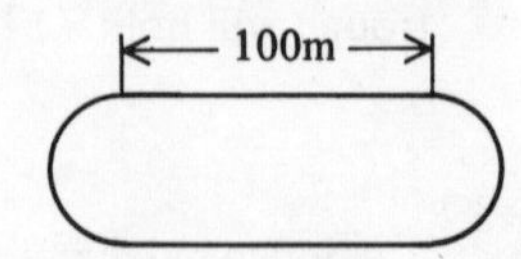

11. A running track is designed so that the length of four complete circuits is 1500 m. It has two parallel straight sections of 100 m and semi-circular ends. Find the distance between the straight sections.

12. A certain square is of side 10 cm. Calculate the radius of:
    (a) the largest circle that can be drawn inside the square,
    (b) the smallest circle that can be drawn outside the square,
    (c) the circle of area equal to that of the square.

Cylinders and spheres

A cylinder is a prism having a circular cross section.

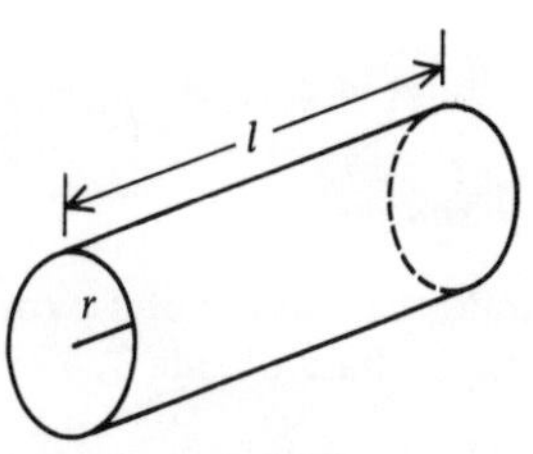

For a right cylinder of radius r and length l

$$\text{volume } V = \pi r^2 l$$

$$\text{curved surface area } S = 2\pi rl$$

$$\text{total surface area } T = 2\pi rl + 2\pi r^2 = 2\pi r(l+r)$$

For a sphere of radius r

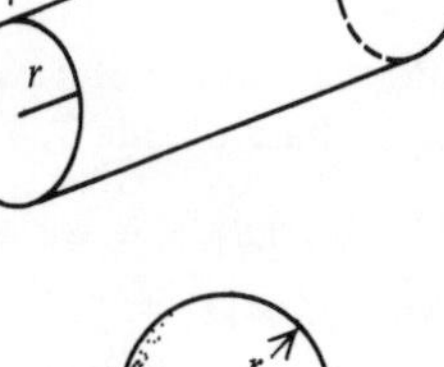

$$\text{volume } V = \tfrac{4}{3}\pi r^3$$

$$\text{surface area } S = 4\pi r^2$$

Exercise 2.3b

1. Calculate the volume and curved surface area of a cylinder 10 m long and of radius 70 cm. (Take π to be $3\frac{1}{7}$).
2. Calculate the volume and total surface area of a solid cylinder of radius 30 cm and length 40 cm. (Take π to be $3\frac{1}{7}$.)
3. A rectangle 10 cm × 12 cm is rolled to form a cylinder of length 12 cm. Calculate the radius and volume of the cylinder.
4. An open cylinder of length 21 cm and radius 3 cm is made from thin metal sheet of mass 0·5 g per cm^2. Calculate:
   (a) the surface area of the cylinder; (b) the mass of the cylinder.
5. A cylindrical metal casting 40 cm long has an internal radius r cm and external radius R cm. Show that the volume of the metal is $40\pi(R+r)(R-r)$ cm^3.
   If $R = 4$, $r = 3$, and $\pi = \frac{22}{7}$, find the volume of metal used in the casting.
6. A cylindrical metal pipe has walls 2 mm thick. The internal radius is 2 cm. Find the volume of metal in 2 m of pipe.
7. Water flows through a pipe of radius 2 cm with a speed of 1 m/s.
   Calculate: (a) the volume of 1 m of pipe
   (b) the rate of flow of the water in litres per minute.
8. Water flows through a pipe of diameter 10 cm at a speed of 2 m/s. Calculate the volume of water delivered in 1 minute.
9. Assuming $\pi = 3$, find the approximate values for the volume and surface area of a sphere of radius:
   (a) 2 cm (b) 3 mm (c) 2·2 cm
10. Assuming $\pi = 3\frac{1}{7}$, calculate the volume and surface area of a sphere of radius:
    (a) 7 cm (b) 2·1 cm (c) 100 cm
11. A concrete post is made in the shape of a hemisphere surmounted on a cylinder each of radius 15 cm. If the total height of the post is 1·50 m, calculate its volume.
12. Calculate the surface area (not including the base) of the concrete post of question 11.
13. A sphere of radius 3 cm fits exactly inside a cylindrical pipe. Calculate the volume of the section of pipe containing the sphere.
14. A concrete block measures 80 cm × 80 cm and 1·20 m high. A hemisphere of radius 40 cm is hollowed out of the top of the block. Calculate the remaining volume.

2.4 Pyramids

The volume of a pyramid $= \frac{1}{3}$(base area) $\times$ height.

Example

Calculate the volume of a pyramid given that a vertex is 6 cm above a triangular base of side 3, 4, 5 cm.

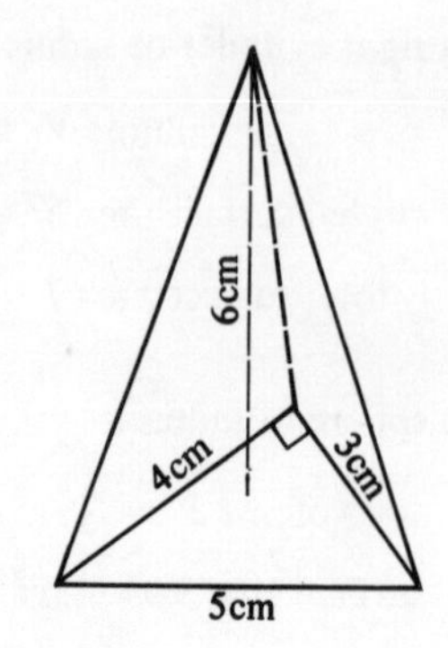

The triangle base is right-angled (Pythagoras)

Area of base $= \frac{1}{2}.3.4$
$= 6\text{ cm}^2$

Volume of Pyramid $= \frac{1}{3}.6.6$
$= 12\text{ cm}^3$

Exercise 2.4

1. The diagram shows a pyramid inside a cube of side a with its vertex at the centre of the cube.
   (a) How many such pyramids can fill the cube?
   (b) What is the volume of the cube?
   (c) Use answers to (a) and (b) to write down the volume of the pyramid.
   (d) Show that your answer to (c) agrees with the formula $\frac{1}{3}$(base area) $\times$ height.

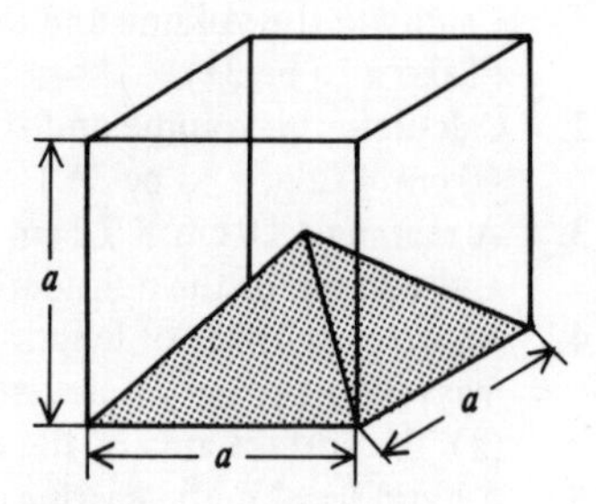

2. The diagram shows the net of a square-based pyramid with the measurements as shown.
   If the pyramid is constructed, calculate:
   (a) the area of the base
   (b) the vertical height (use Pythagoras)
   (c) the volume
   (d) the area of each triangular face
   (e) the total surface area.

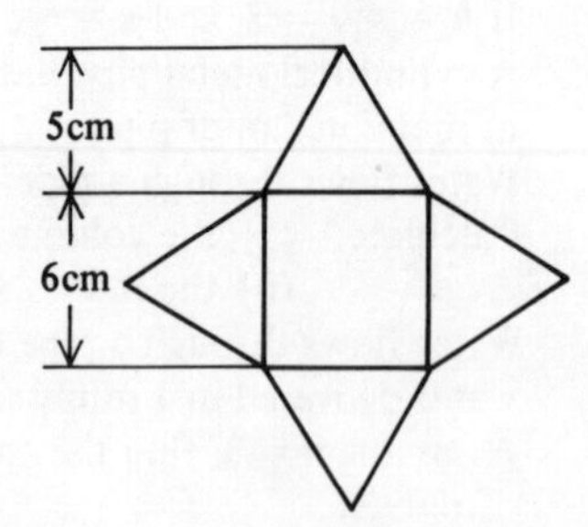

3. An obelisk of height 2 m is in the shape of a pyramid with a square base of side 80 cm. Calculate its volume in m^3.
4. The great pyramid of Cheops has a base area of 52 500 m^2 and a height of 146 m. Calculate its volume, giving your answer to three significant figures.
5. VABCD is a pyramid with V vertically above the centre O of a rectangle ABCD. AB = 6 cm, BC = 8 cm and VA = 13 cm.
   (a) Calculate the length OA.
   (b) Draw the triangle VOA and calculate the length VO.
   (c) Calculate the volume of the pyramid.

6. A certain square-based pyramid of height 10 cm has a volume of 100 cm^3. What is the length of the side of the square base?
(a) 10 cm (b) 17·3 cm (c) 5·48 cm (d) 30 cm (e) 15 cm

7. The diagram shows a child's model of a church. The pyramid VXYZW is symmetrically mounted at one end of a rectangular block. M is the mid-point of XY and N is the centre of the rectangle XYZW.
Using the measurements shown, calculate the lengths of VM and VN and find the total volume of the model.

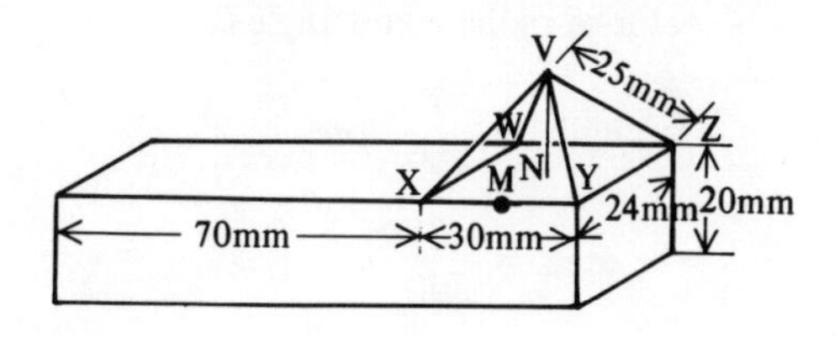

8. An octahedron VXYZWT fits exactly inside a rectangular box with the measurements as shown. XYZW is a square not at the centre of the box. V and T are not at the centres of the end square faces of the box.
(a) Calculate the volume of the pyramid VXYZW assuming it has a height of h cm.
(b) Calculate the total volume of the octahedron.

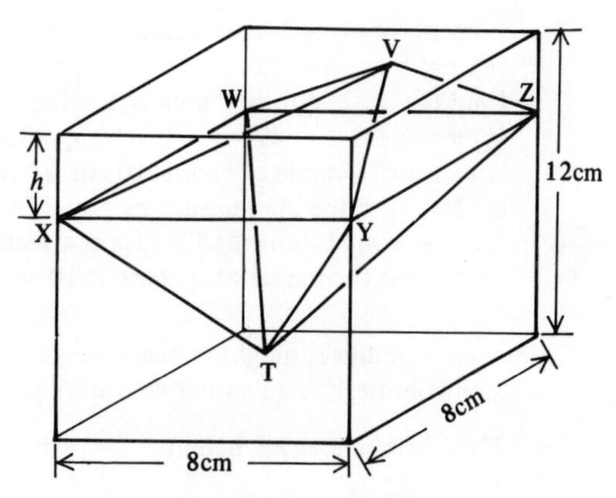

9. The diagram shows a model in which four identical pyramids are mounted at the corners of a rectangular block. The dimensions in centimetres are as shown. Calculate:
(a) the area of each vertical face of the model
(b) the volume of each pyramid
(c) the total volume of the model.

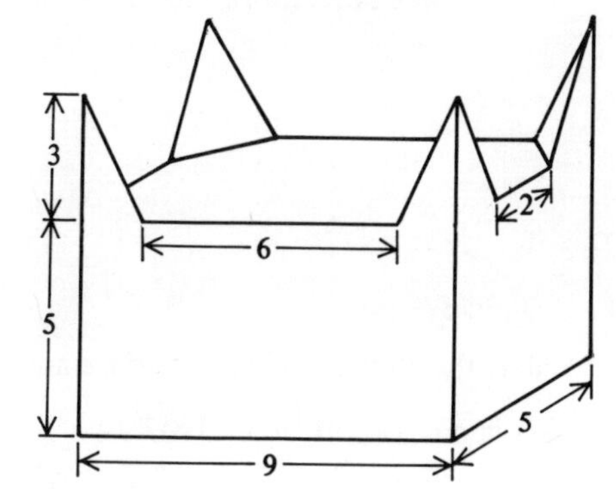

10. VABC is a regular tetrahedron of side 2 cm. M is the mid-point of BC and AG $= \frac{2}{3}$AM. Use Pythagoras' Theorem to find the lengths AM and VG and thus calculate:
(a) the area of triangle ABC
(b) the volume of the tetrahedron.

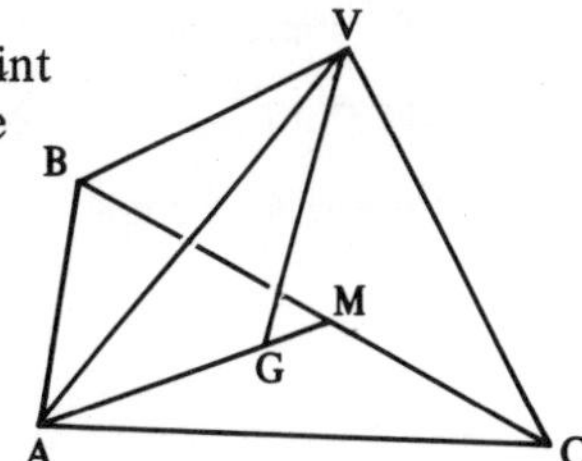

2.5 Sectors and cones

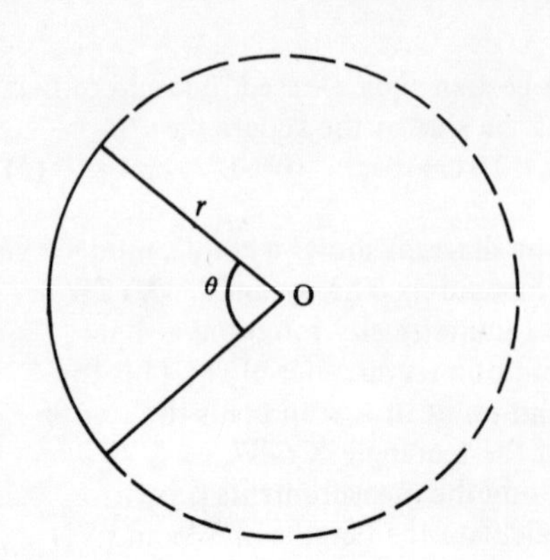

A sector is a part of a circle.
For a sector of radius r and angle θ,

$$\text{arc length} = \frac{\theta}{360} \cdot 2\pi r$$

$$\text{area} = \frac{\theta}{360} \cdot \pi r^2$$

Exercise 2.5a

1. Find the arc length and area of a sector of a circle with
   (a) $r = 5$ cm, $\theta = 28°$ (b) $r = 10$ cm, $\theta = 70°$ (c) $r = 7$ cm, $\theta = 105°$
2. Find the arc length and area of a sector of a circle with
   (a) $r = 8$ cm, $\theta = 42°$ (b) $r = 15$ cm, $\theta = 154°$ (c) $r = 14$ cm, $\theta = 10°$
3. A sector of a circle of radius 10 cm is 'rolled up' to form a cone. If the angle of the sector is 110°, find the curved surface area of the cone and the circumference of its base.
4. A paper cone is 'unrolled' to form a sector of a circle. The slant height of the cone is 21 cm and the radius of its base is 10 cm. Find the arc length of the sector and its angle.

For a cone of radius r, height h, slant height l, curved surface area S and volume V, we can derive the following formulae:

$$V = \tfrac{1}{3}(\text{base area}) \times \text{height} = \tfrac{1}{3}\pi r^2 h$$

$$S = \text{area of sector of radius } l = \pi r l$$

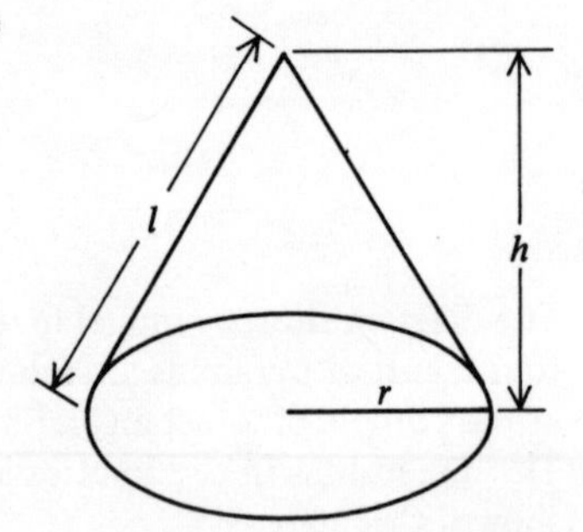

Example

Calculate the volume and total surface area of a cone of radius 8·2 cm and height 12·6 cm.

$$\text{Volume} = \tfrac{1}{3}\pi r^2 h = \tfrac{1}{3} \times 3{\cdot}14 \times 8{\cdot}2^2 \times 12{\cdot}6 = 887 \text{ cm}^3$$

$$l^2 = r^2 + h^2 = 8{\cdot}2^2 + 12{\cdot}6^2 = 226; \text{ so } l = \sqrt{226} = 15{\cdot}0$$

$$S = \pi r l = 3{\cdot}14 \times 8{\cdot}2 \times 15{\cdot}0 = 387 \text{ cm}^2$$

$$\text{Area of base} = \pi r^2 = 3{\cdot}14 \times 8{\cdot}2^2 = 335 \text{ cm}^2$$

So total surface area of cone $= 722 \text{ cm}^2$

Exercise 2.5b

1. Calculate the volume of a cone of radius 4 cm and height 10 cm.
2. Calculate the volume of a cone of radius 30 mm and height 60 mm.
3. A cone has a base diameter of 8 cm and a height of 3 cm. Calculate:
   (a) its slant height (b) its volume (c) its curved surface area.
4. A cone has a vertical height of 12 cm and slant height of 13 cm. Calculate:
   (a) its radius (b) its volume (c) its curved surface area.
5. Which has the greater volume: a cone with radius 2 cm and height 3 cm or a cone with radius 3 cm and height 2 cm?
6. A cone has a volume of 66 cm^3. Find its radius if its height is 7 cm. (Take π to be $\frac{22}{7}$.)
7. A cylinder of height 4 cm and radius 2 cm has a cone of height 4 cm and radius 2 cm mounted on top of it. Find the total volume of the shape.
8. Calculate the total surface area of the shape in question 7.
9. A metal sphere of radius 2 cm is melted down and recast in the shape of a cone of radius 2 cm. Find the height of the cone.
10. A metal cone of height h and radius r is melted down and recast in the shape of a hemisphere of the same radius.
    Which of the following is a true statement?
    (a) $h = r$ (b) $2h = r$ (c) $h = 2r$ (d) $h = r/\pi$ (e) $2\pi r = h$
11. A sphere of radius 3 cm fits exactly inside a cone of radius 6 cm. The diagram shows a cross section.
    (a) By using similar triangles, show that
    $$\frac{6}{3} = \frac{3+x}{y} = \frac{6+y}{x}.$$
    (b) Find the values of x and y.
    (c) State the height of the cone.
    (d) Show that the volume of the sphere is three-eighths of the volume of the cone.

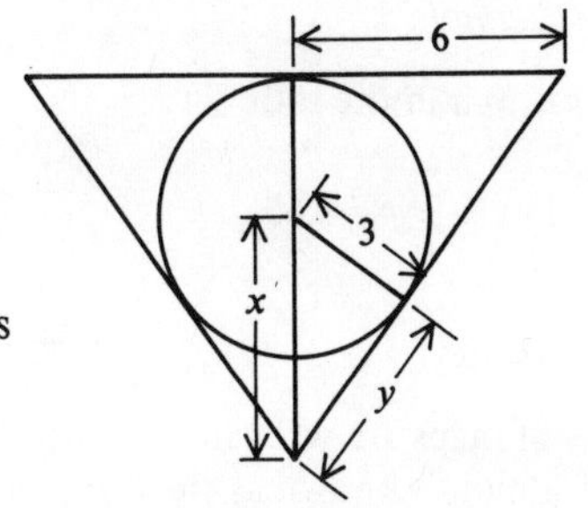

12. A cylinder fits entirely within a cone of radius r and height h as shown. The height of the cylinder is half that that of the cone. Find:
    (a) the radius of the cylinder as a fraction of the radius of the cone.
    (b) the volume of the cylinder as a fraction of the volume of the cone.

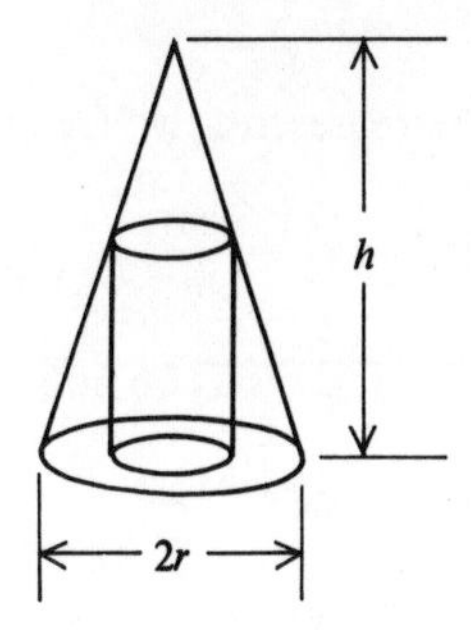

2.6 Ratio and scale

Ratio

Ratio is a comparison of two quantities; it is written either as $a:b$ or as a fraction $\frac{a}{b}$.
Always express the *two* quantities of a ratio in the same units.

Remember:
the ratio of *areas* of similar shapes is equal to the ratio of the *squares* of corresponding lengths;
the ratio of *volumes* of similar shapes is equal to the ratio of the *cubes* of corresponding lengths.

Example 1

Find, as simply as possible, the ratio of 60 mm to 1·60 m.

60 mm : 1·60 m = 6 cm : 160 cm = 3 : 80

Example 2

Two cubes are of length 2 cm and 5 cm. Find the ratio of their volumes in the form $1:n$.

Ratio of volumes = $(2\text{ cm})^3:(5\text{ cm})^3$ = 8 : 125 = 1 : 15·625

Example 3

Increase 3 m in the ratio 2 : 7.

$$3\text{ m} \times \frac{7}{2} = 10{\cdot}5\text{ m}$$

Scales

Scales of maps are written as a ratio or as a fraction. For example, 1 : 50 000 or $\frac{1}{50\,000}$.
The fraction is known as the *representative fraction* (R.F.).

Example 4

Find the representative fraction of a map drawn to a scale of 1·5 cm to 1 km.

$$\text{R.F.} = \frac{1{\cdot}5\text{ cm}}{1\text{ km}} = \frac{15\text{ mm}}{1000 \times 1000\text{ mm}} = \frac{3}{200\,000}$$

The scale is 1 : 66 666.

Example 5

Find the actual area of a lake shown by an area of 2 cm² on a map of scale 1 : 50 000.

$$\text{Area of } 2\text{ cm}^2 \text{ increased in the ratio } (1:50\,000)^2 = 2 \times 50\,000^2\text{ cm}^2$$

$$= \frac{2 \times 50\,000 \times 50\,000}{100\,000 \times 100\,000}\text{ km}^2$$

$$= 0{\cdot}5\text{ km}^2$$

Exercise 2.6

1. Find, as simply as possible, the ratio:
(a) 8 cm to 24 cm (b) 8 mm to 1 cm (c) 25 cm to 2 m (d) 8 m to 2 km
2. Find as a ratio in the form $1:n$
(a) 4 cm to 10 cm (b) 3 cm to 10 m (c) 6 m to 216 km (d) 50 cm to 1 km
3. Increase 5 cm in the ratio 2:5.
4. Increase 8 m in the ratio 16:21.
5. Increase 25 mm in the ratio 5:12.
6. Increase 3 km in the ratio 12:25.
7. Decrease 40 mm in the ratio 8:5.
8. Decrease 1 km in the ratio 4:3.
9. Decrease 2·8 cm in the ratio 7:4.
10. Decrease 6·25 m in the ratio 5:3.
11. A map is drawn to a scale of 1:25 000. Find the actual length of a line on the map of length: (a) 1 cm (b) 2·3 cm (c) 3 mm.
12. A map is drawn to a scale of 1:40 000. Find the length on the map of a straight road of length: (a) 1 km (b) 400 m (c) 600 m.
13. Find the representative fraction (or scale) of a map in which 1·5 km appears as 3 cm.
14. Find the representative fraction (or scale) of a map in which 45 mm represents 9 km.
15. Find the actual area of a lake which appears on a 1:25 000 map as an area of 5 cm^2.
16. Find the actual area of a forest which appears on a 1:10 000 map as an area of 8·2 cm^2.
17. Find the area of an orchard on a 1:40 000 map if its actual area is 80 000 m^2.
18. Find the area of an airfield on a 1:50 000 map if its actual area is 3 km^2.
19. Two maps are drawn of the same park. On the first a lake appears with an area of 2 cm^2 and on the second with an area of 8 cm^2. If the scale of the first map is 1:25 000, what is the scale of the second map?
20. Two plans are drawn of a building site. The first is to a scale of 1:500 in which the building site has an area of 40 cm^2. The second is more detailed and the building site has an area of 1000 cm^2. Find the scale of the second plan.
21. P and Q are two points which appear on a 1:10 000 map at a distance of 1·8 cm apart. P lies on the 400 m contour line and Q on the 450 m contour line. Find:
(a) the difference in height between P and Q
(b) the horizontal distance from P to Q
(c) the actual straight line distance from P to Q (by Pythagoras' Theorem).

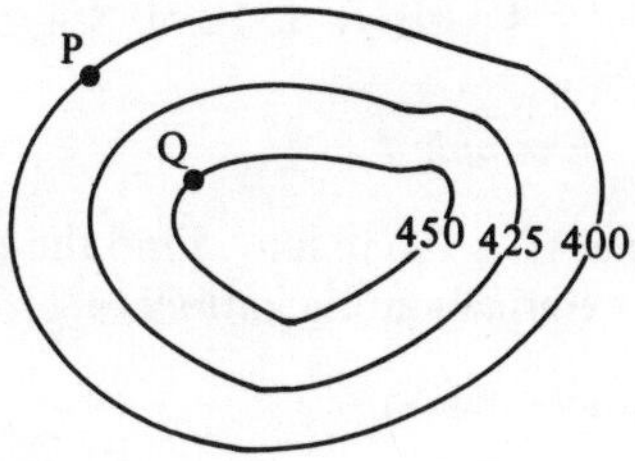

P
Q
450 425 400

22. A straight road from A to B has a slope of 1 in 4: it rises 1 metre for every 4 metres travelled up the road. If the distance along the road from A to B is 600 m, find:
(a) the difference in vertical height between A and B
(b) the horizontal distance between A and B
(c) the distance between A and B shown on a 1:50 000 map.

2.7 Ratio and proportional parts

Example 1

Divide a line 20 cm long into 7 equal parts.

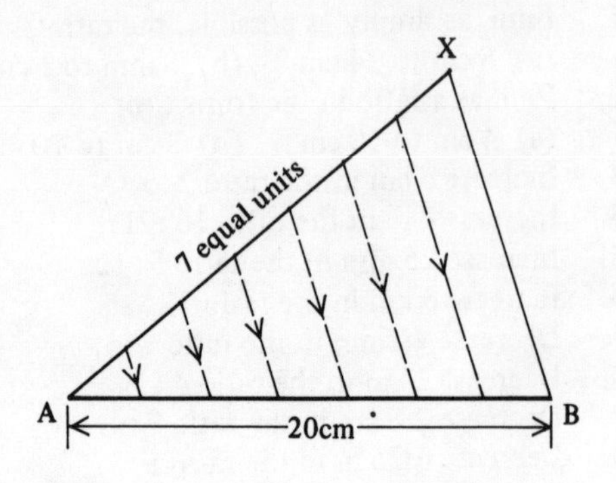

Construction

Draw AB 20 cm long.
Draw AX at any angle to AB 7 units long; say 14 cm.
Join XB.
Draw parallels to XB every unit along AX.

Example 2

Divide a line 16 cm long in the ratio 3 : 5.

$3 + 5 = 8$. Divide 16 cm into 8 equal parts.
3 parts $= 3 \times 2$ cm $= 6$ cm
5 parts $= 5 \times 2$ cm $= 10$ cm

The parts are 6 cm and 10 cm.

Example 3

Divide £2500 between three men in the proportions 1 : 3 : 4.

$1 + 3 + 4 = 8$. Divide £2500 into 8 equal parts.
1 part = £312·50
3 parts = £312·50 × 3 = £937·50
4 parts = £312·50 × 4 = £1250

Example 4

A line AB is 10 cm long. Find the positions of points X and Y such that they divide AB internally and externally in the ratio 9 : 4.

Internal division

AX : XB = 9 : 4

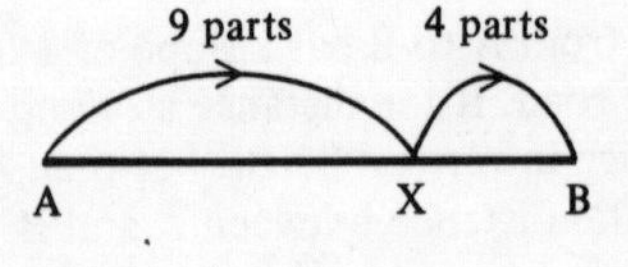

Divide AB into 13 parts
Each part is $\frac{10}{13} \approx 0{\cdot}77$ cm

So AX = 9 × 0·77 = 6·9 cm
XB = 4 × 0·77 = 3·1 cm

External division

AY : YB = 9 : 4; so AB = 5 parts.

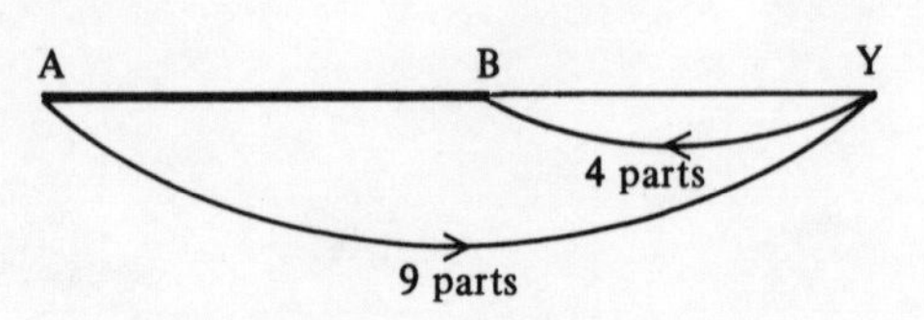

Divide AB into 5 parts
Each part is $\frac{10}{5} = 2$ cm

So AY = 9 × 2 = 18 cm
YB = 4 × 2 = 8 cm

Exercise 2.7

For questions 1–6 you may use any construction, including ruler and set square, to draw parallel lines.

1. Draw a line AB of length 15 cm. Use a construction to divide AB into 7 equal parts. Measure $\frac{3}{7}$ of AB.
2. Draw a line AB of length 20 cm. Use a construction to divide AB into 11 equal parts. Measure $\frac{4}{11}$ of AB.
3. AB = 12 cm. Find, by construction, the point X that divides AB internally in the ratio 3:4. Measure AX.
4. AB = 14 cm. Find, by construction, the point X that divides AB internally in the ratio 2:3. Measure AX.
5. AB = 9 cm. Find, by construction, the point Y that divides AB externally in the ratio 10:1. Measure BY.
6. AB = 11 cm. Find, by construction, the point Y that divides AB externally in the ratio 9:2. Measure BY.
7. Calculate the length of each part of a line:
   (a) 1 m long divided in the ratio 7:13
   (b) 80 cm long divided in the ratio 7:9
   (c) 1·20 m long divided in the proportions 1:2:3
   (d) 78 cm long divided in the proportions 4:3:6.
8. Calculate, to the nearest penny, each part of £1 divided in the proportions
   (a) 2:3 (b) 4:7 (c) 3:2:5 (d) 1:2:6 (e) 3:4:8
9. Three sons are left a legacy of £25 000 to be divided between them in proportion to their ages which are 21, 25 and 29 years. Find how much each son receives.
10. Two men put £5000 and £9000 into an enterprise which produced a profit of £1050. If the profit is divided in the ratio of their original investment, how much did each man receive?
11. ABC is a triangle in which AB = 8 cm, BC = 12 cm and CA = 10 cm.
    X divides AB in the ratio 5:3.
    Y divides AC in the ratio 4:6.
    Draw the triangle and find the point Z where YX meets CB produced.
    (a) Find, by drawing, the ratio BZ:ZC
    (b) Find the value of the product $\frac{AX}{XB} \cdot \frac{BZ}{ZC} \cdot \frac{CY}{YA}$

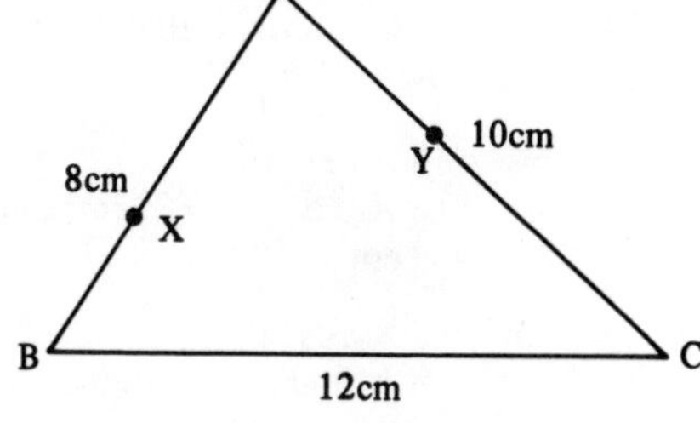

12. Using the same measurements of a triangle ABC as in question 11, let CX and BY meet at T. Let AT produced meet BC at S.
    (a) Find, by drawing, the ratio BS:SC
    (b) Find the value of the product $\frac{AX}{XB} \cdot \frac{BS}{SC} \cdot \frac{CY}{YA}$

Part 3: Trigonometry

3.1 Sine and cosine

If P moves in a circle centre O, and OP = 1 unit, then

$$\text{vertical displacement} = PN = \text{sine } \theta$$
$$\text{horizontal displacement} = ON = \text{cosine } \theta$$

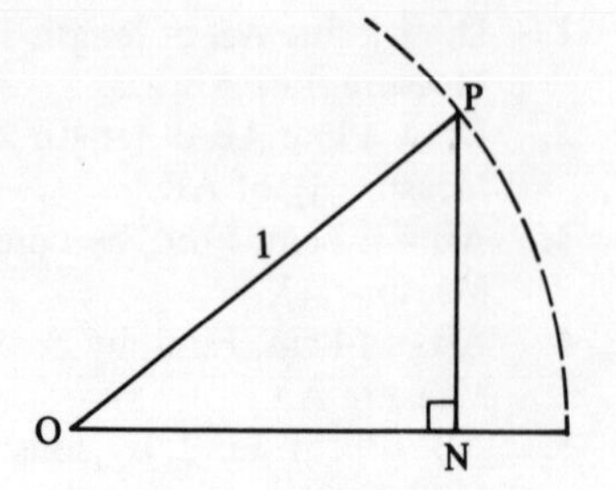

OP is the *hypotenuse* of the triangle NOP. If OP $\neq$ 1, then you multiply the sine and cosine values by the length OP.

Note: In triangle ABC with $\hat{B} = 90°$

$$BC = AC \times \sin \hat{A}$$

so $\sin \hat{A} = \dfrac{BC}{AC} = \dfrac{\text{side opposite}}{\text{hypotenuse}}$

$$AB = AC \times \cos \hat{A}$$

so $\cos \hat{A} = \dfrac{AB}{AC} = \dfrac{\text{side adjacent}}{\text{hypotenuse}}$

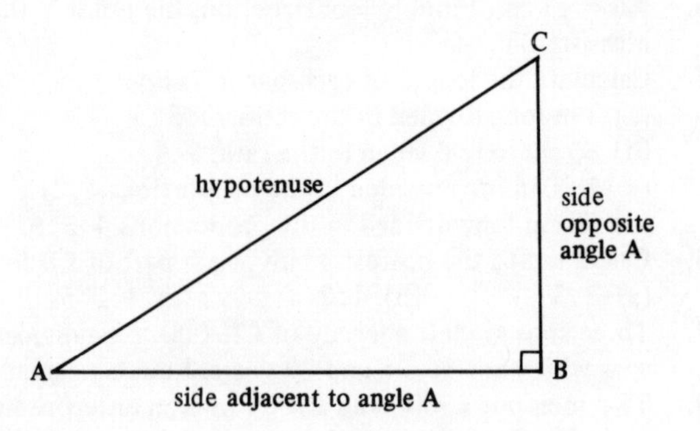

Example 1

The hypotenuse of a right-angled triangle is 8 cm long. One of the angles is 30°. Find the lengths of the other two sides.

$$\begin{aligned} BC &= 8 \times \sin 30° \\ &= 8 \times 0{\cdot}5 \quad \text{(from tables)} \\ &= 4 \text{ cm} \end{aligned}$$

$$\begin{aligned} AB &= 8 \times \cos 30° \\ &= 8 \times 0{\cdot}866 \quad \text{(from tables)} \\ &= 6{\cdot}9 \text{ cm} \end{aligned}$$

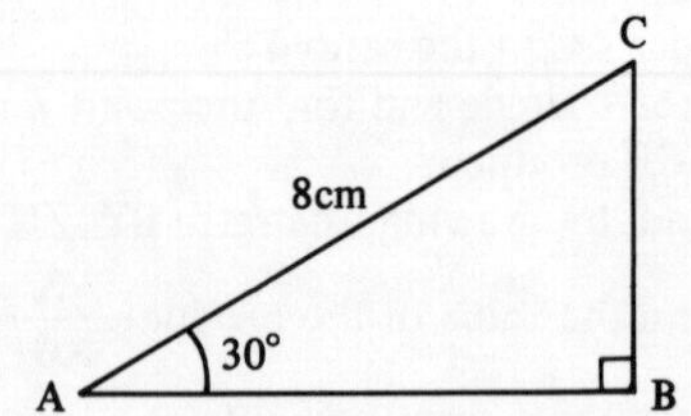

Example 2

A plank 7 m long reaches 5 m up a vertical wall. Find the angle between the plank and the horizontal.

$$\begin{aligned} \sin \theta &= \tfrac{5}{7} = 0{\cdot}714 \\ \theta &= 45{\cdot}6° \quad \text{(from tables)} \end{aligned}$$

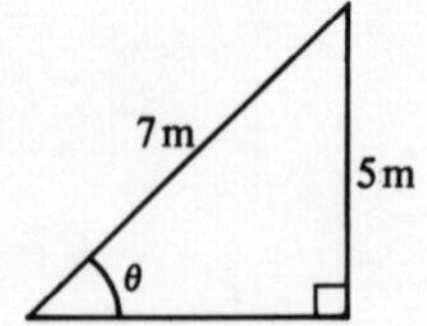

Exercise 3.1

1. Use tables to find the sine of the angle to three significant figures.
   (a) 30° (b) 45° (c) 81° (d) 10·5° (e) 74·5° (f) 64·2°
2. Use tables to find the cosine of the angle to three significant figures.
   (a) 60° (b) 45° (c) 54° (d) 41·8° (e) 8·7° (f) 16·7°
3. Use tables to find, to the nearest 0·1°, the angle whose sine is:
   (a) 0·5 (b) 0·866 (c) 0·766 (d) 0·009 (e) 0·887 (f) 0·434
4. Use tables to find, to the nearest 0·1°, the angle whose cosine is:
   (a) 0·5 (b) 0·515 (c) 0·299 (d) 0·91 (e) 0·905 (f) 0·218
5. From the information in the diagram, find the lengths of the unknown sides:
   (a) (b) (c)

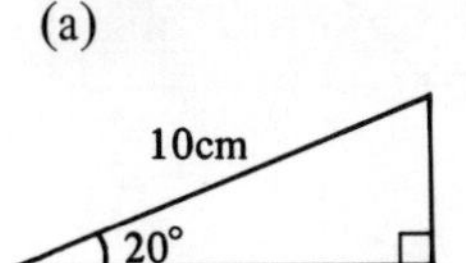

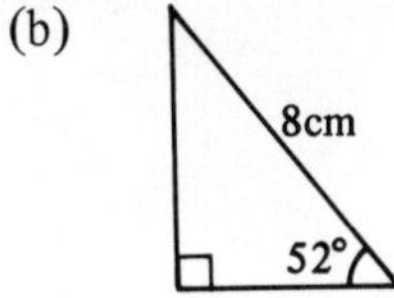

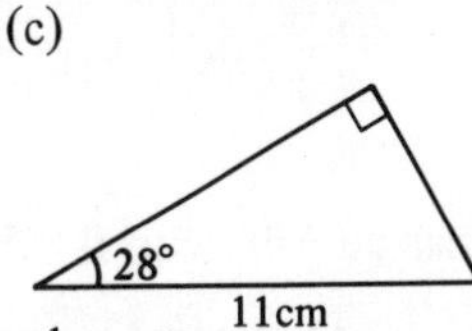

6. From the information in the diagram, find the sizes of the angles:
   (a) (b) (c)

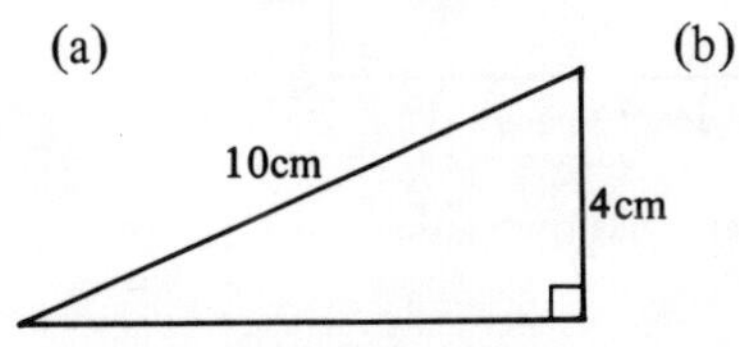

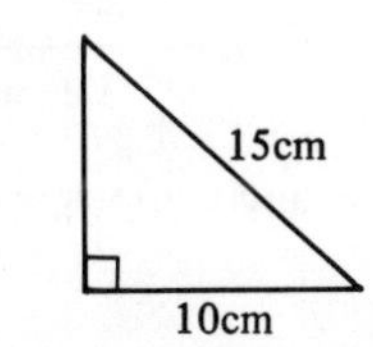

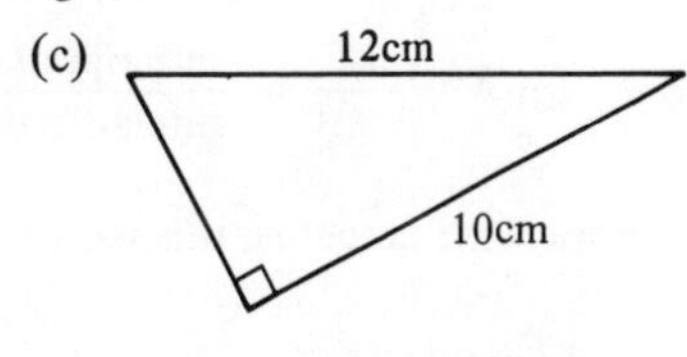

7. A fairground Big Wheel has arms 20 m long. Find the height above the ground of a passenger when his rotor arm is at the following angles to the horizontal:
   (a) 60° below horizontal (b) 10° below horizontal (c) 40° above horizontal
   You may assume that the end of the rotor arm just touches the ground.
8. For the same Big Wheel as in question 7, find the angle of a passenger's rotor arm when he is at the following heights above the ground:
   (a) 20 m (b) 40 m (c) 10 m (d) 26·8 m.
9. A ladder 10 m long leans against a wall at an angle of 70° to the horizontal. It slips 2 m down the wall. Find its new angle of inclination to the horizontal.
10. A pair of step ladders with sides 2·4 m long stands on level ground with the feet of the two parts 2 m apart. Find the angle of inclination of each leg and the height of the top of the steps above the ground.
11. An angle of 66° at the centre of a circle stands on a chord 7·28 cm long. Find the radius of the circle.
12. A ramp inclined at 7° rises 0·3 m. Find the length of the ramp.

3.2 Tangent

If P moves on a line perpendicular to ON and ON = 1 unit, then

$$PN = \tan\theta.$$

If ON $\neq$ 1, then you multiply $\tan\theta$ by the length ON.

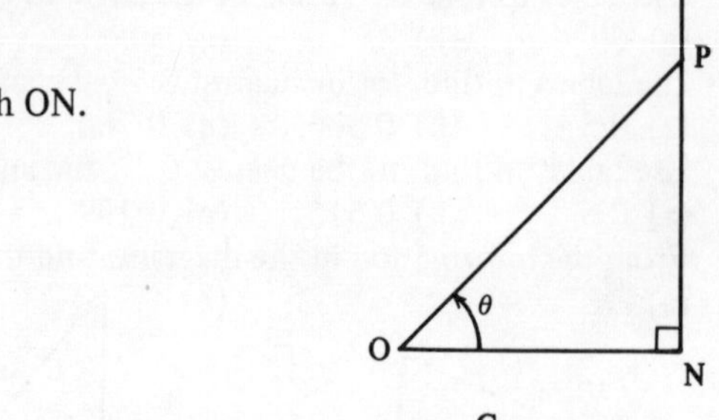

Note: In triangle ABC with $\hat{B} = 90°$

$$BC = AB \times \tan\hat{A}$$

so $$\tan\hat{A} = \frac{BC}{AB} = \frac{\text{side opposite}}{\text{side adjacent}}$$

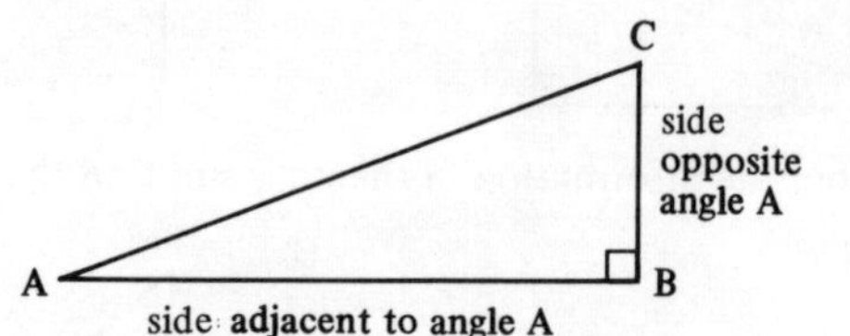

In work with tangents, it is easier to use the angle opposite an unknown side.

Example 1

The foot of a ladder is 1·20 m from the base of a vertical wall. The top of the ladder reaches 5 m up the wall. Find the angle the ladder makes with the horizontal.

$$\tan\theta = \frac{5}{1{\cdot}2} = \frac{50}{12} = 4{\cdot}17$$

so $\theta = 76{\cdot}5°$ (from tables)

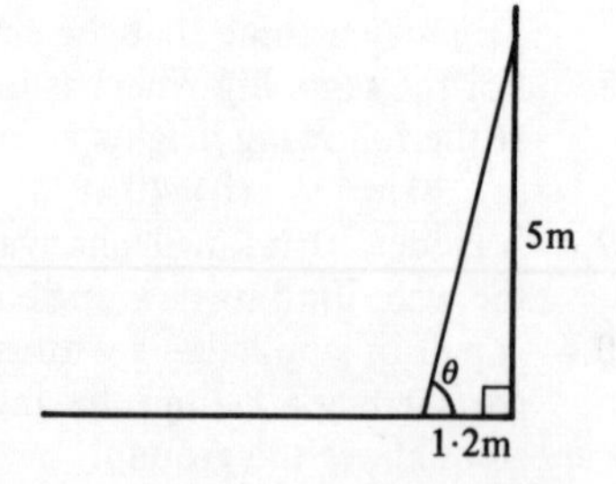

Example 2

A boat is 300 m from the bottom of a vertical cliff. The angle of elevation of the top of the cliff from the boat is 31°. Find the height of the cliff.

$$\frac{h}{300} = \tan 31°$$

so $h = 300 \times \tan 31°$

$= 300 \times 0{\cdot}601$ (from tables)

$= 180\,\text{m}$ (to the nearest metre)

Exercise 3.2

1. ABCD is a square. M is the mid-point of CD. Without using tables state whether the following are true or false:
   (a) $\tan D\hat{B}C = 1$
   (b) $\tan M\hat{B}C = \frac{1}{2}$
   (c) $M\hat{B}C = \frac{1}{2} D\hat{B}C$
   (d) $\sin M\hat{B}C > \tan M\hat{B}C$
   (e) $M\hat{B}C < D\hat{B}M$

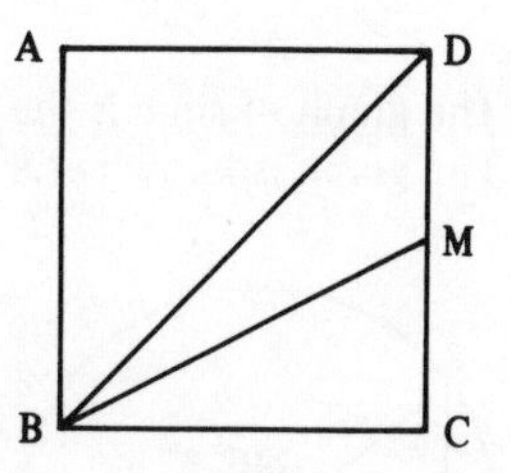

2. ABC is an equilateral triangle of side 2 units. M is the mid-point of BC. Use Pythagoras' Theorem to find the length AM, leaving your answer as a square root. Write down, without using tables:
   (a) $\tan 60°$ (b) $\tan 30°$.
3. Use tables to write down, to three significant figures, the tangent of the angle:
   (a) 40° (b) 17° (c) 65° (d) 80·3° (e) 78·7°
4. From tables find to the nearest 0·1°, the angle whose tangent is:
   (a) 0·404 (b) 1·804 (c) 5·097 (d) 0·378 (e) 7·4
5. From the information in the diagram find the unknown angles:

(a)

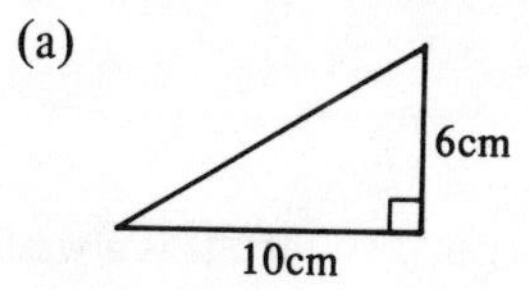

(b)

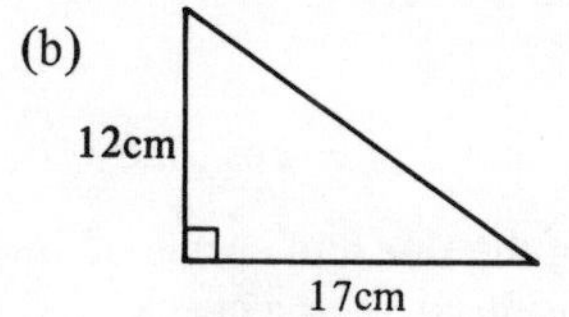

(c)

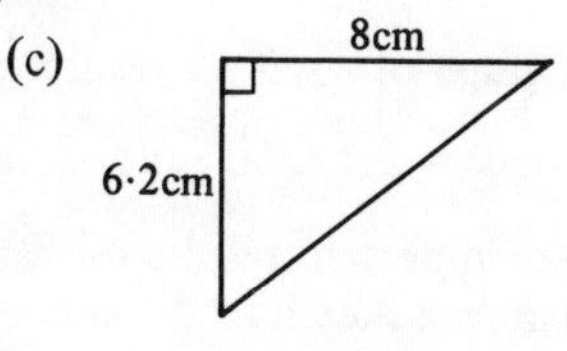

6. From the information in the diagram find the side marked x:

(a)

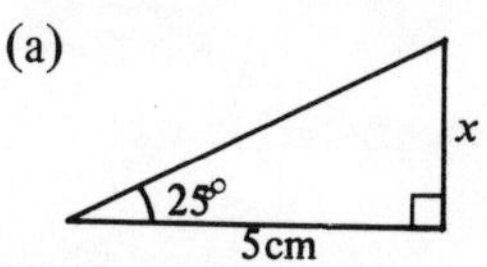

(b)

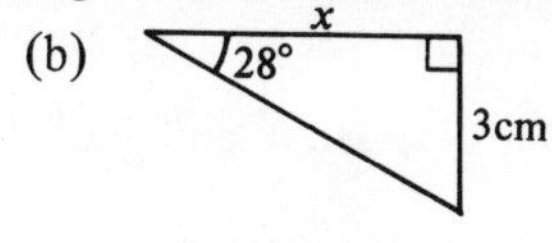

(c)

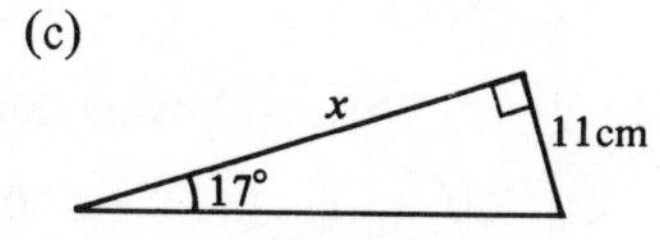

7. From a point 100 m from the foot of a church tower, the angle of elevation of the top of the tower is 18°. Find:
   (a) the height of the tower;
   (b) the angle of elevation of the top of the tower from a point 50 m from the base of the tower.
8. A boy is estimating the height of a cliff 70 m high by taking angle measurements from the beach below.
   (a) Find the angle of elevation he measures from a point 500 m from the bottom of the cliff.
   (b) Find how far he walks towards the cliff if his second measurement is 13°.

3.3 Sine and cosine graphs

The graph of sin θ is the graph of vertical displacements of P as it travels round a unit circle. The graph is known as a *sine wave.* All sine and cosine graphs have a similar appearance.

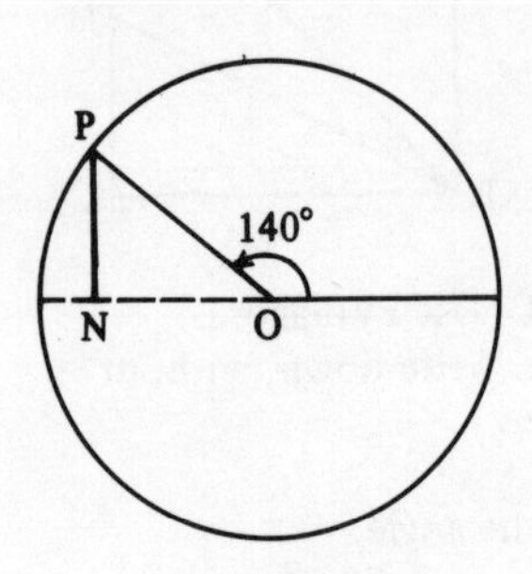

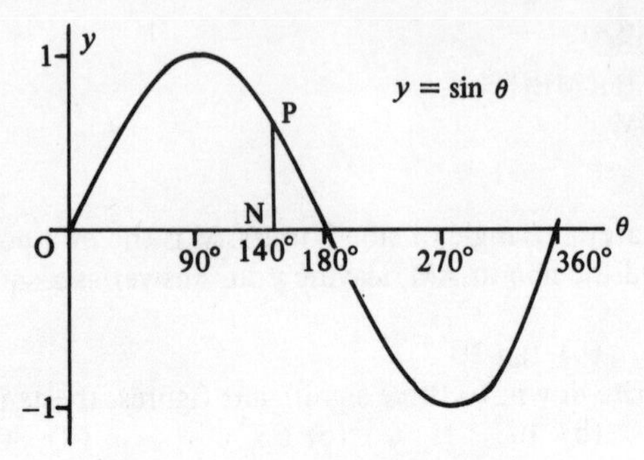

From the symmetry of the graph,

$$\sin\theta = -\sin(360° - \theta) \quad \text{and} \quad \sin\theta = \sin(180 - \theta)$$

From a graph of cos θ the results are

$$\cos\theta = \cos(360 - \theta) \qquad \text{and} \quad \cos\theta = -\cos(180 - \theta)$$

These are important results for finding the sine and cosine of angles larger than 90°. It is always best to draw a sketch of the sine or cosine curve as a check.

Example 1

Find (a) cos 220° (b) two values of x between 0° and 360° if $\cos x = -0{\cdot}4$.

(a) $\cos 220° = \cos(360° - 220°)$
$= \cos 140°$
$= -\cos(180° - 140°)$
$= -\cos 40°$
$= -0{\cdot}766$

(b) $\cos x = -0{\cdot}4$

From tables, $\cos 66{\cdot}4° = 0{\cdot}4$
$x = 180° - 66{\cdot}4°$
$= 113{\cdot}6°$

Also $x = 360° - 113{\cdot}6°$
$= 246{\cdot}4°$

The two values are 113·6° and 246·4°.

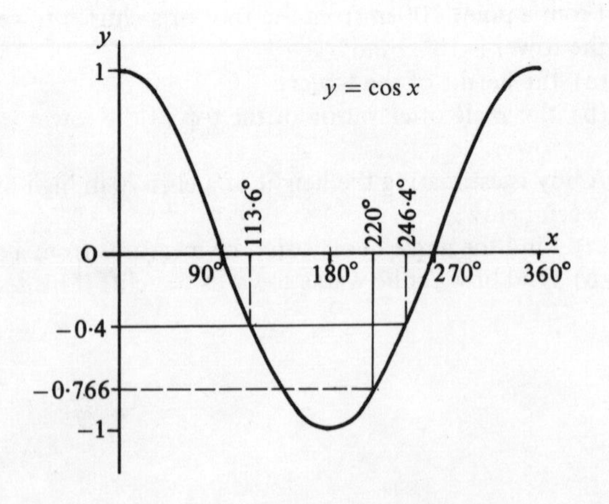

Example 2

Draw the graph of $y = \cos(\frac{1}{2}x)$ for values of x between 0° and 360°. Use the graph to solve the equation $\cos(\frac{1}{2}x) = \frac{x}{90} - 1$.

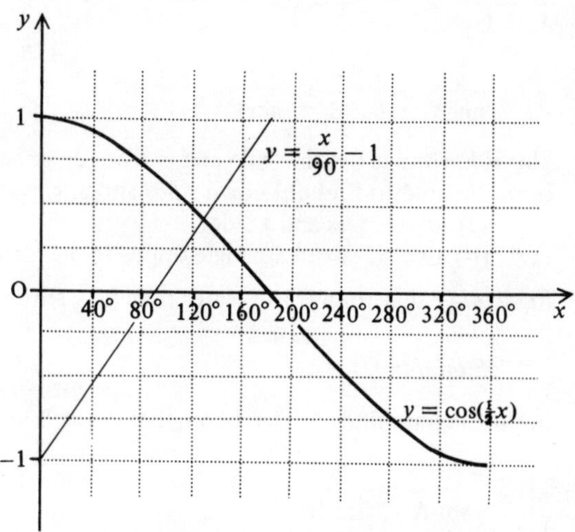

Draw up a table of values at 20° intervals for x from 0° to 360°:

x	0	20	40	60	80
$\frac{1}{2}x$	0	10	20	30	40
$\cos\frac{1}{2}x$	1	0·98	0·94	0·87	0·77

Plot $\cos\frac{1}{2}x$ against x using a suitable scale.

To find a solution of the equation

$$\cos(\tfrac{1}{2}x) = \frac{x}{90} - 1$$

plot the line $y = \frac{x}{90} - 1$ on the same graph.

For a straight line only two points are needed with a third as a check, say (0, −1), (90, 0), (180, 1). The intersection of these two graphs will give the required solution.

Exercise 3.3

1. Draw a sketch of $y = \sin x$ for $0 \leqslant x \leqslant 360°$. Given that $\sin 35° = 0{\cdot}574$, write down the values of the sine of:
   (a) 145° (b) 215° (c) 325°.
2. Draw a sketch of $y = \cos x$ for $0 \leqslant x \leqslant 360°$. Show on your graph that $\cos x = 0{\cdot}3$ has two solutions in this range. Find the solutions from tables.
3. Use tables to write down:
   (a) sin 132° (b) cos 205° (c) sin 187° (d) cos 157°.
4. Use tables to write down:
   (a) cos 310° (b) cos 116° (c) sin 304° (d) sin 211°.
5. Draw your own copy of the graph of Example 2 using a scale of 1 cm to 20° on the x-axis and 4 cm to 1 unit on the y-axis. Hence solve the equation
   $$\cos(\tfrac{1}{2}x) = \frac{x}{90} - 1.$$
6. Draw the graph of $y = 2\sin 2x$ for $0 \leqslant x \leqslant 180°$ using a scale of 1 cm to 10° on the x-axis and 2 cm to 1 unit on the y-axis. Use your graph to solve the equation
   $$2\sin 2x = \frac{x}{30}.$$
7. Draw a graph to find all the solutions between 0° and 360° of the equation
   $$2\cos x = 2 - \frac{x}{240}.$$
8. Draw a graph to find all the solutions between 0° and 360° of the equation
   $$2\sin x = 2 - \frac{x}{90}.$$
9. With the same scales and axes, draw the graphs of $y = 2\cos 2x$ and $y = \sin 3x$ for $0 \leqslant x \leqslant 90°$. Obtain a solution to $2\cos 2x = \sin 3x$ between 0° and 90°.

3.4 Sine rule

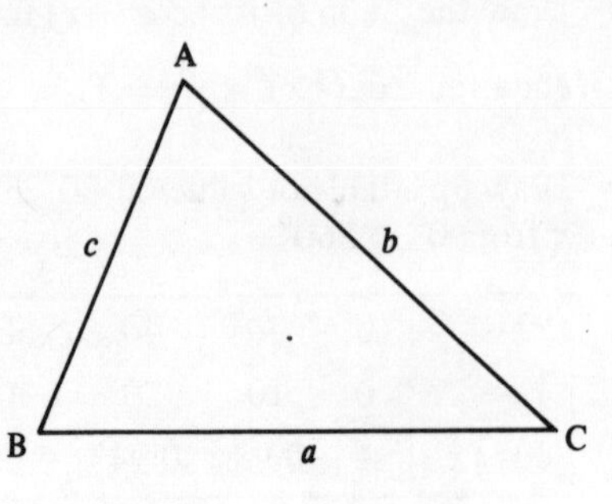

For any triangle ABC with sides a, b, c opposite the angles $\hat{A}, \hat{B}, \hat{C}$

$$\frac{a}{\sin \hat{A}} = \frac{b}{\sin \hat{B}} = \frac{c}{\sin \hat{C}}$$

This is known as the sine rule and is true for any triangle.
It can be used to find unknown sides and angles if you know:

(a) two angles and a side

or (b) two sides and an angle opposite one of these sides.

Remember that for angles greater than 90°, $\sin \theta = \sin (180° - \theta)$.

Example 1

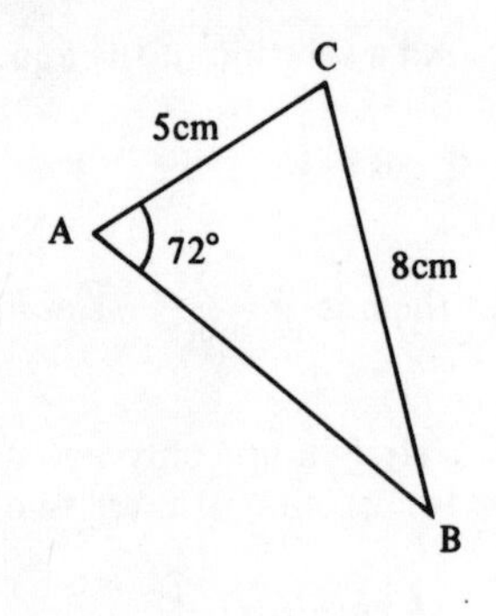

In the diagram find the size of angle B if $a = 8$, $b = 5$, $\hat{A} = 72°$.

$$\frac{a}{\sin \hat{A}} = \frac{b}{\sin \hat{B}}$$

$$\frac{8}{\sin 72°} = \frac{5}{\sin \hat{B}}$$

hence $\sin \hat{B} = \dfrac{5 \sin 72°}{8}$

$$= \frac{5 \times 0{\cdot}951}{8} = \frac{4{\cdot}755}{8} = 0{\cdot}594$$

so $\hat{B} = 36{\cdot}4°$

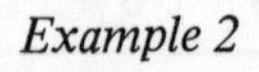

Example 2

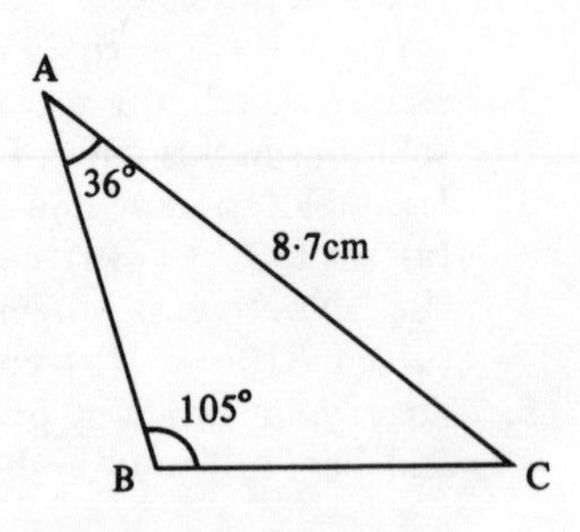

In the diagram find the length of AB.
$AB = c$, $b = 8{\cdot}7$, $\hat{B} = 105°$, $\hat{C} = 180° - (105° + 36°) = 39°$

So $\dfrac{c}{\sin 39°} = \dfrac{8{\cdot}7}{\sin 105°}$

$$c = \frac{8{\cdot}7 \sin 39°}{\sin 105°}$$

$$= \frac{8{\cdot}7 \sin 39°}{\sin 75°}$$

$$= 5{\cdot}67 \text{ cm}$$

No.	Log
8·7	0·940
sin 39°	$\bar{1}{\cdot}799$
	0·739
sin 75°	$\bar{1}{\cdot}985$
5·67	0·754

The calculation is shown on the right. It is best to use log sine tables to look up logs of sin 39°, etc.

Exercise 3.4

In the triangle ABC with sides a, b, c

1. $\hat{A} = 40°, a = 4$ cm, $b = 3$ cm, find $\hat{B}$
2. $\hat{C} = 75°, a = 8$ cm, $c = 9$ cm, find $\hat{A}$
3. $\hat{B} = 18°, b = 2$ cm, $c = 4$ cm, find $\hat{C}$ and $\hat{A}$
4. $\hat{A} = 83°, a = 10$ cm, $b = 9$ cm, find $\hat{B}$ and $\hat{C}$
5. $\hat{A} = 42°, \hat{B} = 64°, a = 16$ cm, find b
6. $\hat{B} = 73°, \hat{C} = 54°, b = 20$ cm, find c
7. $\hat{C} = 36°, \hat{A} = 29°, a = 14$ cm, find b
8. $\hat{A} = 87°, \hat{B} = 46°, a = 8{\cdot}2$ cm, find c
9. $\hat{A} = 104°, \hat{B} = 20°, a = 17{\cdot}3$ cm, find b
10. $\hat{A} = 21°, \hat{C} = 17°, b = 11{\cdot}2$ cm, find a

In questions 11 to 20, find where possible all the remaining sides and angles of the triangle ABC.

11. $\hat{A} = 10°, \hat{B} = 84°, a = 6$ cm
12. $\hat{C} = 25°, \hat{A} = 74°, a = 3$ cm
13. $\hat{B} = 62°, b = 8$ cm, $a = 4$ cm
14. $\hat{A} = 46°, a = 6$ cm, $b = 9$ cm
15. $\hat{A} = 60{\cdot}5°, \hat{B} = 27{\cdot}3°, c = 6{\cdot}2$ cm
16. $\hat{B} = 17°, \hat{C} = 36{\cdot}4°, a = 9{\cdot}3$ cm
17. $\hat{C} = 125°, \hat{B} = 17°, b = 2{\cdot}4$ cm
18. $\hat{A} = 39°, \hat{B} = 16°, c = 4{\cdot}17$ cm
19. $\hat{A} = 58°, a = 11{\cdot}8$ cm, $b = 24{\cdot}2$ cm
20. $\hat{B} = 8°, b = 15$ cm, $c = 42$ cm.

21. The two sides of a step ladder are of length 2 m and 1·85 m. When fully opened, the longer side is inclined at 65° to the horizontal. Find the distance apart of the feet of the step ladder.

22. From a base line AB of length 500 m a surveyor measures the angles XAB and XBA to find the distance of a point X. If he finds $X\hat{A}B = 87°$ and $X\hat{B}A = 85{\cdot}4°$, find the distance XA.

23. To find the width of a river, a boy marks out a length of 20 paces on one bank opposite a tree on the other bank with the tree roughly opposite the centre of his line. He estimates the angles of a sight of the tree from each end of his line to be 40° and 45°. Find the distance of the tree from one end of the line and hence the width of the river to the nearest pace.

24. To measure the height of a cliff from a level beach, two sightings are made of the top of the cliff from points 200 m apart in line with the cliff. Using the information in the diagram, find the length BC and the height of the cliff above sea level.

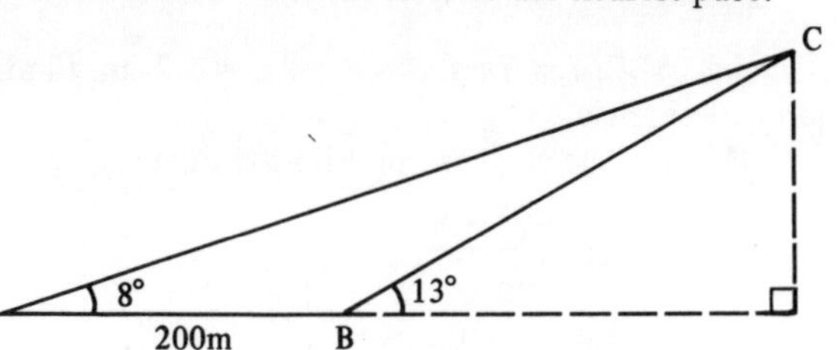

25. A field ABCD is a quadrilateral. A surveyor starts at B and measures the angles $A\hat{B}D = 21°$ and $D\hat{B}C = 54°$. He then measures the distance BD = 150 m and at D measures the angles $A\hat{D}B = 28°$ and $B\hat{D}C = 49°$. Find the lengths of each side of the field.

3.5 Cosine rule

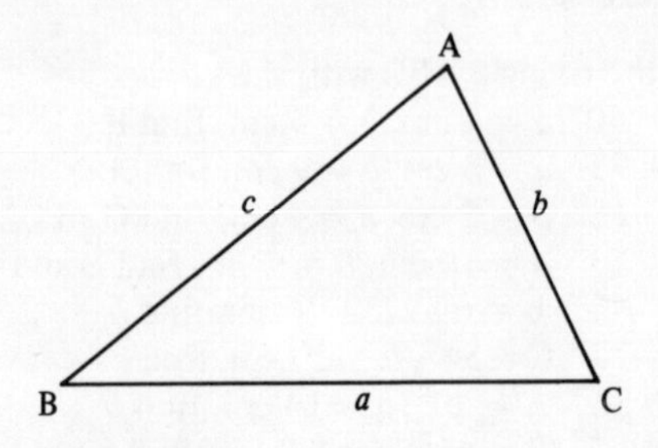

The cosine rule for any triangle ABC is

$$a^2 = b^2 + c^2 - 2bc \cos \hat{A}$$

which can be rearranged as

$$\cos \hat{A} = \frac{b^2 + c^2 - a^2}{2bc}$$

If you want to find other sides or angles, you have to choose the symbols accordingly. You should try to remember the rule by noting its symmetry and it will help if you remember that it starts like Pythagoras' Theorem. The rule is used to find:

(a) a side given the angle opposite and the other two sides;
(b) an angle given all three sides.

Don't forget that for angles greater than 90°, $\cos \theta° = -\cos (180° - \theta°)$.

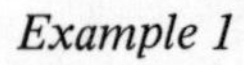

Example 1

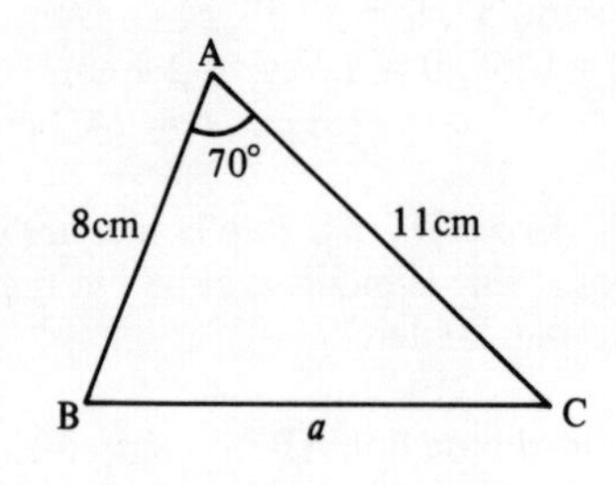

In triangle ABC, $\hat{A} = 70°$, $b = 11$ cm, $c = 8$ cm.
To find a,

$$\begin{aligned} a^2 &= b^2 + c^2 - 2bc \cos \hat{A} \\ &= 11^2 + 8^2 - (2 \times 11 \times 8 \times \cos 70°) \\ &= 121 + 64 - (2 \times 11 \times 8 \times 0{\cdot}342) \\ &= 185 - 60{\cdot}192 \\ &= 124{\cdot}8 \\ a &= \sqrt{124{\cdot}8} = 11{\cdot}2 \text{ cm to three significant figures.} \end{aligned}$$

Example 2

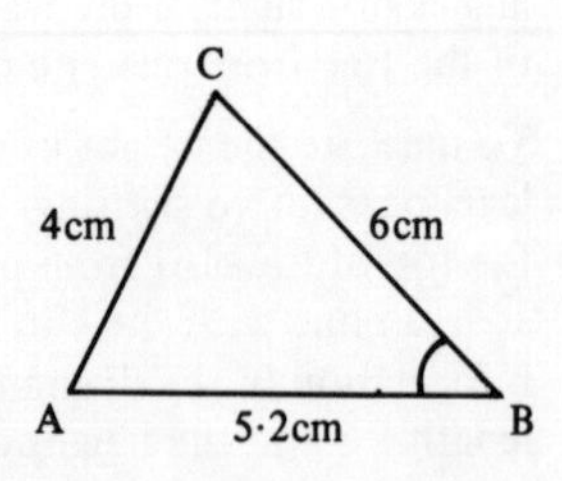

In triangle ABC, $a = 6$ cm, $b = 4$ cm, $c = 5{\cdot}2$ cm. Find the size of the smallest angle.

The smallest angle is opposite the smallest side.

$$\begin{aligned} \cos \hat{B} &= \frac{a^2 + c^2 - b^2}{2ac} \\ &= \frac{6^2 + 5{\cdot}2^2 - 4^2}{(2 \times 6 \times 5{\cdot}2)} \\ &= \frac{36 + 27{\cdot}0 - 16}{62{\cdot}4} = \frac{47{\cdot}0}{62{\cdot}0} \\ \hat{B} &= 40{\cdot}7° \text{ (from log cosine tables)} \end{aligned}$$

No.	Log
47·0	1·672
62·0	1·792
cos 40·7°	$\bar{1}$·880

Example 3

Three sides of a triangle are 3·4 cm, 12·2 cm, 14·7 cm. Find the size of the largest angle.

Let $a = 3{\cdot}4, b = 12{\cdot}2, c = 14{\cdot}7$

The largest angle is $\hat{C}$, opposite 14·7 which is the largest side.

$$\cos C = \frac{a^2 + b^2 - c^2}{2ab}$$

$$= \frac{3{\cdot}4^2 + 12{\cdot}2^2 - 14{\cdot}7^2}{2 \times 3{\cdot}4 \times 12{\cdot}2}$$

$$= \frac{11{\cdot}6 + 149 - 216}{2 \times 3{\cdot}4 \times 12{\cdot}2} = \frac{-55{\cdot}4}{82{\cdot}7}$$

$$\hat{C} = 180° - 47{\cdot}9° = 132{\cdot}1°$$

No.	Log
55·4	1·744
2	0·301
3·4	0·531
12·2	1·086
82·7	1·918
cos 47·9°	$\bar{1}$·826

Exercise 3.5

In questions 1 to 10, find the length of the unknown side of the triangle ABC.

1. $a = 11$ cm, $b = 4$ cm, $\hat{C} = 36°$
2. $b = 6$ cm, $a = 8$ cm, $\hat{C} = 27°$
3. $a = 16$ cm, $c = 21$ cm, $\hat{B} = 80°$
4. $a = 10$ cm, $b = 12$ cm, $\hat{C} = 72°$
5. $a = 1{\cdot}7$ cm, $b = 3{\cdot}2$ cm, $\hat{C} = 24°$
6. $c = 1{\cdot}32$ cm, $a = 2{\cdot}41$ cm, $\hat{B} = 35{\cdot}6°$
7. $b = 1$ cm, $c = 2$ cm, $\hat{A} = 100°$
8. $a = 5$ cm, $b = 4$ cm, $\hat{C} = 156°$
9. $a = 6{\cdot}7$ cm, $b = 8{\cdot}3$ cm, $\hat{C} = 54{\cdot}7°$
10. $b = 21{\cdot}2$ cm, $c = 43$ cm, $\hat{A} = 119°$

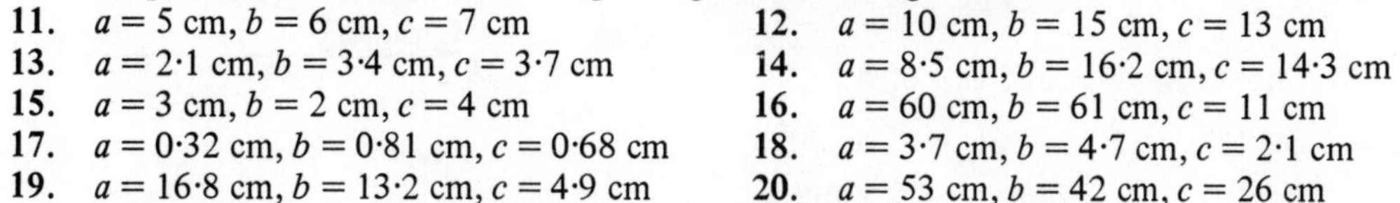

In questions 11 to 20, find the largest angle of the triangle ABC.

11. $a = 5$ cm, $b = 6$ cm, $c = 7$ cm
12. $a = 10$ cm, $b = 15$ cm, $c = 13$ cm
13. $a = 2{\cdot}1$ cm, $b = 3{\cdot}4$ cm, $c = 3{\cdot}7$ cm
14. $a = 8{\cdot}5$ cm, $b = 16{\cdot}2$ cm, $c = 14{\cdot}3$ cm
15. $a = 3$ cm, $b = 2$ cm, $c = 4$ cm
16. $a = 60$ cm, $b = 61$ cm, $c = 11$ cm
17. $a = 0{\cdot}32$ cm, $b = 0{\cdot}81$ cm, $c = 0{\cdot}68$ cm
18. $a = 3{\cdot}7$ cm, $b = 4{\cdot}7$ cm, $c = 2{\cdot}1$ cm
19. $a = 16{\cdot}8$ cm, $b = 13{\cdot}2$ cm, $c = 4{\cdot}9$ cm
20. $a = 53$ cm, $b = 42$ cm, $c = 26$ cm

21. A coastguard finds that the angle between his sights of two boats is 36°. He estimates their distance from him to be 3 km and 5 km. How far are the boats from each other?
22. The lengths of the sides of a triangle are in the proportion 2:5:6. Use the cosine rule to find one of the angles and then the sine rule to find the other angles.
23. A parallelogram has sides of length 4 cm and 7 cm. One of its angles is 60°. Find the length of each diagonal.
24. A golfer attempting a 110 m shot plays at 20° to the correct line and drives the ball 80 m. How far does the ball land from the hole?
25. The sides of a triangle are of length $\sqrt{2}$ cm, $\sqrt{8}$ cm and 3 cm. Find the sizes of all the angles.

3.6 Bearings

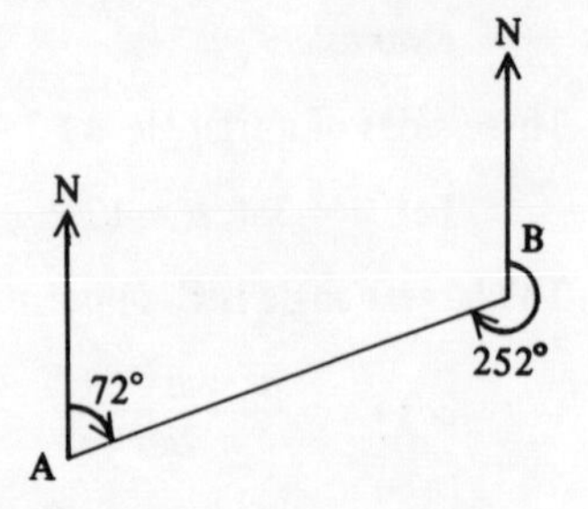

Bearings are directions from North, measured in a clockwise sense.
Bearings are usually given as three figures. The bearing of B from A in the diagram would be written 072°.
The bearing of A from B is 252°. This is also the *back bearing* of B from A.
Notice that bearings and back bearings differ by 180°.

Questions involving navigation and bearings can usually be solved by using right-angled triangles as in Example 1. Sometimes the sine or cosine rule gives a more direct solution (see example 2).

Example 1

A ship sails 7 km on a bearing of 075° from A to B and then 3 km on a bearing of 330° from B to C. Find how far C is north and east of A and the bearing of C from A.

In the diagram ML and AK are drawn perpendicular to the North lines.

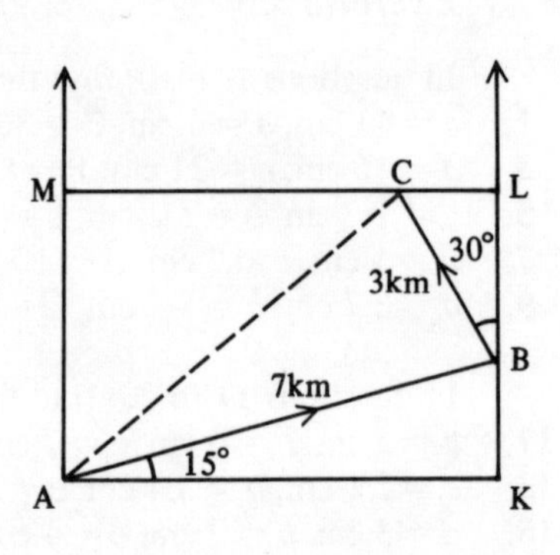

To find the distance of C north of A,

$BK = 7 \sin 15°$ $\quad$ $BL = 3 \cos 30°$

$= 7 \times 0{\cdot}259$ $\quad$ $= 3 \times 0{\cdot}866$

$= 1{\cdot}813$ km $\quad$ $= 2{\cdot}598$ km

$KL = 1{\cdot}813 + 2{\cdot}598 = 4{\cdot}41$ km

To find the distance of C east of A,

$AK = 7 \cos 15°$ $\quad$ $CL = 3 \sin 30°$

$= 7 \times 0{\cdot}966$ $\quad$ $= 3 \times 0{\cdot}5$

$= 6{\cdot}762$ km $\quad$ $= 1{\cdot}5$ km

$CM = AK - CL = 6{\cdot}762 - 1{\cdot}5 = 5{\cdot}26$ km

To find the bearing of C from A,

$$\tan M\hat{A}C = \frac{CM}{AM} = \frac{5{\cdot}26}{4{\cdot}41}$$

$M\hat{A}C = 50°$ (using log tan tables to the nearest 1°).

The bearing of C from A is 050°.

No.	Log
5·26	0·721
4·41	0·644
tan 50°	0·077

Example 2

Two boats sail from port A on bearings of 047° and 325° at speeds of 4 knots and 5 knots respectively. Find their distance apart after 1 hour.

Suppose the first boat reaches B and the second boat reaches C after 1 hour.

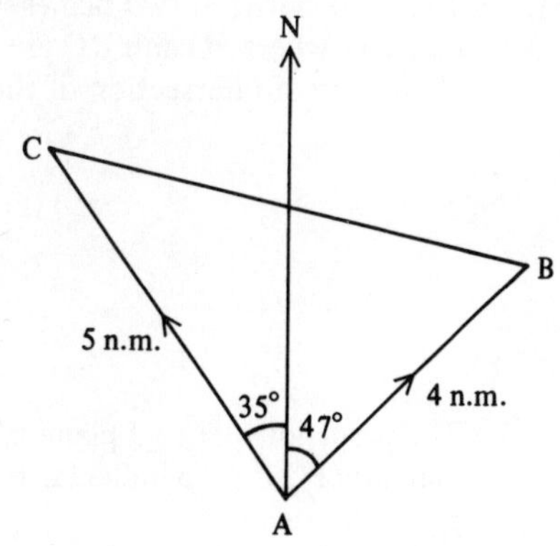

Then AB = 4 nautical miles

AC = 5 nautical miles

$\hat{CAB} = 82°$

Using the cosine rule

$$BC^2 = 5^2 + 4^2 - 2 \times 5 \times 4 \cos 82°$$
$$= 25 + 16 - (40 \times 0{\cdot}139)$$
$$= 41 - 5{\cdot}56 = 35{\cdot}44$$
$$BC = \sqrt{35{\cdot}4} = 5{\cdot}95 \text{ n.m.}$$

Exercise 3.6

1. Two mountain peaks X, Y, lie on an east-west line. From an observer at A, the bearing of X is 027° and the distances AX and AY are equal. Write down the bearings of:

   (a) A from X (b) Y from A
   (c) A from Y (d) Y from X

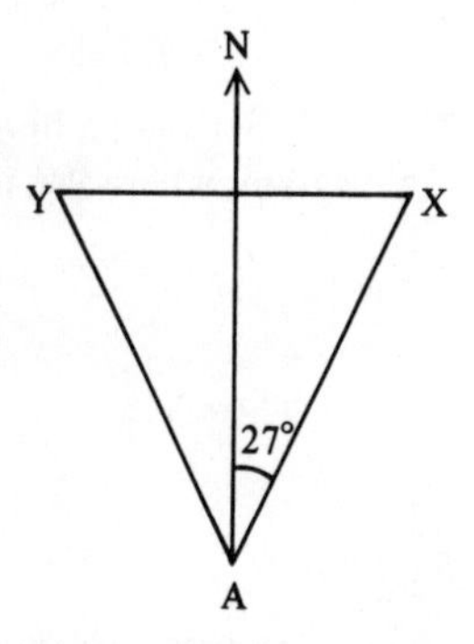

2. Two men set out from the same point. One walks due north and the other on a bearing of 223°. If they walk at the same speed, find the bearing of the more southerly one from the other man after 1 hour.
3. A ship sails 14 km on a bearing of 136°. How far has it travelled south and east of its starting point?
4. An aeroplane flies 20 km on a bearing of 237°. Find how far it has flown south and west of its starting point.
5. One ship sails 26 km on a bearing of 047° while another starting from the same point sails 21 km on a bearing 326°. Find which has achieved the more northerly position.
6. Which aeroplane flies further west: one that flies 200 km on a bearing of 308° or one that flies 160 km on a bearing of 262°?
7. Find the distance apart of the two ships of question 5 when they have reached their final positions.
8. Find the distance apart of the two aeroplanes of question 6 at the ends of their journeys.
9. A ship sails 12 km in the direction 250° and then changes course to 135° and sails a further 18 km. Find the bearing and the distance it would have to travel to return to its starting point.
10. Ship A sails in the direction 073° at 8 knots and ship B leaves from the same port at the same time in the direction 318° at 5 knots. Find their distance apart after 1 hour.

3.7 3-D Trigonometry

1. The angle between two planes is $P\hat{Q}R$ in the diagram, where PQ and RQ are perpendicular to the line of intersection of the planes.

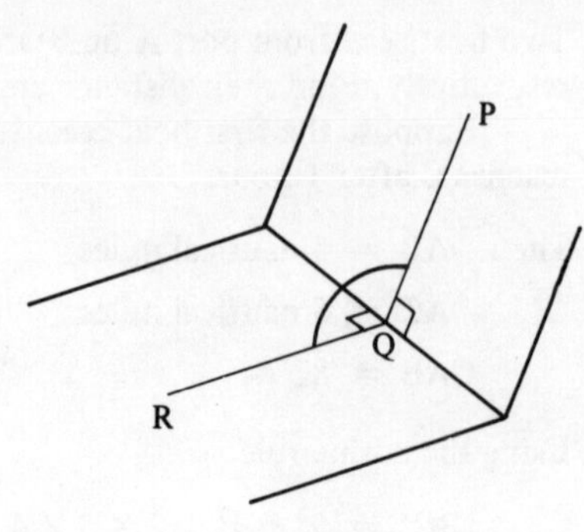

2. The perpendicular to a plane is the line PQ in the diagram; PQ is perpendicular to *any* line of the plane through Q.

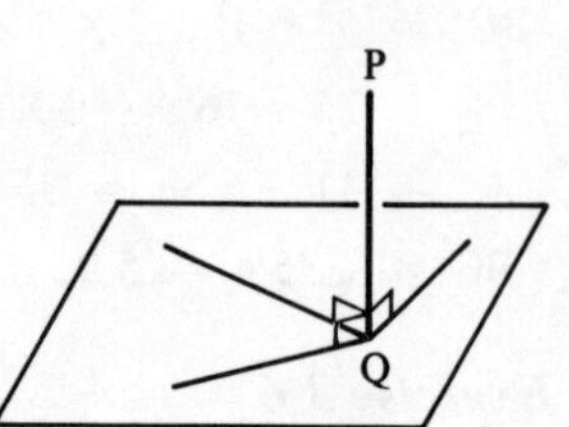

3. The angle between a line l and a plane p is $P\hat{Q}M$ in the diagram, where PM is perpendicular to the plane p.

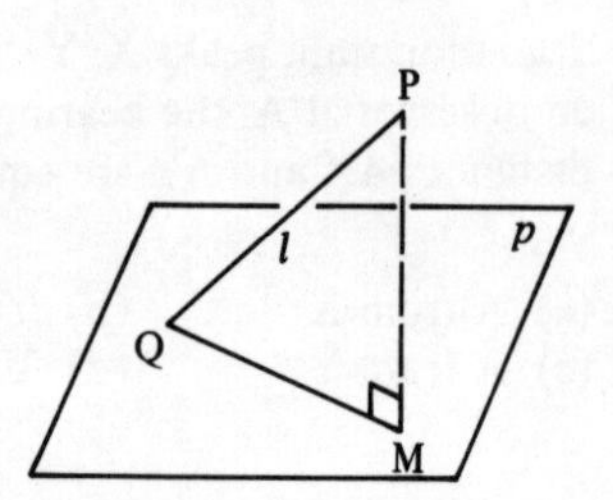

Example

In the diagram, ABCDA′B′C′D′ is a cube. Find the angle between the diagonal A′C and
(a) the edge BC
(b) the face DCC′D′

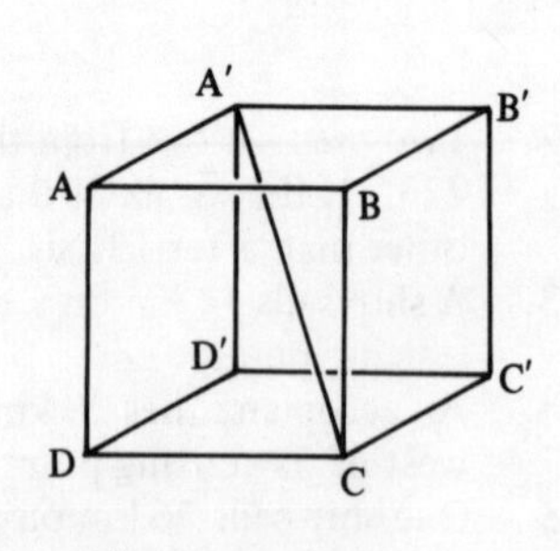

Let the side of the cube be 1 unit.

(a) A′BC is a right-angled triangle.

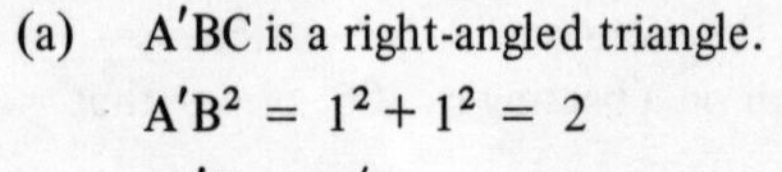

$$A'B^2 = 1^2 + 1^2 = 2$$

$$A'B = \sqrt{2}$$

$$\tan A'\hat{C}B = \frac{\sqrt{2}}{1} = \sqrt{2};\ A'\hat{C}B = 54{\cdot}7°$$

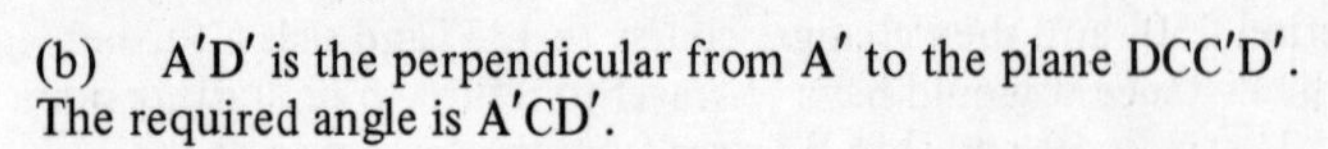

(b) A′D′ is the perpendicular from A′ to the plane DCC′D′.
The required angle is A′CD′.

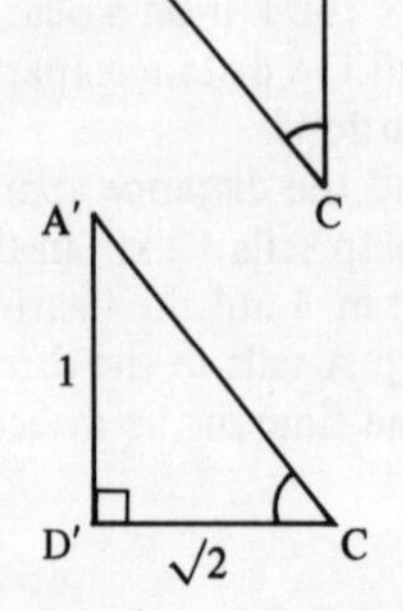

$$\tan A'\hat{C}D' = \frac{1}{\sqrt{2}} = \frac{\sqrt{2}}{2}\quad A'\hat{C}D' = 35{\cdot}3°$$

Exercise 3.7

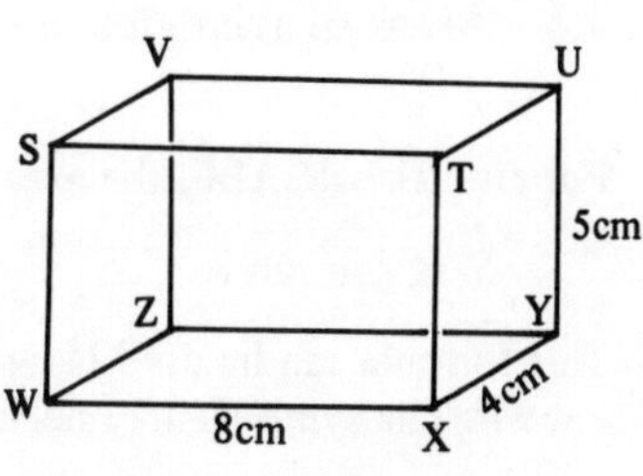

1. The rectangular box STUVWXYZ has dimensions as shown in the diagram.
   Find the angle between the line WU and
   (a) WY (b) UT (c) the plane UTXY
2. In the box STUVWXYZ of question 1, find the angle between the plane WSUY and the plane:
   (a) UTXY (b) VTXZ
3. The pyramid VABCD has a square base ABCD of side 10 cm. The distance from V to each vertex of the base is 15 cm. Find:
   (a) the angles of the triangle VAB
   (b) the length of the perpendicular from A to VB
   (c) the angle between the planes VAB and VBC
   (d) the angle between the planes VAB and ABCD.

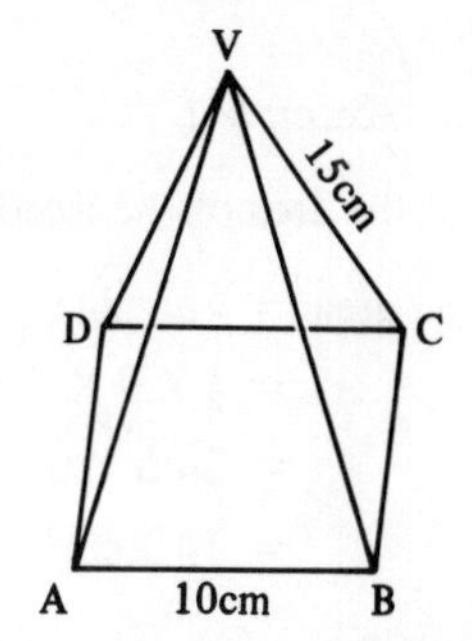

4. In the shape LMNOPQRS, LMNO and PQRS are rectangles. PL = QM = RN = SO = 5 cm. The other lengths are as shown in the diagram. Find the angle between the plane LMNO and the plane:
   (a) LMQP (b) MNRQ

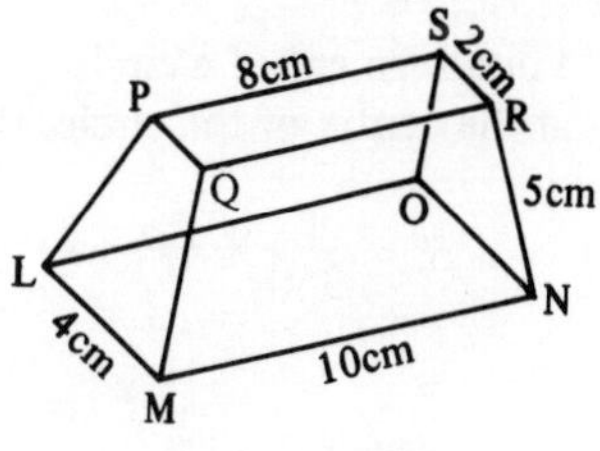

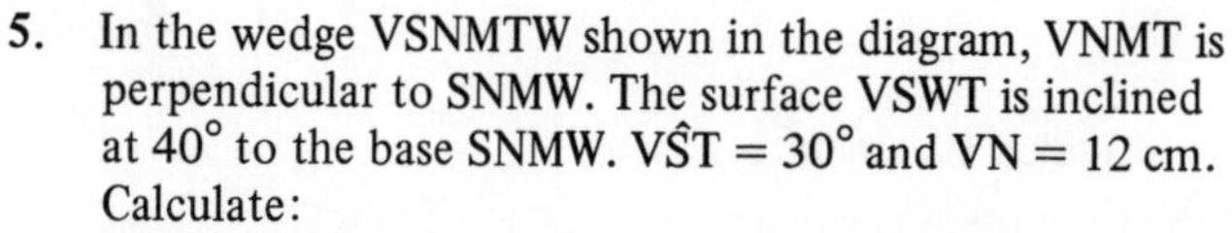

5. In the wedge VSNMTW shown in the diagram, VNMT is perpendicular to SNMW. The surface VSWT is inclined at 40° to the base SNMW. $V\hat{S}T = 30°$ and VN = 12 cm. Calculate:
   (a) the length of VS
   (b) the length of ST
   (c) the angle ST makes with the plane SNMW.

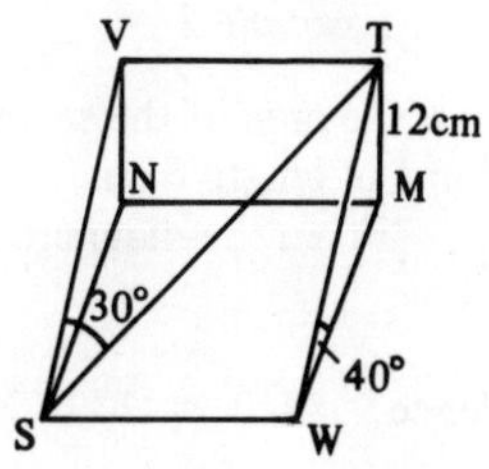

6. A plane hillside slopes at 15° to the horizontal. A straight footpath is inclined at 20° to the line of greatest slope. Calculate the inclination of the footpath to the horizontal.
7. From a point A due south of a church tower, the elevation of the top of the tower is 25°. From a point B due east of the tower, the elevation of the top of the tower is 33°. If the tower is 25 m tall, find the distance AB.

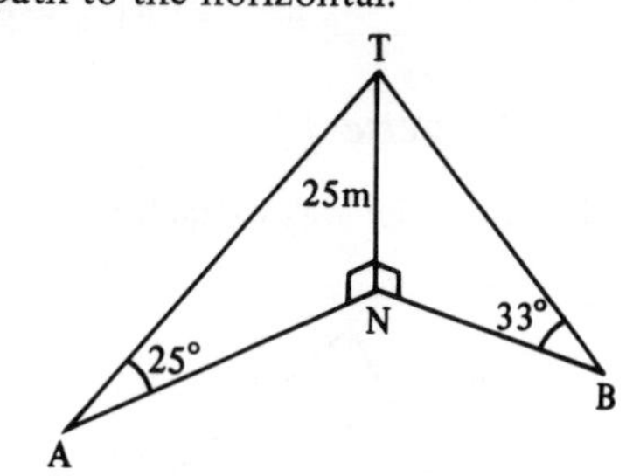

3.8 Areas of triangles and segments

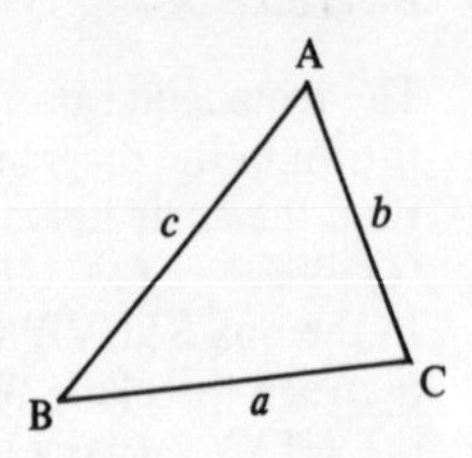

For any triangle ABC, the area S is given by

$$S = \tfrac{1}{2}bc \sin \hat{A}$$

This formula can be used if two sides and the included angle are known. The symbols may need to be rearranged.

Example 1

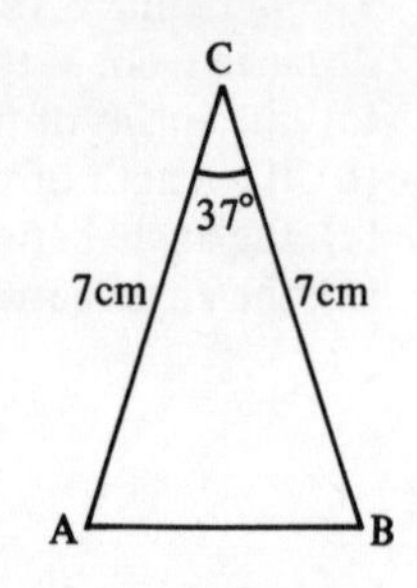

Find the area of the isosceles triangle shown in the diagram.

$$\begin{aligned} \text{area} &= \tfrac{1}{2}ab \sin \hat{C} \\ &= \tfrac{1}{2} \times 7 \times 7 \times \sin 37° \\ &= 24{\cdot}5 \times 0{\cdot}602 \\ &= 14{\cdot}7\ \text{cm}^2 \end{aligned}$$

For a segment of a circle of radius r, cut off by a chord subtending an angle $\theta°$ at the centre of the circle, the area S is given by

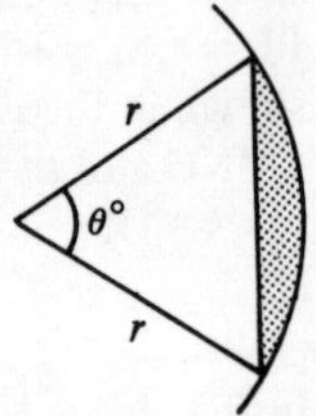

$$S = \frac{\theta}{360} \times \pi r^2 - \tfrac{1}{2}r^2 \sin \theta$$

Example 2

Find the area of the segment of a circle of radius 10 cm cut off by a chord of length 8 cm.

From the diagram, $\sin \text{M}\hat{\text{O}}\text{B} = \frac{4}{10} = 0{\cdot}4$;

$$\text{M}\hat{\text{O}}\text{B} = 23{\cdot}6°$$

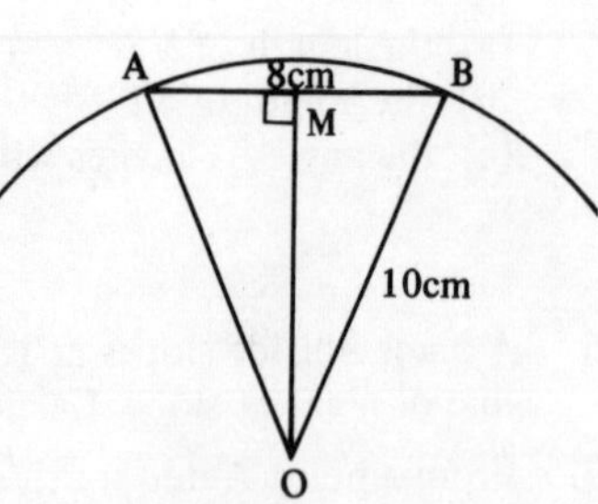

Hence $\text{A}\hat{\text{O}}\text{B} = 23{\cdot}6° \times 2 = 47{\cdot}2°$

$$\begin{aligned} \text{area of segment} &= \frac{47{\cdot}2}{360} \times \pi \times 10^2 - \tfrac{1}{2} \times 10^2 \times \sin 47{\cdot}2° \\ &= 41{\cdot}2 - 36{\cdot}7 = 4{\cdot}5\ \text{cm}^2 \end{aligned}$$

Exercise 3.8

1. Find the area of each triangle.

(a) (b) (c)

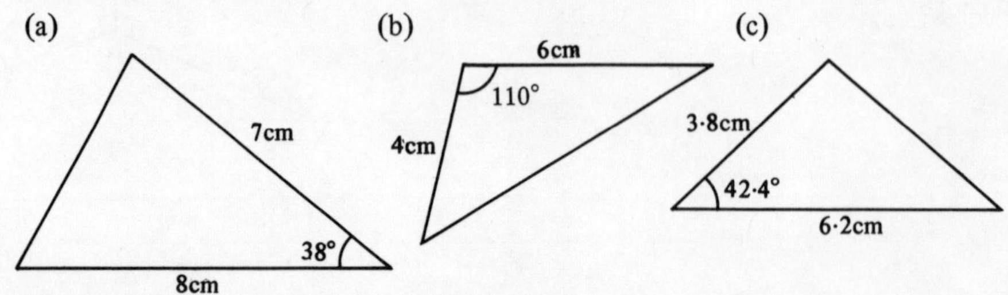

2. Find the area of each triangle:

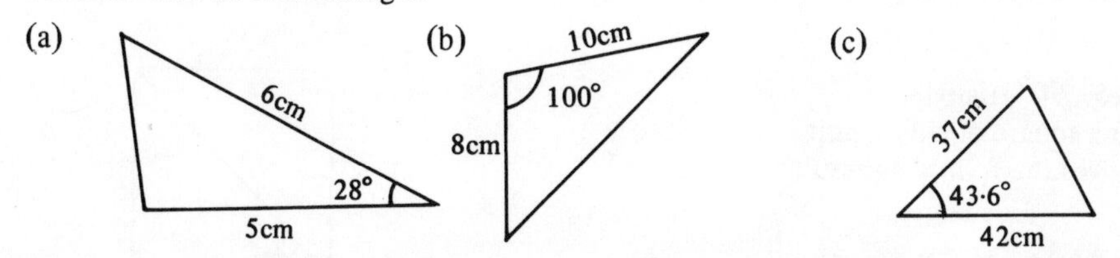

3. ABCD is a parallelogram with AB = 8·1 cm, AD = 4·2 cm and BÂD = 70°. Find the area of the parallelogram.
4. One angle of a rhombus is 142° and its sides are of length 11·2 cm. Find its area.
5. A tent has dimensions as shown in the diagram. Calculate the total area of the canvas needed to make the tent.

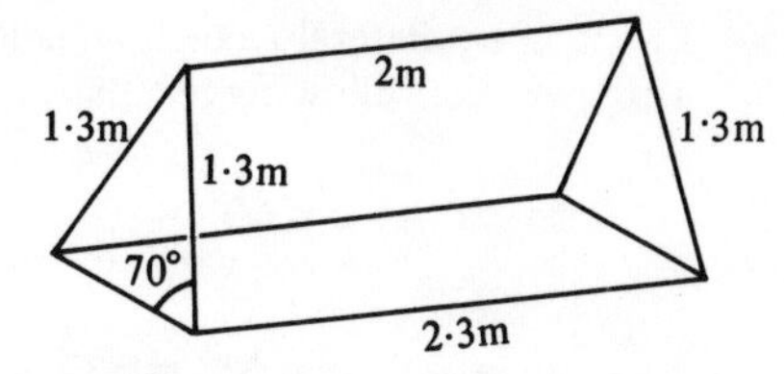

6. A pyramid has a square base of side 9 cm. The length of each slant edge is 15 cm. Find its total surface area.
7. A tetrahedron is made of four equilateral triangles of side 5 cm. Find its surface area.
8. Find the area of a regular pentagon inscribed in a circle of radius 6 cm.
9. Find the areas of a square, an equilateral triangle and a hexagon each of perimeter 12 cm. Which has the largest area?
10. OP is a straight line that rotates in a plane about a centre O; OP = 70 cm and P starts at A. Find the area of the triangle OAP when the size of AÔP is:
(a) 40° (b) 80° (c) 140° (d) 160°.
What would be the size of AÔP to give the maximum area of the triangle OAP?
11. Find the area of a segment of a circle of radius 8 cm cut off by a chord that subtends an angle at the centre of the circle of:
(a) 20° (b) 120°.
12. Find the area of a segment of a circle of radius 15 cm cut off by a chord that subtends an angle at the centre of the circle of:
(a) 30° (b) 130°.
13. Two concentric circles are of radius 3 cm and 5 cm. Find the area of a segment of the larger circle cut off by a tangent to the smaller circle.

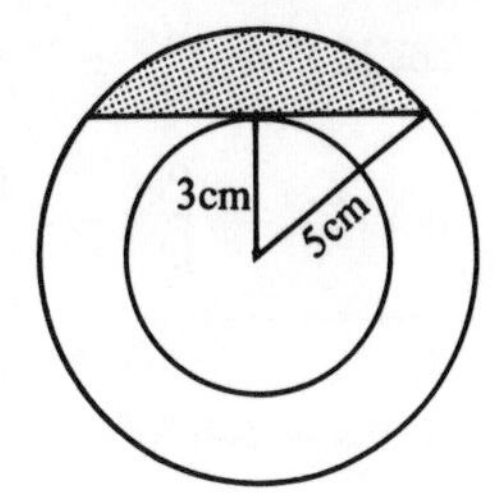

14. A regular hexagon is inscribed in a circle of radius 10 cm. Find the difference in area between that of the hexagon and the circle.
15. Two circles of radius 10 cm and 16 cm have their centres at a distance of 20 cm apart. Find the area of the intersection of the circles.

3.9 Useful results

1. $45°, 45°, 90°$ triangle.
Bisect a square of side 1 unit.
This gives the following results

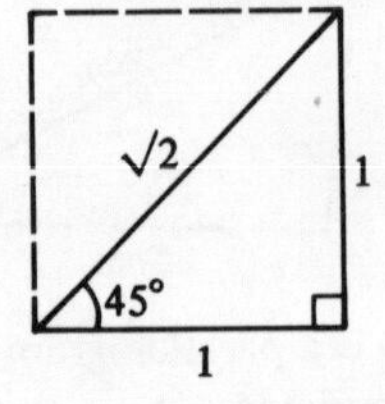

$$\sin 45° = \cos 45° = \frac{1}{\sqrt{2}}$$
$$\tan 45° = 1$$

2. $30°, 60°, 90°$ triangle.
Bisect an equilateral triangle of side 2 units.
This gives the following results.

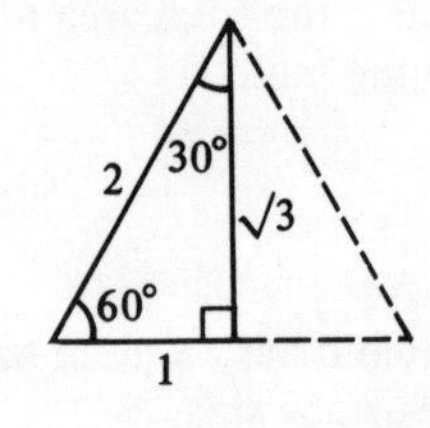

$$\sin 30° = \cos 60° = \frac{1}{2}$$

$$\sin 60° = \cos 30° = \frac{\sqrt{3}}{2}$$

$$\tan 30° = \frac{1}{\sqrt{3}}; \ \tan 60° = \sqrt{3}$$

3. Angles greater than 90°.
Tangent, sine and cosine are found from displacements from the centre. Displacements to the left and below are reckoned as negative (see page 44).

Example 1

Find, without using tables, the sine, cosine, and tangent of 135°.
From the diagram

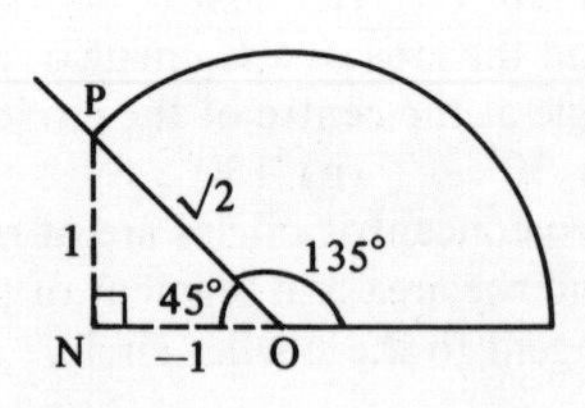

$$\sin 135° = \frac{PN}{OP} = \frac{1}{\sqrt{2}}$$

$$\cos 135° = \frac{ON}{OP} = \frac{-1}{\sqrt{2}} = -\frac{1}{\sqrt{2}}$$

$$\tan 135° = \frac{PN}{ON} = \frac{1}{-1} = -1$$

4. $\sin\theta/\cos\theta = \tan\theta$; $\sin^2\theta + \cos^2\theta = 1$.
In the diagram ABC is any right-angled triangle with $A\hat{B}C = 90°$

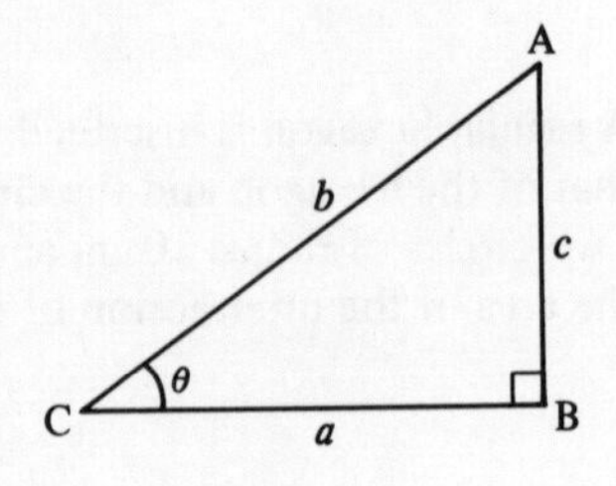

$$\tan\theta = \frac{c}{a}$$

$$\frac{\sin\theta}{\cos\theta} = \frac{c}{b} \div \frac{a}{b} = \frac{c}{a}$$

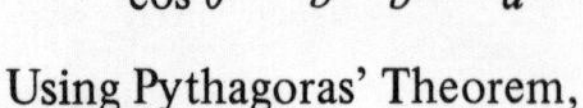

Using Pythagoras' Theorem,

$$c^2 + a^2 = b^2;$$

$$\frac{c^2}{b^2} + \frac{a^2}{b^2} = 1; \qquad \sin^2\theta + \cos^2\theta = 1$$

Example 2

Given $\cos\theta = 0{\cdot}2$, find $\sin\theta$ and $\tan\theta$ without using trigonometric tables.

Method 1

$$\sin^2\theta + \cos^2\theta = 1; \quad \sin^2\theta + 0{\cdot}2^2 = 1; \quad \sin^2\theta + 0{\cdot}04 = 1$$
$$\sin^2\theta = 0{\cdot}96$$
$$\sin\theta = \sqrt{0{\cdot}96} = 0{\cdot}980$$

$$\tan\theta = \frac{\sin\theta}{\cos\theta} = \frac{0{\cdot}980}{0{\cdot}2} = 4{\cdot}90$$

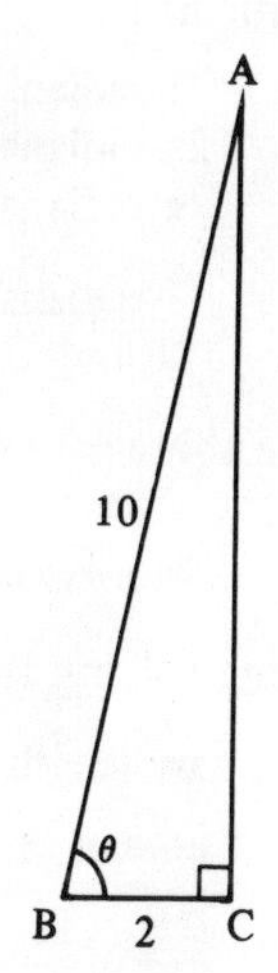

Method 2

Draw a right-angled triangle ABC with $\hat{C} = 90°$ and $\hat{B} = \theta$

$$\cos\theta = 0{\cdot}2 = \tfrac{2}{10}; \text{ let AB} = 10, \text{BC} = 2$$

Then, by Pythagoras' Theorem

$$AC^2 = 10^2 - 2^2 = 100 - 4 = 96$$
$$AC = \sqrt{96} = 9{\cdot}8$$

Hence, $\sin\theta = \dfrac{9{\cdot}8}{10} = 0{\cdot}98$; $\tan\theta = \dfrac{9{\cdot}8}{2} = 4{\cdot}9$.

Exercise 3.9

For these questions do not use trigonometric tables.

1. The arm of a crane whose lower end is at ground level is 8 m long. Find the height of the upper end of the arm when it is inclined to the horizontal at:
   (a) 30° (b) 45° (c) 60°.
2. The displacement s of a particle at time t is given by $s = 2\sin t - 3\cos t$. Find the value of s when the value of t is:
   (a) 60° (b) 90° (c) 135°.
3. Find the value of each of the following, leaving square root signs in your answer:
   (a) sin 120° (b) cos 120° (c) tan 120° (d) sin 135°.
4. Find the value of each of the following, leaving square root signs in your answer:
   (a) sin 150° (b) cos 150° (c) tan 150° (d) cos 135°.
5. If $\hat{A}$ is an acute angle and $\sin\hat{A} = \frac{3}{5}$, find:
   (a) $\cos\hat{A}$ (b) $\tan\hat{A}$.
6. If $\hat{A}$ is an acute angle and $\tan\hat{A} = 2{\cdot}4$, find:
   (a) $\sin\hat{A}$ (b) $\cos\hat{A}$.
7. If $\hat{B}$ is an angle between 90° and 180° and $\tan\hat{B} = -\frac{4}{3}$, find:
   (a) $\sin\hat{B}$ (b) $\cos\hat{B}$.
8. If $\hat{B}$ is an angle between 90° and 180° and $\sin\hat{B} = 0{\cdot}4$, find:
   (a) $\cos\hat{B}$ (b) $\tan\hat{B}$.
9. ABC is a triangle in which $\hat{A} = 30°$, AC = 4 cm and AB = 5 cm. Calculate:
   (a) the area of the triangle (b) the length of BC.
10. XYZ is a triangle in which $\hat{X} = 135°$, XY = 10 cm and XZ = 8 cm. Calculate:
    (a) the area of the triangle (b) the length of YZ.

In questions **11** to **16**, find the value or values of $\sin\theta$.

11. $2\sin^2\theta + \cos^2\theta = 2$
12. $\sin^2\theta - \cos^2\theta = -\frac{1}{2}$
13. $\cos^2\theta + 3\sin\theta = 3$
14. $12\cos^2\theta + \sin\theta = 11$
15. $3\tan\theta + 2\cos\theta = 0$
16. $2\tan\theta - 5\cos\theta = 0$

3.10 Radians

Radians measure angles in terms of the length of arc swept out by the angle. If the length of arc is equal to the radius, the angle is 1 radian.

Remember:

$$\begin{aligned} 1 \text{ radian} &\approx 57^\circ \\ 2\pi \text{ radians} &= 360^\circ \\ \pi \text{ radians} &= 180^\circ \\ \frac{\pi}{2} \text{ radians} &= 90^\circ \end{aligned}$$

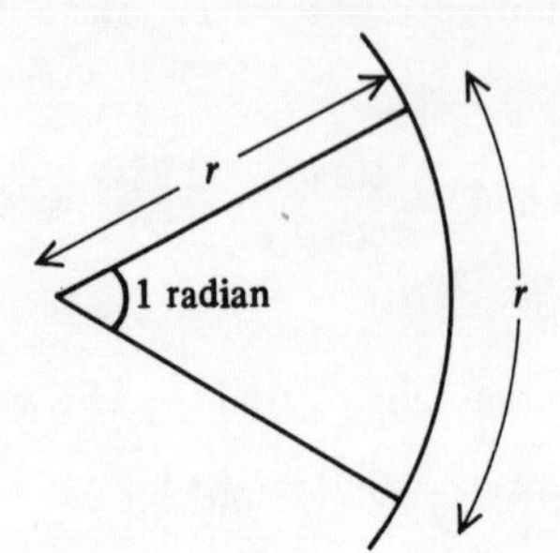

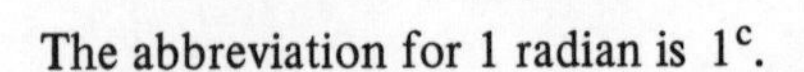

The abbreviation for 1 radian is 1^c.

Sectors and segments

Using radians, the formulas for arc length and area of a sector are:

arc length $s = r\theta$

area of sector $= \frac{1}{2}r^2\theta$

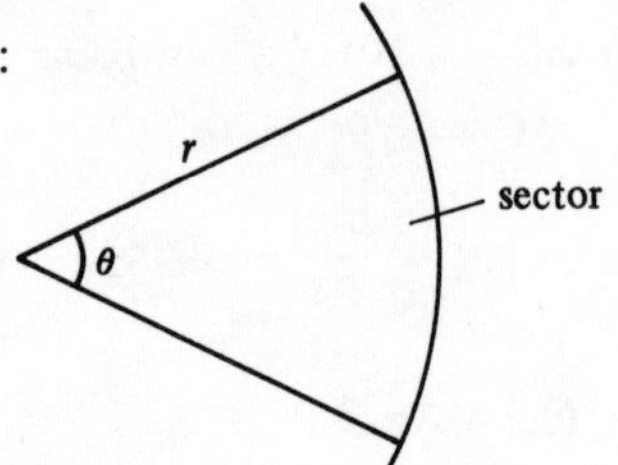

The area of the segment is:

$$\begin{aligned} &\tfrac{1}{2}r^2\theta - \tfrac{1}{2}r^2 \sin\theta \\ = &\tfrac{1}{2}r^2(\theta - \sin\theta) \end{aligned}$$

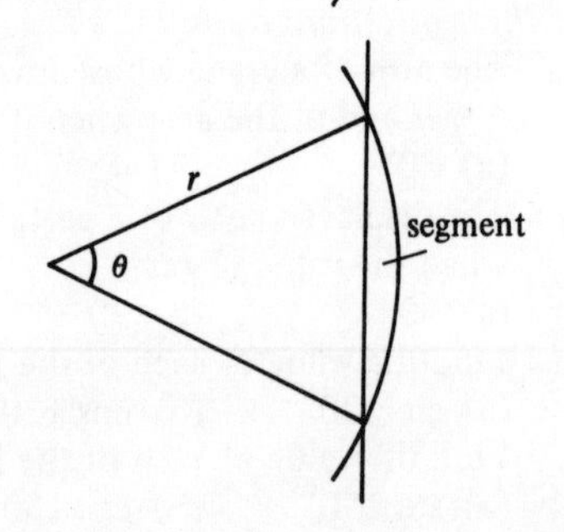

In all these formulas, you must use θ as a *radian.*

Exercise 3.10

1. Write, in degrees, the angle that measures, in radians:

   (a) π (b) $\frac{\pi}{6}$ (c) $\frac{\pi}{3}$ (d) $\frac{2\pi}{3}$ (e) 3π (f) $\frac{3\pi}{4}$

2. Write, in radians, the angle that measures, in degrees:

   (a) 270° (b) 720° (c) 45° (d) 120° (e) 225° (f) 150°

3. Without using tables, find the value of:

   (a) $\sin\frac{\pi}{6}$ (b) $\sin\frac{\pi}{4}$ (c) $\cos\pi$ (d) $\tan\pi$ (e) $\tan\frac{\pi}{4}$

4. Without using tables, find the value of:

   (a) $\cos\frac{\pi}{6}$ (b) $\sin\frac{\pi}{3}$ (c) $\tan 2\pi$ (d) $\cos\frac{3\pi}{2}$ (e) $\tan\frac{3\pi}{4}$

5. Find the value, in radians, of the angle θ between 0 and 2π such that:
(a) $\sin\theta = 1$ (b) $\cos\theta = \frac{1}{2}$ (c) $\cos\theta = -\frac{1}{2}$ (d) $\tan\theta = 1$ (e) $\tan\theta = -1$

6. Find the value, in radians, of the angle θ between 0 and 2π such that:
(a) $\cos\theta = -1$ (b) $\sin\theta = \frac{\sqrt{3}}{2}$ (c) $\sin\theta = -\frac{1}{\sqrt{2}}$ (d) $\sin\theta = -1$ (e) $\sin\theta = -\frac{1}{2}$

7. Use tables to find the value of:
(a) $\sin 1^c$ (b) $\cos 0{\cdot}9^c$ (c) $\tan 0{\cdot}8^c$ (d) $\sin 2^c$ (e) $\cos 2{\cdot}2^c$ (f) $\tan 3^c$

8. Use tables to find the value, in radians, of the angle θ between 0 and π such that:
(a) $\cos\theta = 0{\cdot}4$ (b) $\sin\theta = 0{\cdot}7$ (c) $\tan\theta = -2$
(d) $\cos\theta = -0{\cdot}37$ (e) $\sin\theta = 0{\cdot}18$ (f) $\tan\theta = 1{\cdot}23$

9. A circle has a radius of 10 cm. Find the length of the arc of the circle subtended by an angle at the centre of:
(a) 1^c (b) 2^c (c) $2{\cdot}4^c$ (d) π^c (e) $\left(\frac{\pi}{2}\right)^c$

10. For each part of question 9 find:
(a) the area of the corresponding sector (b) the area of the corresponding segment.

11. A circle centre O has a segment cut off by a chord AB of length 8 cm and 3 cm from O. Find:
(a) the radius of the circle
(b) the size of AÔB in radians
(c) the area of the segment that was cut off.

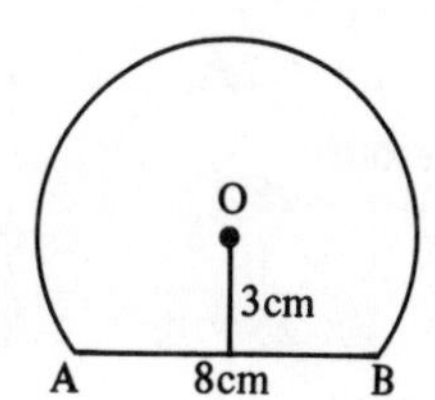

12. Find the difference in area between a hexagon of side 7 cm and its circumscribing circle.

13. 1 nautical mile is the length of arc of a great circle of the earth that subtends an angle of $\frac{1}{60}^{\circ}$ at the centre. Taking the radius of the earth as 6400 km, find:
(a) the size of the angle at the centre in radians
(b) the length in km of 1 nautical mile.

14. The diameter of the Sun is $1{\cdot}4 \times 10^6$ km and its distance from the Earth is $1{\cdot}5 \times 10^8$km. Using these values, find, in radians, the angle that the Sun subtends at an observer on Earth. You may assume that the diameter of the Sun is an arc of a circle with the Earth as centre.

15. A vertical post, height 3·5 m, stands at a distance of 50 m from an observer. Show that the post subtends an angle of approximately 4° at the observer's eye. (You may assume that the post is approximately an arc of a circle.)

Part 4: Algebra

4.1 Brackets and rearranging expressions

Expressions are simplified by collecting like terms together.

i.e. $3a + 4b - 2a + 5b$ $= (3a - 2a) + (4b + 5b)$ $= a + 9b$ or $x + 3y + 7x - 12x$ $= 3y + (x + 7x - 12x)$ $= 3y - 4x$

Brackets are removed by multiplying each term within the bracket by the term outside.

i.e. $3(a + 2b + 5c)$ $= 3a + 6b + 15c$ or $2x(y + 4z) - 3y(z - 2w)$ $= 2xy + 8xz - 3yz + 6yw$

Exercise 4.1a

1. Simplify:
(a) $3x + 5x$ (b) $8a + 2a$ (c) $14b - 2b$ (d) $-7y + 4y$
(e) $8a + 3a - 2a$ (f) $7b - 3b - b$ (g) $4p - 7p - 5p$ (h) $3ab + 5ab + 9ab$
(i) $2a^2 + 7a^2$ (j) $9b^2 - 2b^2$ (k) $x^2 + 3x^2 + 7x^2$ (l) $8y^3 - 2y^3 - 4y^3$

2. Simplify:
(a) $2 \times 3a$ (b) $5b \times 4$ (c) $-3 \times 7y$ (d) $(-6p) \times (-5)$
(e) $4a \times 3b$ (f) $7a \times 2a$ (g) $(-3q) \times 4p$ (h) $(-5y) \times (-3y)$
(i) $3 \times 2b \times 4c$ (j) $2a \times 5b \times 3c$ (k) $3a \times 2a \times a$ (l) $(-2x) \times (5x) \times (-4x)$

3. Simplify:
(a) $6a \div 2$ (b) $8b \div 4b$ (c) $10 \div 2x$ (d) $3 \div 6y$
(e) $5a \div 2b$ (f) $4pq \div 2p$ (g) $9x \div 3xy$ (h) $6ab \div 2ab$
(i) $a^2b \div ab$ (j) $a^2b^3 \div ab^2$ (k) $7ab^4 \div 3a^3b$ (l) $4x^2y^2z^2 \div xy^3z^2$

4. Simplify:
(a) $4x^2 + 7x^2$ (b) $2mn + 5mn - 3mn$ (c) $4l^2m^2 - 7l^2m^2$ (d) $8pqr - 9pqr - 15pqr$
(e) $-3 \times 5m$ (f) $6a \times 11b$ (g) $7a^2 \times 3a^3$ (h) $(-4pq) \times (-2p^2q^3)$
(i) $10m^2 \div 2$ (j) $7a^2b^3 \div 2ab$ (k) $4x^5 \div 7x^8$ (l) $2lm^2n^3 \div 3l^3m^2n$

5. Collect like terms together and simplify:
(a) $4a + 3a + 7b + 5b$ (b) $7a + 3b + 4a + 2b$ (c) $4a + 9b - 2a - 3b$
(d) $6x - 3y - 2x + 7y$ (e) $x - y - 8x - 9y$ (f) $5xy + 2z - 3xy - 5z$
(g) $3ab + pq - ab + 5pq$ (h) $a^2 + b^2 + 6a^2 - 3b^2$ (i) $4a^3 - a - a^3 + 5a$

6. Remove the brackets:
(a) $2(a + b)$ (b) $3(x + 2y)$ (c) $4(3p - 5q)$ (d) $5a(b + c)$
(e) $3p(q - 2r)$ (f) $5(a - 2b + 3c)$ (g) $2x(y - 2 + 3w)$ (h) $-3(2l + m)$
(i) $-2l(m - 5n)$ (j) $p(p + 3q - 5r)$ (k) $-4y(3x - 2)$ (l) $4x(x^2 - x + 1)$

7. Remove the brackets and simplify:
(a) $3(x + y) + 2(x - y)$ (b) $4(x + y) + 3(2x + 5y)$ (c) $2(5x + 4y) - 3(2x + y)$
(d) $8(a - b) - 3(2a - b)$ (e) $6(x - 3) + 5(x + 2)$ (f) $3a(b - 5) - 5a(2b - 3)$
(g) $x(x^2 + 1) + 2x(x^2 + 5)$ (h) $7a - 4a(b + 3)$ (i) $3xy - 2x(y - 4z)$

8. Remove the brackets and simplify:
(a) $2(x - 3y) + 3(y - 2z) + 4(z - 5x)$ (b) $2a(b + c) - 3b(c - a) + 4c(a + b)$
(c) $3m(2 + n) + 5n(1 - 3m) + 12mn$ (d) $xy(x^2 + xy + y^2) - x^2(y^2 - xy - x^2)$

Expressions with two brackets can be simplified by multiplying each term in one bracket by each term in the other bracket.

i.e. $(a+b)(c+d)$ $= ac + ad + bc + bd$ or $(x-2)(x+5)$ $= x^2 - 10 + 5x - 2x$ $= x^2 + 3x - 10$

Exercise 4.1b

1. Remove the brackets:
(a) $(p+q)(r+s)$ (b) $(l+m)(n-p)$ (c) $(a-b)(c-d)$
(d) $(2p+q)(3r-s)$ (e) $(3l+m)(n+5p)$ (f) $(8a-b)(3c-2d)$
(g) $(l+m)(l+m)$ (h) $(p+q)(p-q)$ (i) $(a-b)(a-b)$
(j) $(2p+3q)(2p+3q)$ (k) $(3a-b)(3a+b)$ (l) $(4l-3m)(4l-3m)$

2. Remove the brackets:
(a) $(x+1)(x+2)$ (b) $(x+3)(x+5)$ (c) $(x-1)(x+4)$
(d) $(x-3)(x-4)$ (e) $(1-x)(2-x)$ (f) $(3-x)(4+x)$
(g) $(2x+1)(3x+5)$ (h) $(3x+2)(4x+1)$ (i) $(7x-1)(2x+5)$
(j) $(4x-3)(2x-3)$ (k) $(2p+q)(3p+2q)$ (l) $(5l-2m)(7l+4m)$

3. Remove the brackets:
(a) $3(a+2b)(c+4d)$ (b) $5(x+4)(x+2)$ (c) $2(3a+b)(2a+5b)$
(d) $(p+q)(p+q+r)$ (e) $(x+2y)(x+y-3)$ (f) $(l-m)(2l-3m+4n)$

4. Remove the brackets and simplify:
(a) $(x+2)(x+3)+(x-1)(x-2)$ (b) $(x+7)(x-3)+(x-5)(x-4)$
(c) $(x-2)(x-5)-(x+1)(x+4)$ (d) $(x+4)(x-2)-(x-3)(x+7)$

5. Remove the brackets and simplify:
(a) $(a+b)(c+d)+(a-b)(c-d)$ (b) $(p+q)(r-s)+(p-q)(r+s)$
(c) $(l+m)(l+m)-(l-m)(l-m)$ (d) $(x+y)(x-y)-(x-y)(x+y)$

6. Remove the brackets and simplify:
(a) $(2x+1)(3x-2)+(3x+1)(2x-3)$
(b) $(5x-1)(4x-3)-(3x-2)(7x-1)$
(c) $(2a+3b)(5c+4d)+(3a-4b)(2c-7d)$
(d) $(4p+3q)(p+5q)-(2p+5q)(3p-2q)$

Exercise 4.1c

1. Simplify:
(a) $ab\left(\frac{1}{a}+\frac{1}{b}\right)$ (b) $pq\left(\frac{2}{p}-\frac{3}{q}\right)$ (c) $xyz\left(\frac{1}{xy}+\frac{2}{yz}+\frac{3}{zx}\right)$
(d) $l^2m\left(\frac{m}{l}+\frac{l}{m}\right)$ (e) $x^3y^2\left(\frac{3x}{y^3}-\frac{2y^2}{x}\right)$ (f) $a^3b^2c\left(\frac{a}{bc}-\frac{2b}{a^2c}-\frac{5c}{b^3a}\right)$

2. Simplify
(a) $(x+y)\left(\frac{1}{x}+\frac{1}{y}\right)$ (b) $(3a+2b)\left(\frac{2}{a}-\frac{3}{b}\right)$ (c) $(5p-2q)\left(\frac{4}{p}-3+\frac{1}{q}\right)$
(d) $(a^2+b)\left(\frac{b}{a^2}+\frac{a}{b^2}\right)$ (e) $(x^3-y^2)\left(\frac{x}{y^3}-\frac{y}{x^2}\right)$ (f) $\left(\frac{1}{p}-\frac{1}{q}\right)\left(\frac{p}{3}-\frac{pq}{6}+\frac{q}{2}\right)$

4.2 Factors

$2m$ is common to each of the terms $2lm$, $6mn$ and $10mp$
so $2lm + 6mn + 10mp$ can be written as $2m(l + 3n + 5p)$.

$2m$ is one factor of $2lm + 6mn + 10mp$,
$(l + 3n + 5p)$ is the other factor.

$x^2 + 3x = (x + 3)x$, so $(x + 3)$ and x are the factors of $x^2 + 3x$

$ax + bx = (a + b)x$, so $(a + b)$ and x are the factors of $ax + bx$

Exercise 4.2a

1. Copy and complete:
   (a) $3x + 3y = 3(\quad)$ (b) $5a - 5b = 5(\quad)$
   (c) $4x + 4y + 4z = 4(\quad)$ (d) $6a - 6b + 6c = 6(\quad)$
   (e) $2x + 6y = 2(\quad)$ (f) $8a - 4b = 4(\quad)$
   (g) $3x + 6y + 9z = 3(\quad)$ (h) $25a - 10b - 5c = 5(\quad)$
2. Copy and complete:
   (a) $ax + ay = a(\quad)$ (b) $pa - pb = p(\quad)$
   (c) $px + py + pz = p(\quad)$ (d) $ra - rb + rc = r(\quad)$
   (e) $qx + 3qy = q(\quad)$ (f) $5sa - sb = s(\quad)$
   (g) $2tx + 5ty + tz = t(\quad)$ (h) $7la - 4lb - lc = l(\quad)$
3. Copy and complete:
   (a) $px + qx = (\quad)x$ (b) $as - bs = (\quad)s$
   (c) $px + qx + rx = (\quad)x$ (d) $ra - sa + ta = (\quad)a$
   (e) $3ly + 2my = (\quad)y$ (f) $6fh - 5gh = (\quad)h$
   (g) $4xt + 9yt + zt = (\quad)t$ (h) $2lg - 7mg - 3ng = (\quad)g$
4. Factorize:
   (a) $2a + 2b$ (b) $3a - 3b$ (c) $4x + 12y$ (d) $9p - 6q$
   (e) $px + py$ (f) $ra - rb$ (g) $7sx + 4sy$ (h) $2ta - 7tb$
   (i) $xa + xb + xc$ (j) $la - lb - lc$ (k) $4rx + 5ry + rz$ (l) $pa - 6pb + 8pc$
5. Factorize:
   (a) $lx + mx$ (b) $an - bn$ (c) $7py + 2qy$ (d) $rt - 5st$
   (e) $pt + qt + rt$ (f) $an + bn - cn$ (g) $5lx + mx + 2nx$ (h) $4kg - 2lg - mg$
6. Factorize:
   (a) $x^2 + 3x$ (b) $y^2 - 5y$ (c) $2z^2 + 3z$ (d) $4m^2 - m$
   (e) $x^3 + 2xy$ (f) $4y^2z - y^3$ (g) $ab^2 + a^2b$ (h) $x^2yz^2 - xyz^3$
   (i) $\pi r^2 + 2\pi rh$ (j) $2lm^2 + 8l^2m$ (k) $x^4 + x^3 + x^2$ (l) $32y + 16y^3 + 8y^5$
7. Factorize:
   (a) $abc^2 + ab^2 + a^2b$ (b) $p^3q^2r + p^2q^2r^2 + pq^2r^3$ (c) $7axy + 14bxy + 21cxy$
   (d) $8x^6 + 16x^4 + 48x^3$ (e) $2lmp - lm + 5lm^2$ (f) $f^4g^2 - 6f^2g^3 + 2fg^4$
   (g) $5abcd + 35bcde$ (h) $24k^2lm^2n - 32kl^2m^2n^3$ (i) $16abcx - 28bcdx - 20cdex$
8. Factorize:
   (a) $\frac{p}{4} - \frac{q}{8} + \frac{r}{12}$ (b) $\frac{12a}{x} - \frac{6b}{x} - \frac{24c}{x}$ (c) $\frac{l^2m}{n} + \frac{lm^2}{n^2} + \frac{l^2m^2}{n^3}$

Expressions can often be factorized by first collecting like terms together, and then factorizing each group, before finally finding a bracket which is one of the factors of the original expression.

e.g. $ax + bx + ay + by = (a+b)x + (a+b)y = (a+b)(x+y)$

or, $ax + ay + bx + by = a(x+y) + b(x+y) = (a+b)(x+y)$

Exercise 4.2b

1. Copy and complete:
   (a) $ax - bx + ay - by = (\quad)x + (\quad)y = (\quad)(x+y)$
   (b) $px - py + qx - qy = p(\quad) + q(\quad) = (p+q)(\quad)$
   (c) $ut^2 + vs + us + vt^2 = (\quad)t^2 + (\quad)s = (\quad)(t^2+s)$
   (d) $ut^2 - us + vt^2 - vs = u(\quad) + v(\quad) = (u+v)(\quad)$
   (e) $2lx + mx - my - 2ly = (\quad)x - (\quad)y = (\quad)(x-y)$
   (f) $pr + 3ps - qr - 3qs = p(\quad) - q(\quad) = (p-q)(\quad)$
2. Collect like terms together and then factorize:
   (a) $sx + 5tx + sy + 5ty$
   (b) $am - 2al + bm - 2bl$
   (c) $su^2 + 4tv + tu^2 + 4sv$
   (d) $a^2l - b^2m + b^2l - a^2m$
   (e) $2bx + 3cy + 3cx + 2by$
   (f) $3zw^2 - y^2t + 3zt - y^2w^2$
3. Factorize:
   (a) $ax + bx - cx + ay + by - cy$
   (b) $px^2 - py - pz^3 + qx^2 - qy - qz^3$
   (c) $ls + mt + nt + ms + lt + ns$
   (d) $2fx + gy - 3fz + 2gx + fy - 3gz$
4. Factorize:
   (a) $3m^2x - 4n^3y^2 - 3m^2y^2 + 4n^3x$
   (b) $4ax - 5bx + 5by - 4ay$
   (c) $2ac - 4ad + bc - 2bd$
   (d) $10pq - 5pr - 2sq + sr$

Example

$(x+a)(x+b) = x^2 + (a+b)x + ab$ (sum: $a+b$; product: ab)

$x^2 + 7x + 12 = x^2 + (3+4)x + 12$
$= (x+3)(x+4)$

$(x+a)(x-b) = x^2 + (a-b)x - ab$

$x^2 + 2x - 24 = x^2 + (6-4)x - 24$
$= (x+6)(x-4)$

Exercise 4.2c

1. Copy and complete:
   (a) $x^2 + 5x + 4 = (x+4)(x+\quad)$
   (b) $x^2 + 9x + 14 = (x+2)(x+\quad)$
   (c) $x^2 + 12x - 13 = (x-1)(x+\quad)$
   (d) $x^2 - 10x + 16 = (x-8)(x-\quad)$
   (e) $x^2 + 16x + 64 = (x+8)(x+\quad)$
   (f) $x^2 - 10x + 25 = (x-5)(x-\quad)$
2. Factorize:
   (a) $x^2 + 7x + 10$
   (b) $x^2 - 9x + 18$
   (c) $x^2 + 3x - 18$
   (d) $x^2 + 5x - 14$
   (e) $x^2 + 18x + 81$
   (f) $x^2 - 8x + 16$
   (g) $x^2 + 5x - 24$
   (h) $x^2 - 4x - 21$
3. Factorize:
   (a) $9x^2 + 6x + 1$
   (b) $3x^2 + 4x + 1$
   (c) $5x^2 - 6x + 1$
   (d) $8x^2 - 2x - 1$
   (e) $3x^2 + 5x + 2$
   (f) $5x^2 - 11x + 2$
   (g) $7x^2 - 3x - 4$
   (h) $8x^2 - 10x - 3$

4.3 Difference of two squares

$$\begin{aligned} a^2 - b^2 &= (a-b)a + (a-b)b \\ &= (a-b)(a+b) \\ (7{\cdot}6)^2 - (2{\cdot}4)^2 &= (7{\cdot}6 - 2{\cdot}4)(7{\cdot}6 + 2{\cdot}4) \\ &= 5{\cdot}2 \times 10 \\ &= 52 \end{aligned}$$

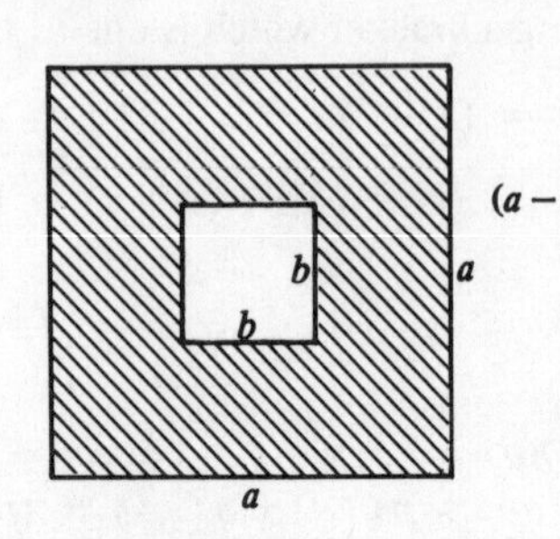

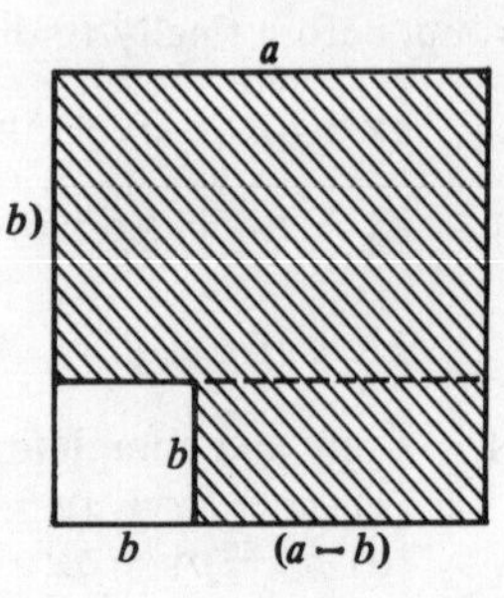

Exercise 4.3a

1. Copy and complete:
   (a) $7^2 - 3^2 = (7-3)(\quad) =$ (b) $14^2 - 6^2 = (14-6)(\quad) =$
   (c) $89^2 - 11^2 = (89-11)(\quad) =$ (d) $28^2 - 18^2 = (28-18)(\quad) =$
   (e) $53^2 - 33^2 = (\quad)(53+33) =$ (f) $72^2 - 42^2 = (\quad)(72+42) =$
2. Copy and complete:
   (a) $(8{\cdot}6)^2 - (1{\cdot}4)^2 = (8{\cdot}6 - 1{\cdot}4)(\quad) =$
   (b) $(5{\cdot}9)^2 - (4{\cdot}1)^2 = (5{\cdot}9 - 4{\cdot}1)(\quad) =$
   (c) $(6{\cdot}2)^2 - (3{\cdot}8)^2 = (6{\cdot}2 - 3{\cdot}8)(\quad) =$
   (d) $(4{\cdot}7)^2 - (2{\cdot}7)^2 = (4{\cdot}7 - 2{\cdot}7)(\quad) =$
   (e) $(5{\cdot}9)^2 - (3{\cdot}9)^2 = (\quad)(5{\cdot}9 + 3{\cdot}9) =$
   (f) $(7{\cdot}5)^2 - (3{\cdot}5)^2 = (\quad)(7{\cdot}5 + 3{\cdot}5) =$
3. First write your answers as a product and then find:
   (a) $8^2 - 2^2$ (b) $12^2 - 8^2$ (c) $53^2 - 47^2$ (d) $91^2 - 71^2$
   (e) $127^2 - 73^2$ (f) $259^2 - 59^2$ (g) $381^2 - 119^2$ (h) $276^2 - 226^2$
4. First write your answer as a product and then find:
   (a) $(5{\cdot}7)^2 - (4{\cdot}3)^2$ (b) $(8{\cdot}1)^2 - (1{\cdot}9)^2$ (c) $(2{\cdot}6)^2 - (1{\cdot}4)^2$ (d) $(5{\cdot}7)^2 - (3{\cdot}7)^2$
   (e) $(6{\cdot}8)^2 - (4{\cdot}2)^2$ (f) $(7{\cdot}9)^2 - (2{\cdot}9)^2$ (g) $(5{\cdot}16)^2 - (4{\cdot}84)^2$ (h) $(8{\cdot}29)^2 - (1{\cdot}71)^2$
5. Copy and complete:
   (a) $p^2 - q^2 = (p-q)(\quad)$ (b) $(l^2 - m^2) = (\quad)(l+m)$
   (c) $p^2 - 4q^2 = (p-2q)(\quad)$ (d) $(9l^2 - m^2) = (\quad)(3l+m)$
   (e) $9p^2 - 4q^2 = (3p-2q)(\quad)$ (f) $(16l^2 - 9m^2) = (\quad)(4l+3m)$
6. Write as the product of two brackets:
   (a) $x^2 - y^2$ (b) $s^2 - t^2$ (c) $4a^2 - b^2$ (d) $l^2 - 9m^2$
   (e) $16a^2 - b^2$ (f) $p^2 - 25q^2$ (g) $16a^2 - 25b^2$ (h) $49p^2 - 16q^2$
7. Write as the product of two brackets:
   (a) $b^2 - c^2$ (b) $81x^2 - y^2$ (c) $p^2 - 100q^2$ (d) $100a^2 - 81b^2$
   (e) $(xy)^2 - z^2$ (f) $a^2 - b^2c^2$ (g) $4p^2 - q^2r^2$ (h) $9l^2m^2 - n^4$
8. Write as the product of two brackets:
   (a) $a^2b^2 - c^2d^2$ (b) $x^2y^2z^2 - w^2$ (c) $16p^2q^2 - 81r^2$ (d) $l^4 - n^4$
   (e) $(a+b)^2 - c^2$ (f) $p^2 - (q+r)^2$ (g) $(x+y)^2 - (y+z)^2$ (h) $(l+m)^2 - (l-m)^2$
9. Find the length of the third side of a right-angled triangle ABC where:
   (a) AC = 13 cm, AB = 12 cm
   (b) AC = 25 cm, AB = 7 cm
   (c) AC = 41 cm, BC = 9 cm

   C
   A B
10. PR is the hypotenuse of a right-angled triangle PQR. Find QR when:
    (a) PR = 61 cm, PQ = 11 cm (b) PR = 85 cm, PQ = 13 cm
    (c) PR = 3·9 cm, PQ = 1·5 cm (d) PR = 3·2 cm, PQ = 1·2 cm

Example

Find the volume of concrete in a pipe 6·3 m long with an external diameter of 1·7 m and an internal diameter of 1·5 m

$V = \pi R^2 h - \pi r^2 h = \pi h (R^2 - r^2)$

i.e. $V = \pi h (R - r)(R + r)$

so $V = \pi \times 6{\cdot}3 \times (1{\cdot}7 - 1{\cdot}5) \times (1{\cdot}7 + 1{\cdot}5)$

$V = \pi \times 6{\cdot}3 \times 0{\cdot}2 \times 3{\cdot}2$

$V \approx 12{\cdot}67$ Taking π as 3·14, the volume of concrete is 12·67 cm³.

Exercise 4.3b

1. Find the area of the shaded part of the diagram on the right where:
   (a) $a = 5{\cdot}7$ cm, and $b = 4{\cdot}3$ cm
   (b) $a = 13{\cdot}2$ cm, and $b = 6{\cdot}8$ cm
   (c) $a = 7{\cdot}9$ cm, and $b = 4{\cdot}7$ cm

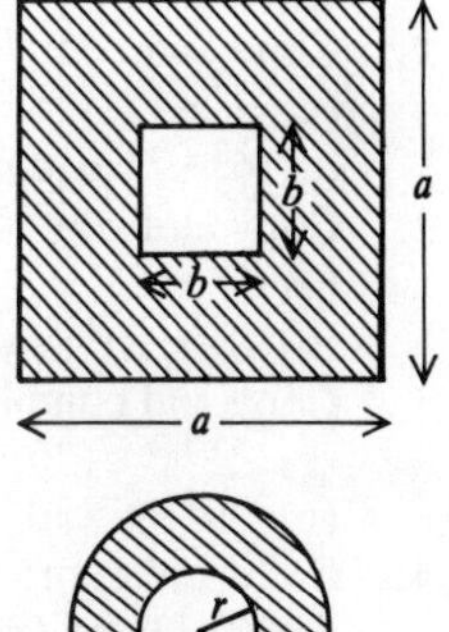

2. Find in terms of π the area of the shaded part of the diagram on the right where:
   (a) $R = 25$ cm, and $r = 5$ cm
   (b) $R = 16{\cdot}8$ cm, and $r = 3{\cdot}2$ cm
   (c) $R = 5{\cdot}9$ cm, and $r = 3{\cdot}7$ cm

3. A plastic duct has a square cross section like that in question 1. Find the volume of plastic where the length of the duct is:
   (a) 18·3 cm, and $a = 3{\cdot}1$ cm, $b = 2{\cdot}1$ cm
   (b) 6·5 m, and $a = 1{\cdot}8$ m, $b = 1{\cdot}3$ m
   (c) 51 cm, and $a = 2{\cdot}6$ cm, $b = 1{\cdot}1$ cm
4. Find the volume of concrete in a pipe, like the one at the top of the page where:
   (a) $R = 1{\cdot}8$ m, $r = 1{\cdot}4$ m, and $h = 7{\cdot}2$ m
   (b) $R = 57$ cm, $r = 43$ cm, and $h = 276$ cm
   (c) $R = 8{\cdot}3$ cm, $r = 6{\cdot}9$ cm, and $h = 45$ cm
5. The volume of a cone is given by $V = \frac{1}{3}\pi r^2 h$. A conical container is made by removing a cone of radius r from a cone of radius R as shown in the diagram. Find the volume of material in the container where:
   (a) $R = 13{\cdot}2$ cm, $r = 9{\cdot}4$ cm, and $h = 24{\cdot}1$ cm
   (b) $R = 47$ cm, $r = 36$ cm, and $h = 215$ cm

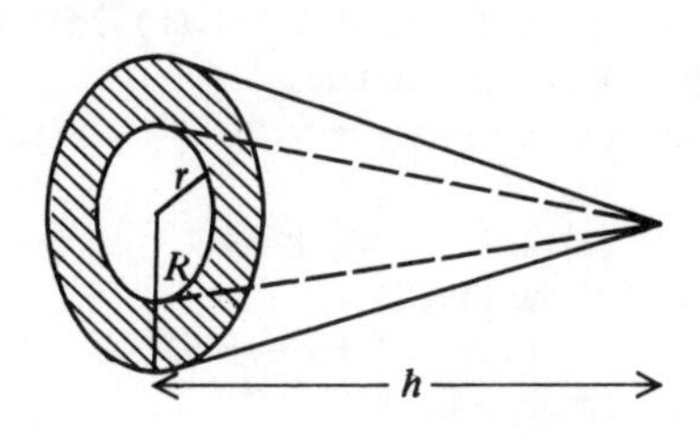

6. In the diagram on the right $A\hat{B}C = B\hat{D}C = 90°$, AC = 15·3 cm, AB = 11·8 cm and BD = 6·5 cm. Use the Theorem of Pythagoras to find the length of BC and hence find the length of DC.

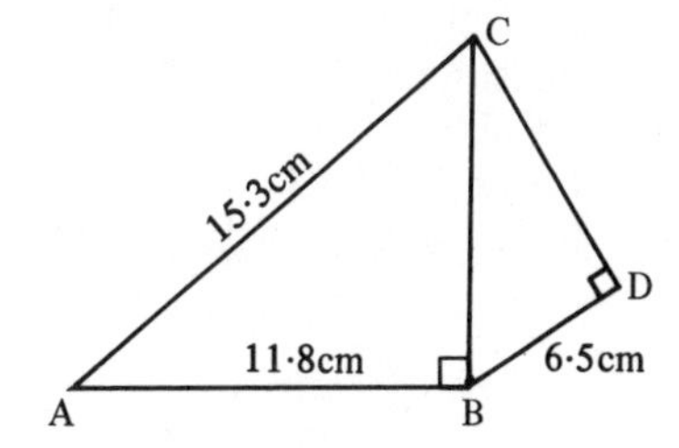

4.4 Completing the square

$$(a+b)^2$$
$$= a^2 + ab + ab + b^2$$
$$= a^2 + 2ab + b^2$$

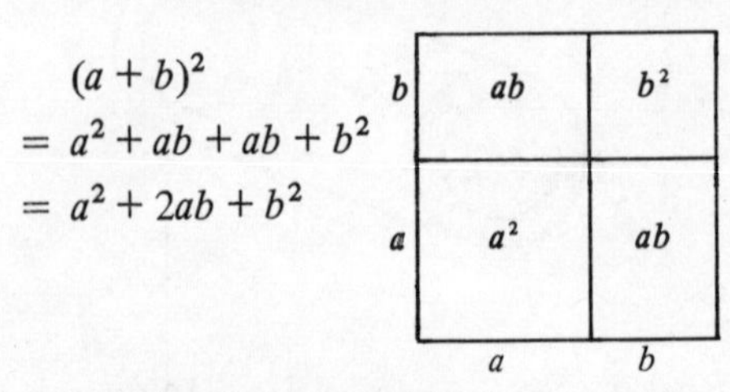

$$(a-b)^2$$
$$= a^2 - 2(ab - b^2) - b^2$$
$$= a^2 - 2ab + b^2$$

$(x+5)^2 = x^2 + 10x + 25$

$(x-7)^2 = x^2 - 14x + 49$

Exercise 4.4a

1. Copy and complete:
(a) $7^2 + 2.7.5 + 5^2 = (\quad + \quad)^2 =$
(b) $8^2 + 2.8.3 + 3^2 = (\quad + \quad)^2 =$
(c) $6^2 - 2.6.4 + 4^2 = (\quad - \quad)^2 =$
(d) $9^2 - 2.9.6 + 6^2 = (\quad - \quad)^2 =$

2. Copy and complete:
(a) $(7+8)^2 = 7^2 + \quad +$
(b) $(5+9)^2 = \quad + \quad +$
(c) $(8-3)^2 = 8^2 - \quad +$
(d) $(11-7)^2 = \quad - \quad +$

3. Copy and complete:
(a) $21^2 = (20+1)^2 = \quad =$
(b) $52^2 = (50+2)^2 = \quad =$
(c) $29^2 = (30-1)^2 = \quad =$
(d) $77^2 = (80-3)^2 = \quad =$

4. Copy and complete:
(a) $(9{\cdot}1)^2 = (9 + 0{\cdot}1)^2 = \quad =$
(b) $(12{\cdot}3)^2 = (12 + 0{\cdot}3)^2 = \quad =$
(c) $(4{\cdot}9)^2 = (5 - 0{\cdot}1)^2 = \quad =$
(d) $(19{\cdot}8)^2 = (20 - 0{\cdot}2)^2 = \quad =$

5. Write without brackets:
(a) $(l+m)^2$ (b) $(x-y)^2$ (c) $(2p+q)^2$ (d) $(a-3b)^2$ (e) $(4m+3n)^2$

6. Write without brackets:
(a) $(x+6)^2$ (b) $(x+11)^2$ (c) $(x-5)^2$ (d) $(x-13)^2$ (e) $(2x+1)^2$
(f) $(5x+1)^2$ (g) $(3x-1)^2$ (h) $(7x-1)^2$ (i) $(3x+5)^2$ (j) $(4x-7)^2$

7. Copy and complete:
(a) $p^2 + 2pq + q^2 = (\quad + \quad)^2$
(b) $l^2 - 2lm + m^2 = (\quad - \quad)^2$
(c) $x^2 + 12x + 36 = (\quad + \quad)^2$
(d) $y^2 - 18y + 81 = (\quad - \quad)^2$

8. Write as a square:
(a) $x^2 + 8x + 16$ (b) $x^2 + 20x + 100$ (c) $x^2 - 14x + 49$ (d) $x^2 - x + \frac{1}{4}$
(e) $9x^2 + 6x + 1$ (f) $25x^2 + 10x + 1$ (g) $16x^2 - 8x + 1$ (h) $36x^2 - 12x + 1$

9. Write as a square:
(a) $x^2 + 6xy + 9y^2$ (b) $4l^2 + 4lm + m^2$ (c) $a^2 + 10ab + 25b^2$
(d) $p^2 - 8pq + 16q^2$ (e) $64x^2 - 16xy + y^2$ (f) $100l^2 - 20lm + m^2$
(g) $4a^2 + 12ab + 9b^2$ (h) $16p^2 - 24pq + 9q^2$

10. Show that:
(a) $(m-n)^2 + (m+n)^2 = 2(m^2+n^2)$ (b) $(m-n)^2 + 4mn = (m+n)^2$

11. Show that $(m^2-n^2)^2 + (2mn)^2 = (m^2+n^2)^2$, and hence write down in terms of m and n the lengths of the sides of a right-angled triangle. By putting $m = 2$ and $n = 1$, find the dimensions of one such triangle.

12. Simplify:
(a) $(x+1)^2 + (x+2)^2 + (x+3)^2$
(b) $(x-3)^2 + (x-1)^2 + (x+1)^2 + (x+3)^2$
(c) $(x+y)^2 + (y+z)^2 + (z+x)^2$
(d) $(2x+y)^2 - (3x-2y)^2 + (x+2y)^2$

To make a perfect square from $x^2 + 6x$, it is necessary to add $(\frac{1}{2}$ coefficient of $x)^2$ i.e. 9

$$x^2 + 6x + (3)^2 = (x + 3)^2$$

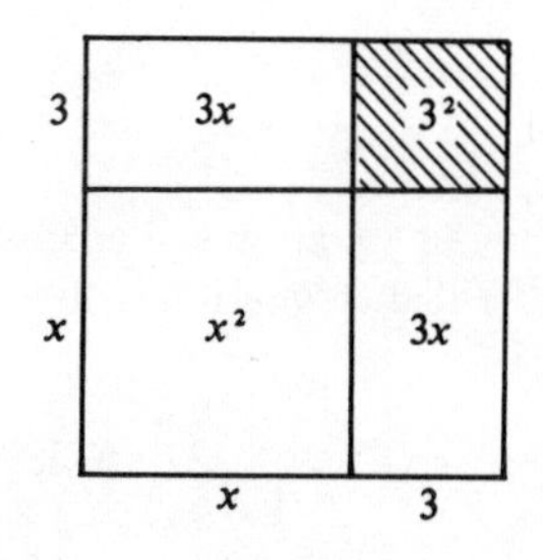

$x^2 + 6x$
$= x^2 + (3x + 3x)$
$= (x + 3)^2 - 3^2$

For example:

$$x^2 + 10x + 28 = (x + 5)^2 + 28 - 5^2 = (x + 5)^2 + 3$$

Exercise 4.4b

1. Copy and complete:
   (a) $x^2 + 4x = (x + \quad)^2 -$
   (b) $x^2 + 8x = (x + \quad)^2 -$
   (c) $x^2 + 12x = (x + \quad)^2 -$
   (d) $x^2 + 20x = (x + \quad)^2 -$
   (e) $x^2 - 6x = (x - \quad)^2 -$
   (f) $x^2 - 10x = (x - \quad)^2 -$
   (g) $x^2 - 4x = (x - \quad)^2 -$
   (h) $x^2 - 14x = (x - \quad)^2 -$
2. Copy and complete:
   (a) $x^2 + 2x + 5 = (x + 1)^2 +$
   (b) $x^2 + 6x + 12 = (x + 3)^2 +$
   (c) $x^2 + 16x + 100 = (x + \quad)^2 +$
   (d) $x^2 + 18x + 100 = (x + \quad)^2 +$
   (e) $x^2 + 10x + 3 = (x + \quad)^2 -$
   (f) $x^2 - 8x + 25 = (x - \quad)^2 +$
   (g) $x^2 - 12x + 40 = (x - \quad)^2 +$
   (h) $x^2 - 16x + 10 = (x - \quad)^2 -$
3. Write in the form $(x + a)^2 + b$:
   (a) $x^2 + 14x$ (b) $x^2 + 22x$ (c) $x^2 - 18x$ (d) $x^2 - 20x$
   (e) $x^2 + 3x$ (f) $x^2 + x$ (g) $x^2 - 3x$ (h) $x^2 - 5x$
4. Write in the form $(x + p)^2 + q$:
   (a) $x^2 + 6x + 17$ (b) $x^2 + 10x + 20$ (c) $x^2 - 8x + 21$ (d) $x^2 - 24x - 100$
   (e) $x^2 + 3x + 5$ (f) $x^2 + 5x + 3$ (g) $x^2 - 3x + 7$ (h) $x^2 - x + 4$

Example

Find the minimum value of $x^2 + 6x + 15$
$x^2 + 6x + 15$ can be written as $(x + 3)^2 + 15 - 9 = (x + 3)^2 + 6$.
When $x = -3$, $(x + 3)^2 = 0$; so the required minimum value is 6.

Exercise 4.4c

1. Write the expression in the form $(x + a)^2 + b$, and hence find the minimum value.
   (a) $x^2 + 4x + 12$ (b) $x^2 + 10x + 30$ (c) $x^2 + 16x + 70$ (d) $x^2 + 2x$
   (e) $x^2 - 2x + 5$ (f) $x^2 - 6x + 15$ (g) $x^2 - 10x + 18$ (h) $x^2 - 3x$
2. Write in the form $(ax + b)^2 + c$, and hence find the minimum value of:
   (a) $4x^2 + 12x + 11$ (b) $9x^2 + 12x + 5$ (c) $16x^2 + 8x + 1$ (d) $25x^2 + 20x$
   (e) $4x^2 - 16x + 20$ (f) $49x^2 - 56x + 19$ (g) $36x^2 - 12x + 1$ (h) $81x^2 - 18x$
3. Find the maximum value of:
   (a) $9 - 4x - x^2$ (b) $11 - 6x - x^2$ (c) $20 + 8x - x^2$ (d) $24x - 9x^2$

4.5 Linear equations

It is important to remember that an equation is like a balance. Any process applied to one side must also be applied to the other if the equation is to remain in balance. This is shown in the four examples below.

(1)
$$\begin{aligned} x+5 &= 11 \\ x+5-5 &= 11-5 \\ x &= 6 \end{aligned}$$

(2)
$$\begin{aligned} 3x &= 39 \\ \tfrac{1}{3}\times 3x &= \tfrac{1}{3}\times 39 \\ x &= 13 \end{aligned}$$

(3)
$$\begin{aligned} 7x-6 &= 50 \\ 7x-6+6 &= 50+6 \\ 7x &= 56 \\ \tfrac{1}{7}\times 7x &= \tfrac{1}{7}\times 56 \\ x &= 8 \end{aligned}$$

(4)
$$\begin{aligned} 12x-21 &= 7x+9 \\ 12x-7x-21 &= 7x-7x+9 \\ 5x-21 &= 9 \\ 5x-21+21 &= 9+21 \\ 5x &= 30 \\ \tfrac{1}{5}\times 5x &= \tfrac{1}{5}\times 30 \\ x &= 6 \end{aligned}$$

Exercise 4.5a

1. Solve:
(a) $x+5=19$ (b) $x+4=12$ (c) $x+73=95$ (d) $x+7=3$ (e) $x+23=-4$

2. Solve:
(a) $x-5=19$ (b) $x-4=12$ (c) $x-67=82$ (d) $x-4=-9$ (e) $x-6=-5$

3. Solve:
(a) $3x=12$ (b) $5x=60$ (c) $13x=117$ (d) $8x=4$ (e) $15x=3$

4. Solve:
(a) $\frac{x}{2}=5$ (b) $\frac{x}{7}=6$ (c) $\frac{x}{19}=2$ (d) $\frac{x}{3}=-4$ (e) $\frac{x}{22}=-3$

5. Solve:
(a) $2x+5=11$ (b) $3x+8=29$ (c) $7x+15=92$ (d) $4x-3=29$ (e) $12x-23=25$

6. Solve:
(a) $\frac{x}{2}+5=11$ (b) $\frac{x}{3}+8=29$ (c) $\frac{x}{6}+11=29$ (d) $\frac{x}{8}-5=15$ (e) $\frac{x}{11}-8=3$

7. Solve:
(a) $2(3x+5)=52$ (b) $5(4x+1)=185$ (c) $7(2x-3)=133$ (d) $3(9x-4)=15$
(e) $3(2x+9)=15$ (f) $6(5x+11)=36$ (g) $4(2x+1)=24$ (h) $8(5x-2)=32$

8. Solve:
(a) $2x+1=x+7$ (b) $9x+4=6x+10$ (c) $5x-7=3x+11$ (d) $12x-23=8x-3$
(e) $4x+3=2x-9$ (f) $8x-1=5x-22$ (g) $7x+4=2x+7$ (h) $6x-13=4x-19$

9. Solve:
(a) $7x+5=11x+1$ (b) $5x+14=7x+4$ (c) $4x-3=9x-23$
(d) $2x-15=11x+12$ (e) $4x+3=38-x$ (f) $7x-5=43-3x$
(g) $19-3x=41-4x$ (h) $43-2x=7-9x$

10. Solve:
(a) $\frac{x}{2}+\frac{x}{4}=15$ (b) $\frac{2x}{3}+\frac{x}{9}=42$ (c) $\frac{x}{5}-3=\frac{x}{10}+4$ (d) $\frac{4x}{7}-5=1-\frac{3x}{14}$

11. Solve:

(a) $2(3(4x + 5) + 6) = 18$ (b) $5(3x + 4) - 16 = 2(7x - 5) + 43$

12. Solve:

(a) $\frac{x-5}{3} + \frac{x-2}{2} = 4$ (b) $\frac{2x+1}{7} - \frac{3x+2}{16} = 1$

Example 1

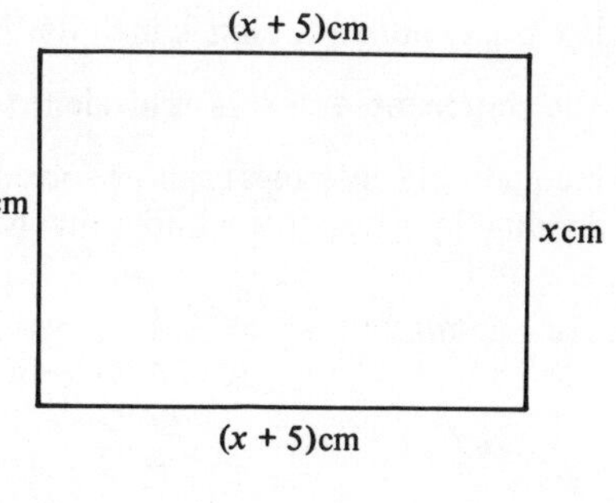

The perimeter of a rectangle is 78 cm. One side is 5 cm longer than the other. Find the length of each side.

Let the length of one side be x cm. Then the length of the other side is $(x + 5)$ cm.

So $x + x + (x + 5) + (x + 5) = 78.$

i.e. $4x + 10 = 78;$

$4x = 68;$

$x = 17.$

The lengths of the sides of the rectangle are 17 cm and 22 cm.

Example 2

Find three consecutive numbers so that their sum is 96.

Let the first number be n, then the second is $(n + 1)$ and the third is $(n + 2)$.

So $n + (n + 1) + (n + 2) = 96.$

i.e. $3n + 3 = 96;$

$3n = 93;$

$n = 31.$

The numbers are 31, 32, and 33.

Exercise 4.5b

1. The perimeter of a rectangle is 98 cm. One side is 7 cm longer than the other. Form an equation and hence find the length of each side.
2. The perimeter of a rectangle is 72 cm. One side is three times the length of the other. Form an equation and hence find the length of each side.
3. Find three consecutive numbers so that their sum is 78.
4. Find three consecutive even numbers so that their sum is 108.
5. A rectangular room is 2·3 m longer than it is wide. Its perimeter is 17 m. Form an equation and hence find the length of each wall.
6. A rectangular room is 3·1 m longer than it is wide. It has one doorway, 1·1 m wide and 25·5 m of skirting board are required. By forming an equation, find the length of each wall. Does the position of the doorway affect your answer?
7. Two boys have £3·30 in cash between them. One gives 60 pence to the other and finds he now has twice as much money as his friend. Form an equation by letting x pence be the amount one boy had at the start. Hence find how much each has then.
8. Two girls have 72 photos of pop stars between them. One gives 11 to the other and finds she now has half the number her friend has. Form an equation by letting n be the number of photos one girl had at the start. Hence find how many each has now.

4.6 Linear inequalities (orderings)

If x is less than 4, we write $x < 4$.
$x \leqslant 4$ means x is less than or equal to 4.

If x is a positive whole number and if $T = \{x | x < 4\}$ then $T = \{1, 2, 3\}$
If x is any number, then T includes numbers such as -1, $1\frac{1}{2}$, 0·1 etc. (i.e. all numbers less than 4).

The statement $4 > x$ is equivalent to $x < 4$.

Inequalities (orderings) can be solved in a similar way to equations, except that multiplying or dividing by a negative number involves a change in the inequality sign (see Example 2).

Example 1

$5x - 4 > 3x - 16$	
$2x - 4 > -16$	subtracting $3x$ from each side
$2x > -12$	adding 4 to each side
$x > -6$	dividing both sides by 2

Example 2

$16 - 5x \geqslant 1 - 2x$	
$-5x + 2x \geqslant 1 - 16$	rearranging
$-3x \geqslant -15$	collecting like terms
$x \leqslant 5$	dividing by -3, so reverse the inequality sign.

Example 3 $\dfrac{16}{x} < 8$

Here $x > 2$ or x may be any negative number.

The number line

A number line is a useful way of showing a result and also of finding which numbers satisfy inequalities such as $x < -5$. For example:

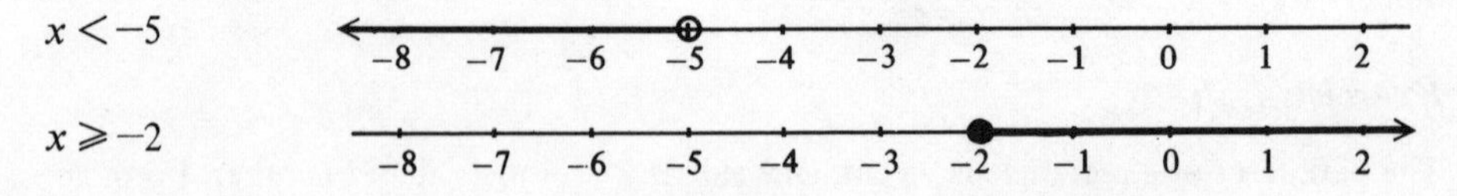

Note: The filled-in circle ● for $x \geqslant -2$ is used to show that x can equal -2.
The empty circle ○ for $x < -5$ is used to show that x is less than -5 but not equal to -5.

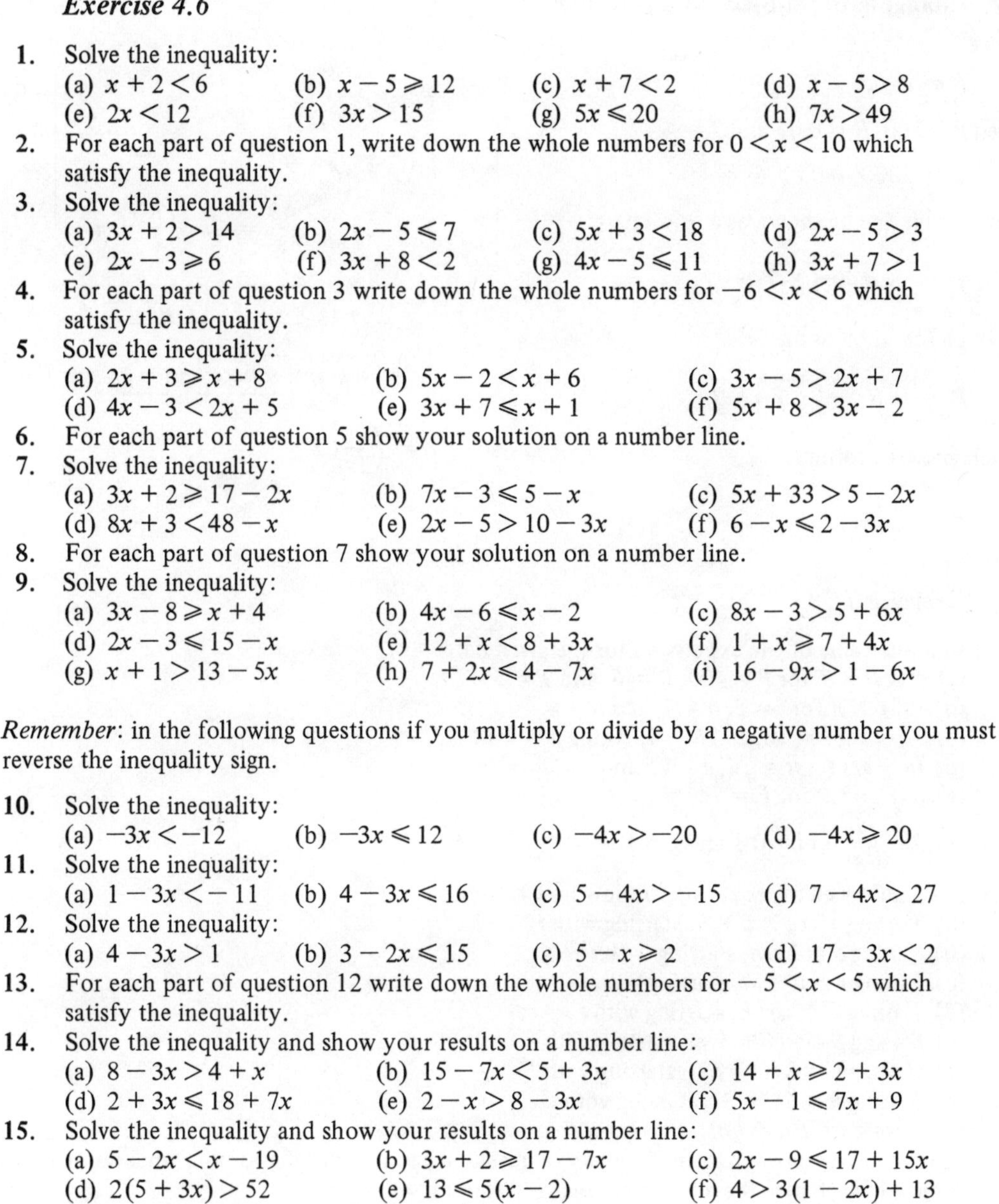

Exercise 4.6

1. Solve the inequality:
   (a) $x+2<6$ (b) $x-5\geqslant 12$ (c) $x+7<2$ (d) $x-5>8$
   (e) $2x<12$ (f) $3x>15$ (g) $5x\leqslant 20$ (h) $7x>49$
2. For each part of question 1, write down the whole numbers for $0<x<10$ which satisfy the inequality.
3. Solve the inequality:
   (a) $3x+2>14$ (b) $2x-5\leqslant 7$ (c) $5x+3<18$ (d) $2x-5>3$
   (e) $2x-3\geqslant 6$ (f) $3x+8<2$ (g) $4x-5\leqslant 11$ (h) $3x+7>1$
4. For each part of question 3 write down the whole numbers for $-6<x<6$ which satisfy the inequality.
5. Solve the inequality:
   (a) $2x+3\geqslant x+8$ (b) $5x-2<x+6$ (c) $3x-5>2x+7$
   (d) $4x-3<2x+5$ (e) $3x+7\leqslant x+1$ (f) $5x+8>3x-2$
6. For each part of question 5 show your solution on a number line.
7. Solve the inequality:
   (a) $3x+2\geqslant 17-2x$ (b) $7x-3\leqslant 5-x$ (c) $5x+33>5-2x$
   (d) $8x+3<48-x$ (e) $2x-5>10-3x$ (f) $6-x\leqslant 2-3x$
8. For each part of question 7 show your solution on a number line.
9. Solve the inequality:
   (a) $3x-8\geqslant x+4$ (b) $4x-6\leqslant x-2$ (c) $8x-3>5+6x$
   (d) $2x-3\leqslant 15-x$ (e) $12+x<8+3x$ (f) $1+x\geqslant 7+4x$
   (g) $x+1>13-5x$ (h) $7+2x\leqslant 4-7x$ (i) $16-9x>1-6x$

Remember: in the following questions if you multiply or divide by a negative number you must reverse the inequality sign.

10. Solve the inequality:
    (a) $-3x<-12$ (b) $-3x\leqslant 12$ (c) $-4x>-20$ (d) $-4x\geqslant 20$
11. Solve the inequality:
    (a) $1-3x<-11$ (b) $4-3x\leqslant 16$ (c) $5-4x>-15$ (d) $7-4x>27$
12. Solve the inequality:
    (a) $4-3x>1$ (b) $3-2x\leqslant 15$ (c) $5-x\geqslant 2$ (d) $17-3x<2$
13. For each part of question 12 write down the whole numbers for $-5<x<5$ which satisfy the inequality.
14. Solve the inequality and show your results on a number line:
    (a) $8-3x>4+x$ (b) $15-7x<5+3x$ (c) $14+x\geqslant 2+3x$
    (d) $2+3x\leqslant 18+7x$ (e) $2-x>8-3x$ (f) $5x-1\leqslant 7x+9$
15. Solve the inequality and show your results on a number line:
    (a) $5-2x<x-19$ (b) $3x+2\geqslant 17-7x$ (c) $2x-9\leqslant 17+15x$
    (d) $2(5+3x)>52$ (e) $13\leqslant 5(x-2)$ (f) $4>3(1-2x)+13$
16. Solve the inequality and show your results on a number line:
    (a) $\frac{10}{x}>2$ (b) $\frac{8}{x}<16$ (c) $\frac{3}{x}\geqslant 12$ (d) $\frac{18}{x}<3$

4.7 Changing the subject of a formula

$$I = P \times R \times T$$

when $P = 300$, $R = 0{\cdot}05$ and $T = 4$

$$I = 300 \times 0{\cdot}05 \times 4 = 60$$

The formula can be shown as a flow chart:

$P \rightarrow$ [$\times R$] $\rightarrow$ [$\times T$] $\rightarrow I$

P can be found by using

$P \leftarrow$ [$\div R$] $\leftarrow$ [$\div T$] $\leftarrow I$

which gives the formula

$$P = \frac{I}{T \times R}$$

$$T = 2\pi\sqrt{\frac{l}{g}}$$

when $l = 2{\cdot}45$, $g = 9{\cdot}8$

$$T = 2\pi\sqrt{\frac{2{\cdot}45}{9{\cdot}8}} = 2\pi\sqrt{\frac{1}{4}} = \pi$$

$l \rightarrow$ [$\div g$] $\rightarrow$ [$\sqrt{\ }$] $\rightarrow$ [$\times 2\pi$] $\rightarrow T$

$l \leftarrow$ [$\times g$] $\leftarrow$ [square] $\leftarrow$ [$\div 2\pi$] $\leftarrow T$

so $l = \left(\frac{T}{2\pi}\right)^2 \times g$

Exercise 4.7a

1. Find the value of the expression for the given data:
   (a) $P \times R \times T$ for $P = 400$, $R = 6$, and $T = 5$
   (b) $l \times b \times h$ for $l = 5$, $b = 7$, and $h = 4$
   (c) $2\pi rh$ for $r = 10$, and $h = 4$, taking $\pi = 3{\cdot}14$
   (d) $(u + at)$ for $u = 15$, $a = 25$ and $t = 2$
   (e) $\frac{5}{9}(F - 32)$ for $F = 78$
   (f) $2\pi\sqrt{\frac{l}{g}}$ for $l = 0{\cdot}1$ and $g = 10$
2. Draw a flow chart, as above, to show how to find:
   (a) V using $V = l \times b \times h$, starting with l
   (b) S using $S = 2\pi rh$, starting with r
   (c) v using $v = u + at$, starting with t
   (d) V using $V = \pi r^2 h$, starting with r
   (e) S using $S = \pi r(2h + r)$, starting with h
   (f) C using $C = \frac{5}{9}(F - 32)$, starting with F
   (g) F using $F = \frac{9}{5}C + 32$, starting with C
   (h) s using $s = t(u + \frac{1}{2}at)$, starting with a
3. Use the inverse flow chart for each part of question 2 to re-write the formula, making as the subject in each case the letter you originally started with.
4. $V = \pi h(R^2 - r^2)$.
   (a) Find V, when $h = 2$ cm, $R = 7{\cdot}5$ cm and $r = 2{\cdot}5$ cm, taking $\pi = 3{\cdot}14$
   (b) Draw a flow chart to show how to find V starting with R
   (c) Use the inverse flow chart to re-write the formula with R in terms of V, h, r, and π.

Example

Make F the subject of the formula in $C = \frac{5}{9}(F-32)$

multiply both sides by 9: $9C = 5(F-32)$

remove brackets: $9C = 5F - 160$

add 160 to both sides: $9C + 160 = 5F$

divide both sides by 5: $\frac{9}{5}C + 32 = F$

so $F = \frac{9}{5}C + 32$

Exercise 4.7b

1. Rearrange the formula and make as subject the given letter:
   (a) l in $A = l \times b$ (b) r in $C = 2\pi r$ (c) I in $V = IR$
   (d) L in $S = \pi r L$ (e) f in $v = u + ft$ (f) x in $y = mx + c$
   (g) T in $P = \dfrac{RT}{V}$ (h) R in $I = \dfrac{E}{R}$ (i) m in $P = \dfrac{m(v-u)}{t}$
   (j) r in $V = \pi r^2 h$ (k) u in $v^2 = u^2 + 2as$ (l) x in $r^2 = (x-a)^2 + y^2$
   (m) g in $T = 2\pi\sqrt{\dfrac{l}{g}}$ (n) v in $\dfrac{1}{f} = \dfrac{1}{u} + \dfrac{1}{v}$ (o) x in $y = \dfrac{c}{x+a}$
2. Make r the subject of the formula in:
   (a) $S = 4\pi r^2$ (b) $V = \frac{4}{3}\pi r^3$ (c) $A = \frac{1}{2}h(R+r)$ (d) $V = \pi h(R^2 - r^2)$
3. Make t the subject of the formula in:
   (a) $s = \dfrac{(u+v)t}{2}$ (b) $u = v - ft$ (c) $R = \dfrac{PV}{mt}$ (d) $T = k(\alpha t + \beta)$
4. Make x the subject of the formula in:
   (a) $y = m(x-a)$ (b) $\dfrac{x}{a} + \dfrac{y}{b} = 1$ (c) $(x-a)y = c$ (d) $y = 1 - \dfrac{1}{x}$
5. Make R the subject of the formula in:
   (a) $A = P\left(1 + \dfrac{R}{100}\right)$ (b) $Rt = 2R + t$ (c) $V = \dfrac{R}{R-r}$ (d) $K = \dfrac{R+1}{R+2}$

Exercise 4.7c

Write down the correct result (or results) from those given:

1. If $a = \dfrac{b}{x}$ then x is: (a) ab (b) $\dfrac{a}{b}$ (c) $\dfrac{b}{a}$ (d) $a+b$
2. If $P = \dfrac{mv^2}{2g}$ then v is: (a) $\sqrt{2Pg - m}$ (b) $\sqrt{\dfrac{2gP}{m}}$ (c) $\dfrac{2g}{m}\sqrt{P}$ (d) $\dfrac{\sqrt{2gP}}{m}$
3. If $A = 2\pi r(r+h)$ then h is: (a) $A - 2\pi r^2$ (b) $\dfrac{A - 2\pi r^2}{2\pi r}$ (c) $\dfrac{A - 2\pi r}{r}$ (d) $\dfrac{A}{2\pi r} - r$
4. If $y = \dfrac{3x+1}{x+2}$ then x is: (a) $\dfrac{3y+1}{y+2}$ (b) $\dfrac{1-2y}{3+y}$ (c) $\dfrac{2y-1}{3-y}$ (d) $\dfrac{y-3}{1-2y}$

4.8 Simultaneous equations

$$\left.\begin{aligned} 3x + 2y &= 8 \\ 5x + y &= 11 \end{aligned}\right\}$$ is a pair of simultaneous equations.

We require to find values of x and y which make both equations true at the same time.
i.e. if $x = 2$ and $y = 1$ then $3 \times 2 + 2 \times 1 = 8$ and $5 \times 2 + 1 = 11$

Example 1 $3x + 2y = 8$ (i)
$5x + y = 11$ (ii)

To make the coefficients of y the same in each equation, multiply (ii) by 2.

(ii) × 2 $10x + 2y = 22$ (iii)

To eliminate the y terms, subtract equation (i) from equation (iii)

(iii) − (ii) $7x + 0 = 14$ i.e. $x = 2$

Substituting $x = 2$ into equation (ii) gives $10 + y = 11$ i.e. $y = 1$

Example 2 $2x + 5y = 16$ (i)
$3x - 4y = 1$ (ii)

To make the coefficients of y the same in each equation:

(i) × 4 $8x + 20y = 64$ (iii)
(ii) × 5 $15x - 20y = 5$ (iv)

To eliminate the y terms, add equation (iii) to equation (iv)

(iii) + (iv) $23x = 69$ i.e. $x = 3$

Substituting $x = 3$ into equation (i) gives

$6 + 5y = 16$ or $5y = 10$ i.e. $y = 2$

Exercise 4.8a

1. Solve the simultaneous equations. Check your results by substitution.
   (a) $4x + y = 8$, $3x + y = 10$ (b) $7x + 2y = 26$, $3x - 2y = 14$ (c) $5x + 3y = 12$, $8x + 3y = 9$
2. Solve the simultaneous equations. Check your results by substitution.
   (a) $x + 5y = 26$, $x + 2y = 14$ (b) $5x + 7y = 18$, $3y - 5x = 22$ (c) $4x - 7y = 41$, $4x - 3y = 29$
3. Solve the simultaneous equations. Check your results by substitution.
   (a) $5x + 3y = 27$, $2x + y = 10$ (b) $7x + 4y = 2$, $3x - y = 9$ (c) $6x + 5y = 35$, $x - 2y = 3$
4. Solve the simultaneous equations. Check your results by substitution.
   (a) $4x + 3y = 23$, $2x + 5y = 29$ (b) $2x + 3y = 28$, $3x + 2y = 27$ (c) $7x - 4y = 37$, $2x + 3y = -6$
   (d) $2x + 7y = 25$, $7x + 2y = 20$ (e) $4a + 3b = 22$, $5a - 4b = 43$ (f) $3p + 3q = 15$, $2p + 5q = 14$

Example 3

An alternative way of solving simultaneous equations is by substitution of one of the variables.

$$3x + 5y = 21 \quad \text{(i)}$$
$$2x + y = 7 \quad \text{(ii)}$$

We can rewrite equation (ii) as $y = 7 - 2x$ and then substitute this value of y into equation (i) giving:

$$3x + 5(7 - 2x) = 21$$

i.e. $3x + 35 - 10x = 21$, or $14 = 7x$, so $x = 2$.

Substituting $x = 2$ gives $y = 7 - 2 \times 2$, so $y = 3$.

Exercise 4.8b

1. Solve the simultaneous equations by substituting one of the variables:
   (a) $3x + 4y = 23$, $y = 2x + 3$ (b) $5x - 2y = 1$, $y = 3x - 2$ (c) $2x + 3y = 8$, $x + 4y = 9$
2. Solve the simultaneous equations by substituting one of the variables:
   (a) $7x - 2y = 8$, $3x + y = 9$ (b) $4x + 3y = 31$, $2y = x - 5$ (c) $5x + 6y = 8$, $2x + 3y = 5$
3. Check your results for questions 1 and 2 by using the elimination method instead.
4. Solve the simultaneous equations by whichever method you prefer:
   (a) $8x - 2y = 26$, $3x + y = 15$ (b) $7x + 3y = 15$, $x - 2y = 7$ (c) $9x - 5y = 47$, $3x + 2y = 1$
   (d) $5x - 6y = 2$, $2x + 3y = 17$ (e) $4x + 5y = 16$, $6x + 2y = 13$ (f) $5 + 3x = 2y$, $3y + 7 = 2x$
   (g) $x + y = 3 = 2x - y$ (h) $3x - 7 = y = 3 - 2x$

Example 4 3 pencils and 2 rubbers cost 40 pence whereas 2 pencils and 5 rubbers cost 67 pence. Find the cost of each.

Using p pence for the cost of one pencil and r pence for the cost of one rubber, we can write this information as a pair of simultaneous equations.

$$3p + 2r = 40 \quad \text{(i)}$$
and
$$2p + 5r = 67 \quad \text{(ii)}$$

These can now be solved giving $p = 6$ and $r = 11$, so the price of one pencil is 5 pence and the cost of one rubber is 11 pence.

Exercise 4.8c

1. Find two numbers such that their sum is 77 and their difference is 25.
2. If 7 pencils and 5 rubbers cost £1·16, whereas 5 pencils and 3 rubbers cost 76 pence, find the cost of each.
3. If 2 adult's tickets and 5 children's tickets at the circus cost £3·50, whereas 3 adult's tickets and 4 children's tickets cost £3·85, find the cost of each type of ticket.
4. A man takes 4 hours to cover a two-stage journey. On one occasion he averages 60 km/h and 80 km/h for the two stages whereas on another occasion he averages 45 km/h and 100 km/h. Find the length of each stage.

4.9 Quadratic equations

Any equation of the form $ax^2 + bx + c = 0$ is a quadratic equation. There are three ways of finding the solution, namely:
(i) by factorizing; (ii) by completing the square; (iii) by using the formula.

Method 1 Factorizing.

	Solve	$x^2 - 6x + 8 = 0$
$x^2 - 6x + 8 = 0$ will factorize	so	$(x-2)(x-4) = 0$
One of the two brackets is zero	so	$(x-2) = 0$ *or* $(x-4) = 0$
	hence	$x = 2$ *or* $x = 4$

Exercise 4.9a

1. Write down the two solutions of the quadratic equation:
(a) $(x-3)(x-4)=0$ (b) $(x-5)(x+2)=0$ (c) $(x+3)(x-7)=0$
(d) $(x+2)(x+9)=0$ (e) $(2x-1)(x-3)=0$ (f) $(3x-1)(x+4)=0$
(g) $(4x-1)(5x-1)=0$ (h) $x(x-3)=0$ (i) $x(3x+1)=0$
(j) $(7x+1)(4x+1)=0$ (k) $(2x-3)(3x-2)=0$ (l) $(7x-2)(3x+4)=0$
2. Write down the two solutions of the quadratic equation:
(a) $(5-x)(2-x)=0$ (b) $(4+x)(3-x)=0$ (c) $(x+2)(1-3x)=0$
3. First factorize the L.H.S. of the equation and hence solve:
(a) $x^2-6x+5=0$ (b) $x^2-5x+6=0$ (c) $x^2-7x+10=0$
(d) $x^2+8x+12=0$ (e) $x^2+4x+3=0$ (f) $x^2+10x+16=0$
(g) $x^2+6x-7=0$ (h) $x^2+5x-14=0$ (i) $x^2+3x-28=0$
(j) $x^2-5x-6=0$ (k) $x^2-6x-16=0$ (l) $x^2-10x-39=0$
4. First factorize the L.H.S. of the equation and hence solve:
(a) $2x^2-3x+1=0$ (b) $6x^2-5x+1=0$ (c) $8x^2-6x+1=0$
(d) $7x^2+8x+1=0$ (e) $10x^2-7x+1=0$ (f) $12x^2-7x+1=0$
(g) $2x^2-9x+4=0$ (h) $3x^2-16x+5=0$ (i) $4x^2-7x+3=0$
(j) $2x^2-5x-3=0$ (k) $3x^2-x-2=0$ (l) $4x^2+5x-6=0$
5. First factorize the L.H.S. of the equation and hence solve:
(a) $12-7x+x^2=0$ (b) $12-x-x^2=0$ (c) $10-3x-x^2=0$
(d) $2-5x-3x^2=0$ (e) $3+5x-2x^2=0$ (f) $3+10x-8x^2=0$
6. The three sides of a right-angled triangle are as shown in the diagram on the right.
Form a quadratic equation and then by factorizing, find the length of the three sides.

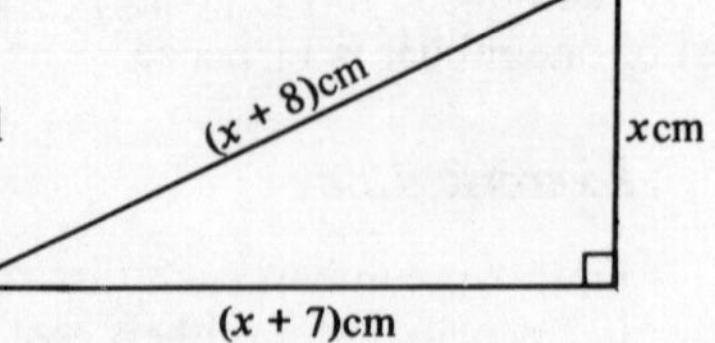

7. The sum of the squares of three consecutive numbers is 110. Show this information as a quadratic equation and hence find the numbers.
8. One side of a rectangle is 10 cm longer than the other. If the area of the rectangle is 56 cm^2, show this information as a quadratic equation and hence find the length of each side.

Method 2 Completing the square

Solve $x^2 + 6x + 4 = 0$

$x^2 + 6x$ can be written as $(x+3)^2 - 9$
$x^2 + 6x + 4$ can be written as $(x+3)^2 - 9 + 4 = (x+3)^2 - 5$
The equation becomes $(x+3)^2 - 5 = 0$, or $(x+3)^2 = 5$.

So $(x+3) = +\sqrt{5}$ *or* $(x+3) = -\sqrt{5}$

i.e. $x = -3 + \sqrt{5}$ *or* $x = -3 - \sqrt{5}$ are the required solutions.

Exercise 4.9b

1. Solve the equation by first taking the square root of each side:
   (a) $(x-2)^2 = 25$ (b) $(x-5)^2 = 64$ (c) $(x-3)^2 = 16$
   (d) $(x+4)^2 = 1$ (e) $(x+6)^2 = 49$ (f) $(x+2)^2 = 4$
2. Solve the equation by first completing the square:
   (a) $x^2 - 6x + 8 = 0$ (b) $x^2 - 12x + 11 = 0$ (c) $x^2 - 14x + 24 = 0$
   (d) $x^2 + 8x + 12 = 0$ (e) $x^2 + 10x - 39 = 0$ (f) $x^2 + 2x - 63 = 0$
3. Solve the equation by first taking the square root of each side:
   (a) $(x-1)^2 = 5$ (b) $(x-4)^2 = 7$ (c) $(x-6)^2 = 11$
   (d) $(x+2)^2 = 3$ (e) $(x+5)^2 = 2$ (f) $(x+1)^2 = 13$
4. Solve the equation by first completing the square:
   (a) $x^2 - 6x + 1 = 0$ (b) $x^2 - 16x + 3 = 0$ (c) $x^2 - 4x + 1 = 0$
   (d) $x^2 + 2x - 7 = 0$ (e) $x^2 + 6x - 12 = 0$ (f) $x^2 + 14x - 3 = 0$

Method 3 Using the formula

Solve $2x^2 - 5x + 1 = 0$

$$x = \frac{-b \pm \sqrt{b^2 - 4ac}}{2a}.$$

In the above equation
$a = 2, b = -5$, and $c = 1$

Substituting the values of a, b and c into the formula gives:

$$x = \frac{5 \pm \sqrt{25 - 4 \times 2 \times 1}}{2 \times 2} \quad \text{i.e. } x = \frac{5 \pm \sqrt{17}}{4}, \text{ so } x = 2{\cdot}281 \text{ or } 0{\cdot}219$$

Exercise 4.9c

1. Evaluate the formula $\dfrac{-b \pm \sqrt{b^2 - 4ac}}{2a}$ when:
   (a) $a = 3,\ b = -7$ and $c = 2$ (b) $a = 5,\ b = -8$ and $c = 3$
   (c) $a = 3,\ b = 4$ and $c = -2$ (d) $a = 7,\ b = 11$ and $c = 2$
2. Solve the equation by using the above formula:
   (a) $x^2 - 4x + 3 = 0$ (b) $x^2 - 9x + 14 = 0$ (c) $2x^2 - 7x + 3 = 0$
   (d) $x^2 + 7x + 10 = 0$ (e) $x^2 + 8x - 20 = 0$ (f) $4x^2 + 11x - 3 = 0$
3. Solve the equation by using the above formula:
   (a) $x^2 - 5x + 1 = 0$ (b) $x^2 - 7x + 3 = 0$ (c) $x^2 + 4x - 9 = 0$
   (d) $x^2 + 6x + 2 = 0$ (e) $3x^2 - 2x - 6 = 0$ (f) $5x^2 - 9x + 1 = 0$
4. The length of a room is 4 metres longer than its width. The height of the room is 3 metres and its volume is $1000\ m^3$. Find the length and width of the room.

4.10 General algebraic problems

Example 1

The same number is added to both the numerator and denominator of the fraction $\frac{13}{24}$. The resulting fraction is $\frac{3}{4}$.

Find the number added.

This problem can be represented by the equation $\dfrac{13+x}{24+x}=\dfrac{3}{4}$.

The equation can be re-written as $4(13+x)=3(24+x)$ so $x=20$.

Exercise 4.10a Linear equations

1. The same number is added to both the numerator and denominator of the fraction $\frac{15}{31}$. If the resulting fraction is $\frac{5}{6}$, find the number.
2. A number is added to the numerator of the fraction $\frac{3}{8}$ and double this number is added to the denominator. If the resulting fraction is $\frac{7}{15}$, find the number.
3. One week a man buys 5 gallons of petrol. Each week for the next three weeks he buys the same amount but he pays a penny per gallon more than in the previous week. If his total bill at the end of the four weeks is £14·60, find the price per gallon of petrol in the last week.
4. A householder can pay for his gas in one of two ways:
   (i) a fixed charge of £3·80 plus 17p per unit, or
   (ii) a fixed charge of £1·50 plus 19p per unit.
   Find how many units are used if it makes no difference which method is used. Which method is more advantageous if more than this number of units is used?
5. A man can choose to pay cash for a car in which case he gets a discount of 5%. Alternatively he can pay in instalments over one year, in which case he has to pay an extra 7% for hire purchase charges. If the difference between the two methods is £177, find the quoted price of the car.
6. During a particular month a salesman lunches at three different restaurants. The cost of the set meal in the second is £1·50 more than in the first. In the third the set meal is half what it costs in the second restaurant. If during the month he has 7 set meals in the first restaurant, 3 in the second, and 8 in the third, and the total bill is £41·30, find the cost of a meal in each restaurant.

Example 2

A ferry with a vehicle deck 112·5 m long can carry 17 cars and 6 caravans end on. Another ferry with a vehicle deck 264 m long can carry 40 cars and 14 caravans end on. Use this information to find the average car and average caravan length of the vehicles carried.

If c metres is the average length of a car, and v metres that of a caravan then:

	$17c + 6v = 112{\cdot}5$	(i)
	$40c + 14v = 264$	(ii)
(i) × 14	$238c + 84v = 1575$	(iii)
(ii) × 6	$240c + 84v = 1584$	(iv)
(iv) − (iii)	$2c = 9$	

so $c = 4{\cdot}5$ and hence $v = 6$.

i.e. The average length of a car is 4·5 m and that of a caravan 6 m.

Exercise 4.10b Simultaneous equations

1. A car park can hold a maximum of 400 cars, or cars and coaches. A coach occupies a space equivalent to 5 cars. The charges per day are 40p for a car and 75p for a coach. When the car park was full the takings on a particular day came to £150. Assuming the vehicles stayed all day, find the number of cars and numbers of coaches in the car park.
2. At Christmas time a woman buys 180 stamps. Some of these cost 7p and others 9p. The total cost of the stamps was £13·78. Find the number of each type bought. How would your answer be altered if for the same amount of money only 178 stamps had been bought?
3. The total area of four sides of a rectangular box of height 3 cm is 72 cm^2. If the length were to be increased by 10% and the breadth decreased by 20%, this area would remain unchanged. Find the length and breadth of the box.
4. A sum of money is made up of an equal number of 10p and 50p pieces. When the number of 50p pieces is increased by 12 and the number of 10p pieces is halved, the sum of money is increased by £4·50. Find the original sum of money.
5. The wages of 12 men and 5 boys come to £148·80 per day. At a similar rate 10 men and 3 boys would earn £116·16 per day. Assuming each man and boy work an 8 hour day, find the hourly rate each is paid. If the hourly rate for men was increased by 10% whilst that of the boys was only increased by 5%, find the cost of employing each group above for one day.

Example 3

A man makes a daily journey of 40 km. When he increases his normal speed by 5 km/h, he finds he takes 2 minutes less time than usual. Find his normal speed.

Let x km/h be the man's normal speed.
The time taken for a journey of 40 km is $40/x$ h.

When his speed is increased by 5 km/h, an equation can be formed.

$$\frac{40}{x} - \frac{40}{x+5} = \frac{2}{60} = \frac{1}{30}$$

This can be rewritten as

$$30 \times 40\,(x + 5 - x) = x\,(x + 5)$$

or $$x^2 + 5x - 6000 = 0$$

Factorizing this equation gives

$$(x - 75)(x + 80) = 0 \qquad \text{i.e. } x = 75 \text{ or } x = -80$$

The man's normal speed is 75 km/h. (A negative speed is not realistic.)

Exercise 4.10c Quadratic equations

1. The hypotenuse of a right-angled triangle is 1 cm longer than one side and 8 cm longer than the other. Find the length of each side.
2. If one edge of a cube is increased in length by 3 cm, and a second edge is decreased in length by 2 cm, the volume of the cuboid formed is 55 cm^3 more than the volume of the cube. Find the length of each edge of the original cube.
3. A man makes a trip of 180 km. On his return journey his average speed is reduced by 10 km/h and he takes 15 minutes longer. Find his average speed on the outward journey.

Part 5: Algebra and graphs

5.1 Straight line graphs

The graph on the right shows part of the line $y = 2x + 1$.

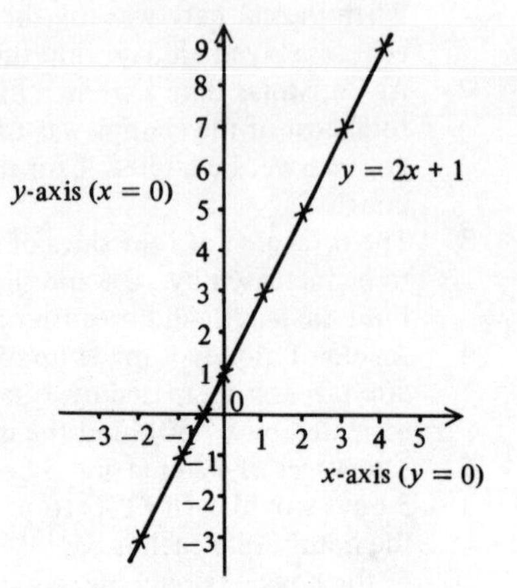

A table of values for this graph is

x	-2	-1	0	1	2	3	4
y	-3	-1	1	3	5	7	9

Each pair of values in the table are the co-ordinates of a point on the line i.e. the points $(-2, -3), (-1, -1), (0, 1), (1, 3), (2, 5), (3, 7), (4, 9)$

The equation $y = 2x + 1$ describes this set of ordered pairs.

Exercise 5.1a

1. Represent each set of ordered pairs as a graph:
   (a) $(0, 2), (1, 3), (2, 4), (3, 5), (4, 6), (5, 7), (6, 8)$
   (b) $(-2, -4), (-1, -2), (0, 0), (1, 2), (2, 4), (3, 6), (4, 8)$
   (c) $(0, -1), (1, 2), (2, 5), (3, 8), (4, 11), (5, 14), (6, 17)$
   (d) $(-1, 7), (0, 6), (1, 5), (2, 4), (3, 3), (4, 2), (5, 1), (6, 0)$
2. For each part of question 1 write down the (x, y) equation which describes the set of ordered pairs.
3. Copy and complete the table of values for each equation:

x	-2	-1	0	1	2	3	4	5	6
y									

   (a) $y = x + 3$ (b) $y = 3x$ (c) $y = 2x + 3$ (d) $y = 7 - x$
4. For each part of question 3 show the information as a graph. In what ways are these four graphs similar to those you drew in question 1?
5. Write down the (x, y) equation which describes the set of ordered pairs:
   (a) $(-2, 3), (-1, 4), (0, 5), (1, 6), (2, 7), (3, 8), (4, 9), (5, 10), (6, 11)$
   (b) $(-3, -12), (-2, -8), (-1, -4), (0, 0), (1, 4), (2, 8), (3, 12), (4, 16), (5, 20)$
   (c) $(-1, 6), (0, 5), (1, 4), (2, 3), (3, 2), (4, 1), (5, 0), (6, -1)$
   (d) $(0, -2), (1, 1), (2, 4), (3, 7), (4, 10), (5, 13), (6, 16)$
   (e) $(-3, 16), (-2, 14), (-1, 12), (0, 10), (1, 8), (2, 6), (3, 4), (4, 2), (5, 0)$
   (f) $(-2, 2), (0, 3), (2, 4), (4, 5), (6, 6), (8, 7)$
6. Complete a table of values as in question 3 for each equation:
   (a) $y = 5x - 3$ (b) $y = 4 - 3x$ (c) $y = \frac{1}{2}x + \frac{1}{2}$ (d) $y = 3(x - 2)$
7. For each part of questions 5 and 6 draw the corresponding graph. Which graphs are similar?

The graph on the right shows the line $y = mx + c$.

The c refers to the *intercept* on the y-axis. i.e. the line passes through the point $(0, c)$.

The m refers to the *gradient* of the line. i.e. the line rises m units for each unit increase in the direction of the x-axis.

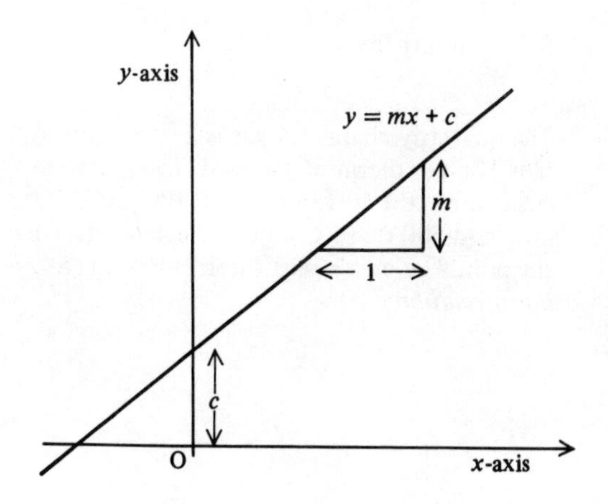

Exercise 5.1b

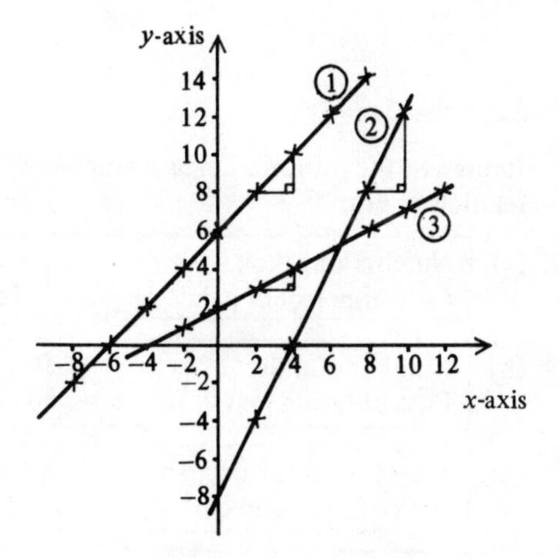

1. Write down the co-ordinates of the marked points on each line in the graph on the right.
   Use this information to write down the (x, y) equations of each line.
2. For each line in the graph on the right write down:
   (a) the value of c, the intercept
   (b) the value of m, the gradient.
3. Use the information in question 2 to write down the (x, y) equation of each line. Check your results with question 1.
4. Use the values of m and c in each equation to *sketch* the graph of:
   (a) $y = 3x + 4$ (b) $y = 5x - 2$ (c) $y = \frac{1}{2}x + 4$ (d) $y = -x + 3$
5. Write down the (x, y) equation of a line parallel to $y = 2x + 3$ which passes through:
   (a) $(0, 0)$ (b) $(0, 1)$ (c) $(0, 5)$ (d) $(0, -3)$ (e) $(0, \frac{3}{2})$
6. Write down the (x, y) equation of a line which passes through the point $(0, 2)$ and has a gradient of:
   (a) 3 (b) 5 (c) $\frac{1}{2}$ (d) -2 (e) 0
7. Write down the (x, y) equation of a line which:
   (a) passes through $(0, 5)$ and has a gradient of 3
   (b) passes through $(0, -2)$ and has a gradient of 4
   (c) passes through $(0, 1)$ and has a gradient of -1
8. Find the gradient of a line which passes through:
   (a) $(0, 2)$ and $(1, 5)$ (b) $(0, 4)$ and $(2, 8)$
   (c) $(0, -1)$ and $(1, 3)$ (d) $(0, -3)$ and $(2, 5)$
   Hence write down the (x, y) equation of each line.

5.2 Linear laws

The quarterly charge for gas is a fixed sum of £5 plus 12p per therm of gas used. The graph on the right shows the cost of using 0, 100, 200, 300, 400, and 500 therms. A graph such as this where the points lie on a straight line represents a *linear relation.*

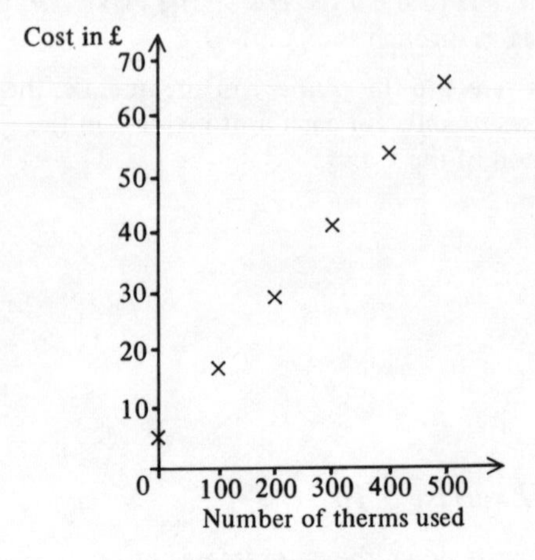

Exercise 5.2a

1. Represent the information as a graph and hence find whether the graph represents a linear relation or not.

(a)

Number of gallons	1	2	3	4	5	6
Cost of petrol in pence	75	150	225	300	375	450

(b)

Number of miles	0	1	2	3	4	5	6
Cost of hiring taxi in pence	50	62	74	86	98	110	122

(c)

Length of side of square in cm	1·2	1·7	2·1	2·5	3·0	3·6
Area of square in cm^2	1·44	2·89	4·41	6·25	9·00	12·96

(d)

Time for journey in hours	1	2	3	4	5	6	7
Distance travelled in km	32	64	100	128	160	200	224

2. In those parts of question 1 where the graph represents a linear relation, write down the (x, y) equation which describes the information.
3. The information given represents a linear relation. Show this on a graph and hence complete the table.

(a)

Time for journey in hours	1	2	3	4	5	6	7
Distance travelled in km	83	166		332		498	

(b)

Number of hectares	0	1	2	3	4	5	6	7
Cost of spraying in £	5	22·50		57·50			110	

(c)

Time taken in hours	0	1	2	3	4	5	6	7
Distance from home in km		172		116		60	32	

4. For each part of question 3 write down the (x, y) equation which describes the information.
5. A garage sells petrol at the rate of 82 pence per gallon. There is a reduction of 10p per gallon for whole gallons only. Show on a graph: (a) the cost of whole gallons only (b) the cost of buying $\frac{1}{2}$, 1, $1\frac{1}{2}$, 2, $2\frac{1}{2}$, 3, $3\frac{1}{2}$, 4, $4\frac{1}{2}$, 5, and $5\frac{1}{2}$ gallons. Which of these represents a linear relation?

The graph on the right shows a set of measurements collected during an experiment. If a line can be drawn through, or close to all the points it can be used to find the *Linear Law* which the data may satisfy, by reading off the gradient and the intercept on the y-axis.

Note: When drawing the line it is usually best to make sure that the same number of points lie on each side of the line.

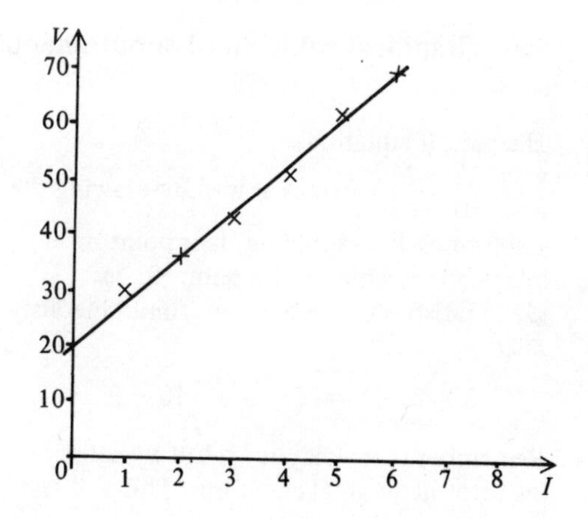

Exercise 5.2b

1. The data in the graph at the top of the page approximately satisfies the relation $V = E + IR$. Use the values of the gradient and intercept on the y-axis to find E and R.
2. Show the information given below on a graph. Draw a line to show how the information approximates to a linear law. Use this line to state the likely law connecting s and t.

(a)

t	1·2	2·5	2·9	3·4
s	16·4	34·1	39·5	46·4

(b)

t	1·8	2·3	3·1	3·8	4·7
s	6·2	8·1	10·8	13·4	16·5

3. The values of C and n are connected by the linear relation $C = I + Pn$. The data below was collected in an experiment. Show the information on a graph and hence find the most likely values for I and P.

n	2·4	3·4	4·3	5·7	7·4	9·5	11·9	15·0
C	22·6	27·9	32·7	40·1	49	60·1	72·7	89

4. Find, by drawing a graph, which of the values given below seems to indicate that an error was made when collecting data satisfying the linear law $E = 0{\cdot}51\ W + 3$.

W	12	15	19	23	28	31	36
E	9·2	10·6	11·1	14·7	17·3	18·8	21·4

Use your graph to find a more likely value for the wrong value of E.

5. Find, by drawing a graph, which two of the values given below seem to indicate that an error was made when collecting the data.

W	0·2	0·5	0·9	1·1	1·6	1·9	2·4	2·7
E	8·8	7·7	6·4	6·0	3·9	2·8	1·1	0·6

Assuming the data fits the linear law $E = aW + b$, find from your graph the most likely values of a and b and hence find a better value for the two inaccurate results.

5.3 Graphical solution of simultaneous equations

The pair of equations

$$\left.\begin{aligned} y &= 3x-2 \\ y &= 10-x \end{aligned}\right\}$$ can be solved by drawing the graph of each and finding their point of intersection, which is the point (3, 7).

(3, 7) satisfies both equations simultaneously since

$$3 \times 3 - 2 = 7 \quad \text{and} \quad 10 - 3 = 7$$

Remember to check your result by substituting into the original equations. This will test the accuracy of your graphs.

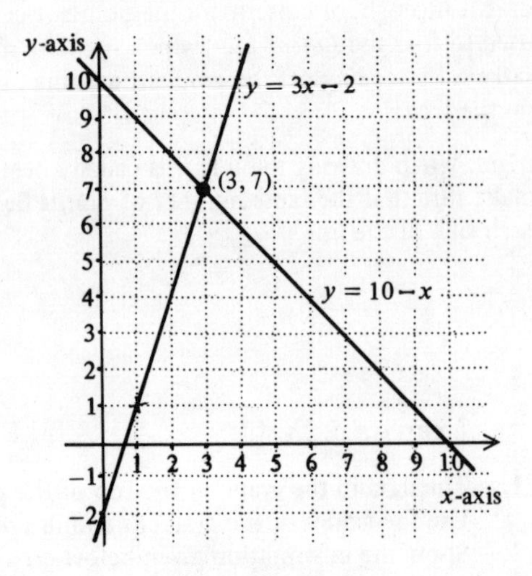

Exercise 5.3a

1. Write down the (x, y) equation which describes each set of ordered pairs. Check that the common ordered pair satisfies both equations.
   (a) {(0, 0), (1, 2), (2, 4), (3, 6), (4, 8), (5, 10)},
   {(0, 6), (1, 5), (2, 4), (3, 3), (4, 2), (5, 1), (6, 0)}
   (b) {(0, 3), (1, 5), (2, 7), (3, 9), (4, 11), (5, 13)},
   {(0, −2), (1, 1), (2, 4), (3, 7), (4, 10), (5, 13)}
2. Show each set of ordered pairs on the same graph. Find the point of intersection of the two lines. Write down the (x, y) equation for each line. Show that the common point satisfies both equations.
   (a) {(1, 6), (2, 5), (3, 4), (4, 3), (5, 2), (6, 1)},
   {(1, 4), (2, 7), (3, 10), (4, 13), (5, 16), (6, 19)}
   (b) {(0, 1), (1, 3), (2, 5), (3, 7), (4, 9), (5, 11)},
   {(0, −4), (1, 0), (2, 4), (3, 8), (4, 12), (5, 16)}
3. Draw each pair of lines on the same graph. Find the point of intersection. Show this point satisfies both equations.
   (a) $y = 2x$, $y = 8 - x$ (b) $y = 2x - 4$, $y = x + 1$ (c) $y = 3x - 2$, $y = x + 4$ (d) $y = 4x - 3$, $y = 2x + 2$
4. Draw each pair of lines on the same graph. Find the point of intersection. Show this point satisfies both equations.
   (a) $y = 5x - 7$, $y = 5 - x$ (b) $y = 2x + 3$, $y = 13 - 3x$ (c) $y = 5x - 4$, $y = 3x - 1$ (d) $y = 7 - x$, $y = 12 - 3x$
5. Copy the graph at the top of this page. Draw the line through (0, 6) parallel to $y = 10 - x$, and also the line through (0, 2) parallel to $y = 3x - 2$. Find the co-ordinates of the vertices of the parallelogram formed in this way. Check that each vertex satisfies the equations of two sides of the parallelogram.

Example

A hiker leaves town A and takes 2 hours to reach town B, 8 km away. After sightseeing, he returns to town A.
A second hiker leaves town B at the same time as the first hiker left town A. He walks towards town A but stops to visit a church halfway between the two towns. He then returns to town B.

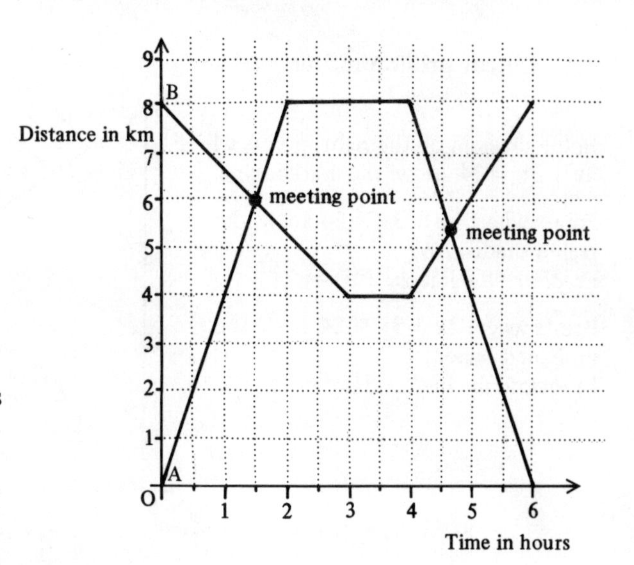

The above information is shown in the graph. The horizontal part of each hiker's trip is where they are sightseeing. They meet on the road between A and B, $1\frac{1}{2}$ hours after setting off, 6 km from A, and again on their return journeys.

Exercise 5.3b

1. Using the graph above find when and where the two hikers meet for the second time, and answer the following questions.
   (a) How long did the first hiker spend sight-seeing in town B?
   (b) How long did the second hiker spend visiting the church?
   (c) Find the average walking speed of each hiker for each part of their trip.
   (d) Find the average walking speed of each hiker for the day, excluding stops.
2. A bus leaves Stratford at 8 a.m. and travels to Bristol 70 miles away. On the journey it stops at Gloucester for 1 hour. Gloucester is 40 miles from Stratford, and for this part of the trip the bus averages 30 m.p.h. The bus arrives at Bristol at 11.05 a.m. A car leaves Stratford at 8.35 a.m. and using the same route as the bus, stops at Cheltenham, 30 miles from Stratford, at 9.05 a.m. Having spent $1\frac{1}{4}$ hours in Cheltenham the car averages 60 m.p.h. on the remainder of the journey to Bristol.
   Show this information on a graph and answer the following questions:
   (a) Find the time the bus arrives at Gloucester and its average speed from Gloucester to Bristol.
   (b) Find the average speed of the car between Stratford and Cheltenham and the time the car finally arrives in Bristol.
   (c) Find at what distance from Stratford the car overtakes the bus. At what times does this happen?
3. Central heating oil may be bought in one of three ways: 18p per litre, a fixed charge of £3 and 12p per litre, or a fixed charge of £6 and 10p per litre. Draw a graph to show the costs of oil by each of the three methods of purchase over the range 0 to 200 litres. Find when it is cheapest by each method.

5.4 Linear programming

In the diagram on the right, points where $2x + 3y = 24$ are *on* the marked line.

Points where $2x + 3y > 24$ are in the region shaded /////, i.e. *above* the line $2x + 3y = 24$.

Points where $2x + 3y < 24$ are in the unshaded region, i.e. *below* the line $2x + 3y = 24$.

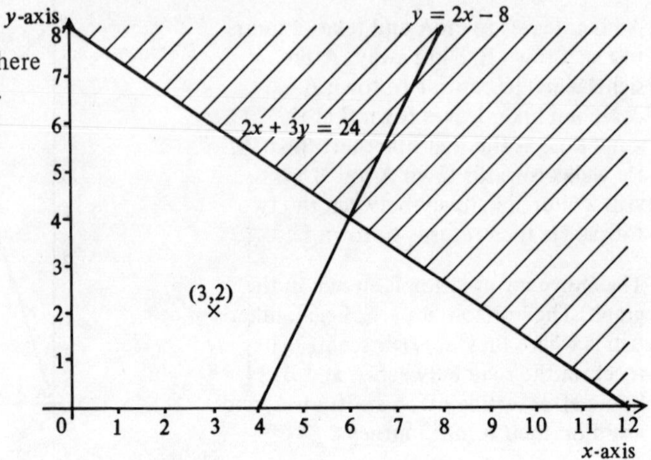

In order to find which region in the above diagram corresponds to $y < 2x - 8$ any point such as (3, 2) is used as a check. Substituting gives $2 > 2.3 - 8$ so the point (3, 2) is in the region where $y > 2x - 8$. The region where $y < 2x - 8$ is therefore on the opposite side of the line $y = 2x - 8$, i.e. on the right of this line.

Exercise 5.4a

1. Draw the given line and then shade the required region. In each case find whether (3, 1) is in the required region or not.
   (a) line $x + y = 7$, region $x + y < 7$
   (b) line $2x + 3y = 12$, region $2x + 3y > 12$
   (c) line $y = 2x + 1$, region $y > 2x + 1$
   (d) line $y = 3x - 7$, region $y < 3x - 7$
2. Draw the given line and then shade the required region.
   (a) line $3x + 2y = 24$, region $3x + 2y < 24$
   (b) line $y = 2x$, region $y > 2x$
   Shade also the region where both $3x + 2y < 24$ and $y > 2x$. Find whether the points (2, 5) and (6, 1) are in this region or not.
3. Draw the given line and then shade the required region.
   (a) line $y = 3x - 4$, region $y < 3x - 4$
   (b) line $y = x + 2$, region $y < x + 2$
   Shade also the region where both $y < 3x - 4$ and $y < x + 2$. Find whether the points (3, 4) and (4, 6) are in this region or not.
4. Draw the lines $x + 2y = 12$ and $2x + y = 10$. Mark the region where
   (a) $x + 2y < 12$ and $2x + y > 10$ (b) $x + 2y > 12$ and $2x + y < 10$
   (c) $x + 2y < 12$ and $2x + y < 10$ (d) $x + 2y > 12$ and $2x + y > 10$
5. Draw the lines $x + y = 11$, $y = 3x - 1$, and $y = x + 3$. Shade the region where $x + y < 11$, $y < 3x - 1$, and $y > x + 3$. Find whether the points (3, 6) and (2, 7) are in this region or not.

Example 1

Shade on a graph the region where $x + 3y \leqslant 24$ and $3x + y < 21$. Find at what points in this region, with whole number values of x and y, $x + y$ is largest.

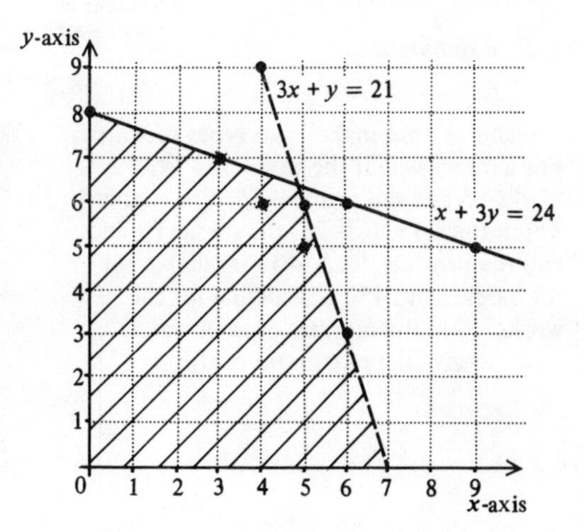

The line $x + 3y = 24$ is shown solid since points *on* the line satisfy $x + 3y \leqslant 24$.
The line $3x + y = 21$ is shown dotted since points on the line do not satisfy $3x + y < 21$.

The points in the shaded region above for which $x + y$ is largest are (3, 7), (4, 6) and (5, 5). The point (5, 6) is excluded since it lies on the dotted line and is therefore not in the region.

Exercise 5.4b

1. Shade on a graph the region where $2x + y \leqslant 10$ and $x + 2y < 12$. Find at what point in this region, with whole number values of x and y, $x + y$ is largest.
2. Shade on a graph the region where $x + 3y < 18$ and $y \geqslant 2x - 10$. Find at what points in this region with whole number values of x and y, $x + y$ is largest. Explain why (6, 4) is excluded from the region.
3. Using the diagram on the right say which region is described by:
   (a) $x + y \leqslant 6$, $x + 3y \geqslant 12$ and $y \leqslant 2x + 3$
   (b) $x + y \geqslant 6$, $x + 3y \geqslant 12$ and $y \leqslant 2x + 3$
   (c) $x + 3y \leqslant 12$, and $y \geqslant 2x + 3$.
4. Using the diagram on the right describe with inequalities the region:
   (a) ⑥ (b) ④ (c) ③

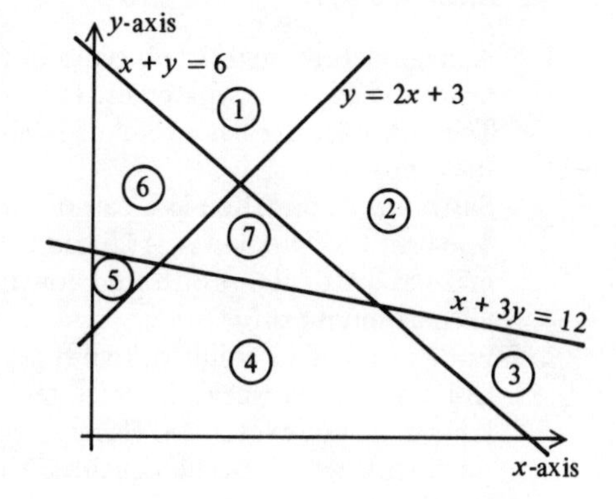

5. Draw the lines $x + y = 7$, $y = 2x + 5$ and $x + 4y = 12$. Mark the region where:
   (a) $x + y \leqslant 7$, $y \leqslant 2x + 5$ and $x + 4y \geqslant 12$
   (b) $x + y \geqslant 7$, $y \leqslant 2x + 5$ and $x + 4y \geqslant 12$
   (c) $x + y \leqslant 7$, $y \geqslant 2x + 5$ and $x + 4y \geqslant 12$
6. Shade on a graph the region where:
   (a) $x + y < 10$, $2x + y > 12$, and $x + 4y \geqslant 16$
   (b) $x + y > 10$, $2x + y > 12$, and $x + 4y \geqslant 16$
   (c) $x + y < 10$, $2x + y < 12$, and $x + 4y \leqslant 16$

Example 2

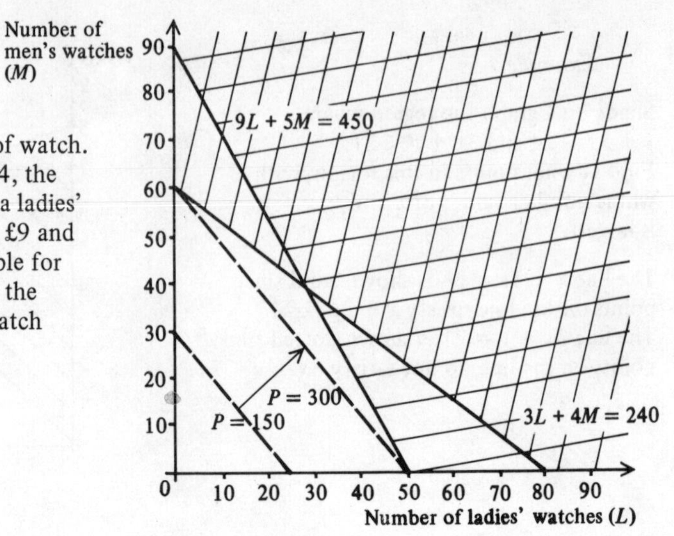

A manufacturer makes two types of watch. For a man's watch the case costs £4, the works £5 and the profit is £5. For a ladies' watch the case costs £3, the works £9 and and the profit is £6. £240 is available for the cases, and £450 is available for the works. Find the number of each watch made to give the maximum profit.

If L and M are the number of ladies' and men's watches made, then for the cases $3L + 4M \leqslant 240$, whilst for the works $9L + 5M \leqslant 450$.
The profit is given by £$(6L + 5M)$. Two profit lines $6L + 5M = 150$ and $6L + 5M = 300$ are shown on the diagram above. These lines are parallel and so the maximum profit will be given by the profit line, which can be drawn in the unshaded region passing through allowable points, furthest to the right. Since only whole number values of L and M are allowable, the solution giving the maximum profit is $L = 28, M = 39$. For these numbers the corresponding profit is £363.

Exercise 5.4c

1. A manufacturer makes two types of product N and L. Type N requires 12 man-hours to make and £5 of raw materials. Type L requires 9 man-hours to make and £8 of raw materials. The manufacturer has a total of 1080 man-hours and £800 for raw materials available for these products.
   Show this information as a pair of inequalities and represent it on a graph.
   The profit on product N is £5 and that on product L is £6. If the manufacturer requires to make at least £600 profit, find how many of each product he must make. How can he make his maximum profit?
2. For a camp of 72 children, two types of tent are available for hire. The large tent sleeps 8 and costs £8 per week, the small tent sleeps 3 and costs £2 per week. The total number of tents must not exceed 18. Using L for the number of large tents and S for the number of small tents write down inequalities to describe the constraints of:
   (a) the number of children (b) the number of tents (c) the cost of hiring.
   If the cost of hiring is to be kept as small as possible, show the information on a graph and find the number of each type that must be hired. What will be the cost of this for the week?
3. A farmer has 100 hectares of land available for planting wheat and/or potatoes, and he is prepared to outlay at most £2700. The initial outlay on each hectare of wheat is £45, whilst that on each hectare of potatoes is £18.
   Show this information as a pair of inequalities and represent it on a graph.
   If the profit on each hectare of wheat is £60 and on each hectare of potatoes £40, find how he should allocate his land to make the maximum profit.
   What is the greatest profit that he could make if he was prepared to use 120 hectares?
   How many hectares must he be prepared to allocate to make it worth growing only potatoes?

Example 3

A firm wants to buy a number of new cars of two types. Type A costs £2000 and £16 per week to run. Type B costs £2400 and £10 per week to run. The firm has £18 000 to spend and wishes to keep its running costs to £120 per week. It must buy at least 4 cars of type A and 2 cars of type B. Find the various possibilities.

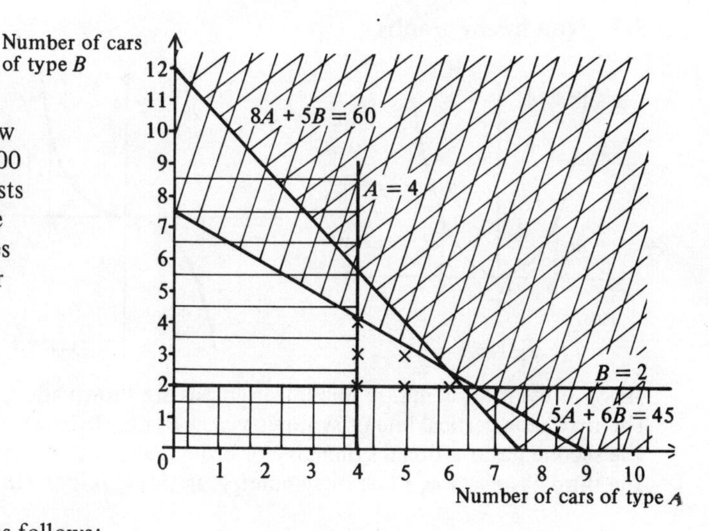

The constraints can be summarized as follows:

Purchase costs: $2000A + 2400B \leqslant 18\,000$ i.e. $5A + 6B \leqslant 45$
Running costs: $16A + 10B \leqslant 120$ i.e. $8A + 5B \leqslant 60$

Also $A \geqslant 4$ and $B \geqslant 2$, where A and B are the number of each car.

From the graph above, we can see the possibilities are 4 cars of type A and 4 of type B, or 5 of A and 3 of B, or 6 of A and 2 of B.

Exercise 5.4d

1. Represent on one graph the set of points (x, y) for which $x \geqslant 2, y \geqslant 10, 4x + y \geqslant 24$ and $3x + 2y \geqslant 36$. Show also the points where $2x + y \leqslant 30$. Hence find the points in this region, with whole number values for x and y, at which $2x + y$ takes its smallest value.
2. A firm with 10 drivers uses vans and lorries to make its deliveries. A van costs £6 per day to run and can carry 600 kg. A lorry costs £15 per day to run and can carry 1100 kg. The firm wants to keep its daily running costs to £90, but be able to deliver at least 6600 kg. Write down inequalities to describe the constraints of:
   (a) drivers (b) running costs (c) carrying capacity.
   Show this information on a graph and find the various possibilities if the firm decides to use at least five vans.
3. A farmer makes a composite feed from two different feeds so that the vitamin content of the final feed is at least:
   Vitamin A–6 units, Vitamin B–8 units, Vitamin C–3 units, Vitamin D–8 units.
   The vitamin content of each feed per kilogram is:

	Vitamin A	Vitamin B	Vitamin C	Vitamin D
Feed 1	3	2	1	1
Feed 2	1	2	1	4

If the costs of each feed per kilogram are 80 pence and 60 pence respectively, find how the farmer should make his feed to keep his costs to a minimum.

5.5 Non-linear graphs

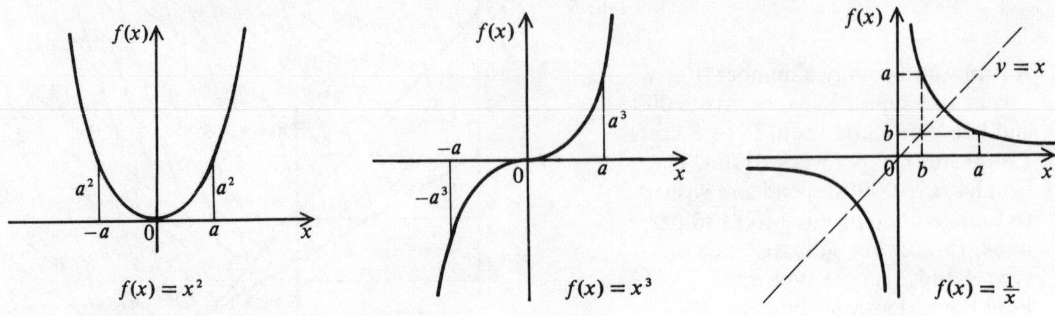

Three of the most common non-linear graphs are shown above.
The first has a vertical line of symmetry, $x = 0$, i.e. $f(a) = f(-a)$.
The second has rotational symmetry of order 2 about 0, i.e. $f(-a) = -f(a)$.
The third has $y = x$ as a line of symmetry. If $f(a) = b$ then $f(b) = a$.

Exercise 5.5a

1. Using the graphs at the top of the page sketch the graph of:
   (a) $f(x) = x^2 + 3$ (b) $f(x) = x^3 + 2$ (c) $f(x) = \frac{1}{x} + 5$
   Describe the transformation which has been applied to each of the original graphs.
2. If a translation of 4 units vertically downwards, was applied to each of the graphs at the top of the page, write down the equation which describes each graph in its new position.
3. Using the graphs at the top of the page, sketch the graph of:
   (a) $f(x) = (x - 3)^2$ (b) $f(x) = (x - 2)^3$ (c) $f(x) = \frac{1}{(x - 5)}$
   Describe the transformation which has been applied to each of the original graphs.
4. If a translation of 4 units to the left was applied to each of the graphs at the top of the page, write down the equation which describes each graph in its new position.
5. Describe the transformation which has been applied to $f(x) = x^2$ if the equation which describes its new position is:
   (a) $f(x) = x^2 + 1$ (b) $f(x) = (x + 1)^2$ (c) $f(x) = x^2 - 1$ (d) $f(x) = (x - 1)^2$
6. Using the four equations in question 5 find for each $f(3)$ and $f(-3)$. For which is $x = 0$ a line of symmetry? Write down the equation of the line of symmetry for the other two.
7. (3, 12), (4, 9), and (6, 6) are three points on $f(x) = 36/x$. Write down the images of these points when they are reflected in the line $y = x$. Check that the images also satisfy $f(x) = 36/x$.
   Write down two more pairs of points which satisfy $f(x) = 36/x$. Use this information to sketch $f(x) = 36/x$.

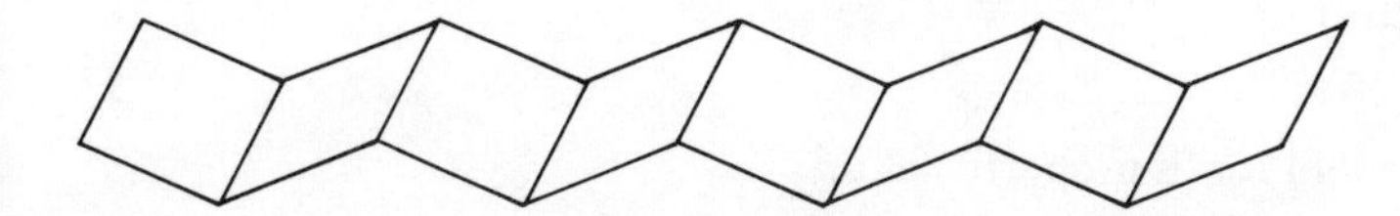

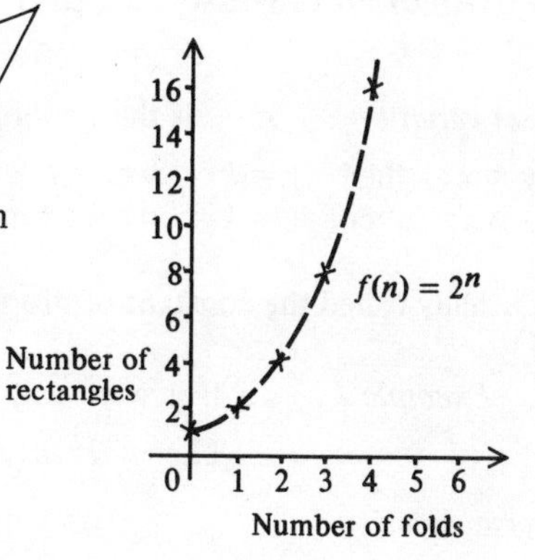

A long strip of paper is folded in half several times, as shown above.
The number of folds and rectangles formed are recorded in the table below.

Number of folds	0	1	2	3	4
Number of rectangles	1	2	4	8	16
Number of fold lines	0	1	3	7	15

The graph shows points which satisfy the equation $f(n) = 2^n$. In this example a dotted line has been used to show the general shape of the graph since points in between those marked have no meaning.

Exercise 5.5b

1. Write down the equation which would show the number of fold lines in the table above for a given number of folds. Describe the connection between the graph for this equation and the graph shown above.
2. Using the numerals 0, 1, and 2 only, write down the number of different numbers that can be formed with 1, 2, 3, 4, or 5 digits which may or may not be the same. For example, using 3 digits, 222, or 010 would be two of the possibilities. Show this information on a graph like the one above and write down the equation which describes the number of numbers which can be made from a given number of digits.
   Draw the corresponding graph if the number zero is excluded.
3. On a 100 km stretch of motorway a car can average any speed between 20 km/h and 100 km/h. Show as a graph the time taken for each possible average speed. Write down the equation which describes this graph and state the domain and corresponding image set (range). Show the corresponding graph if the journey always includes a half-hour coffee stop. Write down the corresponding equation.
4. A printer charges a fixed sum of £10 plus £10 for the first 100 brochures ordered. The charge for the next 100 is £9, and the next 100 is £8 and so on up to 1000 brochures, when the charge remains at £1 per 100. Show this information as points on a graph, joining them with straight lines.
   Write down the equations which describe each of the first three sections of the graph.

5.6 Direct and inverse variation

Direct variation $\propto$ is the symbol for 'is proportional to'

If $y \propto x$ then $y = kx$, where k is a constant
If $y \propto x^2$ then $y = kx^2$, where k is a constant

k is usually called the constant of proportionality.

Example 1 If $y \propto x^2$ and $y = 12$ when $x = 2$, find y when $x = 5$

$12 = k.2^2$ then $k = 3$; so when $x = 5$, $y = 3.5^2 = 75$

Inverse variation

If y is inversely proportional to x then $y \propto \frac{1}{x}$

If y is inversely proportional to x^3 then $y \propto \frac{1}{x^3}$

Example 2 If $y \propto \frac{1}{x^2}$ and $y = 4$ when $x = 3$, find y when $x = 2$

$4 = \frac{k}{3^2}$ then $k = 36$; so when $x = 2$, $y = \frac{36}{2^2} = 9$

Exercise 5.6a

1. If $y \propto x$ and $y = 8$ when $x = 3$, find y when $x = 18$.
2. If $p \propto q$ and $p = 7$ when $q = 5$, find q when $p = 2$.
3. The distance travelled by a car varies directly as the time taken for the journey. Find how far the car travels in $2\frac{1}{2}$ hours if it travels 75 km in $1\frac{1}{2}$ hours.
4. If $y \propto x^2$ and $y = 96$ when $x = 4$, find y when $x = 5$.
5. If $p \propto q^2$ and $p = 45$ when $q = 3$, find q when $p = 80$.
6. The surface area of a cube varies directly as the square of the length of each edge of the cube. Write this statement as an equation and find the constant of proportionality. Hence write down the surface area for an edge length of 4 cm.
7. If $V \propto r^3$ and $V = 6{\cdot}4$ when $r = 4$, find V when $r = 3$.
8. If $y \propto \sqrt{x}$ and $y = 20$ when $x = 25$, find y when $x = 9$.
9. The time of swing of a pendulum varies directly as the square root of its length. Find the time for a length of 100 cm if the time for a length of 16 cm is 0·8 seconds.
10. If y is inversely proportional to x and $y = 8$ when $x = 3$ find:
    (a) y when $x = 4$ (b) x when $y = 10$.
11. If p is inversely proportional to q^2 and $p = 15$ when $q = 2$ find:
    (a) p when $q = 10$ (b) q when $p = 0{\cdot}15$.
12. The current in a circuit is inversely proportional to the resistance. Find the current when the resistance is 48 ohms, if the current is 12 amps when the resistance is 20 ohms.

Example 3

The relationship between s and t is given by $s = at^2 + b$ where a and b are constants.

Experimental results for s and t are

t	0·25	0·5	0·75	1·0	1·25
s	3·13	3·53	4·18	5·10	6·28

By plotting s against t^2 as shown, a straight line can be drawn.

The values of a and b can be found from the gradient and y-intercept.

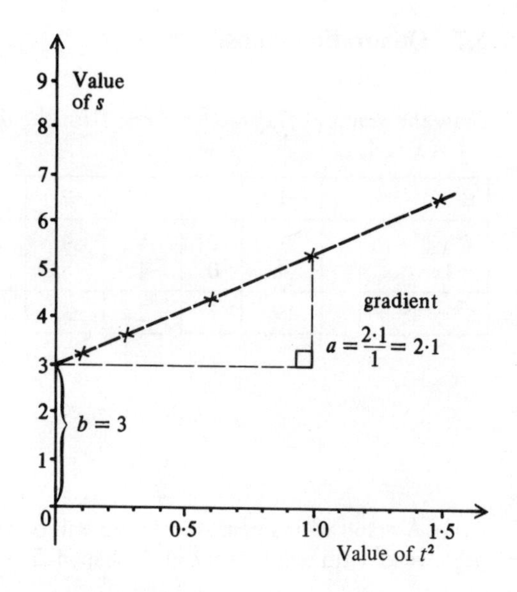

Exercise 5.6b

1. S is proportional to T^2. Experimental results are recorded as follows:

T	1	2	3	4	5
S	3·17	12·81	28·75	51·16	80·41

Draw a graph of S against T^2 and use the gradient of this straight line to find the constant of proportionality.

2. The relationship between x and y is given by $y = k\left(\frac{1}{x^2}\right) + c$. Experimental results are found to be:

x	0·3	0·4	0·5	0·6	0·7	0·8
y	63·2	36·9	24·7	18·0	14·2	11·5

Draw a graph of y against $1/x^2$. Do these points lie on a straight line? Use this line to find the values of k and c.

3. Two quantities v and p are believed to be related, with v being proportional to a power of p. i.e. $v \propto p^n$. Experimental data is collected and shown in the table below.

p	1·1	1·3	1·4	1·7	1·9	2·2	2·5	2·8	3·2	4·0
v	4·61	5·93	6·63	8·87	10·48	13·05	15·81	18·74	22·90	32

Since $v \propto p^n$ we can write $v = k\,p^n$. By taking logs of each side, this equation becomes $\log v = n \log p + \log k$.
A straight line graph can be obtained by plotting $\log v$ against $\log p$.
n will be the gradient of this line and $\log k$ the y-intercept.
Find $\log p$ and $\log v$ for each pair of results in the table. By plotting $\log v$ against $\log p$, find the value of n and also the constant of proportionality. Hence find v when $p = 2$.

5.7 Quadratic graphs

Draw the graph of $f(x) = x^2 - 4x + 1$ for the domain $-1 \leqslant x \leqslant 4$.

x	-1	0	1	2	3	4
x^2 $-4x$	1 4	0 0	1 -4	4 -8	9 -12	16 -16
$x^2 - 4x + 1$	6	1	-2	-3	-2	1

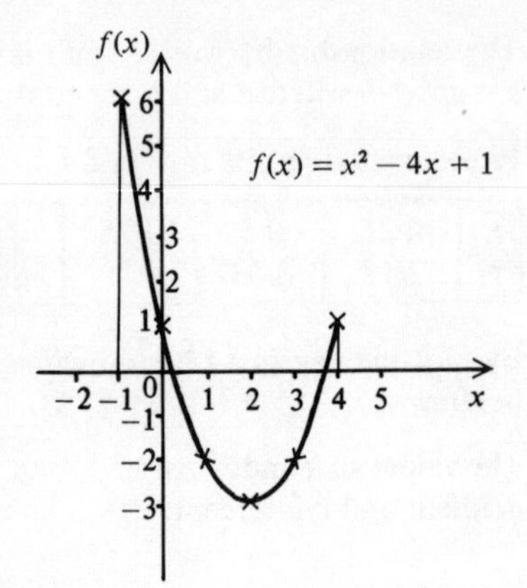

Note: A graph with a positive x^2 term will be a smooth ∪ shaped curve, whilst one with a negative x^2 term will be smooth ∩ shaped. The image set (range) above is $-3 \leqslant f(x) \leqslant 6$.

Exercise 5.7a

1. Write down the image set (range) for the graph drawn at the top of the page. Find $f(\frac{1}{2})$, $f(-\frac{1}{2})$ and $f(3\frac{1}{2})$.
2. For what values of x in the graph above is:
   (a) $f(x) = 1$ (b) $f(x) = -2$ (c) $f(x) = 6$ (d) $f(x) = 0$.
3. By completing a table as above, draw a graph of:
   (a) $f(x) = x^2 + 2x - 3$ for the domain $-4 \leqslant x \leqslant 2$
   (b) $f(x) = x^2 - 4x + 4$ for the domain $-2 \leqslant x \leqslant 4$
   (c) $f(x) = x^2 + 3x + 2$ for the domain $-3 \leqslant x \leqslant 3$
   (d) $f(x) = x^2 - 5x + 4$ for the domain $-1 \leqslant x \leqslant 5$
4. For each graph in question 3, write down the image set (range) and find $f(\frac{1}{2})$.
5. For what values of x in each graph in question 3 is:
   (a) $f(x) = 0$? (b) $f(x) = 2$? (c) $f(x) = 4$? (d) $f(x) = -3$?
6. By completing a table as above, draw a graph of:
   (a) $f(x) = 4 - x^2$ for the domain $-3 \leqslant x \leqslant 3$
   (b) $f(x) = 3x - x^2$ for the domain $-2 \leqslant x \leqslant 4$
   (c) $f(x) = 3 + 2x - x^2$ for the domain $-2 \leqslant x \leqslant 4$
   (d) $f(x) = 3 - 2x - x^2$ for the domain $-4 \leqslant x \leqslant 2$.
7. For each graph in question 6, write down the image set (range) and find $f(\frac{1}{2})$.
8. For what values of x in each graph in question 6 is:
   (a) $f(x) = 0$? (b) $f(x) = 1$? (c) $f(x) = -2$? (d) $f(x) = 3$?
9. What can you say about three of the graphs in question 6? Describe the transformation which would map $f(x) = 4 - x^2$ onto each of the remaining two graphs.
10. Solve the quadratic equation by drawing the associated quadratic graph and finding when $f(x) = 0$. In each case, take as the domain $-5 \leqslant x \leqslant 5$.
    (a) $x^2 - 5x + 6 = 0$ (b) $x^2 + 2x - 8 = 0$
    (c) $x^2 - 5x + 5 = 0$ (d) $x^2 + 2x - 6 = 0$

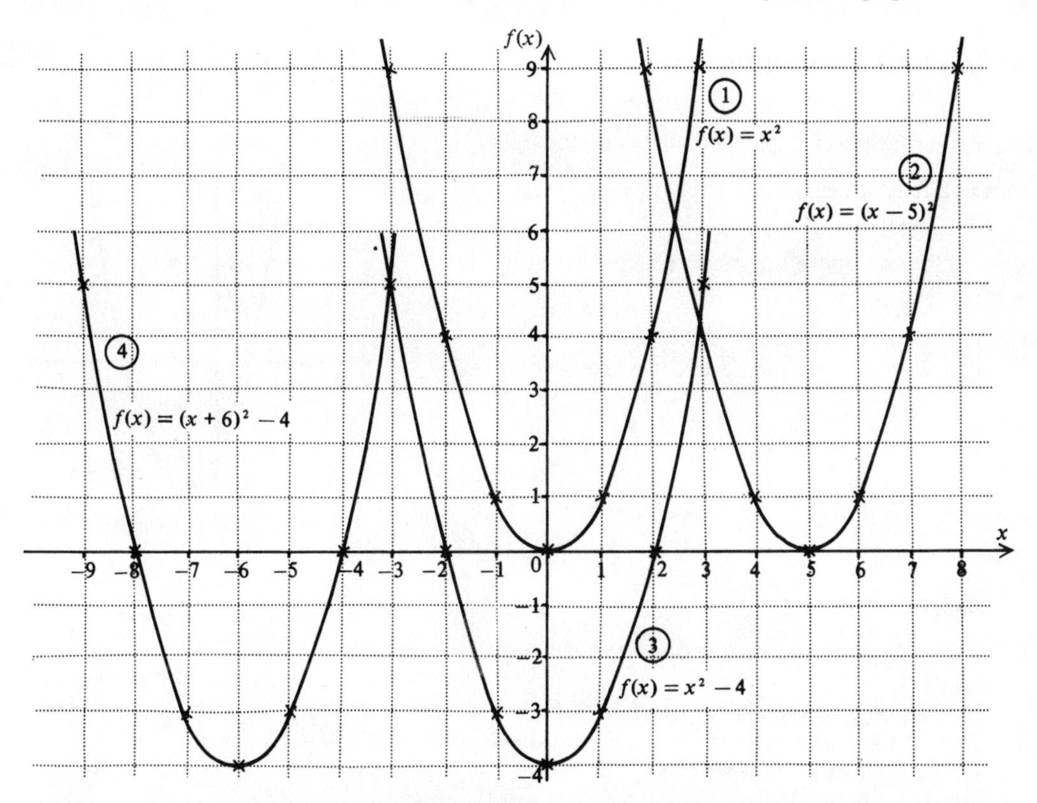

The graph of $f(x) = x^2$ can be translated onto each of the other three graphs.
Under a translation of a upwards the graph becomes $f(x) = x^2 + a$.
Under a translation of b to the right the graph becomes $f(x) = (x - b)^2$.
$f(x) = (x + q)^2 - p$ is the result of a translation of p downwards and q to the left.

Exercise 5.7b

1. For what values of x in each graph above is:
   (a) $f(x) = 0$? (b) $f(x) = 1$? (c) $f(x) = 4$? (d) $f(x) = 5$?
2. Write down the domain and corresponding image set (range) for each of the graphs above.
3. Describe the translation that makes each of graphs ①, ②, and ③ the image of graph ④.
4. Write down the equation of the new graph if $f(x) = x^2$ is translated:
   (a) 3 units upwards (b) 2 units downwards
   (c) 4 units to the right (d) 1 unit to the left
   (e) 2 units to the right and then 5 units upwards
   (f) 3 units to the left and then 1 unit downwards.
5. Write down the equation of the new graph if graph ④ is translated 9 units to the right and 6 units upwards.
6. Use the ideas of translating the graph of $f(x) = x^2$ to sketch:
   (a) $f(x) = x^2 + 4$ (b) $f(x) = (x + 4)^2$ (c) $f(x) = (x - 5)^2 + 2$ (d) $f(x) = (x + 3)^2 - 6$
7. Write $f(x)$ as $(x + a)^2 + b$ and hence sketch the graph of:
   (a) $f(x) = x^2 + 4x$ (b) $f(x) = x^2 - 6x$
   (c) $f(x) = x^2 + 4x + 7$ (d) $f(x) = x^2 - 6x + 5$
   (e) $f(x) = x^2 + 6x + 4$ (f) $f(x) = x^2 - 4x - 1$
8. Sketch the graph of $f(x) = (x + 2)^2 - 3$. Sketch this graph after a translation of 5 units upwards and 5 units to the right. Write down the equation of the new graph.

5.8 Quadratic inequalities

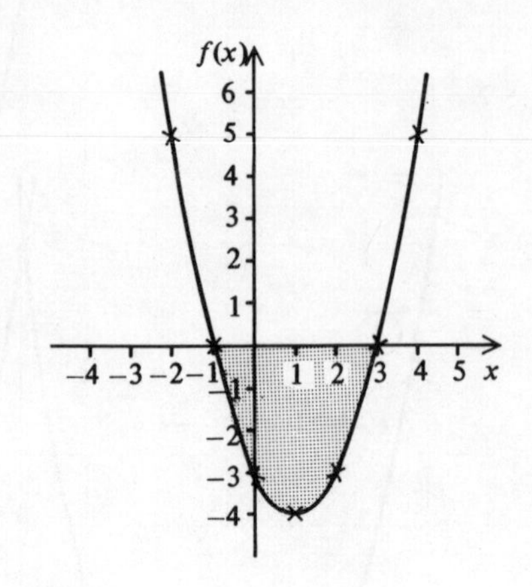

The graph of $f(x) = x^2 - 2x - 3$ is drawn on the right.

$f(x)$ is negative, or below the x-axis, for values of x between -1 and 3.

$f(x)$ is positive, or above the x-axis, when $x > 3$ or when $x < -1$.

So $x^2 - 2x - 3 < 0$ when $-1 < x < 3$
$x^2 - 2x - 3 > 0$ when $x < -1$ or when $x > 3$

Exercise 5.8a

1. Using the graph above write down the values of x for which:
(a) $f(x) < -3$ (b) $f(x) > 5$ (c) $f(x) > -4$ (d) $f(x) = 0$
Find also the values of x for which $x^2 - 2x - 3 < 2$.
2. Draw the graph of $f(x) = x^2 + x - 6$ with the domain $-5 \leqslant x \leqslant 4$.
Find the values of x for which:
(a) $f(x) > 0$ (b) $f(x) < 6$ (c) $f(x) > -6$ (d) $f(x) = 14$
Find also the values of x for which $x^2 + x - 6 > -4$.
3. Draw the graph of $f(x) = x^2 - 2x - 8$ with the domain $-3 \leqslant x \leqslant 5$.
Find the values of x for which:
(a) $f(x) < 0$ (b) $f(x) > -8$ (c) $f(x) \geqslant 7$ (d) $f(x) \leqslant -5$
Find also the values of x for which $x^2 - 2x - 8 \geqslant -9$.
4. Draw the graph of $f(x) = 9 - x^2$ with the domain $-4 \leqslant x \leqslant 4$.
Find the values of x for which:
(a) $f(x) > 0$ (b) $f(x) < 8$ (c) $f(x) \geqslant 5$ (d) $f(x) \leqslant -7$
Find also the values of x for which $9 - x^2 = 7$. Hence find $\sqrt{2}$.
5. Draw the graph of $f(x) = 3 + 2x - x^2$ with the domain $-2 \leqslant x \leqslant 4$.
Find the values of x for which:
(a) $f(x) < 0$ (b) $f(x) = -5$ (c) $f(x) \geqslant 3$ (d) $f(x) \geqslant 4$
Find also the values of x for which $3 + 2x - x^2 \geqslant \frac{1}{2}$.
6. By drawing a quadratic graph, or otherwise, solve the inequality:
(a) $x^2 - 5x + 6 < 0$ (b) $x^2 + 4x - 5 \geqslant 0$ (c) $5 + 4x - x^2 < 0$
(d) $x^2 - 5x + 6 > 2$ (e) $x^2 + 4x - 5 \leqslant -8$ (f) $5 + 4x - x^2 \geqslant 2$
7. By drawing a suitable graph, or otherwise, solve the inequality:
(a) $x^2 - 4x + 1 < 0$ (b) $1 + 3x - x^2 \geqslant 0$ (c) $x^2 - 5x + 4 < 2$
8. Write down two values of x for which $(x-2)(x-4) = 0$. Hence sketch the graph of $f(x) = (x-2)(x-4)$ and write down the solution set for $(x-2)(x-4) \leqslant 0$.
9. Write down two values of x for which $(3-x)(x+4) = 0$. Hence sketch the graph of $f(x) = (3-x)(x+4)$ and write down the solution set for $(3-x)(x+4) < 0$.

The graph of $y = x^2 - 4$ is shown on the right.

The points in the shaded region are where $y > x^2 - 4$.
The points on the curved line are where $y = x^2 - 4$.
The points in the unshaded region are where $y < x^2 - 4$.

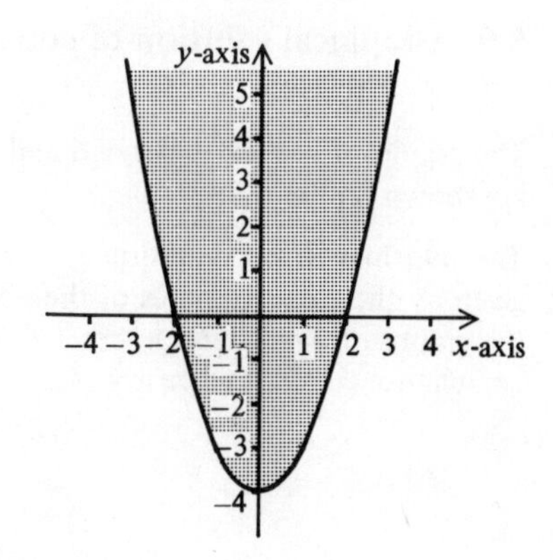

Exercise 5.8b

1. Draw a sketch of the graph above and shade the region where:
   (a) $y > x^2 - 4$ and $y < 0$ (b) $y > x^2 - 4$ and $y > 2$
   (c) $y < x^2 - 4$ and $y > 0$ (c) $y < x^2 - 4$ and $y < -2$
2. Sketch the graph of $y = 3 - x^2$ and shade in the region where:
   (a) $y < 3 - x^2$ and $y > 0$ (b) $y < 3 - x^2$ and $y < -1$
   (c) $y > 3 - x^2$ and $y < 0$ (d) $y > 3 - x^2$ and $y > 1$
3. Sketch the graphs of $y = x^2 - 2$ and $y = x$ on the same piece of paper. Shade the region where: (a) $y > x^2 - 2$ and $y < x$ (b) $y \leqslant x^2 - 2$ and $y > x$
4. Sketch the graphs of $y = x^2$ and $x + y = 4$ on the same piece of paper. Shade the region where: (a) $y < x^2$ and $x + y > 4$ (b) $y \geqslant x^2$ and $x + y \leqslant 4$
5. The graphs of $y = 6/x$ and $x + y = 7$ are shown on the right.
   Copy the graph and shade the region where:
   (a) $y > 6/x$ and $x + y < 7$
   (b) $y < 6/x$ and $x + y \geqslant 7$
   Describe the region shaded on the graph, including the heavy lines.

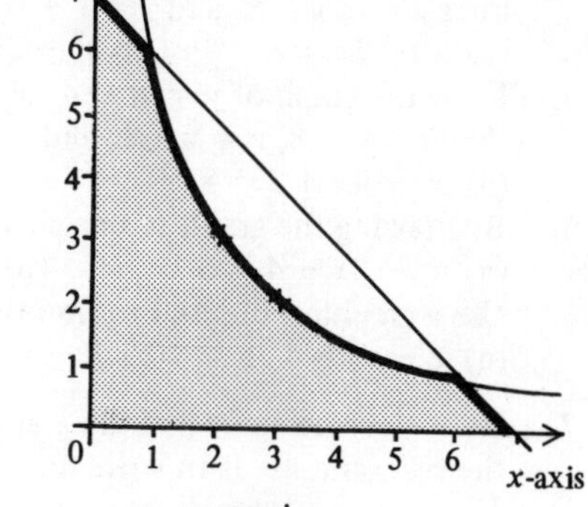

6. The graphs of $y = (x - 1)^2$ and $y = 4 - x^2$ are shown on the right.
   Copy the graph and shade the region where:
   (a) $y \leqslant 4 - x^2$ and $y > (x - 1)^2$
   (b) $y > 4 - x^2$ and $y < (x - 1)^2$
   Describe the region shaded on the graph, including the heavy line.

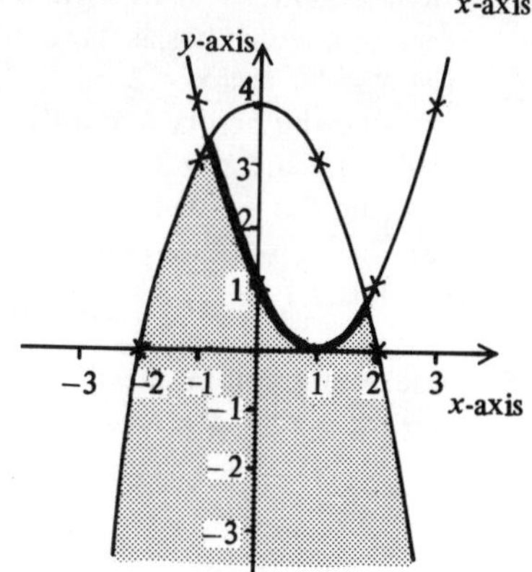

5.9 Graphical solution of equations

The graphs of $y = x^2, y = x + 6$ and $y = x + 2$ are shown on the right.

The solutions of the equation $x^2 = x + 6$ are given by the x co-ordinates of the points where the graphs of $y = x^2$ and $y = x + 6$ meet, i.e. when $x = -2$ or when $x = 3$.

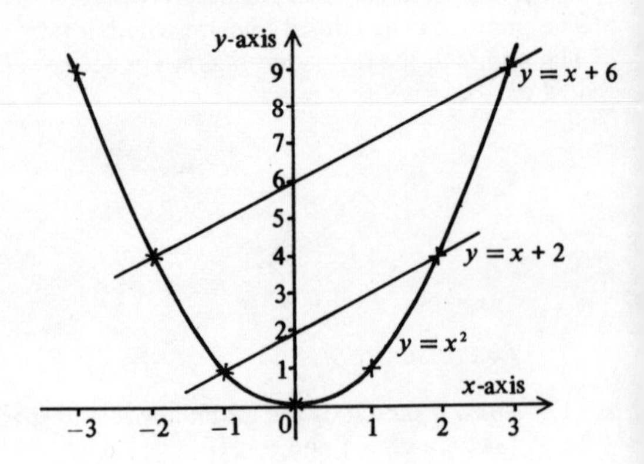

Exercise 5.9a

1. Draw your own copy of the graph above, and use it to write down the solutions of:
(a) $x^2 = x + 2$ (b) $x^2 = 6$ (c) $x^2 = 2$ (d) $x + 6 = 7$.
2. Show the lines $y = x, y = 6 - x$ and $y = 2 - x$ on your graph in question 1, and hence write down the solutions of:
(a) $x^2 = x$ (b) $x^2 = 6 - x$ (c) $x^2 = 2 - x$ (d) $6 - x = x + 2$.
3. Draw the graph of $y = x^2 - 2x$ with the domain $-3 \leqslant x \leqslant 5$. Draw on the same graph the lines $y = 8, y = x$, and $y = x + 4$. Hence write down the solutions of:
(a) $x^2 - 2x = 8$ (b) $x^2 - 2x = x$ (c) $x^2 - 2x = x + 4$.
4. Draw the graph of $y = x^2 - 4x + 3$ with the domain $-1 \leqslant x \leqslant 5$. Draw on the same graph the lines $y = 8, y = x - 1$, and $y = 7 - x$. Hence write down the solutions of:
(a) $x^2 - 4x + 3 = 8$ (b) $x^2 - 4x + 3 = x - 1$ (c) $x^2 - 4x + 3 = 7 - x$.
5. By drawing the graph of one curve and three lines solve:
(a) $x^2 - 3x - 4 = 6$ (b) $x^2 - 3x - 4 = x + 1$ (c) $x^2 - 3x - 4 = 4 - x$.
6. Use a graphical method to find the solution set for:
(a) $8 - x^2 = 4$ (b) $8 - x^2 = x - 4$ (c) $8 - x^2 = 6 - x$.
7. Draw your own copy of the graph on the right and use it to write down an equation which has as its solutions:
(a) $x = -1$ and $x = 2$
(use the points A and B)
(b) $x = 0$ and $x = 3$
(use the points C and D)
(c) the x co-ordinates of the points E and F.

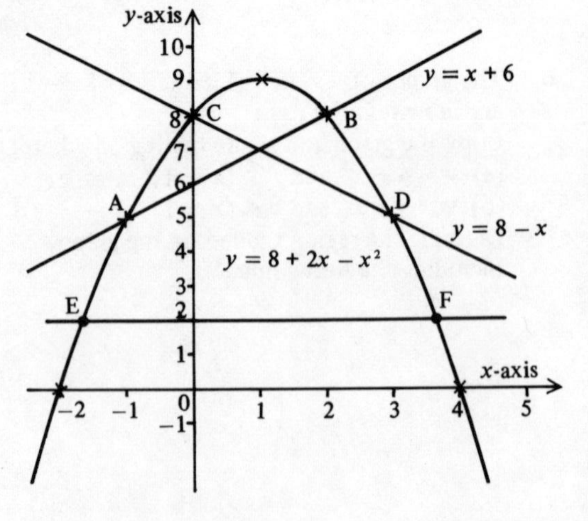

8. Use your graph in question 7 to solve:
$8 + 2x - x^2 = 4 - x$.

Example

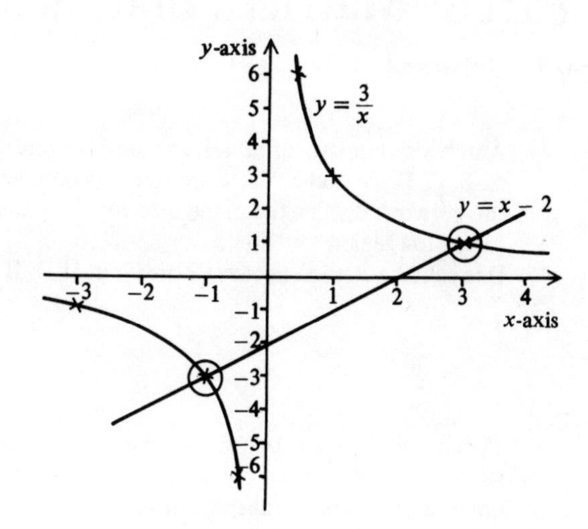

Use a graphical method to solve

$$\frac{3}{x} = x - 2$$

Hence write down the solution set for $x^2 - 2x - 3 = 0$.

The graph of $y = \frac{3}{x}$ meets $y = x - 2$ at the points where $x = 3$ and $x = -1$.

$\frac{3}{x} = x - 2$ can be re-written as $x^2 - 2x - 3 = 0$. Its solution set is $\{-1, 3\}$.

Exercise 5.9b

1. Draw the graphs of $y = \frac{9}{x}$, $y = x$, and $y = 3x$. Hence find the solution set for:
   (a) $\frac{9}{x} = 3$ (b) $\frac{9}{x} = x$ (c) $\frac{9}{x} = 3x$.
2. Draw the graphs of $y = \frac{10}{x}$, $y = x + 3$, and $y = x - 3$. Hence find the solution set for:
   (a) $\frac{10}{x} = x + 3$ (b) $\frac{10}{x} = x - 3$.
3. Show that your results for question 2 can be used to find the solution sets for:
   (a) $x^2 + 3x - 10 = 0$ (b) $x^2 - 3x - 10 = 0$
4. Draw the graphs of $y = x^2 - 1$ and $y = 2x - x^2$ using a scale of 5 cm to 1 unit on the x-axis. Hence find, as accurately as you can, the solution set for $2x^2 - 2x - 1 = 0$.
5. Make an accurate graph of the curve $y = x^3$ shown on the right.
   Use your graph to find the solution set for:
   (a) $x^3 = x$
   (b) $x^3 = 4x$
   (c) $x^3 = 9x$

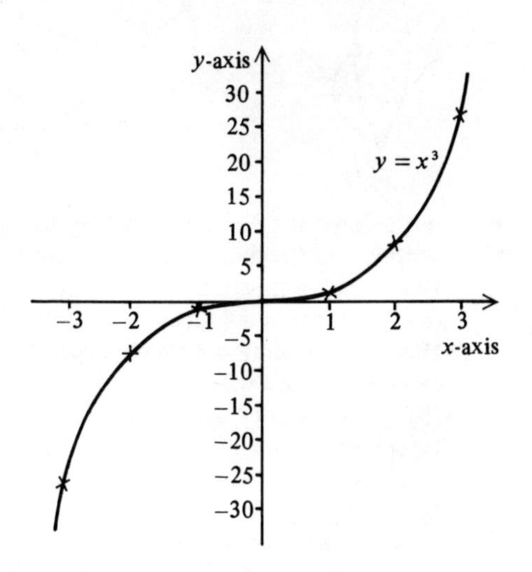

6. Use your graph in question 5 to find the solution set for:

   $$x^3 = 7x - 6$$

   Show that this equation can also be written as

   $$(x - 1)(x^2 + x - 6) = 0$$

   Hence find the solution set for $x^2 + x - 6 = 0$.

Part 6: Matrices and vectors

6.1 Networks

A network consists of *nodes*, *arcs* and *regions.* In the diagram, A, B, C, D are nodes; there are five arcs connecting them and the network divides the plane into three regions. (Do not forget the region 'outside'.)
D is called a *2-node* because 2 arcs join at D. B is a *3-node.*

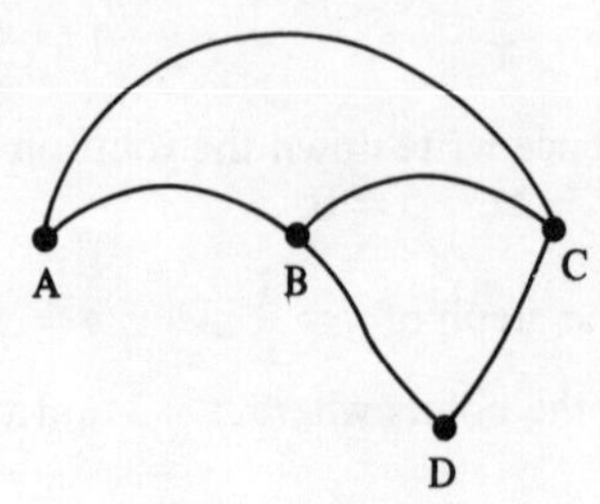

Two networks are *topologically equivalent* if they can be transformed into each other by stretching but without breaking an arc. Points on an arc will remain in the same order under a topological transformation.

Exercise 6.1a

1. Which of these networks are topologically equivalent?

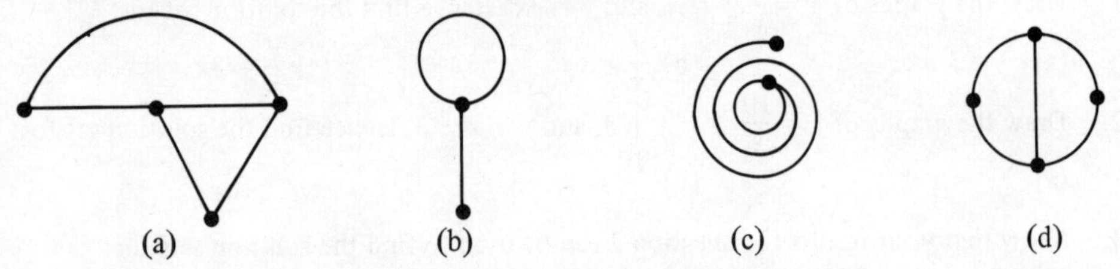

2. Which of these networks are topologically equivalent?

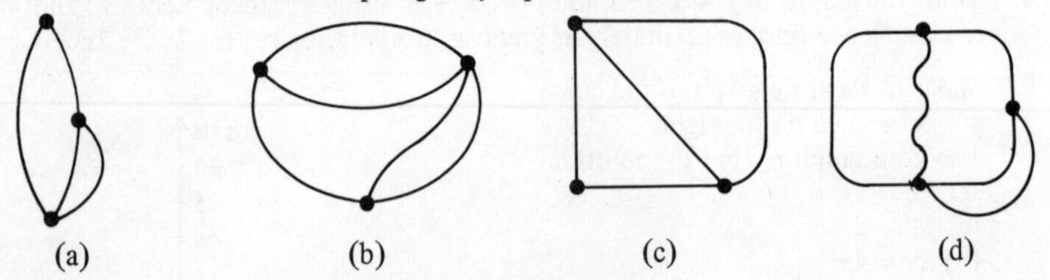

3. Write down the number of nodes, arcs and regions in each network of question 1.
4. Write down the number of nodes, arcs and regions in each network of question 2.
5. Draw a network with: (a) 2 nodes and 3 arcs (b) 3 nodes and 4 arcs (c) 1 node and 2 arcs.
   How many regions are there in each network?
6. Draw a network with: (a) two 1-nodes and one 2-node (b) two 3-nodes and one 2-node (c) one 1-node and one 3-node.
   How many regions are there in each network?
7. Is it possible to draw a network containing:
   (a) 1 node, 1 arc, 2 regions? (b) 3 nodes, 2 arcs, 2 regions?
   Draw the network if it is possible.
8. Using three arcs, design a network with:
   (a) the greatest possible number of regions
   (b) the greatest possible number of nodes.

Euler's relation In any network the number of regions (R), arcs (A) and nodes (N) is connected by the relation $R - A + N = 2$.
A similar relation holds for the network that outlines a solid polyhedron (e.g. a cube). Faces (F), edges (E) and vertices (V) of the solid are connected by the relation $F - E + V = 2$.

Traversibility A network is traversible (can be travelled over) if it can be completely drawn without a break and without repeating any arc.
A network is only traversible if the number of odd nodes is zero or two.

Exercise 6.1b

1. Show that Euler's relation holds for the networks in Exercise 6.1a question 1.
2. Show that Euler's relation holds for the networks in Exercise 6.1a question 2.
3. Use Euler's relation to find the number of regions in each network of Exercise 6.1a question 5.
4. Use Euler's relation to find the number of regions of a network if:
   (a) there are the same number of arcs and nodes
   (b) there are more nodes than arcs. (What is the smallest possible number of regions?)
5. A square-based pyramid has its top removed by a cut parallel with the base. State the number of faces, edges and vertices of the solid and verify Euler's relation.

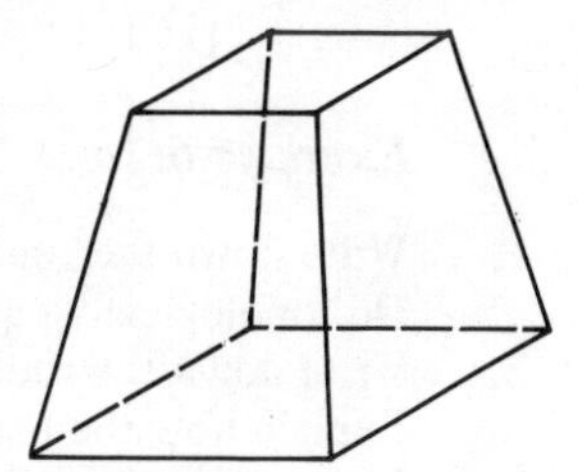

6. (a) A cube ABCDA′B′C′D′ is sawn in half through B, B′, D′, D. Draw a sketch of the half cube and write down the number of its faces, edges and vertices. Show that Euler's relation holds.
   (b) Draw a sketch in the case where the cube is sawn in half through A′, C and the mid-points of DD′ and BB′. Are the number of faces, edges and vertices the same in this half cube as in the other half cube?

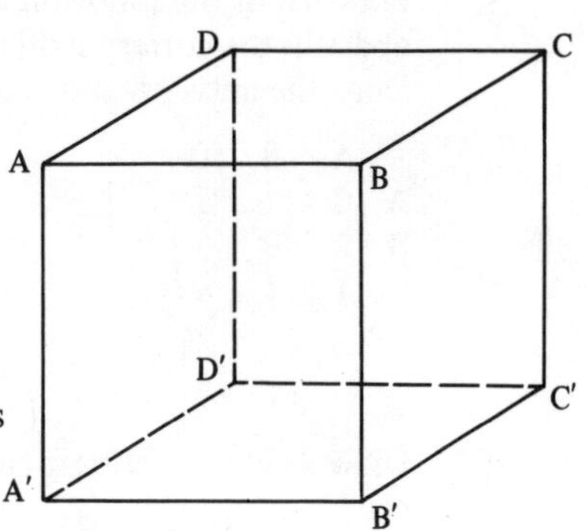

7. Which of the following networks, if any, are traversible?

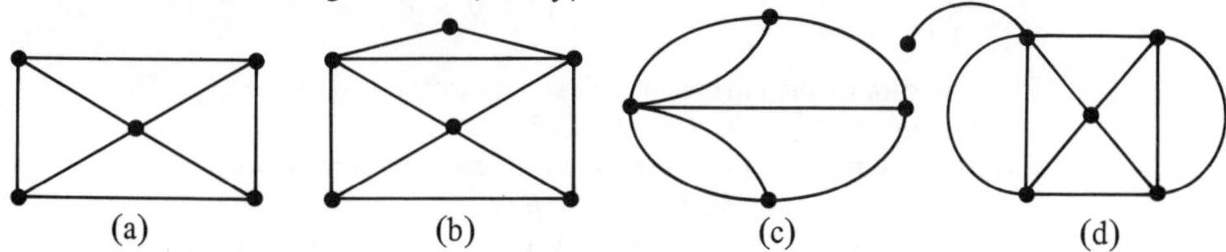

8. Find the number of odd and even nodes of the network of edges of:
   (a) a cube (b) an octahedron.
   Explain why one is traversible and the other is not.

A network may be described by a *route matrix*. Each entry gives the number of arcs from node to node. The matrices below refer to the networks above them. Notice that there are no one-way routes, so in each matrix

(a) there is symmetry about the leading diagonal
(b) all entries in the leading diagonal are zero unless an arc returns to the same node
(c) non-zero entries in the leading diagonal must be even.

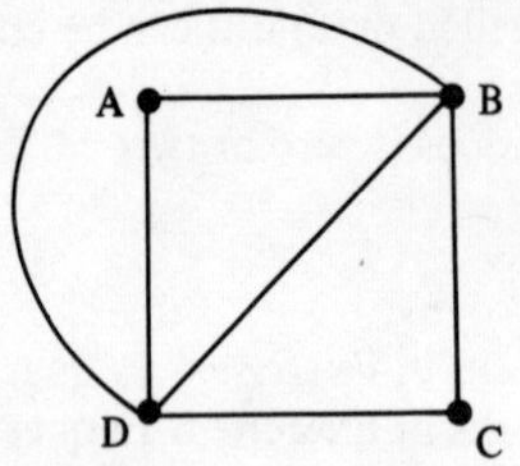

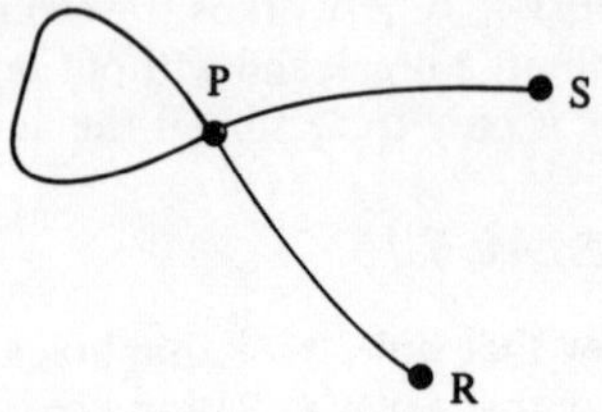

$$\textit{From}\quad \begin{array}{c} \\ A \\ B \\ C \\ D \end{array} \overset{\textit{To}}{\begin{array}{c} \begin{array}{cccc} A & B & C & D \end{array} \\ \begin{pmatrix} 0 & 1 & 0 & 1 \\ 1 & 0 & 1 & 2 \\ 0 & 1 & 0 & 1 \\ 1 & 2 & 1 & 0 \end{pmatrix} \end{array}} \qquad \textit{From}\quad \begin{array}{c} \\ P \\ Q \\ R \end{array} \overset{\textit{To}}{\begin{array}{c} \begin{array}{ccc} P & Q & R \end{array} \\ \begin{pmatrix} 2 & 1 & 1 \\ 1 & 0 & 0 \\ 1 & 0 & 0 \end{pmatrix} \end{array}}$$

Exercise 6.1c

1. Write down the route matrix for each network in questions 1 and 2 of Exercise 6.1a. Do topologically equivalent networks have the same route matrix?
2. Write down the route matrix for each network in question 7 of Exercise 6.1b. Explain how the matrix entries show whether a network is traversible.
3. For each of the following route matrices, write down the number of odd and even nodes in the corresponding network and so state whether the network is traversible. Draw the network and check for traversibility.

$$\begin{array}{c} \\ A \\ B \\ C \end{array} \begin{array}{c} \begin{array}{ccc} A & B & C \end{array} \\ \begin{pmatrix} 0 & 2 & 1 \\ 2 & 0 & 1 \\ 1 & 1 & 0 \end{pmatrix} \end{array} \qquad \begin{array}{c} \\ P \\ Q \end{array} \begin{array}{c} \begin{array}{cc} P & Q \end{array} \\ \begin{pmatrix} 2 & 1 \\ 1 & 0 \end{pmatrix} \end{array} \qquad \begin{array}{c} \\ X \\ Y \\ Z \\ W \end{array} \begin{array}{c} \begin{array}{cccc} X & Y & Z & W \end{array} \\ \begin{pmatrix} 2 & 2 & 0 & 1 \\ 2 & 2 & 1 & 0 \\ 0 & 1 & 2 & 2 \\ 1 & 0 & 2 & 2 \end{pmatrix} \end{array}$$

(a) (b) (c)

4. Draw a network corresponding to the matrix:

(a) $\begin{pmatrix} 0 & 0 & 1 & 0 \\ 0 & 0 & 1 & 1 \\ 1 & 1 & 0 & 1 \\ 0 & 1 & 1 & 0 \end{pmatrix}$ (b) $\begin{pmatrix} 0 & 2 & 1 \\ 2 & 2 & 0 \\ 1 & 0 & 0 \end{pmatrix}$ (c) $\begin{pmatrix} 0 & 3 & 1 \\ 3 & 0 & 2 \\ 1 & 2 & 0 \end{pmatrix}$

5. Find the sum of the entries in each matrix at the top of the page. How is this related to the number of arcs in the network?
6. Explain why none of these matrices can represent a network:

(a) $\begin{pmatrix} 2 & 1 \\ 1 & 1 \end{pmatrix}$ (b) $\begin{pmatrix} 0 & 3 \\ 2 & 0 \end{pmatrix}$ (c) $\begin{pmatrix} 2 & 2 \\ 1 & 0 \end{pmatrix}$

If one-way-only routes are permitted, show that the matrices could represent route networks and draw them.

The matrix **R** represents a one-way traffic system between three points A, B and C in a certain town.
The matrix **R**′ is the *transpose* of **R** obtained by interchanging rows with columns.
R′ is the matrix for the network obtained by reversing the traffic flow.

$$\mathbf{R} = \begin{pmatrix} 0 & 0 & 0 \\ 1 & 0 & 1 \\ 1 & 1 & 0 \end{pmatrix}$$

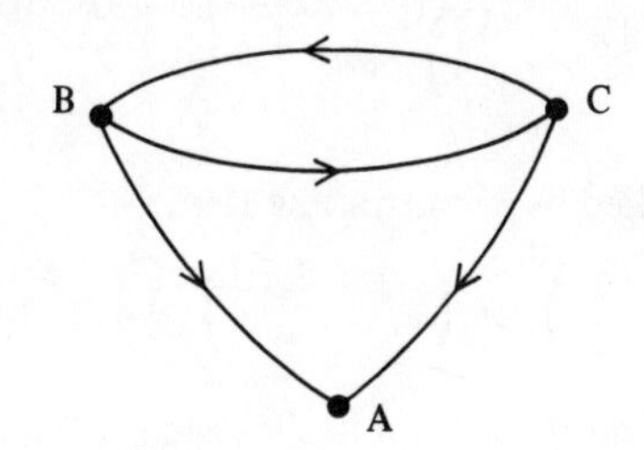

$$\mathbf{R}' = \begin{pmatrix} 0 & 1 & 1 \\ 0 & 0 & 1 \\ 1 & 1 & 0 \end{pmatrix}$$

Exercise 6.1d

1. Draw the network corresponding to **R**′.
2. If the road C to A becomes a two-way street, draw the network and the corresponding route matrix.
3. Write down the matrix corresponding to the network:

(a)

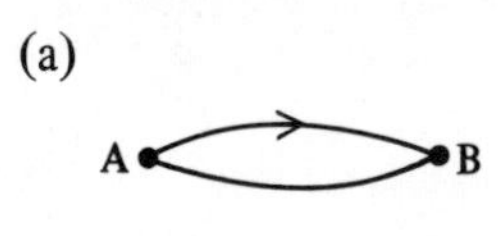

(b)

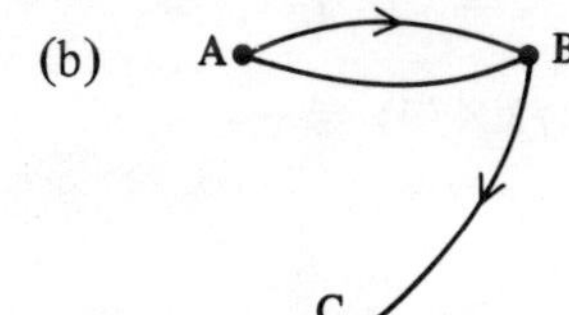

(c)

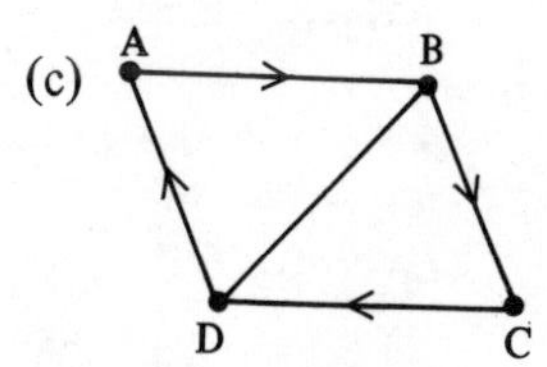

4. Write down the matrix corresponding to the network:

(a)

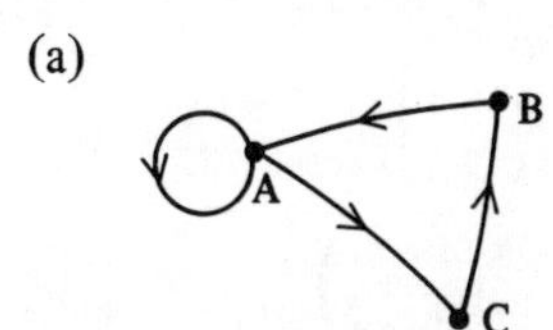

(b)

A B

(c)

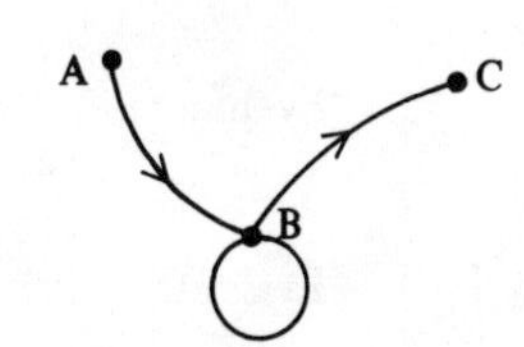

5. The arrow graph shows the relation *is larger than* on the set {0, 4, 10}. Write down the matrix **M** for this arrow graph and its transpose **M**′. What relation does **M**′ represent?

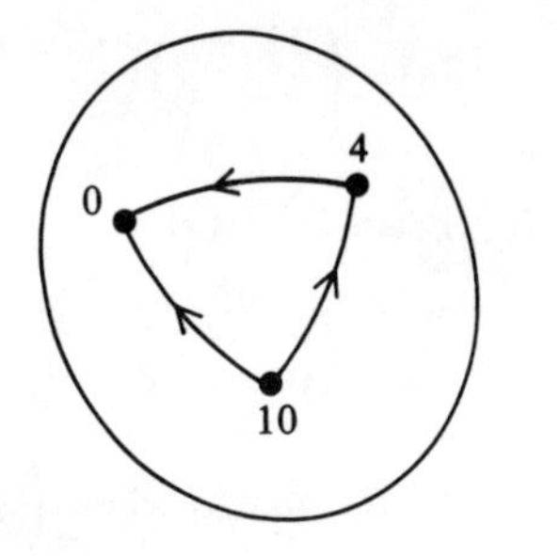

6. Write down the matrix **X** showing the relation *is a multiple of* on the set {3, 5, 9, 15, 45}. Write down the transpose matrix **X**′ and draw its associated arrow graph. What relation does **X**′ represent?
7. Write down the transpose of each of these matrices:

$$\mathbf{A} = \begin{pmatrix} 1 & 0 \\ 1 & 0 \end{pmatrix}, \quad \mathbf{B} = \begin{pmatrix} 3 & 2 \\ -1 & 0 \end{pmatrix}, \quad \mathbf{C} = \begin{pmatrix} 0 & -4 \\ 2 & 1 \end{pmatrix}, \quad \mathbf{D} = \begin{pmatrix} 5 & 2 \\ 3 & -1 \end{pmatrix}$$

Find the sums **A** + **A**′, **B** + **B**′, **C** + **C**′ and **D** + **D**′. What is special about the answers?

8. Let $\mathbf{M} = \begin{pmatrix} a & b \\ c & d \end{pmatrix}$. Write down **M**′ and prove that $(\mathbf{M} + \mathbf{M}')' = \mathbf{M} + \mathbf{M}'$.

6.2 Matrix multiplication

A matrix can be multiplied by a number like this:

$$6\begin{pmatrix}2 & -7\\3 & -2\end{pmatrix} = \begin{pmatrix}12 & 42\\18 & -12\end{pmatrix} \quad \textit{and} \quad 3\begin{pmatrix}1 & 0 & 0\\0 & 2 & 0\\4 & 0 & 3\end{pmatrix} = \begin{pmatrix}3 & 0 & 0\\0 & 6 & 0\\12 & 0 & 9\end{pmatrix}$$

A matrix can be multiplied by a matrix like this:

$$\begin{pmatrix}\boxed{4 \quad 2}\\3 \quad 5\end{pmatrix}\begin{pmatrix}-2 & \boxed{1}\\3 & \boxed{8}\end{pmatrix} = \begin{pmatrix}-2 & \boxed{20}\\9 & 43\end{pmatrix}; \quad 4 \times (1) + 2 \times (8) = 20$$

Each row of the matrix on the left multiplies each column of the matrix on the right *term by term.*

The same method is used for other than 2×2 matrices but the row length of the left matrix must equal the column length of the right matrix. For example:

$$\begin{pmatrix}2 & 3 & 1\\4 & 0 & -2\end{pmatrix}\begin{pmatrix}4 & 1 & 7\\-2 & -1 & -8\\5 & 4 & 0\end{pmatrix} = \begin{pmatrix}7 & 3 & -10\\6 & \boxed{-4} & 28\end{pmatrix}$$

$$4 \times (1) + 0 \times (-1) + -2 \times (4) = -4$$

Exercise 6.2

1. Evaluate $2\begin{pmatrix}4 & 2\\1 & 0\end{pmatrix} + 3\begin{pmatrix}-1 & 2\\4 & -8\end{pmatrix} - 6\begin{pmatrix}1 & 8\\-1 & 4\end{pmatrix}$.

2. Evaluate $7\begin{pmatrix}1 & 0 & 3\\2 & 1 & 0\\4 & -1 & 3\end{pmatrix} + \begin{pmatrix}4 & 1 & 7\\-2 & 4 & 1\\3 & 2 & 6\end{pmatrix} + 2\begin{pmatrix}0 & 0 & 1\\0 & 1 & 1\\1 & 1 & 1\end{pmatrix}$

3. Evaluate: (a) $\begin{pmatrix}2 & 1\\3 & 2\end{pmatrix}\begin{pmatrix}1 & 4\\2 & 3\end{pmatrix}$ (b) $\begin{pmatrix}8 & -2\\5 & 1\end{pmatrix}\begin{pmatrix}4 & 1\\0 & 2\end{pmatrix}$

   (c) $\begin{pmatrix}7 & -1\\3 & 0\end{pmatrix}\begin{pmatrix}-2 & 0\\1 & 5\end{pmatrix}$ (d) $\begin{pmatrix}2 & 3\\1 & 2\end{pmatrix}\begin{pmatrix}2 & -3\\-1 & 2\end{pmatrix}$ (e) $\begin{pmatrix}1 & 0\\0 & 1\end{pmatrix}\begin{pmatrix}4 & 2\\1 & 3\end{pmatrix}$

4. Evaluate: (a) $\begin{pmatrix}5 & 2\\1 & 7\end{pmatrix}\begin{pmatrix}1 & 2\\2 & 1\end{pmatrix}$ (b) $\begin{pmatrix}3 & 2\\-3 & 5\end{pmatrix}\begin{pmatrix}1 & 4\\-6 & 0\end{pmatrix}$

   (c) $\begin{pmatrix}5 & 0\\0 & 5\end{pmatrix}\begin{pmatrix}0 & 4\\4 & 0\end{pmatrix}$ (d) $\begin{pmatrix}0 & 1\\1 & 0\end{pmatrix}\begin{pmatrix}4 & 1\\3 & 2\end{pmatrix}$ (e) $\begin{pmatrix}1 & 0\\0 & -1\end{pmatrix}\begin{pmatrix}-1 & 0\\0 & -1\end{pmatrix}$

5. Find a, b, c, d when $\begin{pmatrix}a & 3\\2 & 6\end{pmatrix}\begin{pmatrix}4 & 7\\-3 & b\end{pmatrix} = \begin{pmatrix}13 & 1\\c & d\end{pmatrix}$

6. Find x, y, z, w when $\begin{pmatrix}2 & 1\\x & y\end{pmatrix}\begin{pmatrix}8 & -3\\0 & -2\end{pmatrix} = \begin{pmatrix}z & w\\-16 & 8\end{pmatrix}$

7. Evaluate: (a) $\begin{pmatrix}3 & 0 & 1\\0 & 1 & 0\\0 & 0 & 3\end{pmatrix}\begin{pmatrix}2 & 0 & 0\\0 & 4 & 0\\0 & 0 & 8\end{pmatrix}$ (b) $(1 \quad 7 \quad -11)\begin{pmatrix}3 & 4\\2 & -1\\1 & 6\end{pmatrix}$

8. Evaluate: (a) $\begin{pmatrix} 4 & 1 \\ 2 & -3 \end{pmatrix}\begin{pmatrix} 0 & 1 & 2 \\ 2 & 1 & 0 \end{pmatrix}$ (b) $\begin{pmatrix} 1 & 4 & 7 \\ 2 & 3 & 0 \\ -1 & 0 & 0 \end{pmatrix}\begin{pmatrix} 0 & 0 & 4 \\ 0 & 3 & -6 \\ 5 & 1 & 1 \end{pmatrix}$

9. Evaluate: (a) $\begin{pmatrix} 1 & 0 & 1 \\ 2 & 1 & 0 \end{pmatrix}\begin{pmatrix} 3 \\ -1 \\ 2 \end{pmatrix}$ (b) $\begin{pmatrix} 1 & 4 & -2 & 0 \\ 3 & -1 & 1 & 4 \end{pmatrix}\begin{pmatrix} 3 & 0 & 3 \\ 1 & -1 & 1 \\ 4 & 2 & 0 \\ -2 & 3 & 2 \end{pmatrix}$

10. Given $\mathbf{A} = \begin{pmatrix} 2 & 1 \\ 1 & -1 \\ 0 & 1 \end{pmatrix}$ and $\mathbf{B} = \begin{pmatrix} 1 & 4 & -2 \\ 5 & 3 & 0 \end{pmatrix}$ evaluate: (a) **A.B** (b) **B.A**

11. Evaluate: (a) $(2 \quad 0 \quad 1 \quad 4)\begin{pmatrix} 3 \\ -1 \\ 4 \\ 2 \end{pmatrix}$ (b) $\begin{pmatrix} 3 \\ -1 \\ 4 \\ 2 \end{pmatrix}(2 \quad 0 \quad 1 \quad 4)$

12. **A** is an $n \times 1$ column matrix and **B** is a $1 \times n$ row matrix. The entries of both matrices are all 1. Describe:
(a) **AB** (b) **BA**.

13. Let $\mathbf{A} = \begin{pmatrix} 1 & 2 \\ 3 & 4 \end{pmatrix}$, $\mathbf{B} = \begin{pmatrix} 3 & 0 \\ 0 & 4 \end{pmatrix}$, $\mathbf{C} = \begin{pmatrix} -2 & 1 \\ \frac{3}{2} & -\frac{1}{2} \end{pmatrix}$

Find: (a) **AB** (b) **BA** (c) **AC** (d) **CA** (e) **BC**

14. Let $\mathbf{P} = \begin{pmatrix} 3 & 4 \\ 1 & 2 \end{pmatrix}$, $\mathbf{Q} = \begin{pmatrix} 0 & 2 \\ 2 & 0 \end{pmatrix}$, $\mathbf{R} = \begin{pmatrix} 2 & -4 \\ -1 & 3 \end{pmatrix}$.

Find: (a) **PR** (b) **RP** (c) **PQ** (d) **QR** (e) **RQ**

15. For **A, B, C** of question 13 find **(AB)C** and **A(BC)**. Are the answers the same?

16. For **P, Q, R** of question 14 find **(PQ)R** and **P(QR)**. Are the answers the same?

17. $\mathbf{X} = \begin{pmatrix} 3 & -5 \\ -1 & 2 \end{pmatrix}$, $\mathbf{Y} = \begin{pmatrix} 4 & a \\ b & 4 \end{pmatrix}$.

Show that $\mathbf{XY} \neq \mathbf{YX}$ in general.
Find particular values of a, b so that $\mathbf{XY} = \mathbf{YX}$.

18. $\mathbf{P} = \begin{pmatrix} 1 & 4 \\ 4 & 1 \end{pmatrix}$. Show that $\mathbf{PP}' = \mathbf{P}'\mathbf{P}$.

19. Let $\mathbf{I} = \begin{pmatrix} 1 & 0 \\ 0 & 1 \end{pmatrix}$, $\mathbf{X} = \begin{pmatrix} 4 & 3 \\ 5 & 2 \end{pmatrix}$.

Show that $\mathbf{X}^2 - 6\mathbf{X} - 7\mathbf{I} = \begin{pmatrix} 0 & 0 \\ 0 & 0 \end{pmatrix}$.

Show further that if $\mathbf{V} = \begin{pmatrix} a \\ b \end{pmatrix}$ then $\mathbf{XV} = x\mathbf{V}$ where x satisfies the equation $x^2 - 6x - 7 = 0$.

20. Let $\mathbf{I} = \begin{pmatrix} 1 & 0 \\ 0 & 1 \end{pmatrix}$, $\mathbf{P} = \begin{pmatrix} 0 & 2 \\ 2 & 0 \end{pmatrix}$. Find $\mathbf{P}^2$, $\mathbf{P}^3$ and $\mathbf{P}^4$. Show that $\mathbf{P}^3 = 4\mathbf{P}$ and $\mathbf{P}^4 = 16\mathbf{I}$.

Suggest an expression for $\mathbf{P}^n$ where n is (a) even
(b) odd.

6.3 Applications of matrix multiplication

$$\mathbf{P} = \begin{array}{c} \\ A \end{array} \begin{array}{cc} B & C \\ (2 & 1) \end{array}$$

$$\mathbf{Q} = \begin{array}{c} \\ B \\ C \end{array} \begin{array}{c} \begin{array}{cc} D & E \end{array} \\ \begin{pmatrix} 1 & 0 \\ 1 & 1 \end{pmatrix} \end{array}.$$

$$\mathbf{R} = \begin{array}{c} \\ A \end{array} \begin{array}{cc} D & E \\ (3 & 1) \end{array}$$

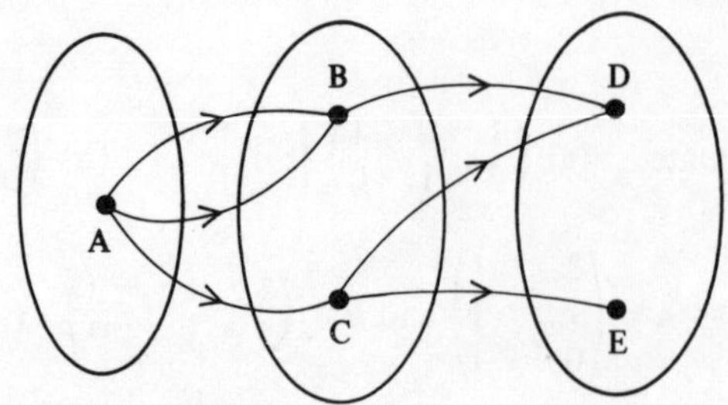

P is the matrix representing the routes from A to B and C.
Q is the matrix representing the routes from B and C to D and E.
R = **P** × **Q** is the matrix representing the routes from A to D and E.

Two matrices can be multiplied if the columns of the first correspond to the rows of the second.

$$\mathbf{S} = \begin{array}{c} \\ X \\ Y \\ Z \end{array} \begin{array}{c} \begin{array}{ccc} X & Y & Z \end{array} \\ \begin{pmatrix} 1 & 0 & 1 \\ 1 & 0 & 1 \\ 1 & 1 & 0 \end{pmatrix} \end{array} \qquad \mathbf{S}^2 = \begin{array}{c} \\ X \\ Y \\ Z \end{array} \begin{array}{c} \begin{array}{ccc} X & Y & Z \end{array} \\ \begin{pmatrix} 2 & 1 & 1 \\ 2 & 1 & 1 \\ 2 & 0 & 2 \end{pmatrix} \end{array}$$

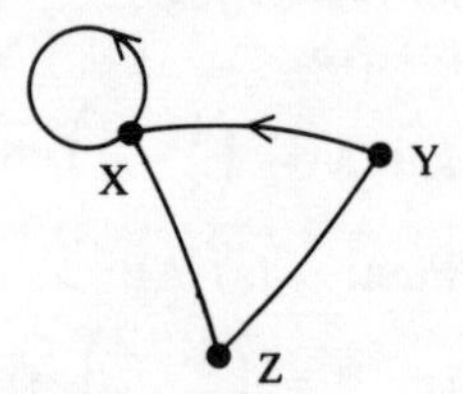

S is the route matrix for the network shown.
$\mathbf{S}^2$ is the *two-stage* matrix giving the number of routes of two parts.

If a square matrix gives information about, for instance, a network, its square gives information about using the network twice.

Exercise 6.3

1. $\mathbf{P} = \begin{array}{c} \\ A \\ B \end{array} \begin{array}{c} \begin{array}{cc} C & D \end{array} \\ \begin{pmatrix} 1 & 1 \\ 0 & 1 \end{pmatrix} \end{array}$, $\mathbf{Q} = \begin{array}{c} \\ C \\ D \end{array} \begin{array}{c} E \\ \begin{pmatrix} 1 \\ 2 \end{pmatrix} \end{array}$ are route matrices for five points A, B, C, D, E.

   Find **PQ** and state the total number of different routes from A and B to E. Draw the network.

2. $\mathbf{P} = \begin{array}{c} \\ A \\ B \\ C \end{array} \begin{array}{c} D \\ \begin{pmatrix} 1 \\ 2 \\ 1 \end{pmatrix} \end{array}$, $\mathbf{Q} = \begin{array}{c} \\ D \end{array} \begin{array}{cc} E & F \\ (1 & 2) \end{array}$ are route matrices for the points A, B, C, D, E.

   Calculate **PQ**, draw the network and state the total number of routes from B to F.

3. Travel from the island of Landy to the island of Nundy has to be via the Mainland. There are two harbours on Landy and one on Nundy; both islands have an airport. Two harbours on the Mainland are used by boats to the islands but only one of the two Landy harbours has connecting services to both Mainland harbours. Two airports on the Mainland handle island traffic.
   (a) Draw a network diagram for journeys between the islands.
   (b) Write down matrices describing the journeys from Landy to the Mainland and for journeys from the Mainland to Nundy.
   (c) Form the product of the matrices and write down the total number of ways journeys can be made from Landy to Nundy by: (i) air (ii) sea.

4. The diagram shows the connections made by a certain relation between the elements of sets A, B and C. Use matrices to describe the connections A to B and the connections B to C. Form the product of the matrices and explain the meaning of the entries in the matrix product.

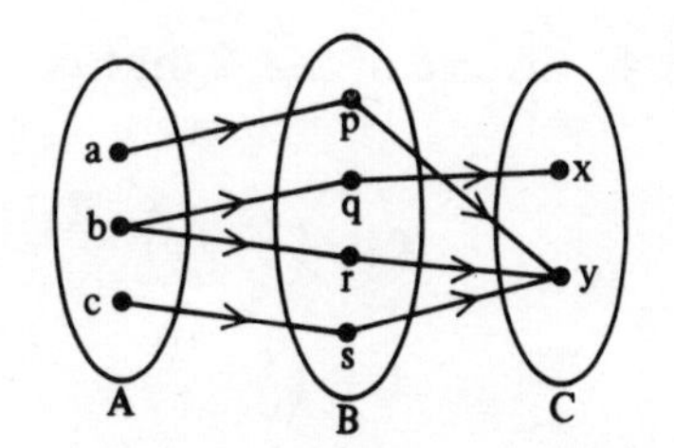

5. The matrix **X** gives the vegetable shopping requirements in kg of a certain housewife on two successive weeks.

$$\mathbf{X} = \begin{matrix} \\ \textit{week 1} \\ \textit{week 2} \end{matrix} \begin{matrix} \textit{Potatoes} & \textit{Carrots} & \textit{Onions} \\ \left(\begin{matrix} 5 \\ 10 \end{matrix}\right. & \begin{matrix} 1 \\ 2 \end{matrix} & \left.\begin{matrix} 1 \\ 1 \end{matrix}\right) \end{matrix}.$$

The prices per kg of these vegetables in two different shops A and B were 12p, 14p, 20p and 10p, 16p, 18p respectively.
Write down a matrix **Y** to show the prices of the vegetables and calculate the product **XY**.
What does the matrix **XY** represent?
Does one of the shops give overall better value for this housewife's shopping list?

6. A shopkeeper holds stocks of two types of screw, S_1, S_2 in three lengths as shown by the matrix M.

$$\mathbf{M} = \begin{matrix} \\ S_1 \\ S_2 \end{matrix} \begin{matrix} \textit{25 mm} & \textit{50 mm} & \textit{75 mm} \\ \left(\begin{matrix} 1000 \\ 1500 \end{matrix}\right. & \begin{matrix} 800 \\ 2100 \end{matrix} & \left.\begin{matrix} 700 \\ 1050 \end{matrix}\right) \end{matrix}$$

The price of each type of screw is the same, namely, 5p, 8p and 12p per ten in order of length.
Write down a price matrix N and calculate **MN**.
Use your answer to calculate the total value of the shopkeeper's stock.

7. Draw the network described by the route matrix **X**.
Calculate $\mathbf{X}^2$ and $\mathbf{X}^3$ and draw the associated network in each case.
What do the networks represent?

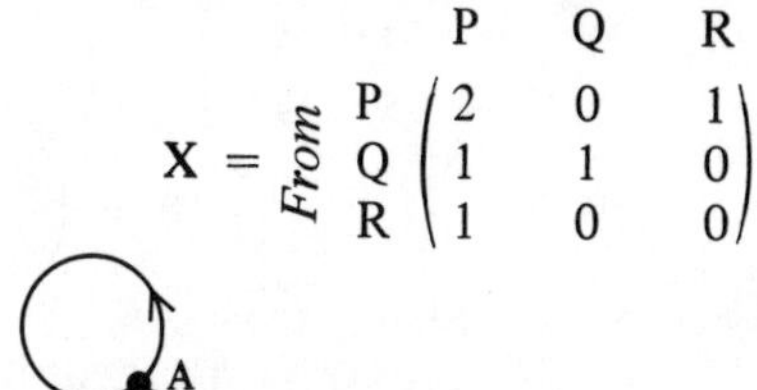

$$\mathbf{X} = \textit{From}\ \begin{matrix} \\ P \\ Q \\ R \end{matrix} \begin{matrix} & \textit{To} & \\ P & Q & R \\ \left(\begin{matrix} 2 \\ 1 \\ 1 \end{matrix}\right. & \begin{matrix} 0 \\ 1 \\ 0 \end{matrix} & \left.\begin{matrix} 1 \\ 0 \\ 0 \end{matrix}\right) \end{matrix}$$

8. Three robots, Fred, George and Harry–F, G, H for short–stand at the vertices A, B, C of a triangle. The network diagram shows the routes they may take. Write down a square matrix **M** showing the possible routes F, G, and H may make.
Find the matrix that gives the number of ways they may reach vertices A, B, C after
(i) two moves (ii) three moves.
In how many different ways may F, G and H return to their starting points after 3 moves, if at all?

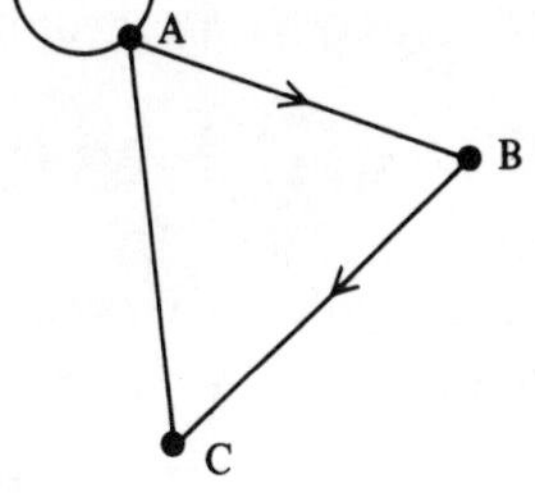

9. Three lines k, l, m in a plane are such that k is perpendicular to both l and m.
Complete the matrix **T** for the relation *is perpendicular to.*

$$\mathbf{T} = \begin{matrix} \\ k \\ l \\ m \end{matrix} \begin{matrix} k & l & m \\ \left(\begin{matrix} \\ \\ \\ \end{matrix}\right. & & \left.\begin{matrix} \\ \\ \\ \end{matrix}\right) \end{matrix}.$$

(a) Calculate $\mathbf{T}^2$.
(b) Draw the relation graph for $\mathbf{T}^2$.
(c) What relation does $\mathbf{T}^2$ describe?

10. Three men A, B, C sit round a table.
Write down a matrix **M** for the relation *is on the left of.*
Calculate $\mathbf{M}^2$ and $\mathbf{M}^3$ and state the relation they describe.

6.4 Inverse of 2 × 2 matrix

$\mathbf{X} = \begin{pmatrix} a & b \\ c & d \end{pmatrix}$ The *determinant* of **X** is $D = ad - bc$

$\mathbf{I} = \begin{pmatrix} 1 & 0 \\ 0 & 1 \end{pmatrix}$ **I** is the *identity matrix* $\mathbf{IX} = \mathbf{X} = \mathbf{XI}$

$\mathbf{X}^{-1} = \frac{1}{D}\begin{pmatrix} d & -b \\ -c & a \end{pmatrix}$ $\mathbf{X}^{-1}$ is the *inverse* of **X**.
$\mathbf{X}^{-1}\mathbf{X} = \mathbf{I} = \mathbf{X}\mathbf{X}^{-1}$

Before changing the terms of the matrix to find the inverse, first find the value of the determinant. If $D = 0$, the matrix has no inverse.

Exercise 6.4

1. Find, if possible, the inverse of the matrix:
(a) $\begin{pmatrix} 5 & 7 \\ 2 & 3 \end{pmatrix}$ (b) $\begin{pmatrix} 4 & 2 \\ 5 & 3 \end{pmatrix}$ (c) $\begin{pmatrix} -2 & 3 \\ -3 & 4 \end{pmatrix}$ (d) $\begin{pmatrix} 6 & 2 \\ 10 & 5 \end{pmatrix}$ (e) $\begin{pmatrix} 3 & 9 \\ 2 & -6 \end{pmatrix}$

2. Find, if possible, the inverse of the matrix:
(a) $\begin{pmatrix} 7 & 4 \\ 5 & 3 \end{pmatrix}$ (b) $\begin{pmatrix} 6 & 11 \\ 2 & 4 \end{pmatrix}$ (c) $\begin{pmatrix} 3 & 12 \\ 2 & 8 \end{pmatrix}$ (d) $\begin{pmatrix} 6 & 5 \\ 5 & 4 \end{pmatrix}$ (e) $\begin{pmatrix} 2 & -1 \\ -2 & 0 \end{pmatrix}$

3. $\mathbf{P} = \begin{pmatrix} 5 & -9 \\ 4 & -8 \end{pmatrix}$, $\mathbf{Q} = \begin{pmatrix} 8 & -9 \\ 4 & -5 \end{pmatrix}$. Calculate **PQ** and use your answer to find $\mathbf{P}^{-1}$ and $\mathbf{Q}^{-1}$.

4. $\mathbf{X} = \begin{pmatrix} 5 & 2 \\ 3 & 1 \end{pmatrix}$, $\mathbf{Y} = \begin{pmatrix} 1 & -2 \\ -3 & 5 \end{pmatrix}$. Calculate **XY** and hence write down the matrix $\mathbf{X}^{-1}$.

5. $\mathbf{A} = \begin{pmatrix} 0 & -1 \\ 2 & 8 \end{pmatrix}$, $\mathbf{B} = \begin{pmatrix} 5 & -3 \\ 2 & -1 \end{pmatrix}$. Find $\mathbf{A}^{-1}$ and $\mathbf{B}^{-1}$.
Calculate: (a) **AB** (b) $(\mathbf{AB})^{-1}$ (c) $\mathbf{B}^{-1}\mathbf{A}^{-1}$
What is significant about the answers (b) and (c)?

6. $\mathbf{U} = \begin{pmatrix} 2 & 4 \\ 3 & 5 \end{pmatrix}$, $\mathbf{V} = \begin{pmatrix} 1 & 0 \\ 7 & 3 \end{pmatrix}$. Write down $\mathbf{U}^{-1}$ and $\mathbf{V}^{-1}$.
Calculate: (a) **UV** (b) $\mathbf{U}^{-1}\mathbf{V}^{-1}$ (c) $\mathbf{V}^{-1}\mathbf{U}^{-1}$
From your answers, which of these statements is true?
$(\mathbf{UV})^{-1} = \mathbf{U}^{-1}\mathbf{V}^{-1}$ *or* $(\mathbf{UV})^{-1} = \mathbf{V}^{-1}\mathbf{U}^{-1}$

7. Find the square of the matrix and hence its inverse:
(a) $\begin{pmatrix} 4 & -5 \\ 3 & -4 \end{pmatrix}$ (b) $\begin{pmatrix} -2 & 0 \\ -4 & 2 \end{pmatrix}$ (c) $\begin{pmatrix} 10 & -7 \\ 13 & -10 \end{pmatrix}$ (d) $\begin{pmatrix} 0 & x \\ x & 0 \end{pmatrix}$

8. Find the inverse of the matrix and use the answer to find the square and cube of the matrix:
(a) $\begin{pmatrix} 5 & -8 \\ 3 & -5 \end{pmatrix}$ (b) $\begin{pmatrix} -7 & 16 \\ -3 & 7 \end{pmatrix}$ (c) $\begin{pmatrix} -6 & 8 \\ -4 & 6 \end{pmatrix}$ (d) $\begin{pmatrix} k & 1-k \\ 1+k & -k \end{pmatrix}$

9. x, y, z, w satisfy the matrix equation $\begin{pmatrix} 2 & 3 \\ 3 & 4 \end{pmatrix}\begin{pmatrix} x & y \\ z & w \end{pmatrix} = \begin{pmatrix} 4 & -1 \\ 2 & 0 \end{pmatrix}$.

Pre-multiply both sides of the equation by the matrix $\begin{pmatrix} -4 & 3 \\ 3 & -2 \end{pmatrix}$ and so find x, y, z, w.

10. What is the inverse of the matrix $\begin{pmatrix} 7 & 10 \\ 3 & 4 \end{pmatrix}$? Use your answer to find the values of

x, y, z, w, that satisfy the matrix equation $\begin{pmatrix} 7 & 10 \\ 3 & 4 \end{pmatrix}\begin{pmatrix} x & y \\ z & w \end{pmatrix} = \begin{pmatrix} 2 & 1 \\ 3 & -1 \end{pmatrix}$.

11. $\mathbf{P} = \begin{pmatrix} 0 & 1 \\ 1 & 0 \end{pmatrix}$, $\mathbf{Q} = \begin{pmatrix} -1 & 0 \\ 0 & -1 \end{pmatrix}$. Find $\mathbf{P}^{-1}$, $\mathbf{Q}^{-1}$, $\mathbf{PQ}$ and $(\mathbf{PQ})^{-1}$.

Show that only four district matrices are produced by products of **P**, **Q** and their inverses.

12. $\mathbf{A} = \begin{pmatrix} 0 & -1 \\ 1 & 0 \end{pmatrix}$. Find $\mathbf{A}^2$, $\mathbf{A}^3$ and $\mathbf{A}^4$.

Hence write down the inverses of $\mathbf{A}$, $\mathbf{A}^2$ and $\mathbf{A}^3$.

13. $\mathbf{R} = \begin{pmatrix} \frac{1}{2} & -\frac{\sqrt{3}}{2} \\ \frac{\sqrt{3}}{2} & \frac{1}{2} \end{pmatrix}$. Find: (a) $\mathbf{R}^{-1}$ (b) $\mathbf{R}^2$ (c) $(\mathbf{R}^2)^{-1}$ (d) $\mathbf{R}^3$ (e) $\mathbf{R}^6$

How many distinct matrices can be produced from products of **R** and its inverse?

14. Let $\mathbf{A} = \begin{pmatrix} 1 & 1 \\ k & k^3 \end{pmatrix}$. If **A** does not have an inverse, show that k can have one of three values.

15. Let $\mathbf{M} = \begin{pmatrix} 6 & 5 \\ 1 & 2 \end{pmatrix}$. Find $\mathbf{M}'$ and $\mathbf{M}^{-1}$. Is it true that $(\mathbf{M}')^{-1} = (\mathbf{M}^{-1})'$?

16. Let $\mathbf{A} = \begin{pmatrix} a & b \\ c & d \end{pmatrix}$ be any matrix.

Write down its inverse, its transpose and the inverse of its transpose.
Is the inverse of the transpose equal to the transpose of the inverse?

6.5 Equations and matrices

Matrices can be used to solve simultaneous equations.

Example 1

Solve: $\left.\begin{aligned} 2x + y &= 1 \\ 5x + 3y &= -1 \end{aligned}\right\}$

Rewrite as a matrix equation: $\begin{pmatrix} 2 & 1 \\ 5 & 3 \end{pmatrix}\begin{pmatrix} x \\ y \end{pmatrix} = \begin{pmatrix} 1 \\ -1 \end{pmatrix}$

The inverse of $\begin{pmatrix} 2 & 1 \\ 5 & 3 \end{pmatrix}$ is $\begin{pmatrix} 3 & -1 \\ -5 & 2 \end{pmatrix}$.

Pre-multiply the equation by the inverse matrix.

$$\begin{pmatrix} 3 & -1 \\ -5 & 2 \end{pmatrix}\begin{pmatrix} 2 & 1 \\ 5 & 3 \end{pmatrix}\begin{pmatrix} x \\ y \end{pmatrix} = \begin{pmatrix} 3 & -1 \\ -5 & 2 \end{pmatrix}\begin{pmatrix} 1 \\ -1 \end{pmatrix}$$

$$\begin{pmatrix} 1 & 0 \\ 0 & 1 \end{pmatrix}\begin{pmatrix} x \\ y \end{pmatrix} = \begin{pmatrix} 4 \\ -7 \end{pmatrix}$$

$$\begin{pmatrix} x \\ y \end{pmatrix} = \begin{pmatrix} 4 \\ -7 \end{pmatrix}$$

From the matrix equation, $x = 4, y = -7$.

Example 2

Solve: $\left.\begin{aligned} 3x - y &= 2 \\ 2x - 4y &= 5 \end{aligned}\right\}$

Rewrite as a matrix equation: $\begin{pmatrix} 3 & 1 \\ 2 & 4 \end{pmatrix}\begin{pmatrix} x \\ y \end{pmatrix} = \begin{pmatrix} 2 \\ 5 \end{pmatrix}$

The inverse of $\begin{pmatrix} 3 & 1 \\ 2 & 4 \end{pmatrix}$ is $\frac{1}{10}\begin{pmatrix} 4 & -1 \\ -2 & 3 \end{pmatrix}$.

Pre-multiply the matrix equation by the inverse matrix.

$$\tfrac{1}{10}\begin{pmatrix} 4 & -1 \\ -2 & 3 \end{pmatrix}\begin{pmatrix} 3 & 1 \\ 2 & 4 \end{pmatrix}\begin{pmatrix} x \\ y \end{pmatrix} = \tfrac{1}{10}\begin{pmatrix} 4 & -1 \\ -2 & 3 \end{pmatrix}\begin{pmatrix} 2 \\ 5 \end{pmatrix}$$

$$\begin{pmatrix} x \\ y \end{pmatrix} = \tfrac{1}{10}\begin{pmatrix} 3 \\ 11 \end{pmatrix} = \begin{pmatrix} 0{\cdot}3 \\ 1{\cdot}1 \end{pmatrix}$$

From the matrix equation, $x = 0{\cdot}3, y = 1{\cdot}1$.

Exercise 6.5

1. Write down the pair of simultaneous equations given by the matrix equation $\begin{pmatrix} 2 & 1 \\ 1 & 1 \end{pmatrix}\begin{pmatrix} x \\ y \end{pmatrix} = \begin{pmatrix} 7 \\ 3 \end{pmatrix}$.

   Find $\begin{pmatrix} x \\ y \end{pmatrix}$ by multiplying both sides of the matrix equation by the matrix $\begin{pmatrix} 1 & -1 \\ -1 & 2 \end{pmatrix}$ and show that the values you obtain for x and y fit your equations.

2. Write in matrix form the pair of equations $\begin{cases} x + 3y = 1 \\ 4x - y = -2 \end{cases}$.

Show that $\begin{pmatrix} x \\ y \end{pmatrix}$ can be found by multiplying both sides of the matrix equation by $\frac{1}{13}\begin{pmatrix} 1 & 3 \\ 4 & -1 \end{pmatrix}$ and find x and y.

3. Use the method of matrices to solve for x and y:

(a) $\begin{cases} 2x + y = 1 \\ 5x + 3y = 3 \end{cases}$ (b) $\begin{cases} 7x + 3y = -3 \\ 2x + y = 2 \end{cases}$ (c) $\begin{cases} 2x - y = -2 \\ x + 2y = 3 \end{cases}$

4. Use the method of matrices to solve for x and y:

(a) $\begin{cases} 2x + y = 2 \\ 7x + 4y = 3 \end{cases}$ (b) $\begin{cases} 4x - 2y = 3 \\ 3x + y = -2 \end{cases}$ (c) $\begin{cases} 2x - y = 1 \\ 3x - 2y = 2 \end{cases}$

5. Use the method of matrices to solve for u and v:

(a) $\begin{cases} 2u + v = 1 \\ 5u + 3v = 2 \end{cases}$ (b) $\begin{cases} 5u - 2v = 1 \\ 2u + v = 3 \end{cases}$ (c) $\begin{cases} u + 3v = 2 \\ 2u - 4v = 0 \end{cases}$

6. Use the method of matrices to solve for r and s:

(a) $\begin{cases} 4r + 3s = 5 \\ 3r + 2s = 1 \end{cases}$ (b) $\begin{cases} r - 3s = -1 \\ r + 2s = 3 \end{cases}$ (c) $\begin{cases} 4r + 3s = 3 \\ 5r - 5s = 1 \end{cases}$

7. Find why the method of matrices does not work if you try to solve the equations

$\begin{cases} 4x + 6y = 3 \\ 6x + 9y = 2 \end{cases}$

Draw a graph of the lines represented by the equations to show why there is no solution.

8. Show that there are solutions to the matrix equation $\begin{pmatrix} 1 & -3 \\ -2 & 6 \end{pmatrix}\begin{pmatrix} x \\ y \end{pmatrix} = \begin{pmatrix} -1 \\ 2 \end{pmatrix}$

even though matrix methods fail to work.

Draw a graph of the lines represented by the equations in x and y.

9. Use matrix methods to find the solution set of

$\left.\begin{matrix} 3x - 2y = 5 \\ x + y = k \end{matrix}\right\}$

for any value k.

Hence find the particular solutions for $k = 0, k = 5, k = 15$.

10. From matrix methods, find the value of a for which the pair of simultaneous equations

$\left.\begin{matrix} 4x + 2y = 3 \\ 3x + ay = 1 \end{matrix}\right\}$

does not have a unique solution.

Find the particular solution when $a = 1$.

11. Rearrange to form a pair of simultaneous equations and solve by the method of matrices:

(a) $5x + 2y = 3 = x + y$ (b) $x + y + 7 = 2x - y = 8$

12. Rearrange to form a pair of simultaneous equations and solve by the method of matrices:

(a) $r + 2s + 3 = 2r + 3s + 4 = 5$ (b) $8 = 7 - u - v = u + 2v$

6.6 Vectors

Representation

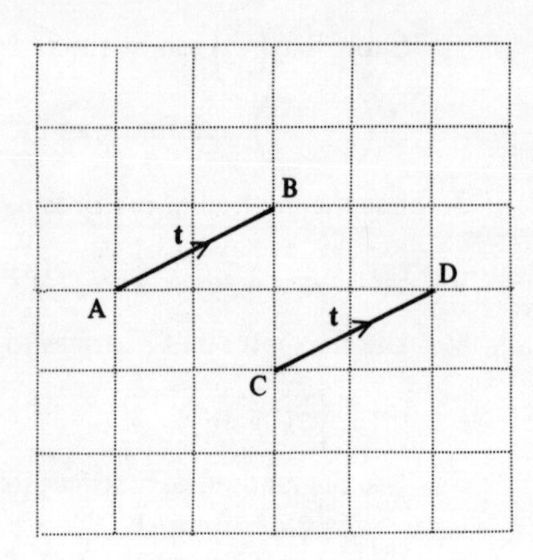

The vector from A to B can be represented in three ways.

$$\overrightarrow{AB} = \mathbf{t} = \begin{pmatrix}2\\1\end{pmatrix}$$

From the diagram, we see that

$$\overrightarrow{CD} = \begin{pmatrix}2\\1\end{pmatrix} = \mathbf{t} = \overrightarrow{AB}.$$

Column vectors are equal if their entries are equal. Equal vectors have the same length and direction.

Length of a vector

If $\overrightarrow{AB} = \mathbf{t} = \begin{pmatrix}x\\y\end{pmatrix}$, the length of $\overrightarrow{AB}$ is $AB = |\mathbf{t}| = \sqrt{(x^2+y^2)}$.

Addition of vectors is by the triangle law.

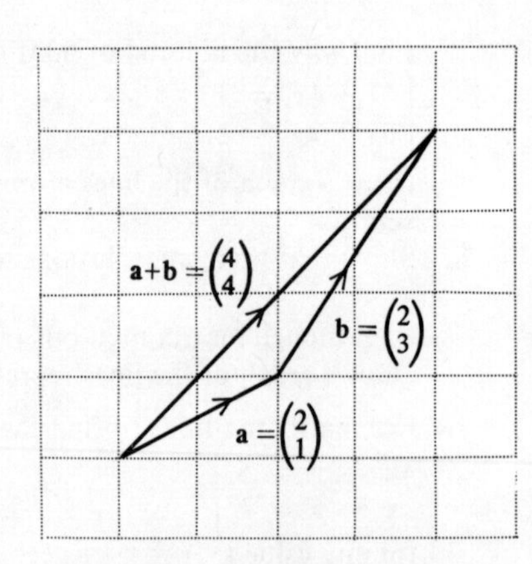

If $\mathbf{a} = \begin{pmatrix}2\\1\end{pmatrix}$, $\mathbf{b} = \begin{pmatrix}2\\3\end{pmatrix}$ then

$$\mathbf{a}+\mathbf{b} = \begin{pmatrix}2\\1\end{pmatrix}+\begin{pmatrix}2\\3\end{pmatrix} = \begin{pmatrix}4\\4\end{pmatrix}.$$

The identity vector is $\mathbf{0} = \begin{pmatrix}0\\0\end{pmatrix}$.

The inverse of $\mathbf{a} = \begin{pmatrix}x\\y\end{pmatrix}$ is $-\mathbf{a} = \begin{pmatrix}-x\\-y\end{pmatrix}$.

Mid-point theorem

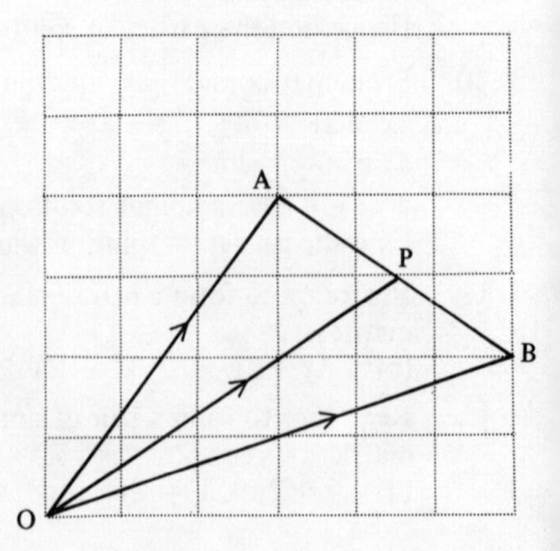

$\overrightarrow{OA} = \begin{pmatrix}3\\4\end{pmatrix}$, $\overrightarrow{OB} = \begin{pmatrix}6\\2\end{pmatrix}$, P is the mid-point of AB.

$$\begin{aligned}\text{Then}\quad \overrightarrow{OP} &= \tfrac{1}{2}(\overrightarrow{OA}+\overrightarrow{OB})\\ &= \tfrac{1}{2}\left(\begin{pmatrix}3\\4\end{pmatrix}+\begin{pmatrix}6\\2\end{pmatrix}\right)\\ &= \begin{pmatrix}4{\cdot}5\\3\end{pmatrix}.\end{aligned}$$

In general, if $\overrightarrow{OA} = \mathbf{a}$, $\overrightarrow{OB} = \mathbf{b}$ and P is the mid-point of AB then $\overrightarrow{OP} = \frac{1}{2}(\mathbf{a}+\mathbf{b})$.

Exercise 6.6

1. Let $\mathbf{a} = \begin{pmatrix}2\\7\end{pmatrix}$, $\mathbf{b} = \begin{pmatrix}3\\-2\end{pmatrix}$, $\mathbf{c} = \begin{pmatrix}-1\\0\end{pmatrix}$.
   Find: (a) $\mathbf{a}+\mathbf{b}$ (b) $\mathbf{b}+\mathbf{c}$ (c) $\mathbf{a}+(\mathbf{b}+\mathbf{c})$ (d) $(\mathbf{a}+\mathbf{b})+\mathbf{c}$
   Are the answers (c) and (d) the same?
   Illustrate the answers by drawing the vectors on a grid.

2. Let $\mathbf{r} = \begin{pmatrix}4\\-2\end{pmatrix}$, $\mathbf{s} = \begin{pmatrix}0\\-3\end{pmatrix}$, $\mathbf{t} = \begin{pmatrix}2\\1\end{pmatrix}$.
   Find $\mathbf{r}+\mathbf{s}$ and $\mathbf{s}+\mathbf{t}$.
   Draw a diagram to show that $(\mathbf{r}+\mathbf{s})+\mathbf{t} = \mathbf{r}+(\mathbf{s}+\mathbf{t})$.

3. Find $\mathbf{x}$ in the equation: (a) $\begin{pmatrix}3\\2\end{pmatrix}+\begin{pmatrix}-1\\3\end{pmatrix} = 2\mathbf{x}$ (b) $\mathbf{x}+\begin{pmatrix}-1\\3\end{pmatrix} = \begin{pmatrix}2\\7\end{pmatrix}$

4. Find $\mathbf{x}$ in the equation:
   (a) $\begin{pmatrix}4\\-2\end{pmatrix}+\mathbf{x} = \begin{pmatrix}0\\0\end{pmatrix}$ (b) $3\mathbf{x}+\begin{pmatrix}1\\-3\end{pmatrix} = \begin{pmatrix}4\\0\end{pmatrix}$

5. Find x, y to fit the vector equation:
   (a) $\begin{pmatrix}3\\y\end{pmatrix}+\begin{pmatrix}2\\7\end{pmatrix} = \begin{pmatrix}x\\-3\end{pmatrix}$ (b) $\begin{pmatrix}x\\7\end{pmatrix}+\begin{pmatrix}3\\y\end{pmatrix} = \begin{pmatrix}2\\2\end{pmatrix}$

6. Find x, y to fit the vector equation:
   (a) $\begin{pmatrix}x\\5\end{pmatrix}+\begin{pmatrix}y\\4\end{pmatrix} = \begin{pmatrix}7\\y\end{pmatrix}$ (b) $\begin{pmatrix}x\\-3\end{pmatrix}-\begin{pmatrix}y\\y\end{pmatrix} = \begin{pmatrix}y\\1\end{pmatrix}$

7. Given $\mathbf{a} = \begin{pmatrix}2\\3\end{pmatrix}$ and $\mathbf{b} = \begin{pmatrix}2\\-1\end{pmatrix}$ write down the inverse of vectors $\mathbf{a}$ and $\mathbf{b}$ and use your answer to find the inverse of $\mathbf{a}+\mathbf{b}$.

8. Given $\mathbf{a} = \begin{pmatrix}1\\2\end{pmatrix}$ and $\mathbf{b} = \begin{pmatrix}-3\\0\end{pmatrix}$ write down the inverse of vectors $\mathbf{a}$ and $\mathbf{b}$. Draw a diagram to show that the inverse of $(\mathbf{a}+\mathbf{b})$ equals the sum of the inverse of $\mathbf{a}$ and the inverse of $\mathbf{b}$.

9. $\overrightarrow{OA} = \begin{pmatrix}2\\7\end{pmatrix}$, $\overrightarrow{OB} = \begin{pmatrix}-1\\10\end{pmatrix}$; P is the mid-point of AB. Find the vector $\overrightarrow{OP}$.

10. A = (3, −1), B = (−1, 6); P is the mid-point of AB. Find $\overrightarrow{OA}$, $\overrightarrow{OB}$ and $\overrightarrow{OP}$.

11. A = (−3, 2), B = (2, 6), X is the mid-point of AB. Write down $\overrightarrow{OA}$, $\overrightarrow{OB}$, $\overrightarrow{OX}$ and $2\overrightarrow{OX}$. Hence write down the co-ordinates of the fourth vertex of the parallelogram of which the other vertices are O, A and B.

12. A = (1, 1), B = (5, 3), C = (3, 5) are three vertices of a parallelogram ABDC. Find the vectors $\overrightarrow{AB}$, $\overrightarrow{AC}$ and hence $\overrightarrow{AD}$. By calculating the lengths of the vectors $\overrightarrow{AB}$ and $\overrightarrow{AC}$, show that ABDC is a rhombus.

13. $\overrightarrow{OP} = \begin{pmatrix}7\\24\end{pmatrix}$, $\overrightarrow{OQ} = \begin{pmatrix}15\\20\end{pmatrix}$. If O, P, Q are three vertices of a parallelogram, find the fourth vertex. Calculate the length of each side.

14. $\mathbf{a} = \begin{pmatrix}3\\4\end{pmatrix}$, $\mathbf{b} = \begin{pmatrix}5\\12\end{pmatrix}$. Calculate:
    (a) $|\mathbf{a}|$ (b) $|\mathbf{b}|$ (c) $|\mathbf{a}|+|\mathbf{b}|$ (d) $|\mathbf{a}+\mathbf{b}|$

6.7 Vector geometry

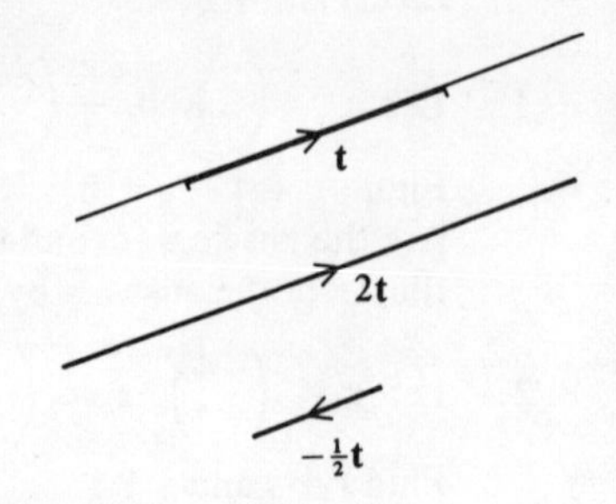

Most vector geometry problems are based on parallel vectors.

2**t** is parallel to **t** and twice its length.

$-\frac{1}{2}$**t** is parallel to **t** in the opposite direction and half its length.

Example 1

ABCD is a parallelogram.
M and N are the mid-points of AB and BC respectively.
Prove that MN $= \frac{1}{2}$ AC.

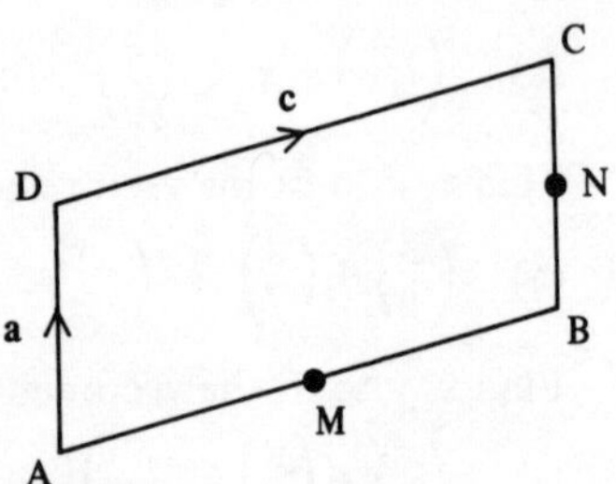

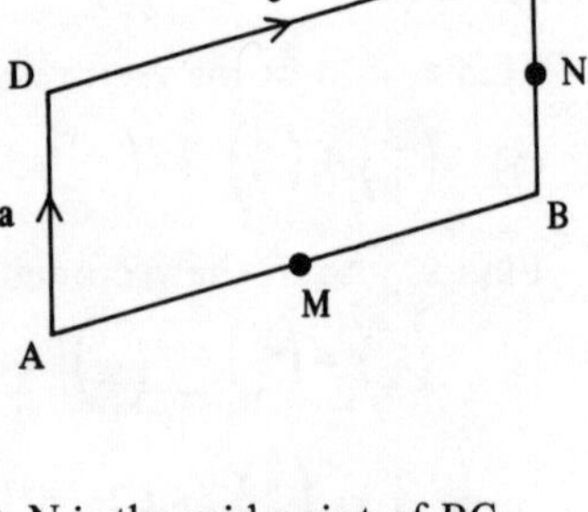

Let $\overrightarrow{AD} = \mathbf{a}$, $\overrightarrow{DC} = \mathbf{c}$ so $\overrightarrow{AC} = \overrightarrow{AD} + \overrightarrow{DC} = \mathbf{a} + \mathbf{c}$.
$\overrightarrow{AB} = \mathbf{c}$ so $\overrightarrow{MB} = \frac{1}{2}\mathbf{c}$
$\overrightarrow{BC} = \mathbf{a}$ so $\overrightarrow{BN} = \frac{1}{2}\mathbf{a}$
Hence $\overrightarrow{MN} = \overrightarrow{MB} + \overrightarrow{BN} = \frac{1}{2}\mathbf{c} + \frac{1}{2}\mathbf{a} = \frac{1}{2}(\mathbf{a} + \mathbf{c})$.
This proves that MN is parallel to AC and MN $= \frac{1}{2}$AC.

Exercise 6.7a

1. ABCD is any quadrilateral. M is the mid-point of AB, N is the mid-point of BC. Let $\overrightarrow{BM} = \mathbf{x}$, $\overrightarrow{BN} = \mathbf{y}$. Prove that $\overrightarrow{MN} = \frac{1}{2}\overrightarrow{AC}$
2. ABCD is any quadrilateral. K, L, M, N are the mid-points of AB, BC, CD, DA respectively.
   Let $\overrightarrow{KB} = \mathbf{x}$, $\overrightarrow{BL} = \mathbf{y}$, $\overrightarrow{ND} = \mathbf{u}$, $\overrightarrow{DM} = \mathbf{v}$.
   Express $\overrightarrow{AC}$ (a) in terms of **x** and **y** (b) in terms of **u** and **v**.
   Prove that $\overrightarrow{KL} = \overrightarrow{NM}$ and that KLMN is a parallelogram.
3. ABC is any triangle.
   M, N, L are the mid-points of AB, BC, CA respectively. G lies on CM and CG = 2 GM.
   Let $\overrightarrow{CA} = \mathbf{a}$, $\overrightarrow{CB} = \mathbf{b}$.
   Use **a**, **b** to find an expression for each of the following vectors:
   $\overrightarrow{CM}$, $\overrightarrow{CG}$, $\overrightarrow{AN}$, $\overrightarrow{AG}$.
   Hence prove that G lies on AN and AG : GN = 2 : 1.
4. ABC is any triangle. X lies on AB and AX $= \frac{1}{3}$AB.
   Z lies on CB and CZ $= \frac{1}{4}$ CB.
   Y lies on CX and CY $= \frac{1}{2}$ CX.
   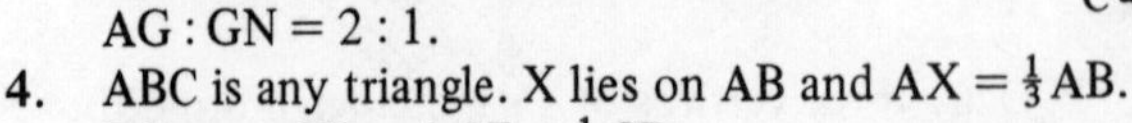
   Let $\overrightarrow{CA} = \mathbf{a}$, $\overrightarrow{CB} = \mathbf{b}$ and express the following vectors in terms of **a** and **b**:
   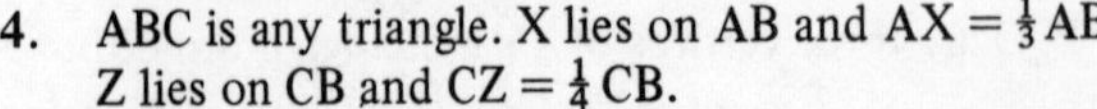
   $\overrightarrow{CX}$, $\overrightarrow{AY}$, $\overrightarrow{CY}$, $\overrightarrow{YZ}$.
   Show that Y must lie on AZ. What fraction of AZ is AY?

Example 2

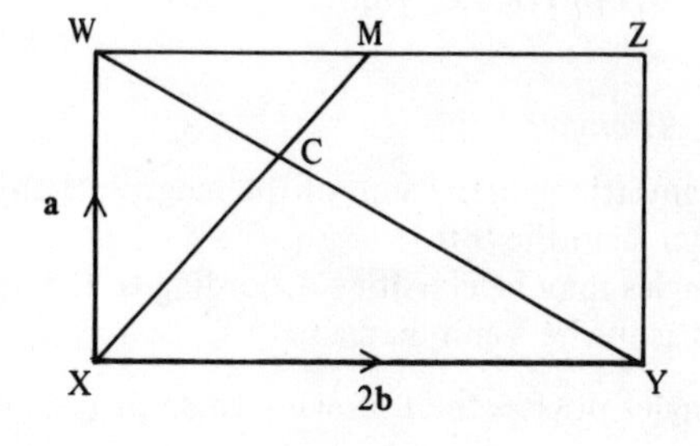

XYZW is a rectangle. M is the mid-point of ZW. YW intersects XM at C. Prove that C divides both lines in the ratio 2 : 1.

Let $\overrightarrow{XW} = \mathbf{a}$, $\overrightarrow{XY} = 2\mathbf{b} \Rightarrow \overrightarrow{WM} = \mathbf{b}$.
Let $\overrightarrow{XC} = k\,\overrightarrow{XM}$, $\overrightarrow{YC} = l\,\overrightarrow{YW}$.
Then $\overrightarrow{XM} = \mathbf{a} + \mathbf{b}$ so $\overrightarrow{XC} = k\mathbf{a} + k\mathbf{b}$.
Then $\overrightarrow{YW} = \mathbf{a} - 2\mathbf{b}$ so $\overrightarrow{YC} = l\mathbf{a} - 2l\,\mathbf{b}$.
$\overrightarrow{XC} = \overrightarrow{XY} + \overrightarrow{YC}$ so

$$k\mathbf{a} + k\mathbf{b} = 2\mathbf{b} + l\mathbf{a} - 2l\,\mathbf{b}.$$

or $$k\mathbf{a} + k\mathbf{b} = l\mathbf{a} + (2 - 2l)\,\mathbf{b}.$$

From the vector equation, since **a** is not parallel to **b**,

$$k = l \quad \text{and} \quad k = 2 - 2l$$

Solving the simultaneous equations gives $k = l = \frac{2}{3}$.
Hence XC : CM = 2 : 1 and YC : CW = 2 : 1.

Exercise 6.7b

1. XYZW is a parallelogram. XY is extended to R so that XY = YR. WR meets YZ at S.
   Let $\overrightarrow{XY} = \mathbf{a}$, $\overrightarrow{XW} = \mathbf{b}$, $\overrightarrow{YS} = k\,\overrightarrow{YZ}$, $\overrightarrow{WS} = l\,\overrightarrow{WR}$.
   Use the fact that $\overrightarrow{ZS} = \overrightarrow{ZW} + \overrightarrow{WS}$ to form a vector equation involving **a** and **b** and hence find k and l.
2. PQRS is a parallelogram. X lies on PQ such that PX = XQ. Y lies on PS such that PY = 2YS. PR meets XY at Z.
   Let $\overrightarrow{PQ} = 2\mathbf{a}$, $\overrightarrow{PS} = 3\mathbf{b}$.
   Let PZ $= k$ PR, YZ $= l$ YX.
   Find $\overrightarrow{PZ}$ and $\overrightarrow{YZ}$ in terms of **a** and **b** and hence find the values of k and l.
3. PQRS is a parallelogram. X lies on QS and QX $= \frac{2}{5}$QS.
   Prove that RX, when extended, cuts QP in the ratio 2 : 1.
4. ABCD is a square. X lies on AB such that AX = 2 XB. Y lies on CD such that CY = 3YD. Find the ratio in which XY divides BD.
5. ABC is a triangle. M is the mid-point of BC and N lies on AC such that AN = 2NC. K is a point on AM such that AK $= k$ AM.
   Let $\overrightarrow{BA} = \mathbf{a}$, $\overrightarrow{BC} = \mathbf{c}$.
   Find $\overrightarrow{BK}$ and $\overrightarrow{BN}$ in terms of **a** and **c** and hence find the value of k for which K lies on BN.
6. ABCDA′B′C′D′ is a cube. M is the mid-point of the face CDD′C′ and N is the mid-point of line A′M.
   Let $\overrightarrow{A'A} = \mathbf{a}$, $\overrightarrow{A'B'} = \mathbf{b}$, $\overrightarrow{A'D'} = \mathbf{c}$. Use **a**, **b**, **c** to find an expression for:
   (a) $\overrightarrow{A'M}$ (b) $\overrightarrow{D'N}$.
   Show that D′N extended meets A′B at X where A′X = XB.

Part 7: Geometry

7.1 Properties of plane shapes

Triangles

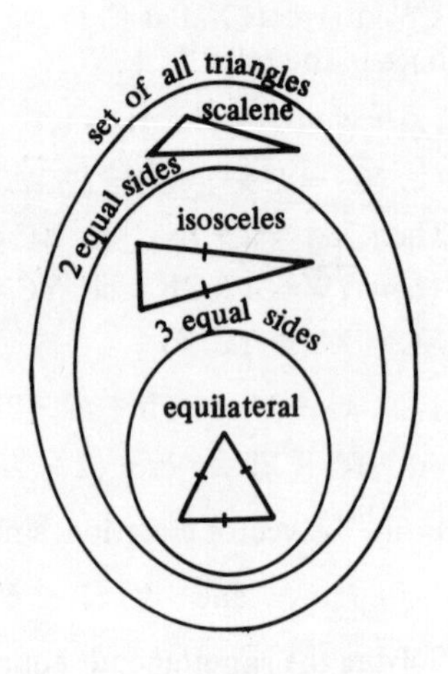

For any triangle, the sum of the lengths of any two sides is greater than the third.
Triangles may be classified according to the lengths of their sides as in the Venn diagram.

Triangles possess the following angle properties:

For *all* triangles: angle sum = 180°
exterior angle = sum of interior opposite angles.

For *isosceles* triangles: base angles are equal

For *equilateral* triangles: all angles equal 60°.

Exercise 7.1a

1. Draw a Venn diagram to show the sets of all triangles, right-angled triangles and isosceles triangles.
2. Draw a Venn diagram to show the sets of all triangles, obtuse-angled triangles, acute-angled triangles and isosceles triangles.
3. Find the value of the lettered angle in the triangles:

(a)

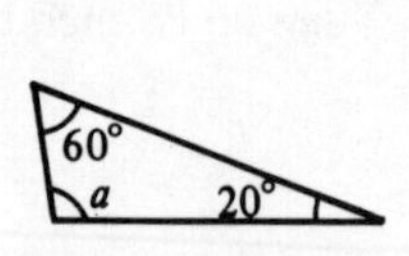

(b)

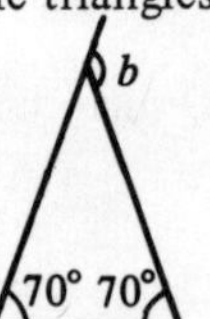

(c)

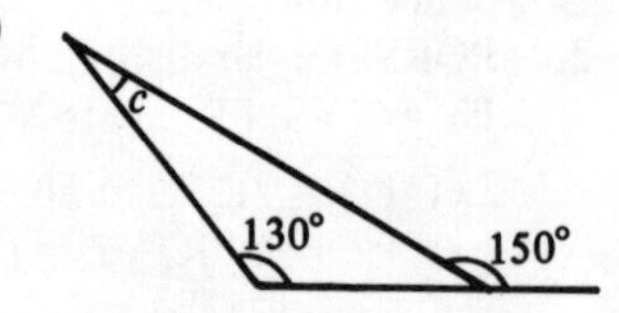

4. Find the value of the lettered angle in the triangles:

(a)

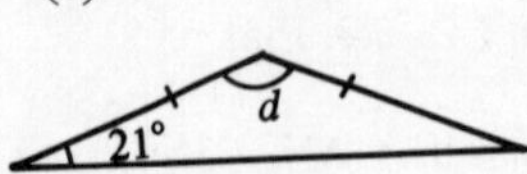

(b)

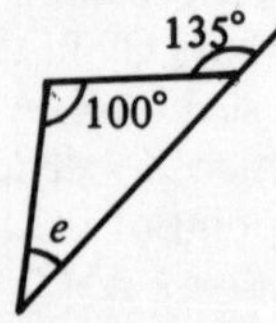

(c)

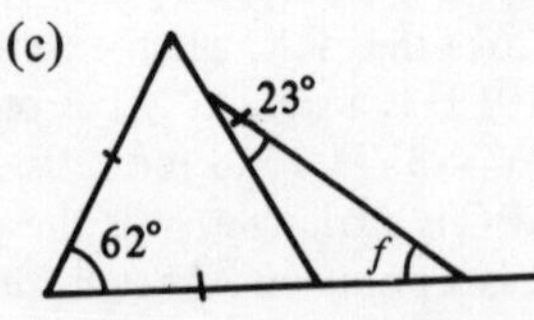

5. Find x if the sizes of the angles of a triangle in degrees are:
(a) 80, 20, x (b) 50, $x + 50$, $2x + 50$ (c) x, $2x$, $3x$.
6. One angle of an isosceles triangle is 40°. Calculate the other two angles, giving both possible answers.
7. ABC is an equilateral triangle. X lies on AC and Y lies on BC produced so that CYX is an isosceles triangle. Calculate the size of $C\hat{X}Y$.
8. ABCD is a square and ABP is an equilateral triangle drawn outside the square. Find the size of $A\hat{P}D$.

Angles of polygons

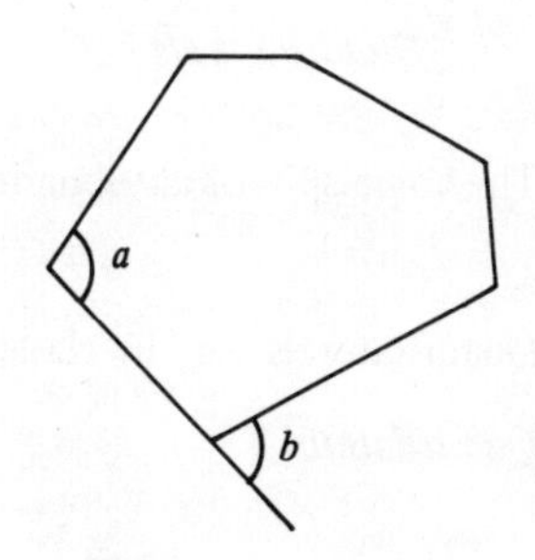

In the diagram, angle a is an *interior* angle and angle b is an *exterior* angle.

For any polygon of n sides

1. The sum of the exterior angles is 360° or 4 right angles.
2. The sum of the interior angles is $(2n - 4)$ right angles.

A polygon with equal sides *and* equal angles is called a *regular* polygon.

Example 1

Three angles of a pentagon are 60°, 70°, and 80°. Find the size of each of the other two angles, given that they are equal.

$n = 5$

Sum of interior angles $= (2n - 4)$ right angles $= 6 \times 90° = 540°$

Sum of given angles $= 60° + 70° + 80° = 210°$

Hence, sum of unknown angles $= 540° - 210° = 330°$
Each unknown angle $= 165°$.

Example 2

Calculate the size of an interior angle of a 9-sided regular polygon.

$n = 9$

Sum of interior angles $= (2n - 4)$ right angles $= 14 \times 90°$

Sum of each angle $= \dfrac{14 \times 90°}{9} = 140°$.

Exercise 7.1b

1. Calculate the sum of the interior angles of a decagon (10 sides).
2. Calculate the sum of the interior angles of a 12-sided polygon.
3. Calculate the size of each interior angle of a regular polygon of:
   (a) 5 sides (b) 10 sides (c) 15 sides.
4. Calculate the size of each interior angle of a regular polygon of:
   (a) 8 sides (b) 12 sides (c) 20 sides.
5. The exterior angle of a certain regular polygon is one-fifth the size of the interior angle. How many sides has the polygon?
6. A certain regular polygon has 20 sides. Calculate the size of an exterior angle. Use your answer to find the size of an interior angle. Find the sum of all the interior angles and show that your answer fits the formula $(2n - 4)$.
7. The angles of a pentagon, in degrees, are x, $2x$, $3x$, $4x$, and $5x$. Find x.
8. Three angles of a hexagon are 90°, 110°, and 110°. Find the size of each of the other angles given that they are equal.

Quadrilaterals

The angle sum of any quadrilateral is 360°.

Quadrilaterals may be classified according to the properties of their sides, angles, or diagonals.

Quadrilateral	*Sides*	*Angles*	*Diagonals*
Trapezium	One pair parallel.	—	—
Kite	Two pairs of adjacent sides are equal.	One pair of opposite angles equal	One diagonal bisects the other at right angles.
Parallelogram	Opposite sides parallel and equal.	Opposite angles equal.	Bisect each other.
Rhombus	All sides equal; opposite sides parallel.	Opposite angles equal.	Bisect each other at right angles.
Rectangle	Opposite sides parallel and equal.	All angles equal 90°.	Bisect each other.
Square	All sides equal; opposite sides parallel and equal.	All angles equal 90°.	Bisect each other at right angles.

By examining the list you see that some shapes possess all the properties of another shape. Thus, a square is a rectangle because the square satisfies all the conditions required of a rectangle. In set language, the set of squares is a subset of the set of rectangles. For example, examine the following Venn diagram.

$R \cap S$ = set of squares

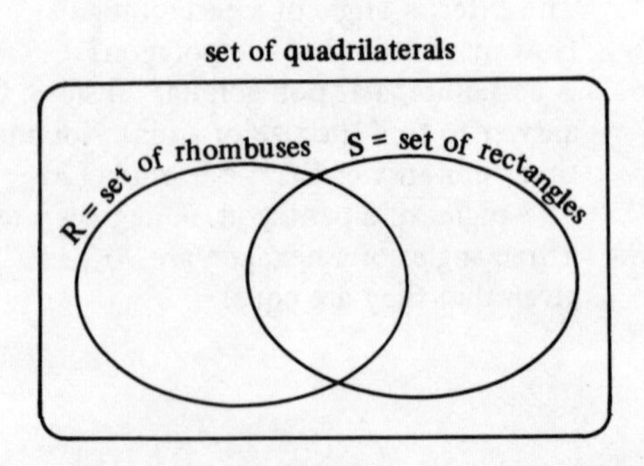

Exercise 7.1c

1. Draw a Venn diagram to show the subsets
   P = $\{x \mid x$ is a parallelogram$\}$,
   R = $\{x \mid x$ is a rhombus$\}$,
   S = $\{x \mid x$ is a square$\}$, of the set Q = $\{x \mid x$ is a quadrilateral$\}$.
2. Draw a Venn diagram to show the sets of quadrilaterals that are trapeziums, parallelograms, and squares.
3. ABCD is a square, ABXY is a rhombus with the side AB common to both figures. Describe the triangle DAY, giving reasons for your answer.
4. ABCD is a rectangle, ABXY is a parallelogram with the side AB common to both figures. Which of the following statements is true?
   (a) CDYX is a parallelogram
   (b) CDYX cannot be a rhombus
   (c) CDYX must be a rectangle
   (d) If CDYX is a rectangle, then so is ABXY.
5. A pair of congruent isosceles triangles are joined along an equal side. Draw diagrams to show that the resulting quadrilateral can be:
   (a) a parallelogram
   (b) a rhombus
   (c) a kite.
6 ABC is a right-angled isosceles triangle with $\hat{B} = 90°$. ACX is an isosceles triangle drawn on the side AC. Describe the triangle ACX if
   (a) ABCX is a square
   (b) ABCX is a kite
   (c) $A\hat{C}X = 105°$.
7. AOC, BOD are diagonals of a quadrilateral, intersecting at O. Name the quadrilateral if:
   (a) OA = OC, OB > OD, and $A\hat{O}B = 90°$
   (b) OA = OC, OB = OD, and $A\hat{O}C > 90°$
   (c) OA = OB = OC = OD.
8. AOC, BOD are two rods that are pivoted at their centre point O. Name the quadrilateral ABCD if:
   (a) AC = BD
   (b) $A\hat{O}B = 90°$
   (c) AO = OB and $A\hat{O}B = 90°$.
9. ABCD is a square, BCXY is a parallelogram drawn on the side BC outside the square and in the same plane. For each of the following statements, state whether:
   (A) it must be true
   (B) it may be true
   (C) it must be false
   (a) BCXY is a rectangle
   (b) ADXY is a parallelogram
   (c) ADXY is a square
   (d) ADXY is a rectangle
   (e) ABY is equilateral.
10. PQRS is a square, RSTU is a rhombus with $R\hat{S}T \neq 90°$. The figures are cut out of card and hinged along the common side RS. For each of the following statements, state whether:
   (A) it must be true
   (B) it may be true
   (C) it must be false
   (a) PST is an isosceles triangle
   (b) PST is an equilateral triangle
   (c) PQUT is a rectangle
   (d) PQUT is a parallelogram
   (e) PQUT is a rhombus.

Symmetry

Plane shapes may possess two types of symmetry.

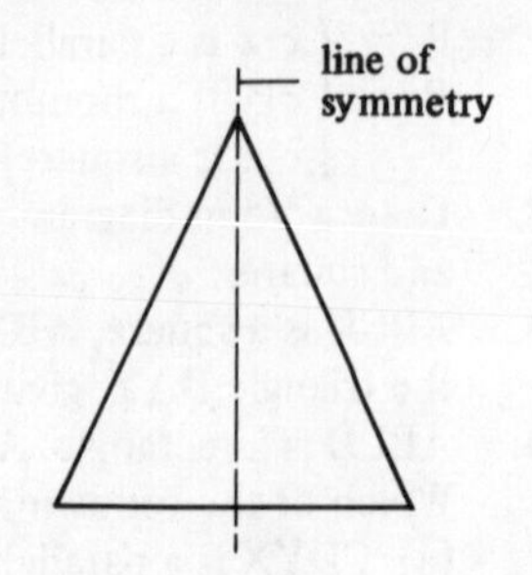

Line symmetry. The isosceles triangle shown has a line of symmetry, shown in the diagram as a broken line. The shape on each side of the line is a mirror image of the shape on the other side.

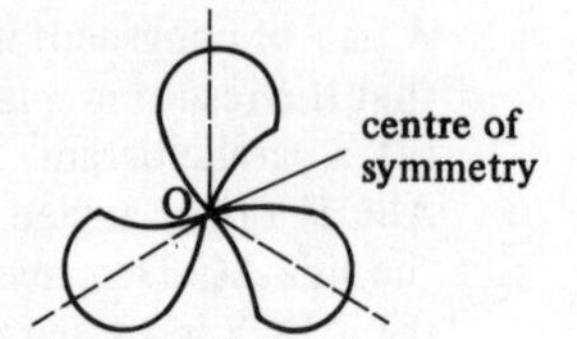

Rotational symmetry. The shape on the right has rotational symmetry. It may be turned about O, its centre of symmetry, into any one of three identical positions. The shape has rotational symmetry of order 3.

If a figure possesses more than one line of symmetry, then the lines of symmetry intersect at the same point which is also a centre of rotational symmetry for the figure.

Exercise 7.1d

1. Draw a sketch of a regular pentagon to show that it possesses five lines of symmetry and rotational symmetry of order 5. Draw a second sketch to show a figure which possesses only rotational symmetry of order 5.
2. Draw a sketch of a regular hexagon and show its lines of symmetry. If a regular polygon of n sides has every line of symmetry passing through a vertex, what can you say about n?
3. AOB is a line of symmetry of a figure. OX is a straight line. Draw a sketch to show lines AOB, OX, and any other lines that the figure must have. If OX lies along another line of symmetry and the figure only possesses two lines of symmetry, show that $X\hat{O}B = 90°$.
4. Two lines of symmetry of a certain figure are inclined at 60° to each other. Show that there must be at least one further line of symmetry.
5. State which of the following statements are true (T) and which are false (F):
   (a) A parallelogram has rotational symmetry of order 2.
   (b) A rhombus has rotational symmetry of order 4.
   (c) A parallelogram which has a line of symmetry is a rectangle.
   (d) A rhombus which has a line of symmetry is a square.
   (e) A square is a rhombus with rotational symmetry of order 4.

6. State which of the following statements are true (T) and which are false (F):
   (a) An isosceles triangle with more than one line of symmetry is an equilateral triangle.
   (b) A right-angled triangle can have at most one line of symmetry.
   (c) A right-angled triangle with more than one line of symmetry is an equilateral triangle.
   (d) A pentagon which has more than one line of symmetry must be regular.
   (e) A hexagon can be drawn with only one line of symmetry.
7. A regular tessellation is constructed of equal squares and equal regular octagons. Draw a sketch of part of the tessellation and show that there exist centres of rotation for the tessellation other than the centres of the squares and octagons.
8. A regular tessellation is constructed of equal equilateral triangles. Draw a sketch of part of the tessellation and show that there exist three different (non-congruent) centres of rotation for the tessellation.

Exercise 7.1e Miscellaneous

1. An equilateral triangle ABC has an isosceles triangle ABX drawn on one side so that X lies inside triangle ABC. Which of the following statements (i) must be true? (ii) may be true? (iii) cannot be true?
   (a) $AB = BC$ (b) $BX < AC$ (c) $BX + XC < BC$
   (d) $AX > BC$ (e) $AX < \frac{1}{2}BC$
2. ABC is an equilateral triangle drawn on the side AB of the square ABXY so that C lies outside the square. Which of the following triangles are isosceles?
   (a) CYX (b) ACY (c) BXY (d) ABC (e) BYC
3. ABCDEF is a regular hexagon. Sides AB, DC are produced to meet at X. Calculate the sum of the interior angles of the pentagon AXDEF.
4. A square has an equilateral triangle drawn on each side. Calculate the sum of the interior angles of the 8-sided star-shaped polygon so formed.
5. In the rectangle ABCD, the diagonals meet at X.
   If $A\hat{D}X = 54°$, find $X\hat{C}B$ and $A\hat{X}B$.
6. In the parallelogram ABCD, $A\hat{B}D = 40°$. Find $B\hat{D}C$.
7. In the rhombus PQRS, $P\hat{Q}S = 70°$. Find $P\hat{Q}R$ and $Q\hat{R}S$.
8. In the rhombus EFGH, the diagonals meet at X and $X\hat{G}H = 24°$.
   Find $F\hat{H}G$, $F\hat{G}H$ and $F\hat{X}G$.
9. Find the angles of a parallelogram in which one angle is four times another.
10. The sizes of the angles of a pentagon are x, x, y, y, y. Show that y cannot be less than $60°$ and find the value of y if $x = 170°$.

7.2 Congruence

Two plane figures are congruent if one fits exactly on the other. All corresponding sides and corresponding angles are equal.

Triangles are congruent if at least three special facts are known. There are four types of congruence:

1. Three sides [S.S.S].

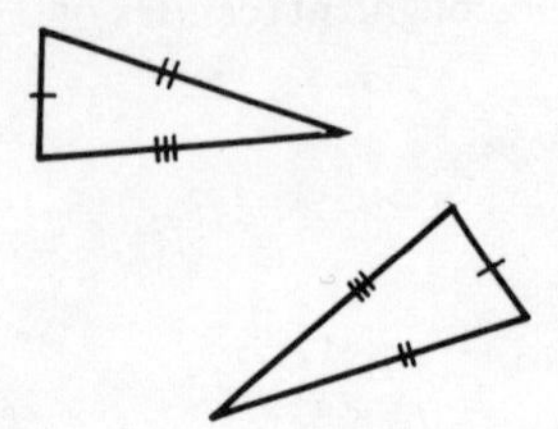

2. Two sides and the angle between them [S.A.S.].

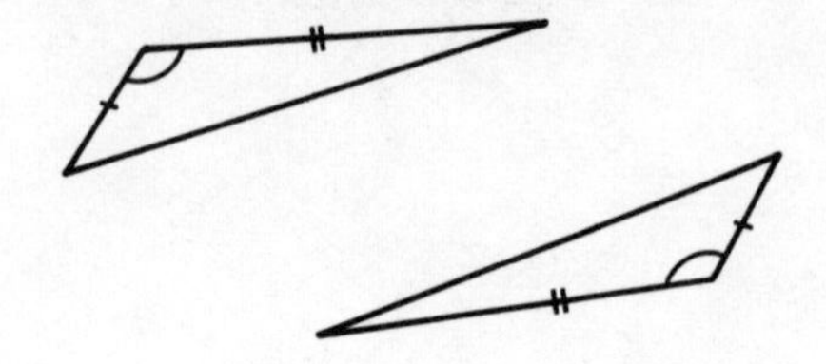

3. Two angles and a corresponding side [A.S.A.].

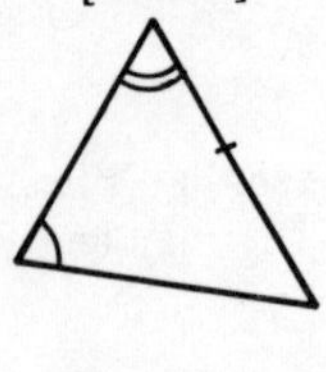

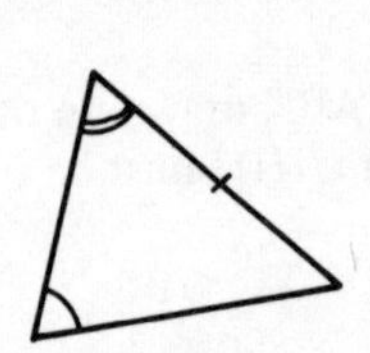

4. (Right-angled triangles only.) Right angle, hypotenuse and one other side [R.H.S.].

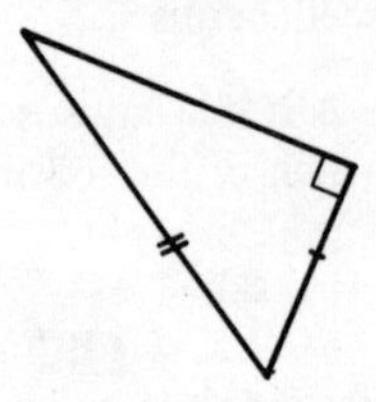

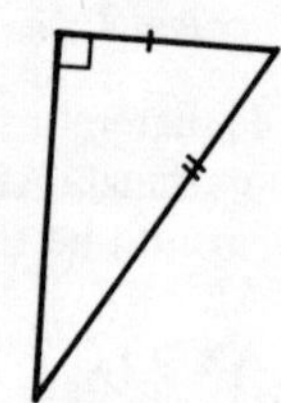

In all questions using congruent triangles, first establish one of these cases of congruence. You can then use the equality of all the other triangle dimensions.

Example 1

In the quadrilateral ABCD,

AB = AD and BC = CD.

Prove that $A\hat{B}C = A\hat{D}C$

Proof. Join AC. In triangles ABC, ADC,

AC = AC common side
AB = AD given
BC = CD given

⇒ triangles ABC, ADC are congruent [S.S.S.]
⇒ $A\hat{B}C = A\hat{D}C$

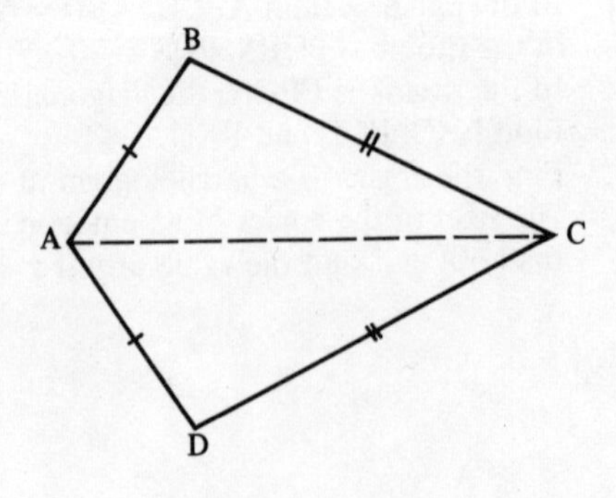

Example 2

ABC is an isosceles triangle with AB = BC. Prove that the perpendicular from A to BC bisects BC.

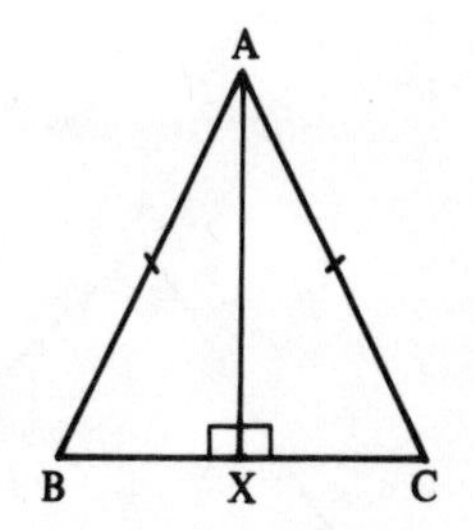

Proof Let the perpendicular from A to BC meet BC at X.

AX = AX	common side
AB = AC	ABC isosceles
$A\hat{X}B = A\hat{X}C$	given

⇒ triangles ABX, ACX are congruent [R.H.S.].

⇒ BX = XC; i.e. X bisects BC.

Exercise 7.2

In each question, first prove the congruency of a pair of triangles.

1. ABC is an isosceles triangle with AB = AC. M is the mid-point of BC. Prove that $A\hat{B}C = A\hat{C}B$.
2. ABC is an isosceles triangle with AB = AC. AX is the bisector of angle BAC meeting BC at X. Prove that X is the mid-point of BC.
3. ABCD is a parallelogram. Prove that $A\hat{B}C = A\hat{D}C$. (First join AC and use alternate angles.)
4. ABCD is a parallelogram. AC meets BD at X. Prove that AX = XC and hence that the diagonals of a parallelogram bisect each other.
5. A, B, and C lie on a circle centre O with $A\hat{O}B = B\hat{O}C$. Prove that AB = BC.
6. A, B, and C lie on a circle centre O with AB = BC. Prove that $A\hat{O}B = B\hat{O}C$.
7. Prove that a quadrilateral with equal sides is a parallelogram.
8. Using the definition that a rhombus is a quadrilateral with equal sides and also the results of questions 7 and 4, prove that the diagonals of a rhombus are perpendicular to each other.
9. ABCD is a rectangle and a point X outside the rectangle is fixed so that BX = XC. Prove that AX = DX.
10. ABCD is a quadrilateral in which $A\hat{B}C = A\hat{D}C = 90°$ and AB = DC. Prove that BC = AD.

7.3 Similarity

Similar triangles

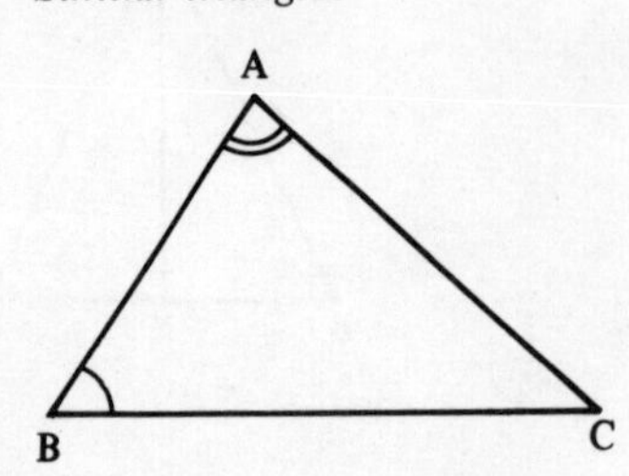

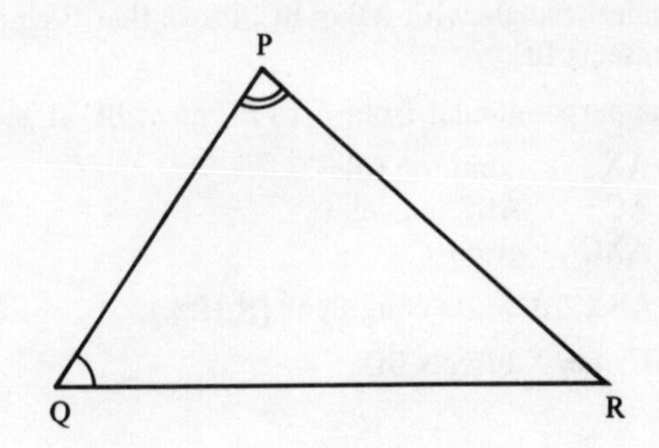

If two triangles are similar then:
- (a) corresponding angles are equal
- (b) the ratio of corresponding sides is the same

In the triangles ABC, PQR above, we have

$$\hat{A} = \hat{P},\ \hat{B} = \hat{Q},\ \hat{C} = \hat{R}$$

and $\dfrac{AB}{PQ} = \dfrac{BC}{QR} = \dfrac{CA}{RP}$

Example 1

The two triangles drawn below are similar with equal angles as marked and lengths in centimetres as shown in the diagram. Find the unknown marked lengths.

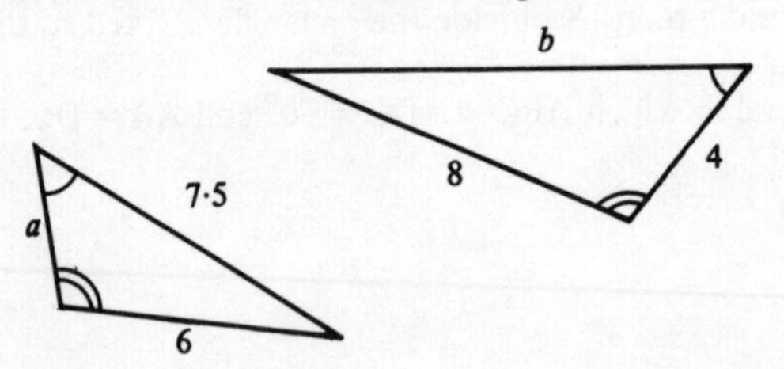

The triangles are similar, so the corresponding sides are in the same ratio.

$$\frac{a}{4} = \frac{6}{8} = \frac{7{\cdot}5}{b}$$

$\Rightarrow \quad a = 3$ and $b = 10$.

Example 2

XY is parallel to BC in the triangle ABC with lengths in centimetres as shown in the diagram.
Prove that triangle AXY is similar to triangle ABC, and calculate the length of BC.

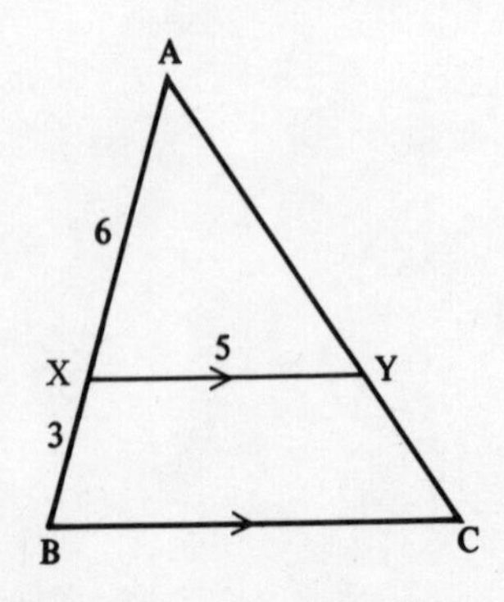

Proof

In triangles AXY, ABC,

$\hat{A}$ is common

$A\hat{X}Y = A\hat{B}C$ (XY is parallel to BC, corresponding angles)

$\Rightarrow$ angles are equal so triangles are similar

$$\Rightarrow \frac{BC}{XY} = \frac{AB}{AX}$$

$$\Rightarrow \frac{BC}{5} = \frac{9}{6}$$

$$\Rightarrow BC = 7{\cdot}5 \text{ cm}$$

Exercise 7.3a

1. Find the unknown marked lengths. Equal angles are marked. The lengths are in centimetres.

(a)

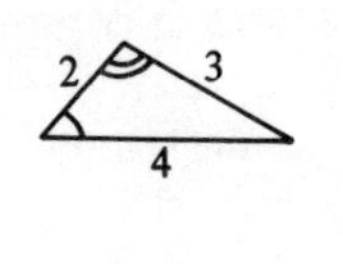

a b 8

(b)

9 10 11

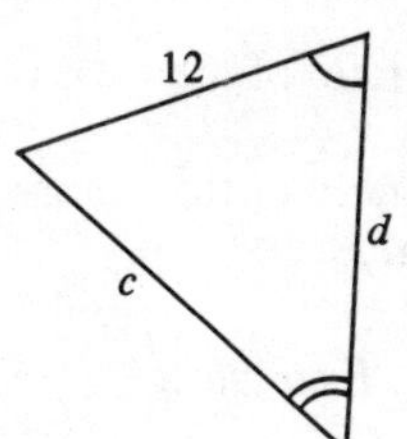

(c)

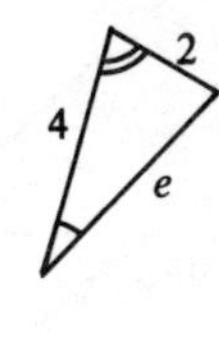

6 6 f

(d)

2 3 h 4 2 g

(e)

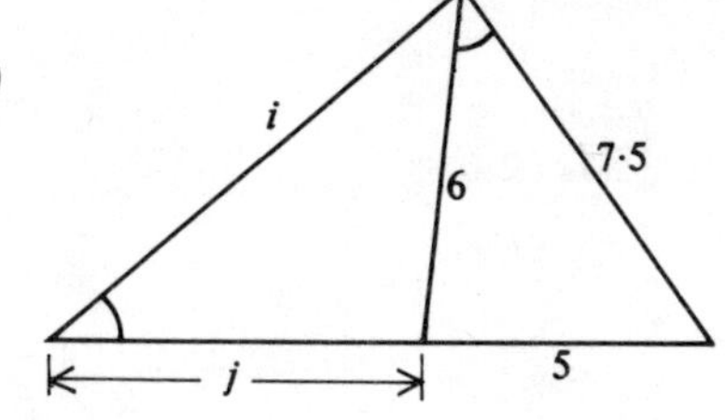

(f)

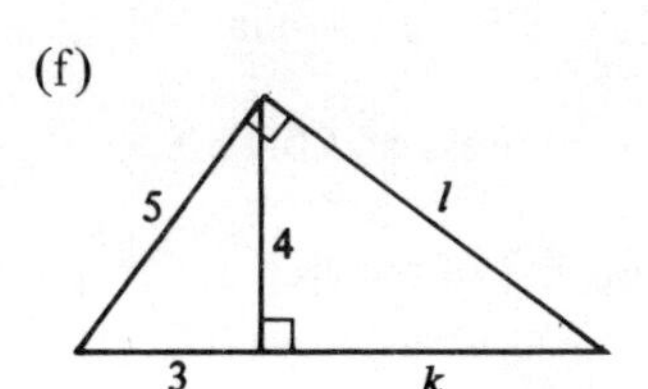

2. Two triangles are similar. The sides of one are 4, 6 and 7 cm. The shortest side of the other is 10 cm. Calculate the lengths of each of the other two sides.
3. PQRS, XYZW are similar quadrilaterals with corresponding equal angles in that order. PQ = 5 cm, RS = 10 cm, SP = 4 cm, XY = 3 cm and YZ = 5 cm. Calculate the lengths of the other sides.
4. From the ground a man sights the top of a tree in line with the top of a vertical pole. The height of the pole is 2 m. It is 3 m from the man and 7 m from the tree. Calculate the height of the tree.

5. In triangle ABC, X lies on BC, Y lies on AC such that XY is parallel to BA; AB = 10 cm, XY = 8 cm and CY = 10 cm. Prove that triangles CXY, CBA are similar and calculate the length AY.
6. ABC is a right-angled triangle with $\hat{B} = 90°$. BN is the perpendicular from B to AC meeting AC at N. Prove that triangles ANB, BNC are similar and hence that $AN.NC = BN^2$.
7. PQRS is a parallelogram. X is a point on RS such that RX:XS = 1:2. Prove that PR cuts QX in the ratio 3:1.
8. ABCD is a parallelogram with AB = 9 cm and AD = 6 cm. E lies on BC and BE = 2 cm. DE meets AB produced at F. Calculate the length BF.
9. XYZW is a rectangle with XY = 10 cm, YZ = 8 cm. R lies on XY and S lies on YZ with XR = 3 cm, YS = 2 cm. WS meets ZR at T. ZR meets WX produced at U. Calculate the length XU and the ratio WT:TS.

Areas and volumes

The ratio of the areas of two similar plane figures is equal to the *square* of the ratio of corresponding sides.
The ratio of the volumes of two similar solids is equal to the *cube* of the ratio of corresponding lengths.

Example 3

In triangle ABC, D lies on BC and $B\hat{A}D = A\hat{C}D$. AB = 6 cm, BC = 9 cm. Calculate the area of triangle ABD as a fraction of the area of triangle ABC.

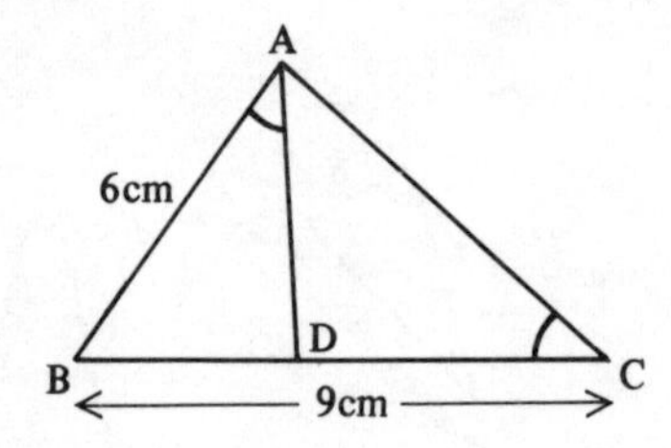

In triangles ABD, CBA

$B\hat{A}D = A\hat{C}B$ given

$A\hat{B}D = C\hat{B}A$ common

⇒ Triangles ABD, CBA are similar.

The ratio of the areas of ABD:CBA $= \left(\frac{AB}{BC}\right)^2 = \left(\frac{6}{9}\right)^2 = \frac{4}{9}$

⇒ Triangle ABD is $\frac{4}{9}$ of the area of triangle ABC.

Exercise 7.3b

1. In triangle ABC, a line parallel to BC cuts AB and AC at X and Y respectively. AX = 4 cm, XB = 3 cm. Calculate the ratio of the areas of triangles AXY and ABC.
2. ABC is a right-angled triangle with AB = 3 cm, BC = 5cm, CA = 4 cm. AX is the perpendicular from A to the line BC at X. Calculate the ratio of the areas of the triangles ABX to AXC.
3. In the diagram, XY, TU, WZ are parallel lines. The lengths of some lines are given in cm. Calculate the lengths of XY and WZ and the ratio of the areas of the triangles TYX and TWZ.

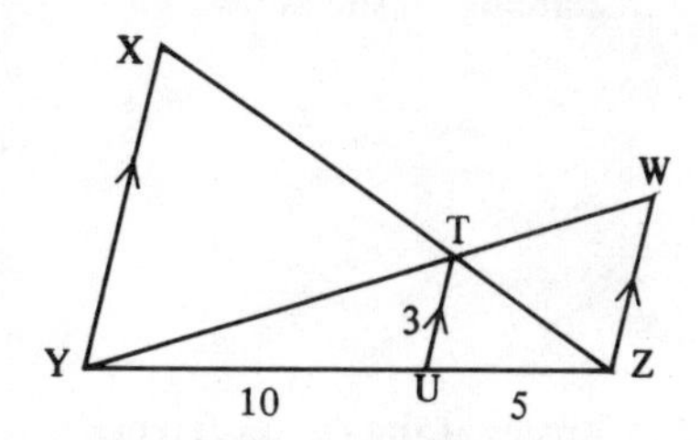

4. In triangle ABC, a line parallel to BC cuts AB at X and AC at Y such that the area of the trapezium BXYC is $\frac{7}{16}$ the area of triangle ABC. Calculate the ratio of the lengths AX to XB.
5. Two similar trays are made of lengths 50 and 70 cm. The width of the smaller tray is 40 cm. Calculate the width of the longer tray and the ratio of the areas of the trays.
6. Two scale models of the same house are produced. One is 7 cm long and 4·9 cm tall. The other is 9·1 cm long. Calculate its height.
7. A scale model is made of a house on a scale 1 : 20. The height of the house is 7 m.
   (a) Find the height of the model.
   (b) The length of the model is 60 cm. Find the length of the house.
   (c) Find the area of the dining room floor of the model if the dining room of the house has a floor area of 16 m^2.
   (d) Find the number of stairs in the house if there are 20 in the model.
   (e) Find the capacity of the oil tank of the model if the house oil tank has a capacity of 1200 litres.
8. A model dinghy is made $\frac{1}{25}$ the size of an actual dinghy. Find the details of the model (in suitable units) that correspond to the following specification of the boat.
   (a) Length 4 m
   (b) Maximum height 6 m
   (c) Deck area 5 m^2
   (d) Number of sails 2
   (e) Stowage capacity 0·5 m^3.

7.4 Angles in circles

The diagrams on the right illustrate the results about angles in circles. They are:

1. The angle at the centre is twice the angle at the circumference standing on the same arc.

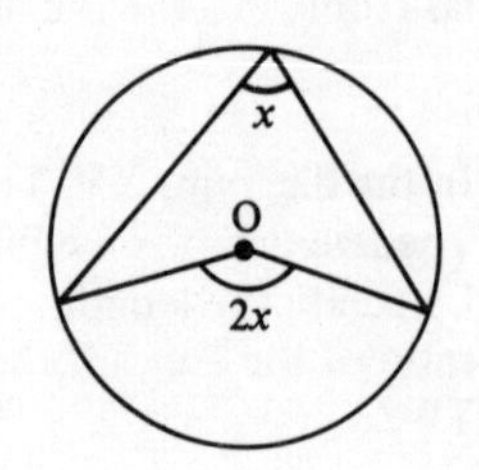

2. Angles at the circumference standing on the same arc (or equal arcs) are equal.

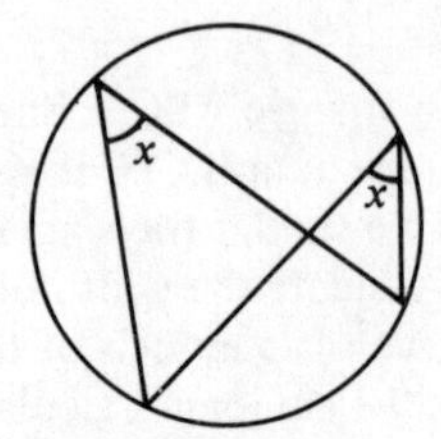

3. Angles in semi-circles are right angles.

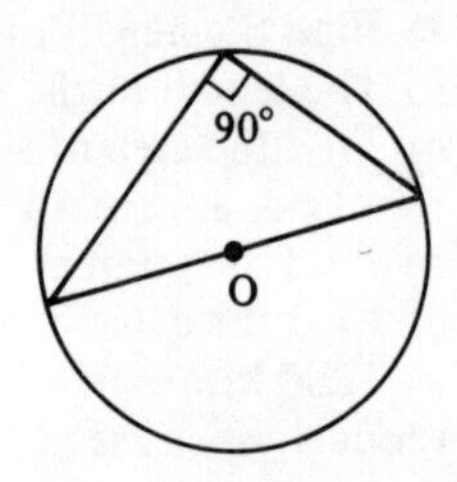

4. Opposite angles of a cyclic quadrilateral are supplementary (add up to 180°).

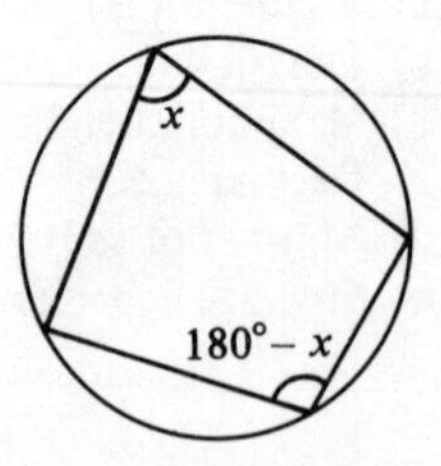

5. The exterior angle of a cyclic quadrilateral equals the interior opposite angle.

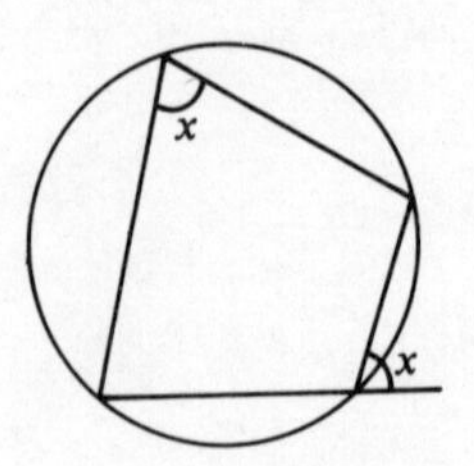

The *converse* of each result is also true.

For example, if opposite angles of a quadrilateral add up to 180°, the quadrilateral is cyclic.

Exercise 7.4a

1. Find the sizes of the angles marked by letters. O is the centre of the circle.

(a)

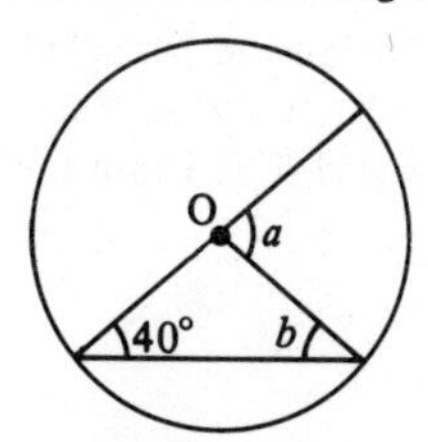

(b)

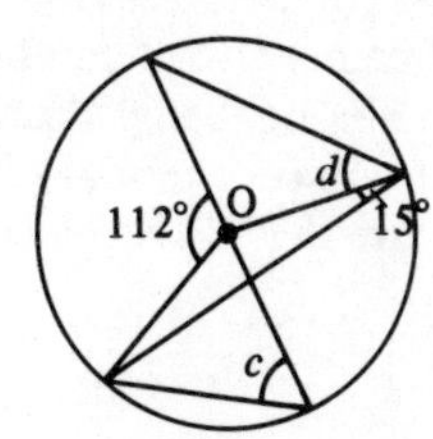

(c)

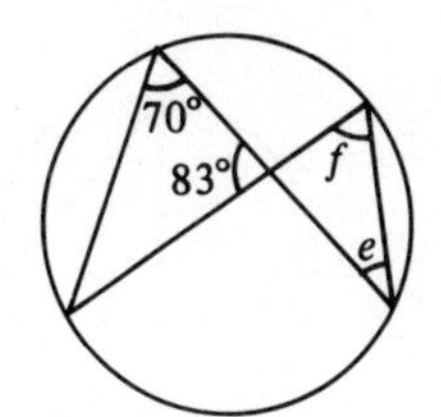

(d)

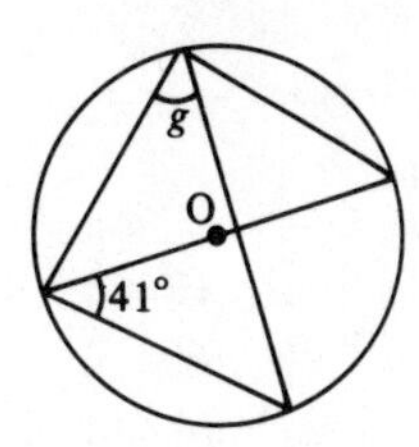

2. Find the sizes of the angles marked by letters. O is the centre of the circle.

(a)

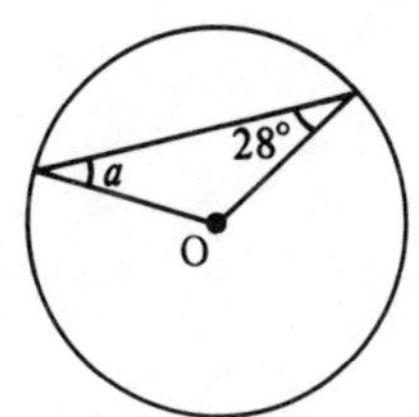

(b)

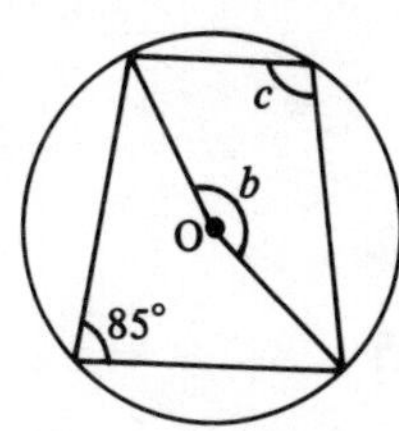

(c)

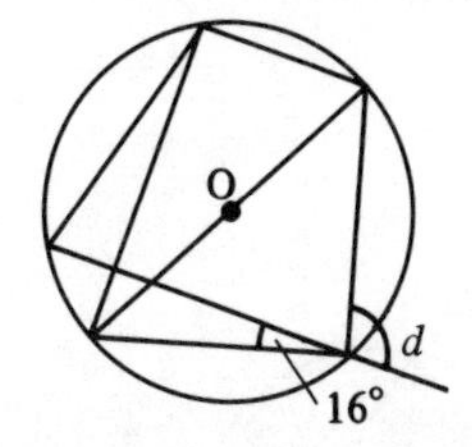

(d)

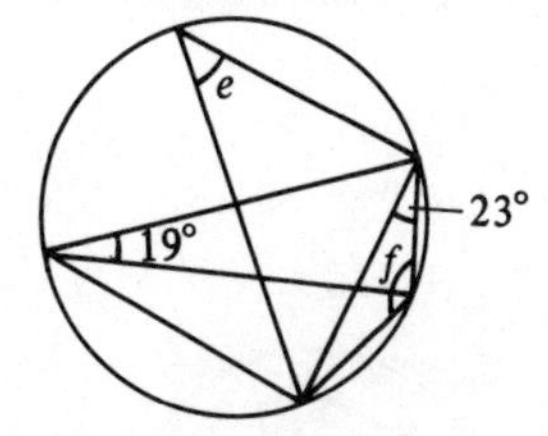

3. A, B, C, D are four points on the circumference of a circle. $A\hat{B}D = 41°$ and AB is parallel to DC. Find $A\hat{C}D$, $B\hat{D}C$, and $B\hat{A}C$.
4. AOB is a diameter of a circle. X and Y are points on the circumference on opposite sides of AOB. AX = XB. Find the sizes of the angles:
   (a) $A\hat{X}B$ (b) $X\hat{A}B$ (c) $X\hat{Y}B$
5. A, B, C, and D are four points on the circumference of a circle. AC meets DB at X. Prove that triangles AXB and DXC are similar.
6. A, B, C, and D are four points on the circumference of a circle. $A\hat{B}D = B\hat{A}C$. Prove that AB is parallel to DC.
7. ABCD is a quadrilateral. Show that it is cyclic if $A\hat{B}D = A\hat{C}D$.
8. ABCD is a cyclic quadrilateral. X lies in the same plane such that $D\hat{X}B = D\hat{A}B$. Prove that $D\hat{X}B + D\hat{C}B = 180°$.
9. The perpendiculars from A and B to the opposite sides of the triangle ABC meet them at M and N respectively. X is the mid-point of AB.
   (a) Prove that $B\hat{A}M = B\hat{N}M$
   (b) Prove that XN = XM.

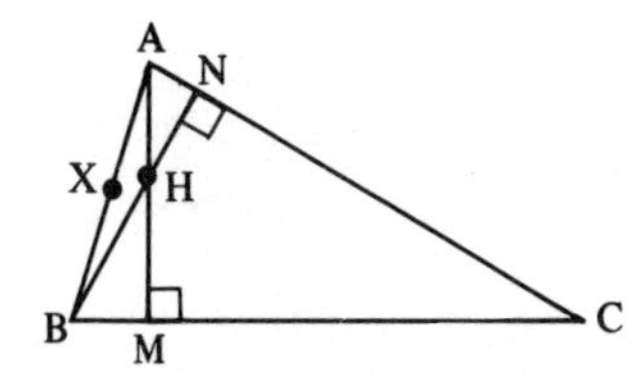

10. ABCD is a quadrilateral. X lies on CD such $X\hat{A}B + B\hat{C}D = 180°$. Prove that $A\hat{X}D = A\hat{B}C$.
11. Prove that a cyclic trapezium has an axis of symmetry.
12. Prove that a cyclic parallelogram is a rectangle.
13. Prove that the circumcentre of an obtuse-angled triangle lies outside the triangle. Where is the circumcentre of a right-angled triangle?
14. ABCDX is a pentagon (not regular). Y is a point on BC. ABYX and XYCD are both cyclic quadrilaterals. Prove that $B\hat{A}X + C\hat{D}X = 180°$.

Alternate segment

AB is a chord of a circle.
SAT is the tangent at A.
P is any point on the major arc AB.
Angle BAT and angle APB are equal.

The angle BAT between the chord and the tangent equals the angle APB in the alternate segment.

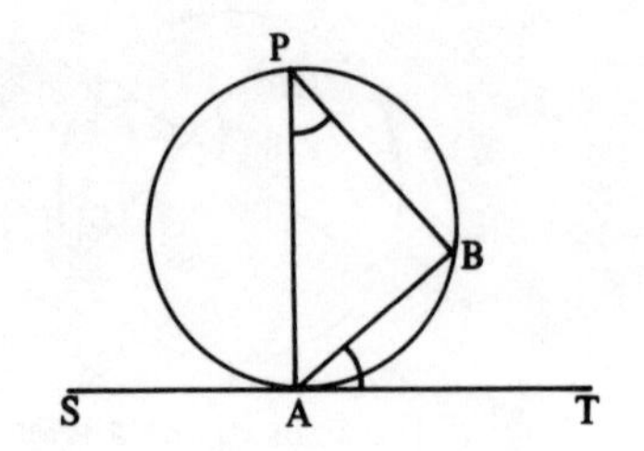

Exercise 7.4b

1. Find the sizes of the angles marked by letters. T and S are points where the tangents touch the circle.

(a)

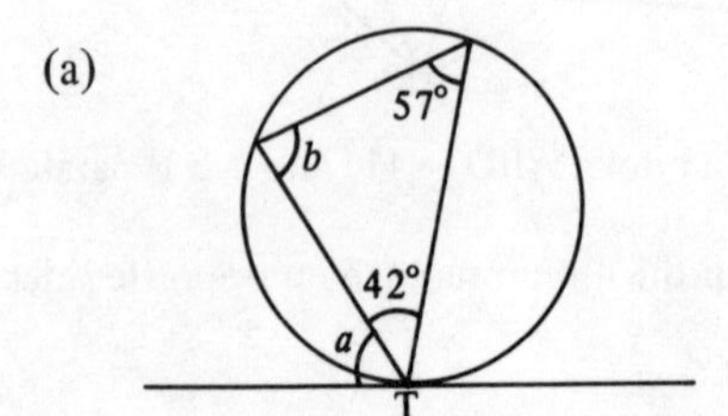

(b)

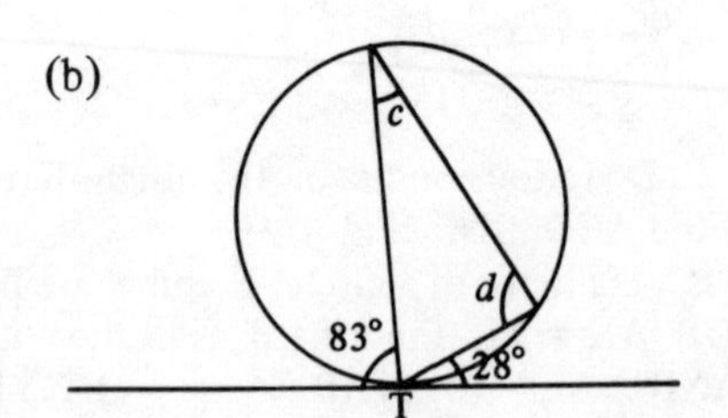

(c)

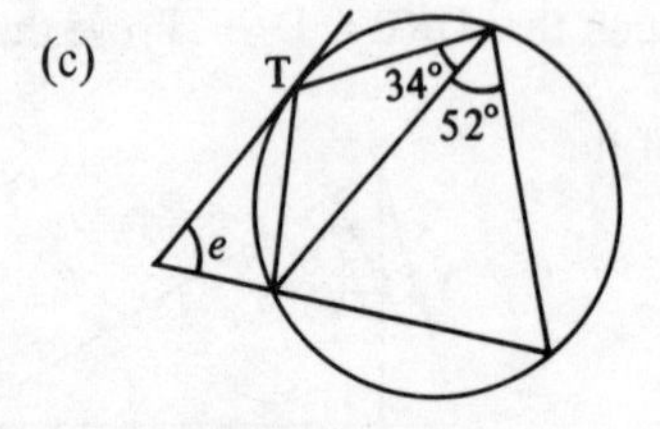

(d)

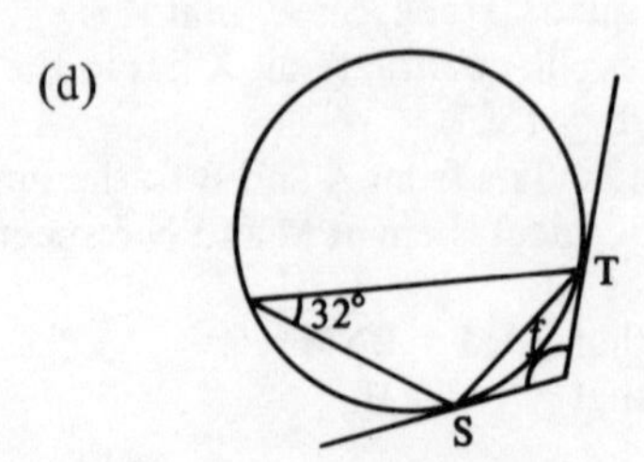

2. Find the sizes of the angles marked by letters. T and S are points where the tangents touch the circles.

(a)

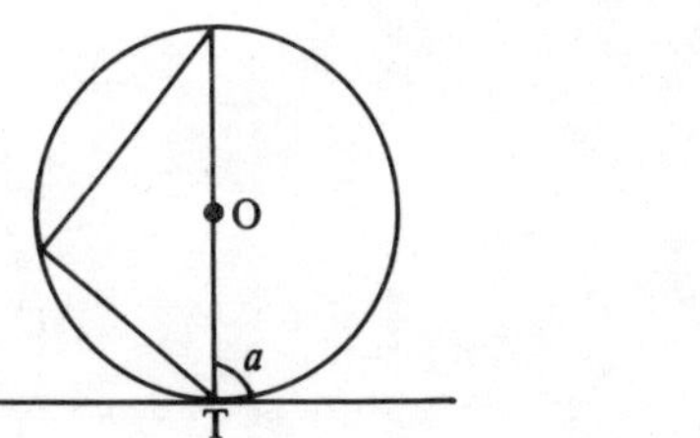

(b)

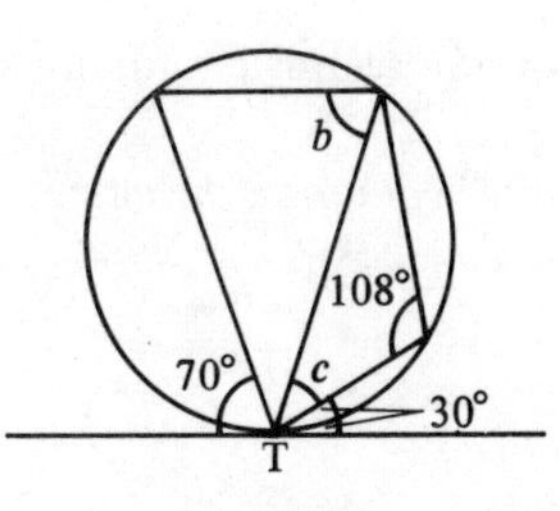

(c)

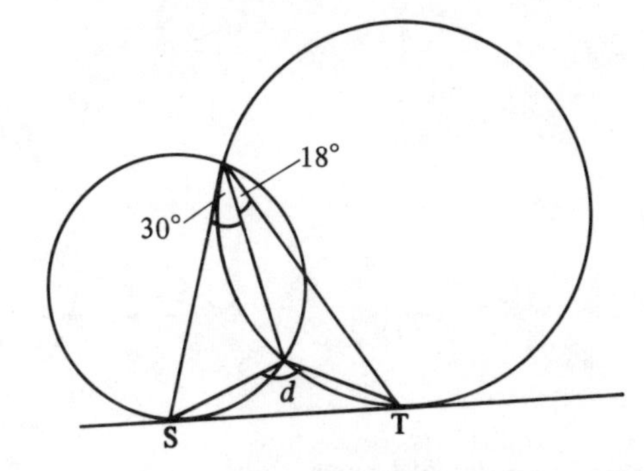

3. ATB is a tangent to a circle at T. Points X, Y lie on the circle such that $X\hat{T}A = 27°$ and $Y\hat{T}B = 72°$.
Calculate the sizes of the angles of the triangle XYT.
4. The angles of a triangle ABC are $\hat{A} = 17°$, $\hat{B} = 64°$, $\hat{C} = 99°$. Find the sizes of the angles that AB and CB make with the tangent to the circumcircle at B.
5. ABCDE is a regular pentagon inscribed in a circle. XAY is a tangent to the circle. Give reasons for the following angle facts:
(a) $D\hat{A}C = C\hat{A}B$ (b) $C\hat{A}B = B\hat{C}A$ (c) $B\hat{C}A = B\hat{A}X$
Hence, by symmetry or otherwise, find the sizes of $B\hat{A}X$ and $E\hat{A}Y$.
6. Prove that if ABCD . . . is a polygon of n sides, the tangent to the circumscribed circle at A makes an angle of $(180/n)°$ with the chord AB.
7. Two circles touch each other externally at T. ATX, BTY are straight lines such that A, B lie on one circle and X, Y on the other. Prove that AB is parallel to XY.
Hint: draw in the common tangent.
8. PQRS is a trapezium with PQ parallel to SR.
RQ and SP meet at X.
Prove that the circumcircles of triangles PQX and SRX have a common tangent at X.

7.5 Chords and tangents

The results for intersecting chords are as follows.

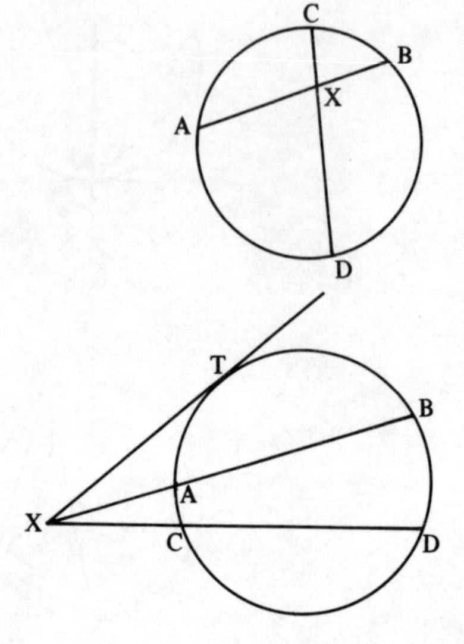

Chords that intersect inside the circle

$$AX.XB = CX.XD$$

Chords that intersect outside the circle

$$AX.XB = CX.XD = XT^2$$

or $$XA.XB = XC.XD = XT^2$$

Note: the distances are always taken from the point of intersection.

Example 1

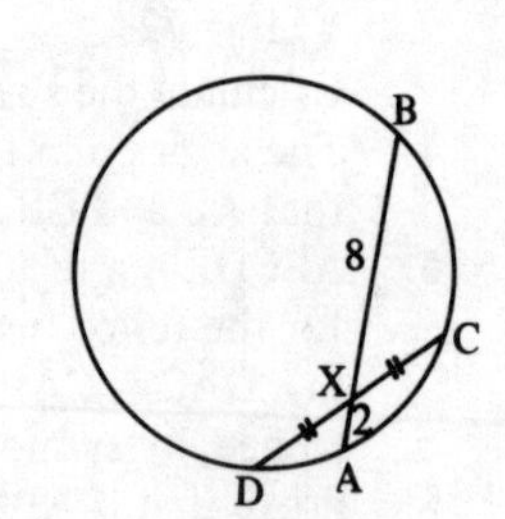

Chord AB of a circle bisects chord CD at X.
$AX = 2$ cm; $XB = 8$ cm. Find the length of CD.

$$CX.XD = AX.XB$$

But $CX = XD$;

so $$CX^2 = 2.8 = 16$$
$$CX = 4$$

Hence $CD = 8$ cm.

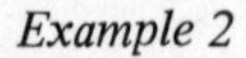

Example 2

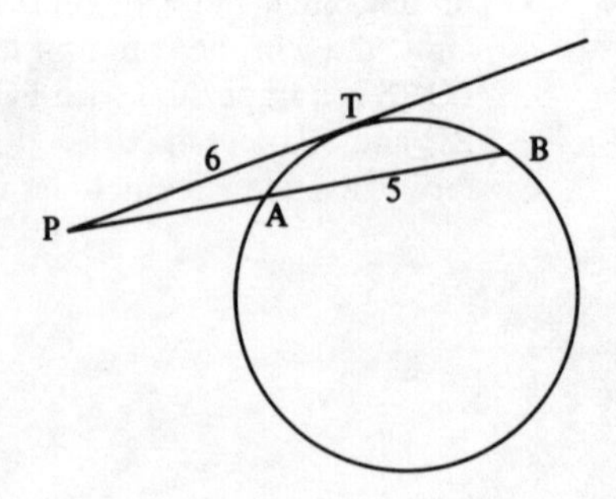

From a point P, a tangent PT is drawn to a circle and a line PAB cuts the circle at A and B.
$PT = 6$ cm; $AB = 5$ cm. Find the length of PA.

Let $$PA = x \text{ cm}$$
$$PA.PB = PT^2$$
$$x(x+5) = 36$$
$$x^2 + 5x - 36 = 0$$
$$(x+9)(x-4) = 0 \quad \Rightarrow x = 4$$

Hence $PA = 4$ cm.

Exercise 7.5

1. Find the lengths marked by letters in the diagrams (not drawn to scale). The lengths may be taken as centimetres.

(a)

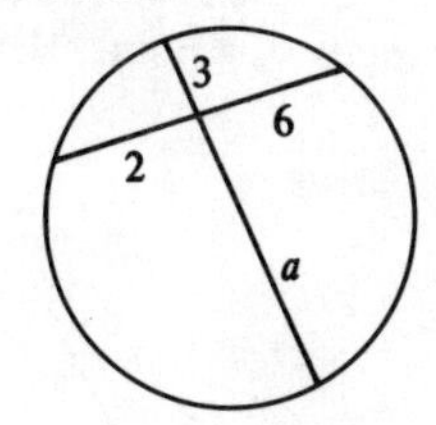

(b)

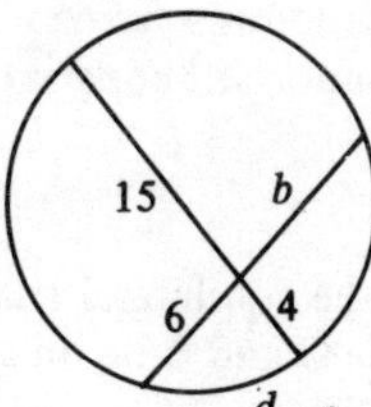

(c)

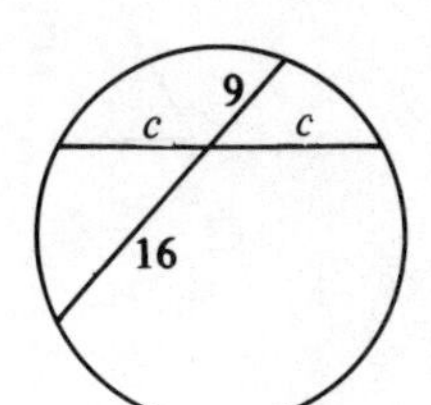

(d)

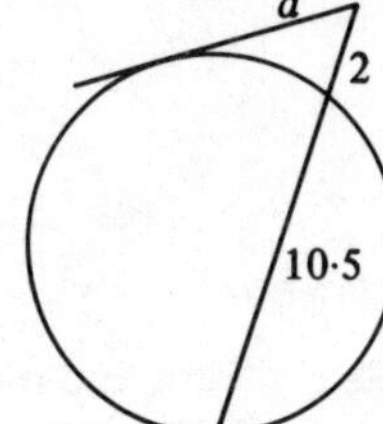

(e)

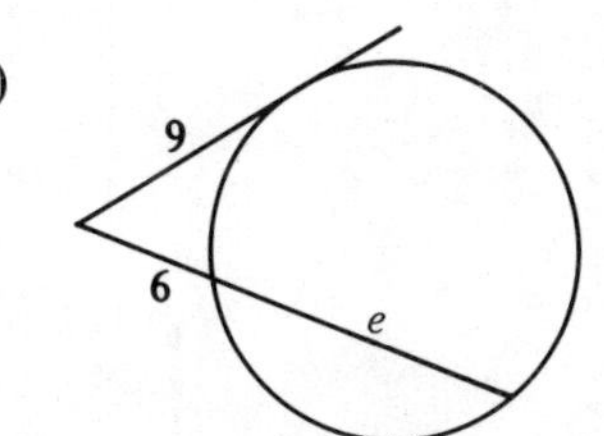

(f)

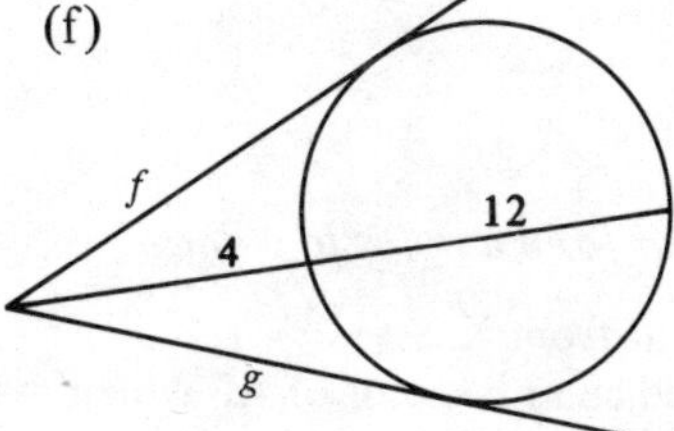

2. Chords AB, DC meet at X outside the circle.
   (a) XA = 10; XB = 4; XC = 3; find XD.
   (b) AB = 5; BX = 4; XD = 12; find CD.
   (c) CD = 7; XC = 3; XB = 5; find AB.

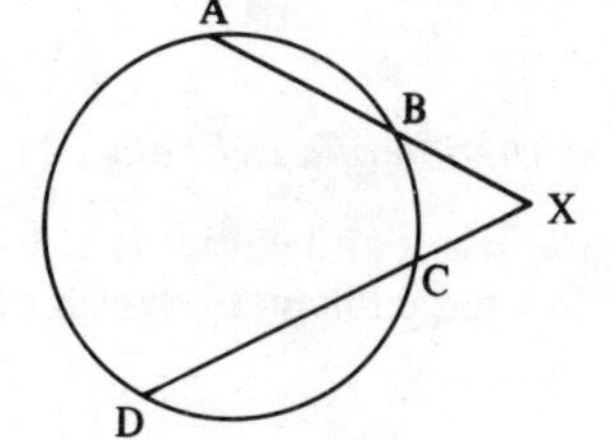

3. Using the same diagram as in question 2
   (a) AB = 13; BX = 3; XC = 4; find CD.
   (b) XB = BA = 6; XD = 18; find CD.
   (c) AB = 11; CD = 1; XB = 3; find XC.

4. The common chord AB of two circles meets a common tangent at C. The common tangent touches the circles at T and S.
   (a) Prove that CT = CS.
   (b) If CT = 8 cm and AB = 12 cm, find the length BC.

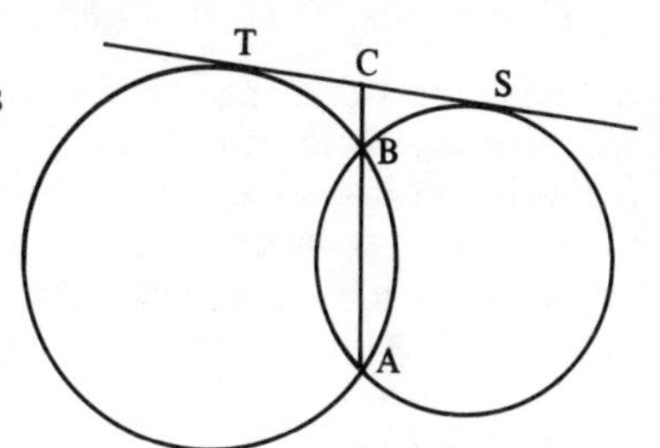

5. A, B, C lie on a circle. The tangent at C meets AB produced at X. XD is the tangent to any other circle through AB touching it at D. Prove that XD = XC.

7.6 Constructions

Geometrical constructions involving ruler and compasses only can be used to produce accurate drawings. The constructions are connected with the idea of the *locus* of a point. Here, to remind you, are the most important constructions.

Angle of 60°

This is based upon the equilateral triangle. A and B lie on a circle centre O while O and B lie on an equal circle centre A.

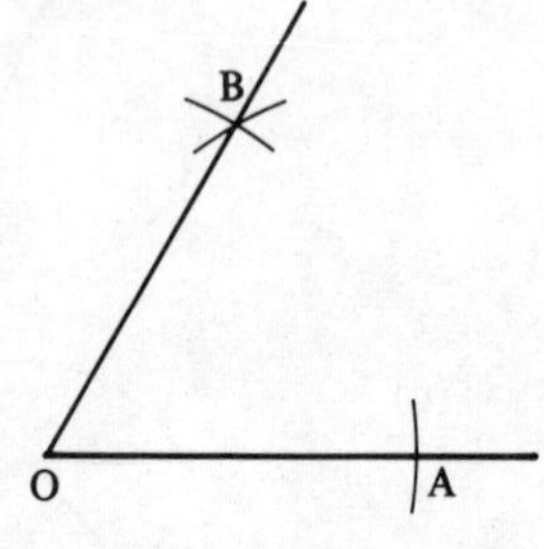

Perpendicular bisector

P and Q lie on a circle centre A and on an equal circle centre B.
PQ is the set of points equidistant from A and B.

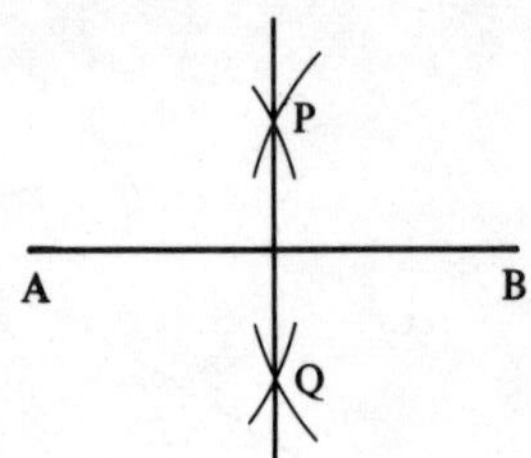

Perpendicular from a point to a line

A, B are equidistant from X.
P is on the perpendicular bisector of AB which must pass through X.
XN is the perpendicular and the shortest distance from X to the line.

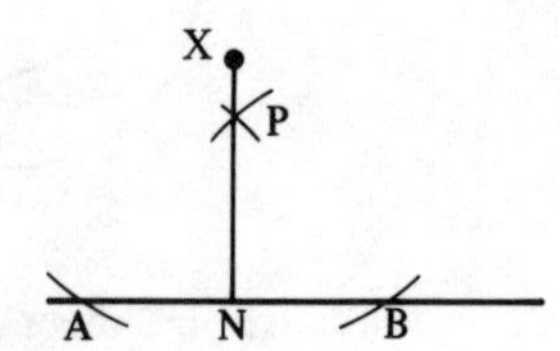

Construction of a right angle (1)

A and B are first found equidistant from O.
X lies on the perpendicular bisector of AB.

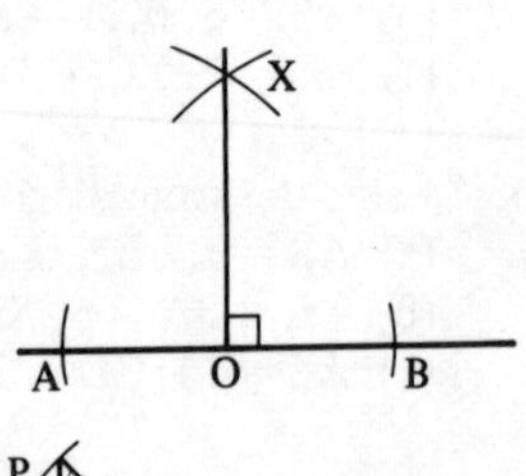

Construction of a right angle (2)

Based on the fact that angles in a semi-circle are right angles.
A is chosen anywhere on the line.
B is equidistant from O and A and is the centre of the semi-circle AOP; ABP is the diameter.
This construction is sometimes preferred when the right angle is at the end of a line.

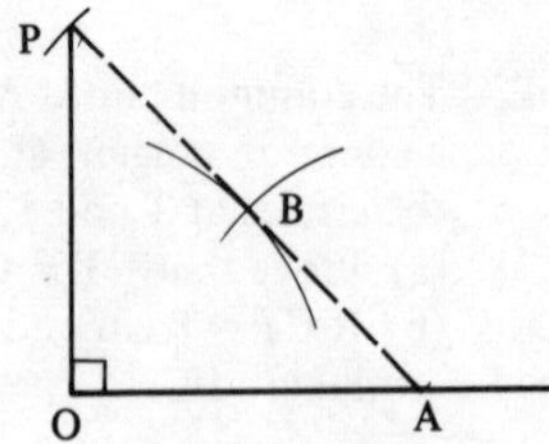

Angle bisector

P lies on the set of points equidistant from the two 'arms' of the angle.

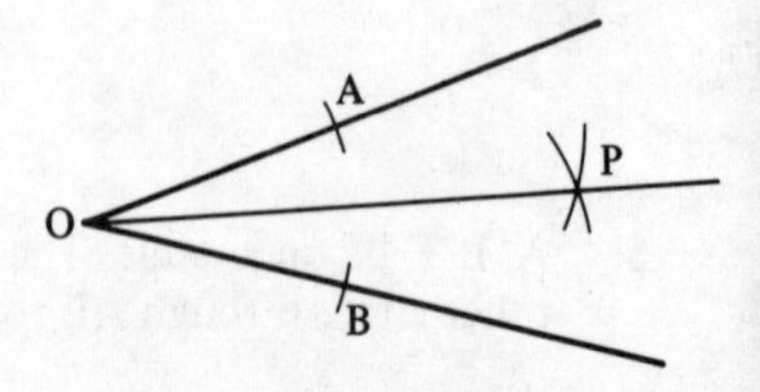

Exercise 7.6

In these questions, use ruler and compasses only.

1. Construct an equilateral triangle of side 6 cm.
2. Construct a right-angled triangle whose shorter sides are 4·5 cm and 5·5 cm.
3. Construct a triangle ABC with AB = 8 cm, BC = 6 cm, AC = 9 cm. Construct the perpendicular from B to meet AC at X. Measure the length BX.
4. A post box and a telephone box are 500 m apart on a straight country road. A farmhouse is 200 m from the post box and 400 m from the telephone box. Using a construction find the shortest distance from the farmhouse to the road.
5. Construct a regular hexagon ABCDEF of side 5 cm. Measure the distance AC.
6. Construct a triangle ABC with AB = 5 cm, $\hat{A} = 90°$, $\hat{B} = 60°$. Measure the length BC.
7. Construct a triangle ABC with dimensions as in question 6. Construct the bisector of angle B to meet AC at X. Measure the length AX.
8. Construct a triangle ABC with AB = 8 cm, BC = 9 cm, CA = 10 cm. Construct the perpendicular bisector of the side CA to meet BC at X. Measure BX.
9. ABC is a triangle with AB = 6·5 cm, BC = 5 cm, CA = 6 cm. Construct the triangle and the perpendicular bisectors of each side. Show that the three perpendiculars meet at a single point X. Draw the circle, with centre X, which passes through A, B and C.
10. Using the same dimensions as in question 9, draw the triangle ABC and construct the bisectors of each angle. Show that these bisectors meet at a single point.
11. Use the constructions for 90°, 60°, and bisecting an angle to construct an angle of:
    (a) 45° (b) 30° (c) 150° (d) 15°.
12. Construct a triangle ABC with AB = 8 cm, $\hat{A} = 60°$, $\hat{B} = 45°$. Measure the lengths BC and CA.

7.7 Locus

The position of a point P is defined in some way. The set of all possible positions that P may take is called the *locus* of P. This will often be a straight line or curved line. But it may be just a single point (or two) or it could be an area.

Problems involving the idea of locus call upon a general knowledge of geometrical facts.

Example 1

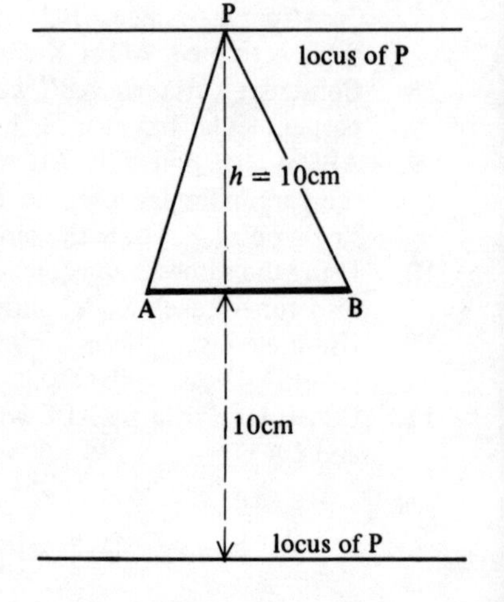

In triangle PAB, AB = 8 cm and the area of the triangle is 40 cm^2. Find the locus of P.

Let the perpendicular height of P above AB be h cm.
Then area of triangle $= \frac{1}{2}.8.h = 40$
$\Rightarrow h = 10.$

The locus of P is the set of points distant 10 cm from AB i.e. a pair of parallel lines.

Example 2

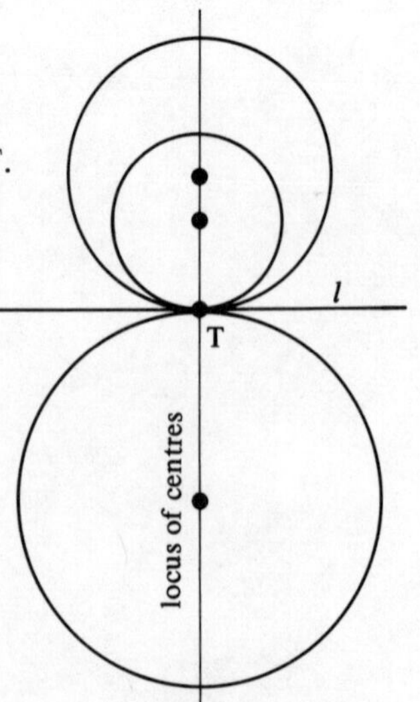

Find the locus of the centre of circles that touch a given line at a given point.

Let the line be l and the point where the circles touch the line be T.

l is a tangent to the circles. The line from the centre of each circle to T is perpendicular to l.

The locus is a line through T, perpendicular to l.

Exercise 7.7

1. Draw AB = 8 cm and construct the locus of P such that PA = PB.
2. Sketch the locus of a point P such that its distance from a given line is constant.
3. Draw an angle ABC = 72°. Construct the locus of P such that P is equidistant from BA and BC.
4. AXB and CXD are two straight lines intersecting at X so that angle CXB = 40°. Construct the locus of P such that P is equidistant from both lines (two possible lines).
5. A triangle ABP has an area of 40 cm^2; AB = 10 cm and P is free to move. Sketch the locus of P.
6. A triangle ABP is right-angled at P; AB = 10 cm. P is free to move. Construct the locus of P.
7. X, Y are fixed points and XY = 8 cm; P is a point free to move. Draw a sketch on the same diagram, to show clearly the locus of P when:
   (a) PX = PY (b) PX = 4 cm (c) PY = 8 cm
   Is it possible for P to satisfy both conditions (a) and (c)?
   Is it possible for P to satisfy both conditions (a) and (b)?
   Is it possible for P to satisfy all three conditions (a), (b) and (c)?
8. X, Y are fixed points and XY = 8 cm.
   Draw a sketch to show three different circles passing through X and Y.
   If P is the centre of a circle passing through X and Y, what is the locus of P?
9. P is a point 1 cm from the circumference of a circle radius 5 cm. Sketch the locus of P (two possible paths).
10. ABCD is a square of side 10 cm. P is a point within the square such that its distance from any corner is greater than 5 cm. Sketch the limit of the locus of P.

For questions **11** *to* **14** *graph paper will be useful.*

11. Ox, Oy are two lines at right angles. The distance of P from Ox is twice its distance from Oy. Draw the locus of P (two lines).
12. P moves so that the sum of its distance from two perpendicular lines is 10 cm. Draw the locus of P (two lines).
13. S is a point 6 cm from a line l. The distance of P from l and S is equal. Draw the locus of P. Hint: draw arcs of circles centre S, radii 3 cm, 4 cm, etc.
14. Using the same information as in question 13, let PS be half the distance of P from l. Construct the locus of P.

7.8 The sphere and earth geometry

Position

The position of a point on the surface of the earth is given by two *angles*, latitude and longitude. If P lies on latitude 60°N and on longitude 25°E, its position is given by (60°N, 25°E).

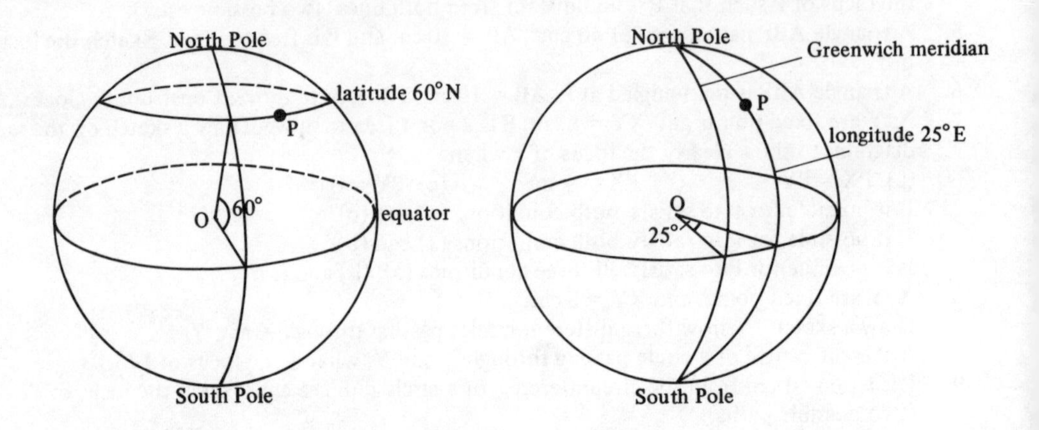

A circle on the surface of the earth with its centre at the centre of the earth is a *great circle.* All lines of longitude and the equator are great circles.

Distance

The shortest distance between two points lies on a great circle. Distances along great circles may be measured in *nautical miles*.

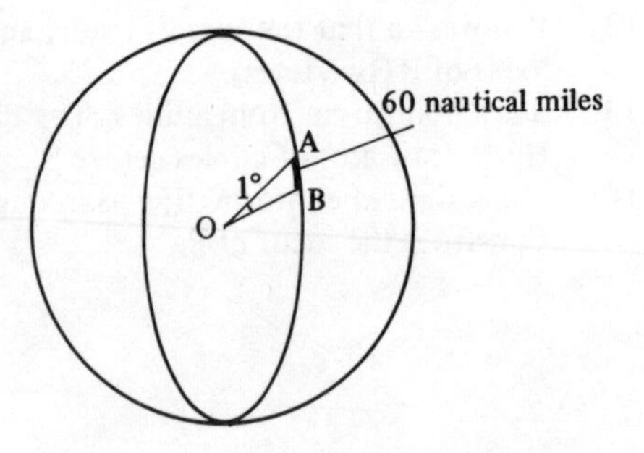

1 nautical mile is the length of arc on a great circle that subtends an angle of 1 minute at the centre.

1 knot is a speed of 1 nautical mile per hour.

60 nautical miles along a great circle subtend an angle of 1 degree at the centre of the earth.

Example 1

Find, in nautical miles, the distance from A (50°N, 27°W) to B (20°S, 27°W).

A, B lie on the great circle longitude 27°W.

Angular distance $= 50° + 20° = 70°$.

Distance $= 70 \times 60 = 4200$ nautical miles.

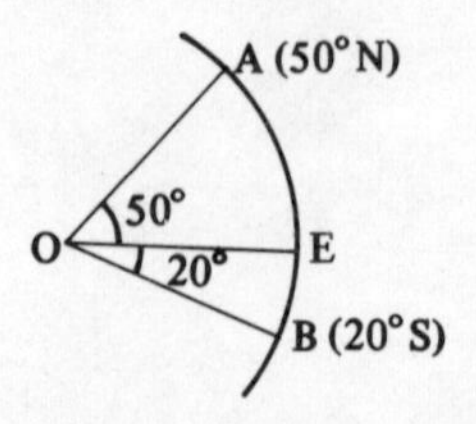

Example 2

Find the distance, in kilometres, from Moscow (56°N, 38°E) to Edinburgh (56°N, 0°W) along the line of latitude 56°N. Take the radius of the earth as 6400 km.

Let X be the centre of the circle of latitude 56°N.

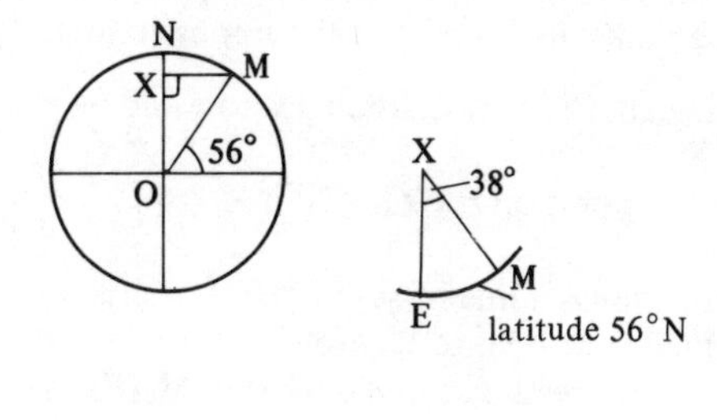

$$XM = OM \cos 56°$$
$$= 6400 \cos 56°$$

$$\text{arc } ME = \frac{38}{360} \times 2\pi \times XM$$
$$= \frac{38}{360} \times 2\pi \times 6400 \cos 56°$$
$$\approx 2370 \text{ km}$$

Exercise 7.8

1. Find the shortest distance, in nautical miles, between:
   (a) (60°N, 10°E) and (45°N, 10°E) (b) (27°N, 15°E) and (27°S, 15°E)
2. Find the shortest distance, in nautical miles, between two points on the equator whose longitudes are 10°W and 17°E.
3. Calculate, in nautical miles:
   (a) the length of the equator (b) the distance from the equator to the North Pole.
4. Find the latitude and longitude of a point:
   (a) 1000 nautical miles due west of (0°, 50°W)
   (b) 2000 nautical miles due north of (10°N, 15°E)
   (c) 3000 nautical miles due south of (10°N, 15°E)
5. The positions of two places A and B are (68°N, 79°E) and (68°N, 101°W) respectively. Find, in nautical miles, the shortest distance from A to B.
6. Find the distance between the places (60°S, 27°W) and (60°S, 153°E):
   (a) along latitude 60°S (b) across the South Pole.
7. Find the speed in knots of a ship that sails from (39°N, 10°W) to (50°N, 10°W) in 55 hours along longitude 10°W.
8. Find the speed in knots of a ship that sails from latitude 20°S to latitude 5°N in 150 hours, assuming it follows the shortest route.

In questions **9** *to* **11** *take the radius of the earth to be 6400 km.*

9. Calculate the radius of the circle of latitude 36°N. Find the distance along latitude 36°N between Malta (36°N, 14°E) and Gibraltar (36°N, 5°W).
10. Calculate the distance along a line of latitude between:
    (a) Cairo (30°N, 31°E) and New Orleans (30°N, 90°W)
    (b) Vancouver (49°N, 123°W) and Volgograd (49°N, 44°E)
    (c) Krakow (50°N, 20°E) and Krasnokamensk (50°N, 118°E).
11. Calculate the lengths in kilometres of a degree of longitude at the equator and a degree of longitude at 60°N. What is the distance in nautical miles of a degree of longitude at 60°N?
12. The radius of the moon is approximately 1700 km. Calculate the distance, along latitude 45°N, between two places on the moon whose positions are (45°N, 60°E) and (45°N, 110°E).

Part 8: Transformation geometry

8.1 Reflections

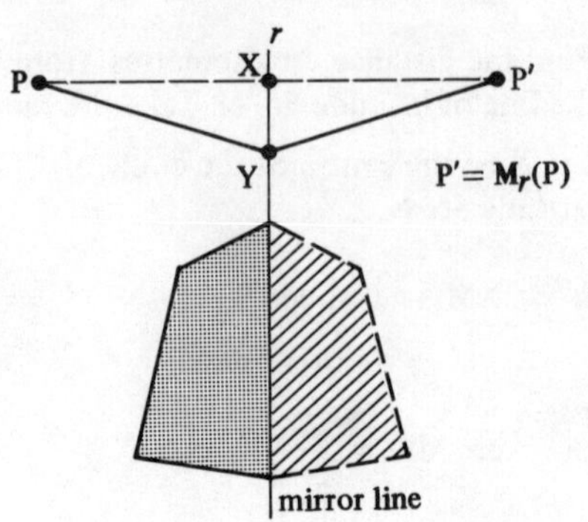

If r is a mirror line, the reflection of a point P is a point P′ equidistant from r on the opposite side.

In the diagram PP′ is at right angles to r and PX = P′X.
Y is another point on r; $P\hat{Y}X = P'\hat{Y}X$.

The mirror line is a line of symmetry for any shape which is its own reflection.
The image of P reflected in r is written $M_r(P)$.

Successive reflections

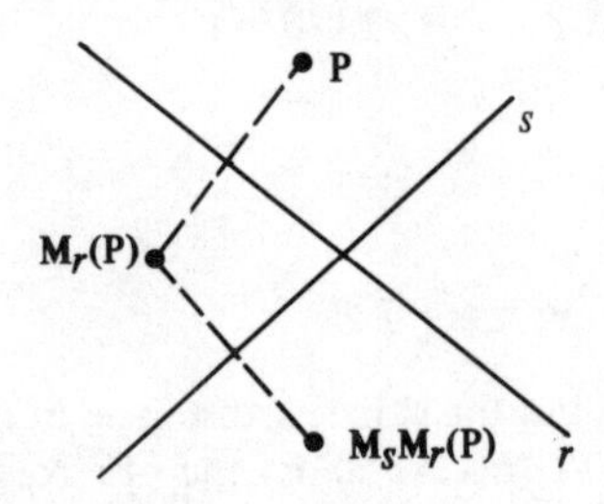

$M_sM_r(P)$ means the image of P reflected first in r then in s.
The order is important.

Exercise 8.1

1. Which of the following shapes can be reflected onto themselves?
   (a) square (b) isosceles triangle (c) parallelogram (d) rhombus
2. An isosceles triangle ABC has AB = BC. What is the quadrilateral produced by ABC and its image reflected in:
   (a) AB? (b) AC?
3. Draw a triangle ABC with AB = 8 cm, BC = 5 cm, CA = 7 cm. Use ruler and compasses to construct the image of ABC reflected in:
   (a) BC (b) the bisector of angle A (c) the perpendicular bisector of BC.
4. Draw a right-angled triangle ABC with AB = 5 cm, BC = 10 cm and $\hat{B} = 90°$. Construct the image of ABC in:
   (a) the perpendicular bisector of BC (b) the bisector of $\hat{B}$ (c) AC.
5. A, B, C, D are four points on a circle, centre O. BD is a diameter.
   Draw a sketch to show the reflection of ABCD in:
   (a) BD
   (b) the tangent at D
   (c) the chord BC.

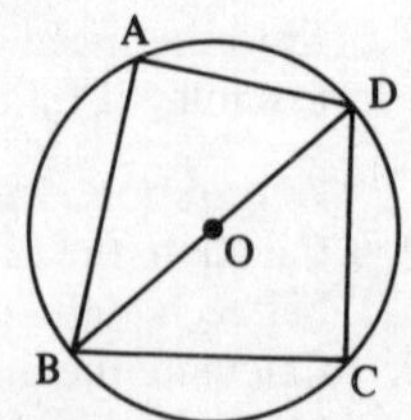

6. S = {letters in the word MATHS}. Draw a Venn diagram to show the subsets:
   (a) H = {letters unchanged by reflection in the horizontal axis}
   (b) V = {letters unchanged by reflection in the vertical axis}.

7. A 12-sided polygon is inscribed in a circle. Sketch the polygon and its image after reflection in a line outside the circle. If the vertices of the polygon represent the hours on a clock, what can you say about the direction of movement of the hour hand in the image?
8. Find the image of the point $(1, -2)$ in the mirror lines:
(a) $y = 0$ (b) $x = 0$ (c) $x = 1$ (d) $y = -2$ (e) $y = x$
9. Find the image of the point $(3, 1)$ in the mirror lines:
(a) $x = 3$ (b) $y = 1$ (c) $x = 0$ (d) $y = 0$ (e) $y = x$
10. If the image of the point $(-3, 7)$ under a certain reflection is $(3, -1)$, find one point on the mirror line. What is the gradient of the mirror line? Write down the equation of the mirror line.
11. Find the equation of the mirror line if the image of $(1, 2)$ is:
(a) $(1, -2)$ (b) $(-1, 2)$ (c) $(2, 1)$ (d) $(-2, -1)$ (e) $(2, 2)$
12. Sketch the graph of $y = x^2$ and its image under a reflection in the line:
(a) $x = 0$ (b) $y = 0$ (c) $y = x$
13. Sketch the graph of $y = 1/x$ and its image under a reflection in the line:
(a) $x = 0$ (b) $y = x$ (c) $y = 1$
14. The triangle with vertices $(0, 1)$, $(0, 3)$, $(1, 1)$ is reflected in the line $y = 1$. Its image is reflected in the line $y = 3$. Draw a diagram to show the two images. Use another diagram to show the images formed if the order of reflection is reversed (first in $y = 3$, then in $y = 1$). Is the final result the same?
15. The triangle with vertices $(1, 1)$, $(1, 4)$, $(2, 4)$ is reflected in the line $y = 1$. Its image is reflected in the line $x = 2$. Show that, if the order of reflections is reversed, the final result is the same.
16. Draw a sketch to show the reflection of a triangle in the lines $x = 1, x = -1, x = 2$ in that order. Find the equation of the mirror line that could replace these three.
17. Reflections in the lines r, s, t in that order are equivalent to a reflection in the line $y = 0$. If the equations of r and s are $y = 1$ and $y = 2$ respectively, find the equation of t.
18. The diagram shows an equilateral triangle divided into four equal equilateral triangles. Draw diagrams to show the reflection that will carry the shaded triangle to:
(a) triangle A (b) triangle B
Hence show that the shaded triangle can be mapped to triangle A by two successive reflections.

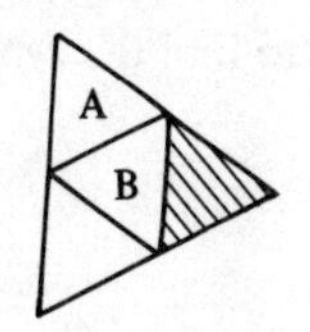

19. The image of P in the mirror line r is P′; PP′ is perpendicular to r. The image of P′ in a mirror line s is P″. Produce an argument to show that if s is perpendicular to r, the order of reflections makes no difference to the final image.

8.2 Rotations

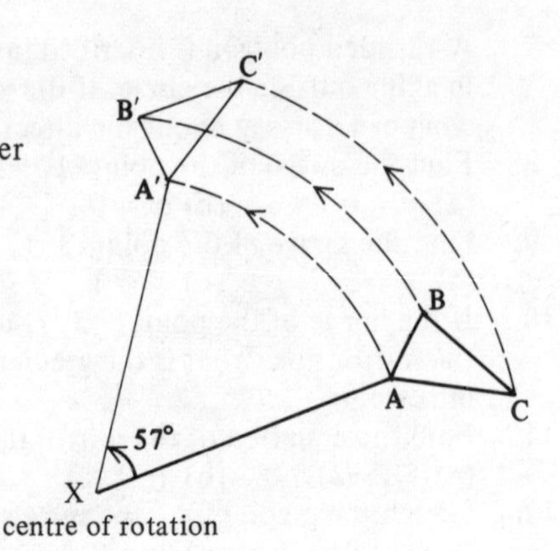

In the diagram, triangle ABC moves to triangle A′B′C′ under a rotation centre X, angle 57°.

A rotation, centre X, angle 57° is written **R** (X, 57°).
Here, **R** (X, 57°)(A) = A′.

Results

1. There is just one fixed point which is the centre of a rotation.
2. Angles of rotation are added.
3. The inverse of **R**(X, θ) is **R**(X, $-\theta$).
4. Two successive reflections in non-parallel lines are equivalent to a rotation.
5. Two successive rotations about the *same* centre are equivalent to another rotation about the *same* centre.
6. Two successive rotations about *different* centres are equivalent to another rotation about a third centre.

Construction

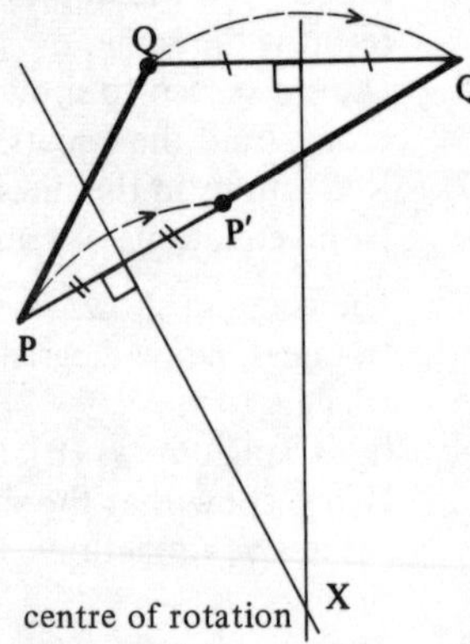

If P′ is the image of P under a rotation, the centre of rotation lies on the perpendicular bisector of PP′.

Two points (P, Q) and their images (P′, Q′) are needed to find the centre of rotation X.

Exercise 8.2

1. A, B, C, D, are vertices of a square, centre O. Which of the following statements are true about the image of A under a rotation?
   (a) **R**(O, 90°)(A) = D (b) **R**(O, 180°)(A) = C
   (c) **R**(D, 180°)(A) = B (d) **R**(B, 45°)(A) = O

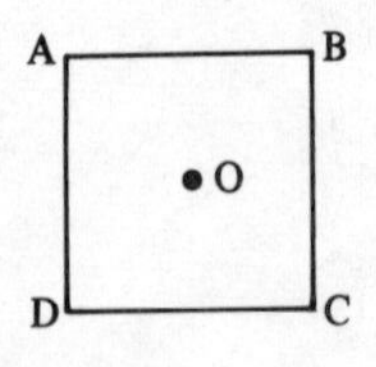

2. Using the diagram of question 1, find the centre and angle of rotation if:
   (a) B → A and C → B (b) B → D and C → C (c) B → D and C → A
3. Draw a sketch of the diagram and show the position of the shaded triangle after a rotation of:
   (a) 180° about X (b) 240° about Y
   (c) 180° about Y followed by 180° about X.

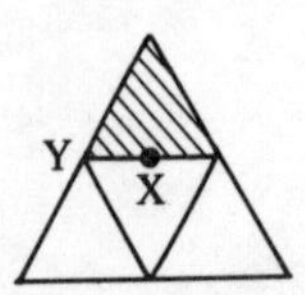

4. The shaded shape in the diagram is rotated 90° about X followed by 90° about Y. Draw a sketch to show the final image and state the centre and angle of the single rotation that can replace these two.

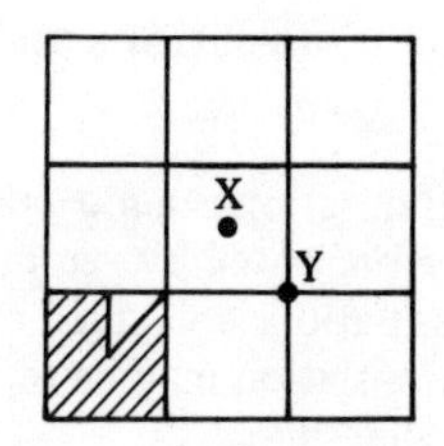

5. Use squared paper to draw the image of the unit square (0, 0), (0, 1), (1, 1), (1, 0) under the following rotations:
   (a) centre (0, 0), angle 90° (b) centre (2, 0), angle 180°
   (c) centre (2, 2), angle 90° followed by centre (2, 0), angle 180°
   (d) centre (2, 0), angle 180° followed by centre (2, 2), angle 90°
6. Find the centre and angle of the single rotation that can replace the pair of rotations in each of (c) and (d) in question 5.
7. Draw a triangle ABC with AB = 10 cm, BC = 5 cm, $A\hat{B}C = 60°$. Draw the image of the triangle under the rotations:
   (a) **R**(B, 60°) (b) **R**(C, 90°) (c) **R**(C, 90°) . **R**(B, 60°)
8. The triangle with vertices (0, 0), (2, 0), (0, 4) is rotated 180° about the point (0, 4). Find which of the following are equivalent transformations:
   (a) A reflection in $x = 2$ followed by a reflection in $y = 4$
   (b) A reflection in $y = 4$ followed by a reflection in $x = 0$
   (c) A rotation of 90° about (0, 0) followed by a rotation of 90° about (−4, 0)
   (d) A rotation of 10° about (0, 4) followed by a rotation of 170° about (0, 4)
9. The vertices of two squares ABCD and XYZW are given by the sets of co-ordinates: (−3, 1), (−3, −1), (−1, −1), (−1, 1) and (1, 3), (1, 1), (3, 1), (3, 3). Find the centre and angle of rotation which maps:
   (a) ABCD to ZWXY (b) ABCD to YZWX (c) ABCD to WXYZ
10. **R** is a rotation such that **R**:(0, 0) → (2, 2) and **R**:(3, 1) → (3, −1). Find:
   (a) the centre and angle of rotation of **R** (b) the image of (1, 1)
   (c) the point that is mapped to (5, 2)

8.3 Translations and glides

A translation is a movement without turning. Translations can be represented by vectors (see section 6.6 on page 112). Translations are added or subtracted using the Triangle Law. A translation may be replaced by two successive reflections in parallel lines.

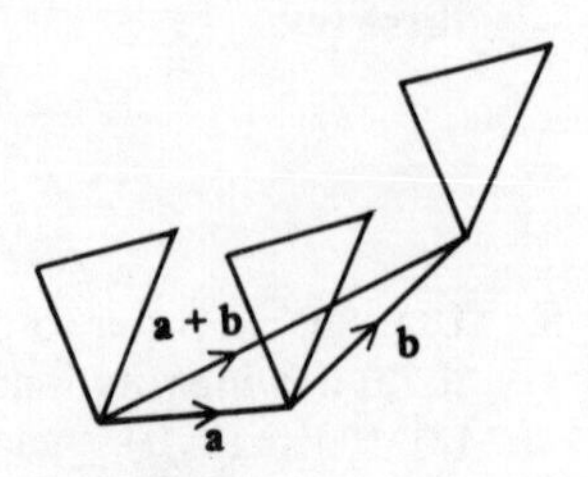

Exercise 8.3a

1. A triangle has vertices (0, 0), (2, 1), (3, −2). Draw the triangle and its image under the translation:
   (a) $\begin{pmatrix}-3\\2\end{pmatrix}$ (b) $\begin{pmatrix}3\\3\end{pmatrix}$
2. A triangle has vertices (1, 1), (1, 4), (−1, 3). Draw the triangle and its image under the translation:
   (a) $\begin{pmatrix}3\\-1\end{pmatrix}$ (b) $\begin{pmatrix}-1\\3\end{pmatrix}$
3. The unit square with vertices (0, 0), (0, 1), (1, 1), (1, 0) is translated by $\begin{pmatrix}4\\-1\end{pmatrix}$, then by $\begin{pmatrix}-2\\3\end{pmatrix}$.
   Draw the square and its two successive images and state the single translation that would return it to its original position.
4. Using the same unit square as in question 3, show the two successive translations $\begin{pmatrix}-2\\0\end{pmatrix}$ and $\begin{pmatrix}3\\-4\end{pmatrix}$. Hence find the inverse of the sum of these translations.
5. If $\mathbf{a} = \begin{pmatrix}2\\3\end{pmatrix}$ and $\mathbf{b} = \begin{pmatrix}-1\\4\end{pmatrix}$, show on squared paper the translations $\mathbf{a} + \mathbf{b}$ and $\mathbf{a} - \mathbf{b}$.
6. If $\mathbf{a} = \begin{pmatrix}0\\5\end{pmatrix}$ and $\mathbf{b} = \begin{pmatrix}-3\\-2\end{pmatrix}$, show on squared paper the translations $\mathbf{a} + \mathbf{b}$ and $\mathbf{a} - \mathbf{b}$.

Isometries

Geometrical transformations which do not change the shape of a figure are called isometries. Reflections, rotations and translations are isometries.
When two isometries are combined, the result is always an isometry.

Glides

A translation and a reflection combined is called a glide. The direction of the translation is parallel to the mirror line which is the glide axis.

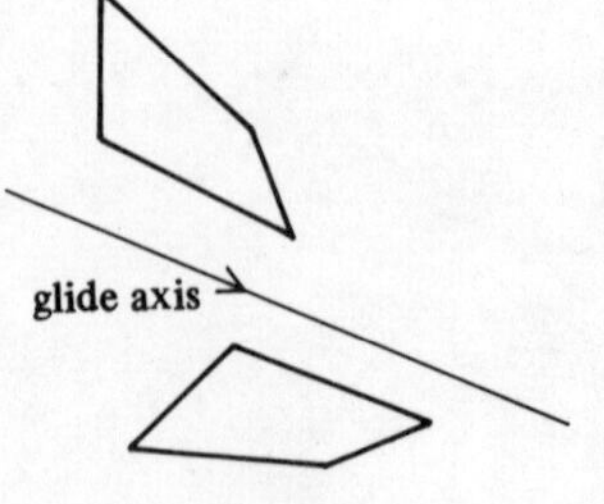

Exercise 8.3b

1. On squared paper draw the triangle with vertices $(-3, 0)$, $(-3, 3)$, $(-1, 3)$. Draw its reflection in $x = 0$ and the reflection of its image in $x = 3$. Describe the combined transformation.
2. Repeat question 1, using mirror lines $x = 1$, then $x = 4$. Is the result the same?
3. Repeat question 1, taking the mirror lines in the reverse order.
4. Repeat question 2, taking the mirror lines in the reverse order.
5. r and s are a pair of parallel mirror lines. A shape is reflected in r and its image is reflected in s. Which of the following statements are true? Draw a sketch to illustrate each answer.
   (a) The combined transformation is equivalent to a translation.
   (b) If the distance between r and s is x cm, the combined transformation is equivalent to a translation of x cm.
   (c) If r is to the right of s, the combined transformation is a translation to the left.
   (d) If the reflections are carried out in the reverse order (s first, then r), the result is a translation of the same distance but in the opposite direction.
6. a and b are two parallel mirror lines. $\mathbf{M}_a$ and $\mathbf{M}_b$ are defined as reflections in a and b respectively. Which of the following are true?
   (a) $\mathbf{M}_a . \mathbf{M}_b = \mathbf{M}_b . \mathbf{M}_a$ (b) $\mathbf{M}_a . \mathbf{M}_b = -\mathbf{M}_b . \mathbf{M}_a$ (c) $\mathbf{M}_a . \mathbf{M}_a = \mathbf{M}_b . \mathbf{M}_b$
7. A triangle has vertices $(0, 0)$, $(3, 0)$, $(3, 1)$. It is rotated through an angle of $90°$ about $(0, 0)$ and its image is rotated through an angle of $-90°$ about $(5, 0)$. Draw the triangle and its two successive images and describe the resultant transformation.
8. Which of the following are true? Draw a sketch to illustrate each answer.
   (a) $\mathbf{R}(X, \theta) . \mathbf{R}(X, \theta) \equiv$ a translation
   (b) $\mathbf{R}(X, \theta) . \mathbf{R}(Y, -\theta) \equiv$ a translation
   (c) $\mathbf{M}_a . \mathbf{M}_b = \mathbf{M}_b . \mathbf{M}_a \Rightarrow a, b$ are perpendicular
   (d) If $\mathbf{T}$ is a translation of x units parallel to $y = 0$, and $\mathbf{M}$ is a reflection in the line $y = 0$, then $\mathbf{T} . \mathbf{M} \neq \mathbf{M} . \mathbf{T}$.
9. A triangle T has vertices $(0, 0)$, $(2, 0)$, $(0, 1)$. A glide of 4 units along $y = 0$ as glide axis takes it to T$'$. Then T$'$ is subject to a glide of 3 units along $x = 2$ as glide axis to T$''$. What single transformation takes T to T$''$?
10. A triangle T has vertices $(-2, 1)$, $(-2, -2)$, $(0, 3)$. It is transformed to T$'$ by a rotation of $180°$ about $(0, 3)$. T$'$ is subject to a glide of -4 units along $x = 2$ as glide axis to T$''$. What single transformation would take T$''$ to T?

8.4 Enlargements

An enlargement is not an isometry. The image is a figure similar to the object.
Under an enlargement:

(a) Angles are unchanged.
(b) Lengths are increased in the same ratio.
(c) Lines and their images are parallel.

Examples

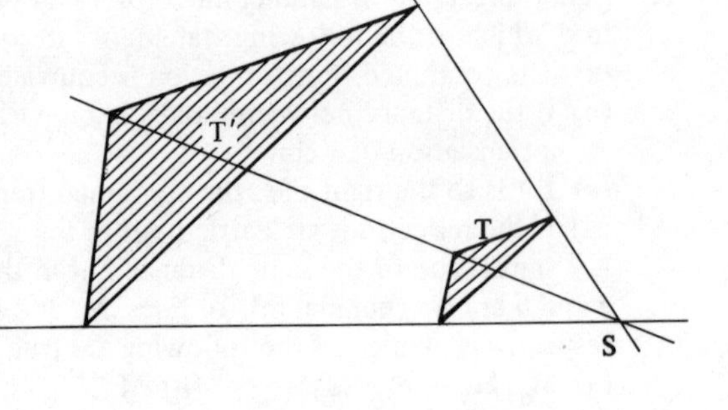

1. T′ is the image of T under an enlargement, centre S, scale factor 3.

   $\mathbf{E}(S, 3)(T) = T'$

2. Fractional enlargements produce a smaller image nearer the centre.

   $\mathbf{E}(S, \frac{1}{3})(T') = T$

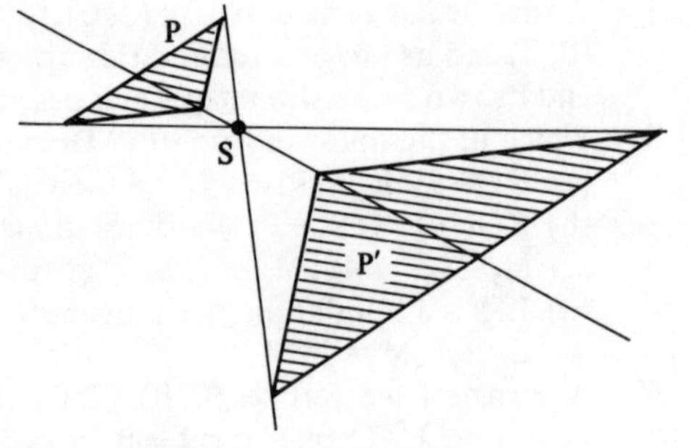

3. Negative enlargements produce an image on the opposite side of the centre of enlargement.

   $\mathbf{E}(S, -2)(P) = P'$
   also $\mathbf{E}(S, -\frac{1}{2})(P') = P$

4. Two successive enlargements are equivalent to another enlargement; the scale factor is the product of the other two scale factors.

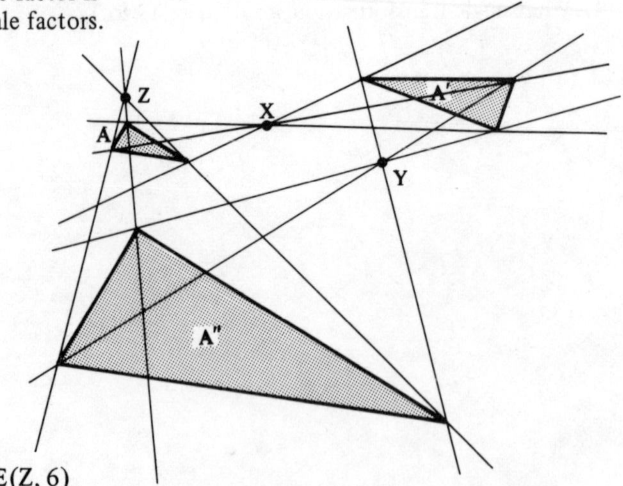

If $\mathbf{E}(X, -2)(A) = A'$
and $\mathbf{E}(Y, -3)(A') = A''$
then $\mathbf{E}(Z, 6)(A) = A''$
i.e. $\mathbf{E}(Y, -3) . \mathbf{E}(X, -2) = \mathbf{E}(Z, 6)$

Exercise 8.4

1. On squared paper draw the triangle with vertices (2, 0), (2, 3), and (3, 3). Taking the origin as the centre of enlargement, draw enlargements of this triangle using a scale factor of:
   (a) 2 (b) −1 (c) $\frac{1}{2}$
2. On squared paper draw the square with vertices (2, 2), (5, 2), (5, −1), and (2, −1). Taking the origin as the centre of enlargement, draw enlargements of this square using a scale factor of:
   (a) 3 (b) $\frac{1}{2}$ (c) −2
3. Using the triangle of question 1, draw the two successive enlargements centre (2, 0) scale factor 2 followed by centre (2, 0) scale factor −2. What single enlargement could replace these two?
4. Using the square in question 2, draw the two successive enlargements centre (2, 0) scale factor −1 followed by centre (0, 0) scale factor 2. Find the centre and scale factor of the single enlargement this represents.
5. PQRS is a square inside a triangle ABC with side SR on BC and Q on AC. By using an enlargement of the square taking C as the centre, find the largest square that can be drawn inside the triangle such that one side lies on BC and the other two vertices lie on AB and AC.

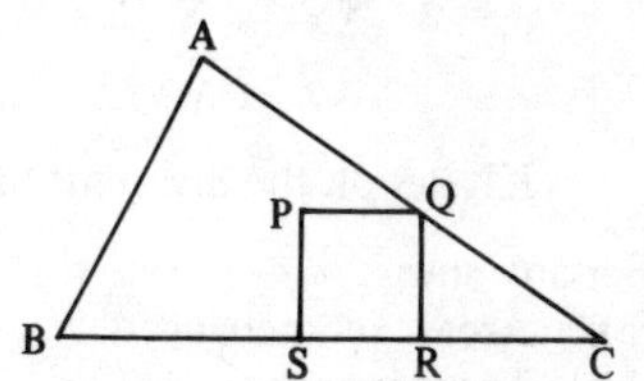

6. ABC is any triangle. BCX is an equilateral triangle with X outside triangle ABC. Taking A as the centre of enlargement, find a smaller equilateral triangle such that each vertex lies on a different side of triangle ABC.
7. The square with vertices O(0, 0), P(0, 1), Q(1, 1), R(1, 0) is transformed to OP′Q′R′ by an anticlockwise rotation of 45° about O followed by an enlargement, centre O scale factor 2. Which of the following are true and which are false?
   (a) R lies on OQ′ (b) OP′Q′R′ is a non-square rhombus
   (c) OQ′ = 2OP (d) OR′ = $\sqrt{2}$ OQ
   (e) area OP′Q′R′ = 4 × area OPQR
8. A triangle has vertices O(0, 0), A(4, 0), B(4, 3). It is transformed to PQR by a translation $\begin{pmatrix} -2 \\ -1 \end{pmatrix}$ followed by an enlargement centre O scale factor −2.

   Find the co-ordinates of P, Q, and R and the area of triangle PQR.

8.5 Shears

Example 1

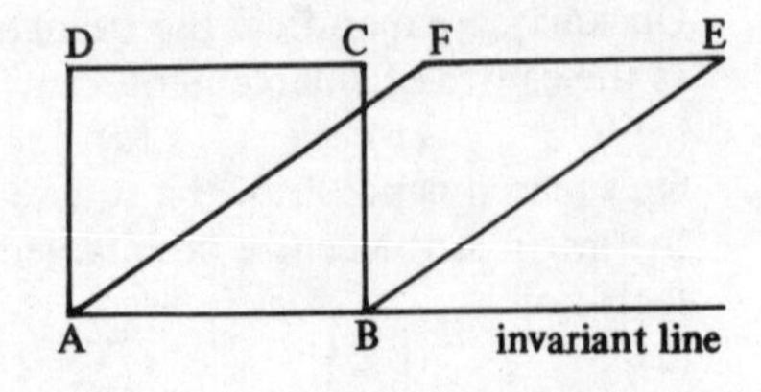

The rectangle ABCD maps to the parallelogram ABEF under a shear.
AB remains fixed.
All other points move parallel to AB.

Example 2

$WXYZ \rightarrow W'X'Y'Z'$ by means of a shear.

area $WXYZ$ = area $W'X'Y'Z'$

$XY = X'Y'$

$WZ = W'Z'$

KL lies on the invariant line.

For any shear:
(a) Areas are unchanged.
(b) There is an invariant line.
(c) All points move parallel to the invariant line.
(d) Distances between points on lines parallel to the invariant line are preserved.

Exercise 8.5

1. XL is the invariant line of a shear; XZ is perpendicular to XL; XZ = 6 cm and Y lies on XZ such that XY = 2 cm. The image of Z under the shear is Z′ where ZZ′ = 4 cm. Draw a diagram of the shear and join XZ′. Hence find the image Y′ and the length YY′.

2. ABC is a straight line with AB = BC. ABXY is a rectangle and BCX′Y′ is a parallelogram of the same area. Construct the invariant line of the shear that maps ABXY to BCX′Y′.

3. T and R are on opposite sides of the invariant line of a shear that maps T to T′ and R to R′. A straight line meets TT′ at X, TR at Y, the invariant line at Z, and T′R′ at W. The shear maps X, Y, Z, W to X′, Y′, Z′, W′ (not shown on the diagram).
Find which statements are true and which false:
(a) XX′ = TT′ (b) Y′ lies on R′T′
(c) W′ lies on TR
(d) Z′ is on the invariant line but at a different position from Z
(e) X′Y′Z′W′ is a straight line.

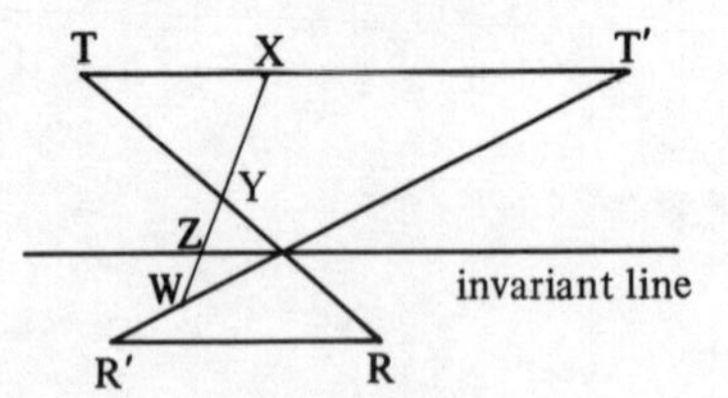

4. XYZW is a rectangle and C is the mid-point of WY. XYZ′W′ is the image of the rectangle under a shear which has XY as the invariant line. C′ is the image of C. Which of the following are true and which false?
(a) W′Z′ = WZ (b) area XCY > area XC′Y
(c) CC′ = XY (d) YC′W is a straight line
(e) XC′ = $\frac{1}{2}$XZ′

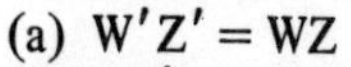

5. Construct the quadrilateral ABCD in which AB = 6 cm, BC = 5 cm, CD = 2 cm, DA = 3 cm, and BD = 6 cm. Draw a line parallel to BD through C to meet AB produced at E. Hence find:
(a) the invariant line of the shear that maps BDC to BDE
(b) a triangle which is equal in area to ABCD
(c) the length AE and the perpendicular distance of D from AB
(d) the area of the quadrilateral ABCD.

6. XYZW is a trapezium with XY = XW = 5 cm, XZ = 7 cm, and W$\hat{X}$Y = X$\hat{Y}$Z = 90°. By means of a shear find a point T on ZY produced such that the area of triangle TWZ equals the area of XYZW. Find also a point S on YZ produced such that the area of triangle YWS equals the area of XYZW. What is the length ST?

7. In the diagram, ABC is a right-angled triangle with A$\hat{B}$C = 90°. ABXY and ACWZ are squares. BST is parallel to AZ; ZUV is parallel to AB.

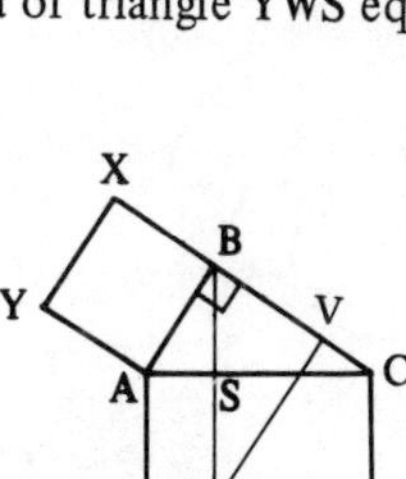

(a) Describe a shear that maps ASTZ to ABUZ
(b) Describe a shear that maps ABUZ to a square.
(c) Describe a transformation that maps the square in (b) to ABXY.
(d) Explain why this proves that area ASTZ equals area ABXY.
(e) Draw a diagram and describe the transformations that map CSTW to the square that can be drawn on BC.
(f) How does this prove the Theorem of Pythagoras?

8. Construct the image of a circle radius 5 cm under a shear where a tangent is the invariant line and the distance of the centre from its image is 2·5 cm.

8.6 Transformation matrices

A matrix can be used to describe transformations in a plane.

Example

The vertices of the quadrilateral ABCD are A(0, −1), B(2, 0), C(2, 2), and D(0, 3). ABCD is transformed to A′B′C′D′ using the matrix $\begin{pmatrix} 0 & 1 \\ 2 & 0 \end{pmatrix}$. Find the co-ordinates of the vertices of the transformed quadrilateral and illustrate by a sketch.

The new co-ordinates are found by evaluating:

$$\begin{pmatrix} 0 & 1 \\ 2 & 0 \end{pmatrix}\overset{A}{\begin{pmatrix} 0 \\ -1 \end{pmatrix}} = \overset{A'}{\begin{pmatrix} -1 \\ 0 \end{pmatrix}} \qquad \begin{pmatrix} 0 & 1 \\ 2 & 0 \end{pmatrix}\overset{B}{\begin{pmatrix} 2 \\ 0 \end{pmatrix}} = \overset{B'}{\begin{pmatrix} 0 \\ 4 \end{pmatrix}}$$

$$\begin{pmatrix} 0 & 1 \\ 2 & 0 \end{pmatrix}\overset{C}{\begin{pmatrix} 2 \\ 2 \end{pmatrix}} = \overset{C'}{\begin{pmatrix} 2 \\ 4 \end{pmatrix}} \qquad \begin{pmatrix} 0 & 1 \\ 2 & 0 \end{pmatrix}\overset{D}{\begin{pmatrix} 0 \\ 3 \end{pmatrix}} = \overset{D'}{\begin{pmatrix} 3 \\ 0 \end{pmatrix}}$$

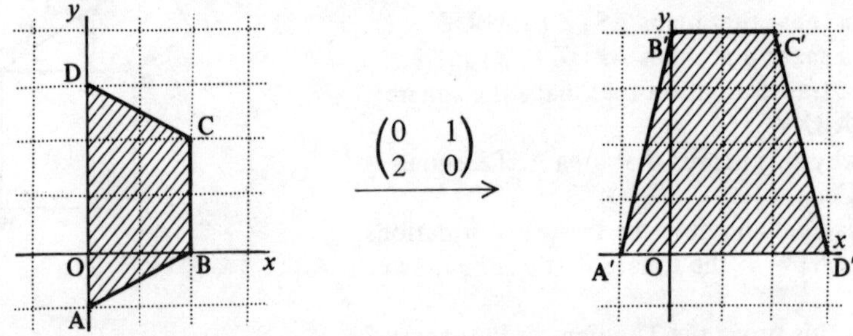

The sketch shows the transformed quadrilateral drawn on the original axes.

Another way of finding the new co-ordinates is to find the new value of the point $\begin{pmatrix} x \\ y \end{pmatrix}$ and substitute.

$$\begin{pmatrix} 0 & 1 \\ 2 & 0 \end{pmatrix}\begin{pmatrix} x \\ y \end{pmatrix} = \begin{pmatrix} y \\ 2x \end{pmatrix}. \qquad \text{For A, } x = 0, y = -1, \text{ so } A' = \begin{pmatrix} -1 \\ 0 \end{pmatrix}.$$

Exercise 8.6a

1. ABCD is the quadrilateral with vertices as in the example above. Find the new vertices A′, B′, C′, D′, and draw a diagram to show the transformation using the matrix:
   (a) $\begin{pmatrix} 1 & 1 \\ 0 & 1 \end{pmatrix}$ (b) $\begin{pmatrix} 0 & 2 \\ 1 & 0 \end{pmatrix}$ (c) $\begin{pmatrix} 0 & 2 \\ 0 & 1 \end{pmatrix}$ (d) $\begin{pmatrix} 1 & 1 \\ 1 & 0 \end{pmatrix}$ (e) $\begin{pmatrix} 0 & 2 \\ 2 & 0 \end{pmatrix}$.

2. Find the transformation of the triangle with vertices (0, 0), (0, 1), (2, 1) using the following matrices. Draw a diagram of the transformed triangle and describe the transformation.
   (a) $\begin{pmatrix} -1 & 0 \\ 0 & 1 \end{pmatrix}$ (b) $\begin{pmatrix} 0 & 1 \\ 1 & 0 \end{pmatrix}$ (c) $\begin{pmatrix} 2 & 0 \\ 0 & 2 \end{pmatrix}$ (d) $\begin{pmatrix} 0 & 1 \\ -1 & 0 \end{pmatrix}$ (e) $\begin{pmatrix} 2 & 0 \\ 1 & 0 \end{pmatrix}$.

Enlargement matrices

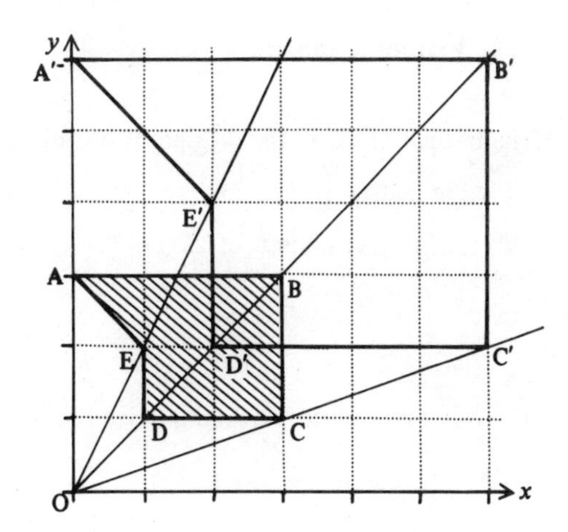

ABCDE is enlarged to A′B′C′D′E′ using centre O, scale factor 2. This enlargement is produced by the matrix $\begin{pmatrix} 2 & 0 \\ 0 & 2 \end{pmatrix}$.

You can check this works for each vertex of the figure.
For example,

$$\begin{pmatrix} 2 & 0 \\ 0 & 2 \end{pmatrix}\begin{pmatrix} 1 \\ 2 \end{pmatrix} = \begin{pmatrix} 2 \\ 4 \end{pmatrix} \text{ so E} \rightarrow \text{E}'.$$

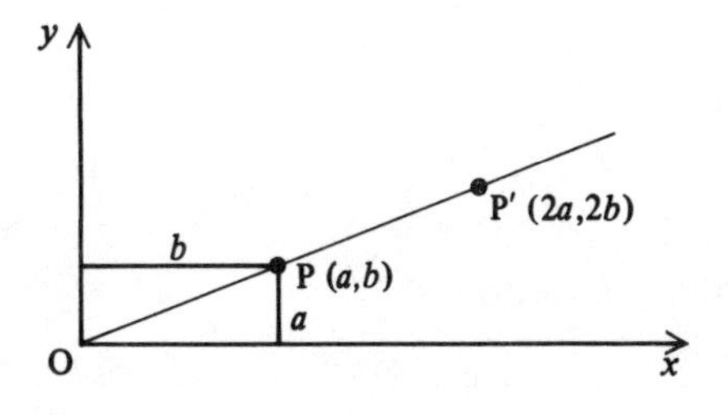

The image of any point P(a, b) when using the enlargement represented by $\begin{pmatrix} 2 & 0 \\ 0 & 2 \end{pmatrix}$ is P′$(2a, 2b)$.
Clearly, P′ lies on OP and OP′ = 2 OP.
This proves that the matrix represents the enlargement described.
In general, the matrix $\begin{pmatrix} k & 0 \\ 0 & k \end{pmatrix}$ produces an enlargement centre O, scale factor k.

Exercise 8.6b

1. ABCD has vertices (1, 1), (0, 2), (2, 3), (3, 1). Find the vertices of the enlarged quadrilateral A′B′C′D′ and draw it on squared paper using:
(a) $\begin{pmatrix} 2 & 0 \\ 0 & 2 \end{pmatrix}$ (b) $\begin{pmatrix} 3 & 0 \\ 0 & 3 \end{pmatrix}$ (c) $\begin{pmatrix} \frac{1}{2} & 0 \\ 0 & \frac{1}{2} \end{pmatrix}$ (d) $\begin{pmatrix} -2 & 0 \\ 0 & -2 \end{pmatrix}$ (e) $\begin{pmatrix} -\frac{1}{2} & 0 \\ 0 & -\frac{1}{2} \end{pmatrix}$
2. ABCD has vertices (1, 1), (1, −2), (−1, −1), (−2, 1). Find the vertices of the enlarged quadrilateral A′B′C′D′ and draw it on squared paper using:
(a) $\begin{pmatrix} 2 & 0 \\ 0 & 2 \end{pmatrix}$ (b) $\begin{pmatrix} 3 & 0 \\ 0 & 3 \end{pmatrix}$ (c) $\begin{pmatrix} \frac{1}{2} & 0 \\ 0 & \frac{1}{2} \end{pmatrix}$ (d) $\begin{pmatrix} -2 & 0 \\ 0 & -2 \end{pmatrix}$ (e) $\begin{pmatrix} -\frac{1}{2} & 0 \\ 0 & -\frac{1}{2} \end{pmatrix}$
3. The unit square with vertices (0, 0), (1, 0), (1, 1), (0, 1) is enlarged using the matrix $\begin{pmatrix} k & 0 \\ 0 & k \end{pmatrix}$. Find the co-ordinates of the enlarged square and draw a sketch to show the length of its sides. What is the relationship between the ratio of the areas of the two squares and the determinant of the matrix?
4. Write down the matrices that represent enlargements centre O, scale factor:
(a) $-\frac{1}{2}$ (b) −3.
Show, by the product of the matrices, that these two successive enlargements are equivalent to an enlargement scale factor $1\frac{1}{2}$.
Illustrate your answer using the unit square (0, 0), (1, 0), (1, 1), (0, 1).

8.7 Isometry matrices

The triangle OAB in the diagram is transformed to OA′B′ using the matrix $\begin{pmatrix} 0 & -1 \\ 1 & 0 \end{pmatrix}$.

This transformation is a rotation of 90° about O.

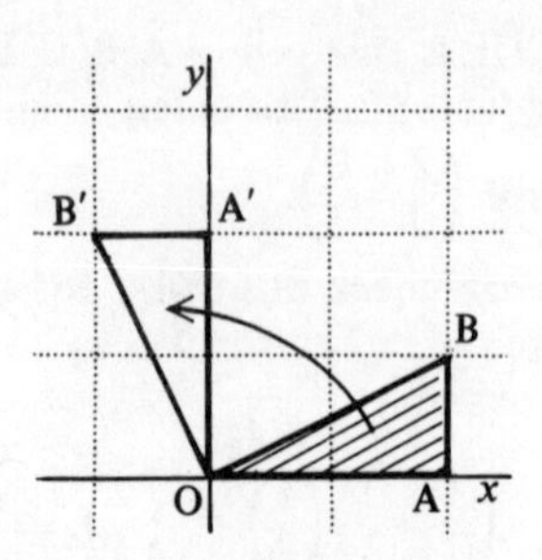

The isometry matrices are summarized in this table:

Rotation about O of:	90°	180°	270°	360° or 0°
	$\begin{pmatrix} 0 & -1 \\ 1 & 0 \end{pmatrix}$	$\begin{pmatrix} -1 & 0 \\ 0 & -1 \end{pmatrix}$	$\begin{pmatrix} 0 & 1 \\ -1 & 0 \end{pmatrix}$	$\begin{pmatrix} 1 & 0 \\ 0 & 1 \end{pmatrix}$
Reflection in:	Ox	Oy	$y = x$	$y = -x$
	$\begin{pmatrix} 1 & 0 \\ 0 & -1 \end{pmatrix}$	$\begin{pmatrix} -1 & 0 \\ 0 & 1 \end{pmatrix}$	$\begin{pmatrix} 0 & 1 \\ 1 & 0 \end{pmatrix}$	$\begin{pmatrix} 0 & -1 \\ -1 & 0 \end{pmatrix}$

Exercise 8.7a

1. The triangle OAB with vertices (0, 0), (1, 1), (2, 3) is transformed to OA′B′. Find which matrix represents the transformation given that A → A′ and B → B′ as follows:
   (a) (1, 1) → (1, −1)
   (2, 3) → (2, −3)
   (b) (1, 1) → (1, 1)
   (2, 3) → (3, 2)
   (c) (1, 1) → (1, −1)
   (2, 3) → (3, −2)
2. Find the matrix which represents the transformation in which:
   (a) (−1, 2) → (2, −1)
   (−3, 2) → (2, −3)
   (b) (−1, 2) → (1, −2)
   (−3, 2) → (3, −2)
   (c) (−1, 2) → (−2, 1)
   (−3, 2) → (−2, 3)
3. Evaluate the product $\mathbf{M} = \begin{pmatrix} 1 & 0 \\ 0 & -1 \end{pmatrix}\begin{pmatrix} 0 & 1 \\ 1 & 0 \end{pmatrix}$.

   Describe the transformation that the matrix **M** represents.
4. Evaluate the product $\mathbf{N} = \begin{pmatrix} 0 & 1 \\ 1 & 0 \end{pmatrix}\begin{pmatrix} 1 & 0 \\ 0 & -1 \end{pmatrix}$.

   Describe the transformation that the matrix **N** represents.
5. Let **X** be a reflection in the x-axis and **R** be a rotation about the origin of 90°. Write down the matrices associated with **X** and **R** and evaluate the products **XR** and **RX**. If P is the point (−1, 4), find **XR**(P) and **RX**(P).
6. Let **Y** be a reflection in the y-axis and **H** be a half turn about the origin. Write down the matrices associated with **Y** and **H**. If P is any point (a, b) in the plane, show that **YH**(P) = **HY**(P) and describe in words the equivalent isometry.
7. Use the triangle OAB shown at the top of the page to illustrate the transformations given by $\begin{pmatrix} 0 & -1 \\ 1 & 0 \end{pmatrix}$ followed by $\begin{pmatrix} 0 & 1 \\ 1 & 0 \end{pmatrix}$.
   (a) Describe the resultant transformation.
   (b) What matrix represents this transformation?
   (c) Show that the product of the two given matrices yields the same result if multiplied in the correct order, but not otherwise.

8. Draw up a multiplication table to show the products of all four rotation matrices with each other. Hence show that the products of two rotations always produces a rotation. Give an example to show that the product of two reflections does not give a reflection.

General rotation matrix

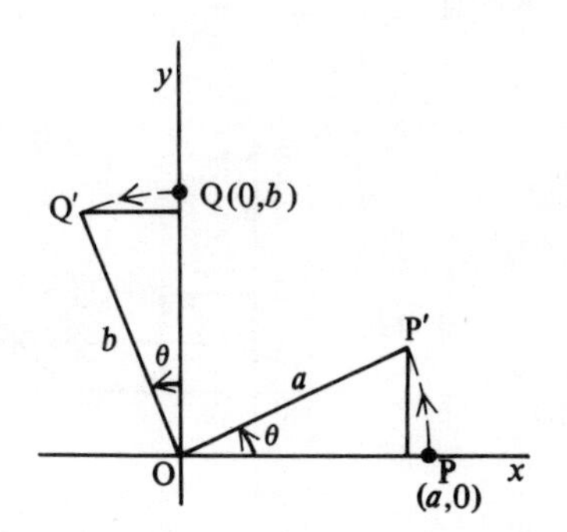

In a rotation of θ about O, P(a, 0) → P′, and Q(O, b) → Q′. From the right-angled triangles, the co-ordinates of P′ and Q′ are

$$P'(a\cos\theta, a\sin\theta), \quad Q'(-b\sin\theta, b\cos\theta).$$

So, after a rotation of θ,

$$\begin{pmatrix} a \\ 0 \end{pmatrix} \to \begin{pmatrix} a\cos\theta \\ a\sin\theta \end{pmatrix} \text{ and } \begin{pmatrix} 0 \\ b \end{pmatrix} \to \begin{pmatrix} -b\sin\theta \\ b\cos\theta \end{pmatrix}$$

This gives for point (a, b)

$$\begin{pmatrix} a \\ b \end{pmatrix} \to \begin{pmatrix} a\cos\theta & -b\sin\theta \\ a\sin\theta & +b\cos\theta \end{pmatrix} = \begin{pmatrix} \cos\theta & -\sin\theta \\ \sin\theta & \cos\theta \end{pmatrix}\begin{pmatrix} a \\ b \end{pmatrix}$$

The matrix $\mathbf{R} = \begin{pmatrix} \cos\theta & -\sin\theta \\ \sin\theta & \cos\theta \end{pmatrix}$ represents a rotation of θ about O.

Exercise 8.7b

1. By putting $\theta = 90°$, show that the matrix **R** above represents the correct matrix for this rotation. Check also using the values $\theta = 180°$ and $\theta = 270°$.
2. Find the rotation matrix for an angle of:
   (a) 45° (b) 60°.
   Use your answers to find the image of the point (2, 1) under these rotations.
3. Find the rotation matrix for an angle of:
   (a) 120° (b) 135° (c) 150° (d) 300°.
4. Find the image of the triangle (0, 0), (1, 2), (2, 2) under the transformation represented by the matrix $\begin{pmatrix} 0{\cdot}6 & -0{\cdot}8 \\ 0{\cdot}8 & 0{\cdot}6 \end{pmatrix}$. Show that this is a rotation and find the angle of rotation to the nearest degree.
5. Find the image of the triangle (0, 0), (0, 4), (4, 3) under the transformation represented by the matrix $\begin{pmatrix} 0{\cdot}8 & 0{\cdot}6 \\ -0{\cdot}6 & 0{\cdot}8 \end{pmatrix}$. Show that this is a rotation and find the angle of rotation to the nearest degree.
6. Write down the rotation matrix for an angle of 30°. Call this matrix **T**. Write down the rotation matrix for an angle of 60°. Call this matrix **S**.
   Find:
   (a) $\mathbf{T}^2$, and check that this matrix is the same as **S**.
   (b) **TS** and **ST**, and show that each represents a rotation of 90°.
   (c) $\mathbf{S}^2$, and write down the angle involved in this rotation matrix.
   What can you say about $\mathbf{T}^3$ and $\mathbf{S}^3$?
7. Find $\begin{pmatrix} \cos\theta & -\sin\theta \\ \sin\theta & \cos\theta \end{pmatrix}\begin{pmatrix} \cos\phi & -\sin\phi \\ \sin\phi & \cos\phi \end{pmatrix}$.
   The resulting matrix represents a rotation of angle $(\theta + \phi)$. Hence write down an expression for: (a) $\cos(\theta + \phi)$ (b) $\sin(\theta + \phi)$.

8.8 Shear matrices

A shear can be represented by a matrix. The effect of two shear matrices are shown in the diagrams:

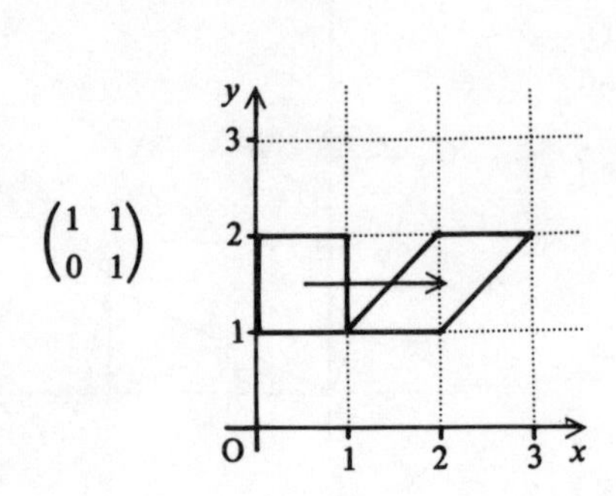

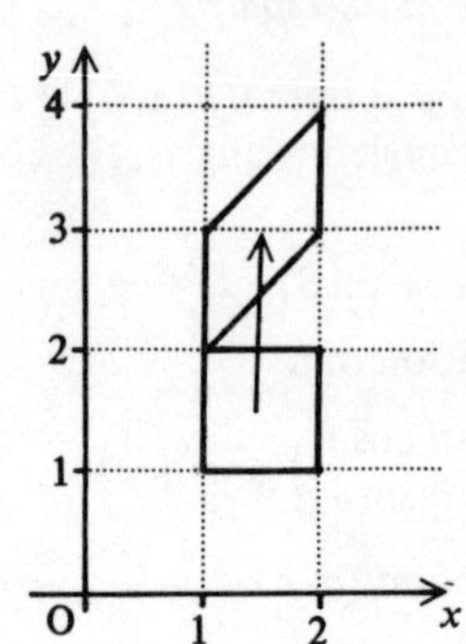

Remember

$\begin{pmatrix} 1 & a \\ 0 & 1 \end{pmatrix}$ is a shear with Ox as the invariant line.

$\begin{pmatrix} 1 & 0 \\ a & 1 \end{pmatrix}$ is a shear with Oy as the invariant line.

Exercise 8.8

1. Draw a diagram to show the image of the following squares under a transformation by the shear matrix $\begin{pmatrix} 1 & 2 \\ 0 & 1 \end{pmatrix}$:
   (a) (0, 0), (1, 0), (1, 1), (0, 1) (b) (2, 2), (2, 3), (3, 3), (3, 2)
   (c) $(1, \frac{1}{2}), (2, \frac{1}{2}), (2, -\frac{1}{2}), (1, -\frac{1}{2})$.

2. Draw a diagram to show the image of the following shapes under a transformation by the shear matrix $\begin{pmatrix} 1 & 0 \\ 2 & 1 \end{pmatrix}$:
   (a) triangle (0, 0), (2, 1), (2, 2) (b) square $(\frac{1}{2}, \frac{1}{2}), (\frac{1}{2}, -\frac{1}{2}), (-\frac{1}{2}, -\frac{1}{2}), (-\frac{1}{2}, \frac{1}{2})$.
   (c) pentagon (1, 1), (0, 2), (−2, 2), (−2, 0), (−1, −1).

3. The square with vertices (0, 0), (1, 0), (1, 1), (0, 1) is transformed by a shear to the parallelogram P shown in the diagram.
   P is transformed by another shear to parallelogram Q.
   Find the two matrices associated with these shears and the matrix associated with the single shear that takes Q back to the original square.

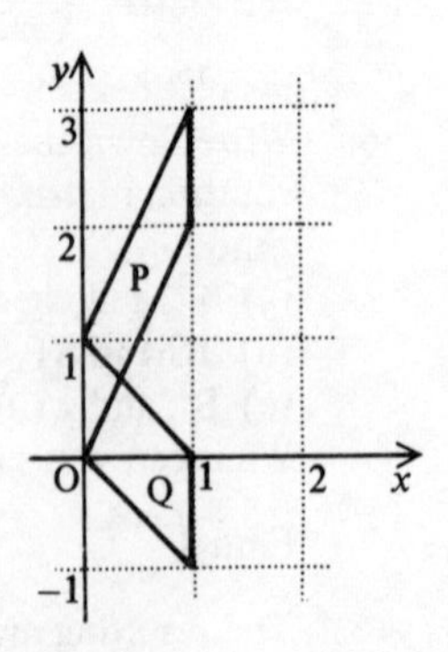

4. Under a certain shear, $(1, \frac{1}{2})$ maps to $(1, 2)$, and $(2, 1)$ maps to $(2, 4)$.
   (a) What is the invariant line of the shear?
   (b) Find the matrix associated with this shear.
5. Transform the square with vertices (0, 0), (1, 0), (1, 1), (0, 1) using the shear matrix $\begin{pmatrix} 1 & \frac{1}{2} \\ 0 & 1 \end{pmatrix}$. Find the matrix for the inverse transformation.
6. $\begin{pmatrix} 1 & 3 \\ 0 & 1 \end{pmatrix}$ and $\begin{pmatrix} 1 & 0 \\ -2 & 1 \end{pmatrix}$ are shear matrices. Draw diagrams to show the transformations of the square with vertices (0, 0), (0, 1), (1, 1), (1, 0) using these two matrices. Find the matrices that represent the inverse transformations.
7. $\mathbf{M} = \begin{pmatrix} 1 & 0 \\ 1 & 1 \end{pmatrix}$, $\mathbf{N} = \begin{pmatrix} 1 & 0 \\ -3 & 1 \end{pmatrix}$, $\mathbf{P} = \begin{pmatrix} 1 & 2 \\ 0 & 1 \end{pmatrix}$ are three shear matrices. T is the triangle with vertices (1, 0), (1, 2), (2, 2).
   Find the vertices of the image of T under the transformations:
   (a) **N**(T) (b) **MN**(T) (c) **P**(T) (d) **NP**(T)
   Draw diagrams to illustrate (b) and (d). Are they both equivalent to a single shear? If so, find the associated shear matrix.
8. Is a shear followed by a shear equivalent to a single shear? Use different combinations of the matrices $\begin{pmatrix} 1 & 0 \\ 1 & 1 \end{pmatrix}$, $\begin{pmatrix} 1 & 0 \\ 2 & 1 \end{pmatrix}$ and $\begin{pmatrix} 1 & 1 \\ 0 & 1 \end{pmatrix}$ to answer the question.
9. Transform the square with vertices (0, 0), (1, 1), (0, 2), (−1, 1) using the matrix $\begin{pmatrix} \frac{1}{2} & -\frac{1}{2} \\ \frac{1}{2} & 1\frac{1}{2} \end{pmatrix}$. Show that the transformation is a shear and find its invariant line.
10. By considering the rectangle with vertices (1, 0), (2, −1), (4, 1), (3, 2), show that the matrix $\begin{pmatrix} 1\frac{1}{2} & \frac{1}{2} \\ -\frac{1}{2} & \frac{1}{2} \end{pmatrix}$ represents a shear. Find the invariant line of the shear.
11. Transform the square with vertices (0, 0), (1, −1), (2, 0), (1, 1) using the matrix $\begin{pmatrix} 4 & 3 \\ -3 & -2 \end{pmatrix}$. Show that the transformation is a shear and find the equation of the invariant line of the shear.
12. Transform the rectangle with vertices (0, 0), (2·5, 0), (2·5, 5), (0, 5) using the matrix $\begin{pmatrix} 0{\cdot}4 & 0{\cdot}3 \\ -1{\cdot}2 & 1{\cdot}6 \end{pmatrix}$. Find:
    (a) two invariant points of the transformation
    (b) the equation of the line on which they lie.
    Show also that if (a, b) is any point on this line, it is unchanged by the transformation.

8.9 General transformations

A matrix alone cannot represent all transformations of the plane. We may need to use a vector transformation as well.

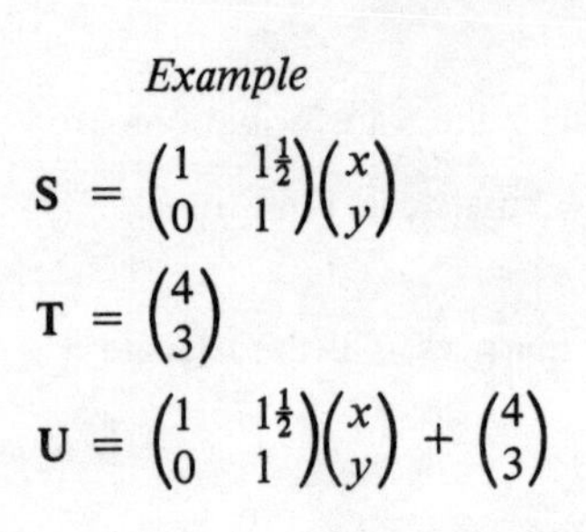

Example

$$\mathbf{S} = \begin{pmatrix}1 & 1\frac{1}{2}\\0 & 1\end{pmatrix}\begin{pmatrix}x\\y\end{pmatrix}$$

$$\mathbf{T} = \begin{pmatrix}4\\3\end{pmatrix}$$

$$\mathbf{U} = \begin{pmatrix}1 & 1\frac{1}{2}\\0 & 1\end{pmatrix}\begin{pmatrix}x\\y\end{pmatrix} + \begin{pmatrix}4\\3\end{pmatrix}$$

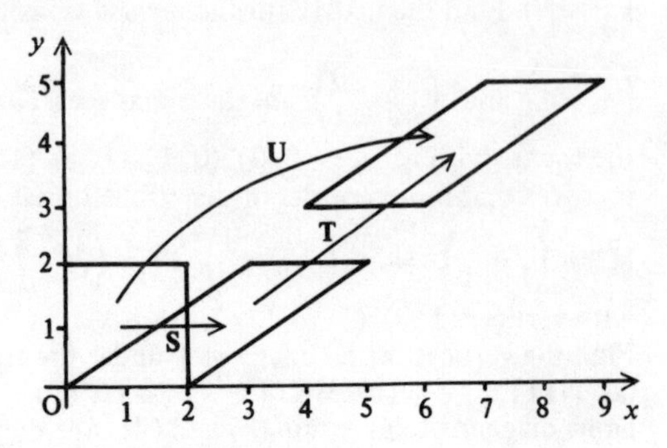

Exercise 8.9a

1. Draw a diagram to show the effect of each of the following transformations on the square with vertices (0, 0), (0, 1), (1, 1), (1, 0). In each case describe in words the single transformation this represents.

   (a) $\begin{pmatrix}-1 & 0\\0 & 1\end{pmatrix}\begin{pmatrix}x\\y\end{pmatrix} + \begin{pmatrix}3\\0\end{pmatrix}$ (b) $\begin{pmatrix}2 & 0\\0 & 2\end{pmatrix}\begin{pmatrix}x\\y\end{pmatrix} + \begin{pmatrix}2\\2\end{pmatrix}$

   (c) $\begin{pmatrix}0 & 1\\-1 & 0\end{pmatrix}\begin{pmatrix}x\\y\end{pmatrix} + \begin{pmatrix}0\\-1\end{pmatrix}$ (d) $\begin{pmatrix}1 & 0\\0 & 1\end{pmatrix}\begin{pmatrix}x\\y\end{pmatrix} + \begin{pmatrix}4\\3\end{pmatrix}$.

2. A rotation of the square with vertices (0, 0), (0, 1), (1, 0), (1, 1) maps it to the square with vertices (−2, 2), (−3, 2), (−2, 3), (−3, 3) in that order. Find:
   (a) the centre of rotation (b) the angle of rotation
   (c) the translation vector that maps the vertex (0, 0) to its image
   (d) the matrix and translation vector that gives the rotation.

3. In the diagram, the square ABCD is enlarged by scale factor 2 to the square CEFG.
   (a) What is the centre of enlargement?
   (b) Find the vertices of the image of ABCD under an enlargement scale factor 2, centre O.
   (c) What translation vector maps this image to CEFG?
   (d) What is the matrix **M** and vector **v** that represents an enlargement scale factor 2, centre (0, 3)?

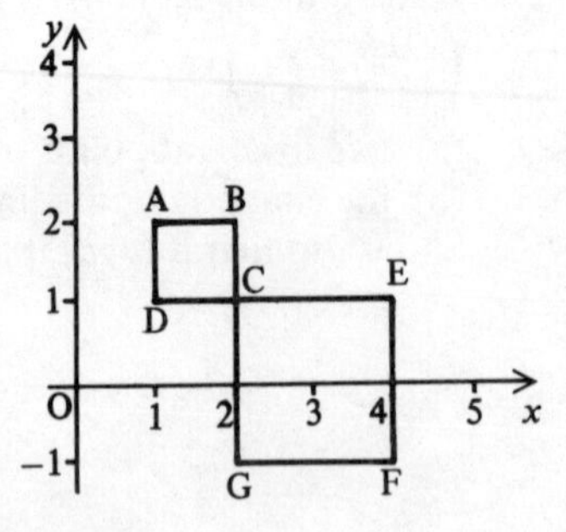

4. Find the matrix **M** and vector **v**, given that $\mathbf{M}\begin{pmatrix}x\\y\end{pmatrix} + \mathbf{v}$ represents:

   (a) a translation $\begin{pmatrix}2\\-1\end{pmatrix}$ (b) a rotation of 180° about the origin

   (c) a reflection in the line $y = x + 1$ (d) an enlargement scale factor −2 centre (4, 3).

Area and matrices

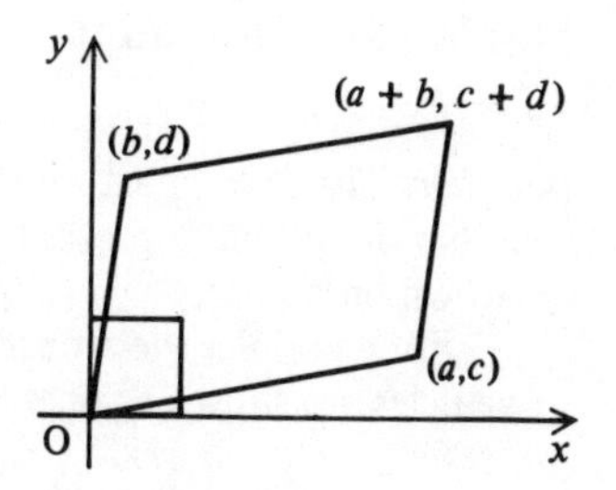

The transformation represented by the matrix $\begin{pmatrix} a & b \\ c & d \end{pmatrix}$ maps the unit square with vertices (0, 0), (1, 0), (1, 1), (0, 1) on to the parallelogram with vertices (0, 0), (a, c), $(a + b, c + d)$, (b, d).

The area of the parallelogram is $ad - bc$. The determinant of any transformation matrix is the area scale factor of the transformation.

Exercise 8.9b

1. Find the image of the square with matrices (0, 0), (1, 0), (1, 1), (0, 1) under the matrix transformation:

   (a) $\begin{pmatrix} 3 & 0 \\ 0 & 3 \end{pmatrix}$ (b) $\begin{pmatrix} 2 & 1 \\ 1 & 0 \end{pmatrix}$ (c) $\begin{pmatrix} 3 & 2 \\ 4 & 3 \end{pmatrix}$ (d) $\begin{pmatrix} -1 & 2 \\ 2 & -1 \end{pmatrix}$.

   In each case, find the area of the image.
2. The triangle with vertices (0, 0), (1, 2), (0, 2) is transformed using the given matrix. Find the area of the image.

   (a) $\begin{pmatrix} 4 & 0 \\ 0 & 4 \end{pmatrix}$ (b) $\begin{pmatrix} -2 & 0 \\ 0 & -2 \end{pmatrix}$ (c) $\begin{pmatrix} 2 & 3 \\ 1 & 4 \end{pmatrix}$ (d) $\begin{pmatrix} 3 & 4 \\ 6 & 8 \end{pmatrix}$.
3. Draw a diagram to show the effect of using the matrix $\begin{pmatrix} -1 & 0 \\ 0 & 1 \end{pmatrix}$ on a shape of your choice. What is the change in area? How is this shown by the determinant of the matrix?
4. Isometries can be described as direct or opposite. Find the determinants of the eight isometry matrices given on page 152. Classify the isometries as direct or opposite and give reasons for your choice.
5. The parallelogram ABCD has vertices (0, 0), (7, 2), (10, 7), (3, 5). Find the transformation that maps the square with vertices (0, 0), (1, 0), (1, 1), (0, 1) to ABCD and hence find the area of the parallelogram.
6. The square with vertices (0, 0), (1, 0), (1, 1), (0, 1) is mapped to ABCD under the transformation matrix $\begin{pmatrix} 5 & -1 \\ 2 & 4 \end{pmatrix}$. Find the vertices of ABCD and its area.
7. A transformation **T** is equivalent to the shear $\begin{pmatrix} 1 & 2 \\ 0 & 1 \end{pmatrix}$ followed by a rotation of 90° about O. Find the single matrix associated with **T** and show that **T** does not change the area.
8. The matrix $\begin{pmatrix} 2 & 3 \\ 3 & a \end{pmatrix}$ represents a transformation in which area remains unaltered. Find the two possible values of a.

Part 9: Statistics and probability

9.1 Display of information

1. *Bar chart* In a bar chart, the length of each bar or column represents the total as shown on a scale.
   For example, the stopping distances of vehicles are shown against a scale of metres.

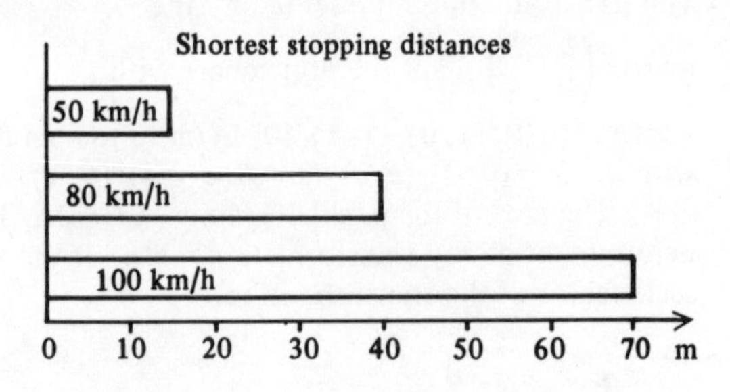

2. *Histogram* In a histogram, the area of each rectangle is in proportion to the total data. Widths may vary according to the information given, but heights will have to be adjusted so that the areas represent totals.

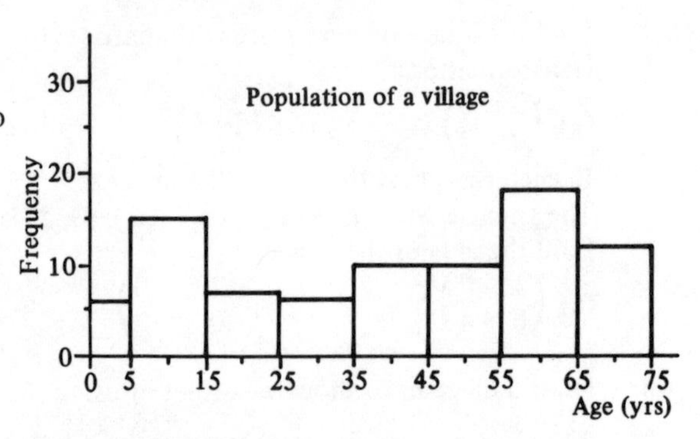

For example, the graph shows the distribution of population in age groups for a certain village. The information is in the table below.

Age (years)	under 5	5–15	15–25	25–35	35–45	45–55	55–65	over 65
Numbers of people	3	15	7	6	10	10	18	12

The height of the 'under 5' column is drawn to 6 on the graph because it is only half the width of the other columns.
For the graph, it is assumed that the 'over 65' group spans 10 years.

3. *Pie chart* A pie chart shows the division of a total into its constituent parts. Each angle at the centre has to be worked out before drawing the diagram.
   For example, the way a schoolboy spends a school day is shown.

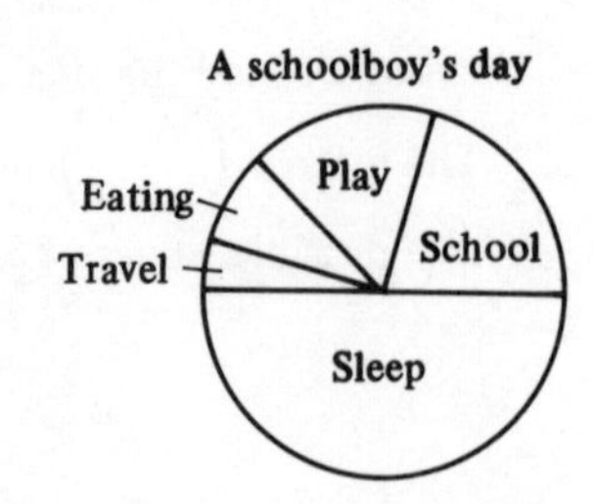

Activity	sleep	school	travel	eating	play
Hours	12	5	1	2	4

First find the angles at the centre.
The total of 24 hours is given by 360°; so 1 hour is given by $\frac{360^\circ}{24} = 15^\circ$.

Exercise 9.1

1. The column graph shows the total number of cars, lorries, vans, and motor-cycles counted on a busy road in 1 hour. From the graph, find:
   (a) the total number of vehicles
   (b) how many more cars there were than vans
   (c) what percentage of the total vehicles were lorries.

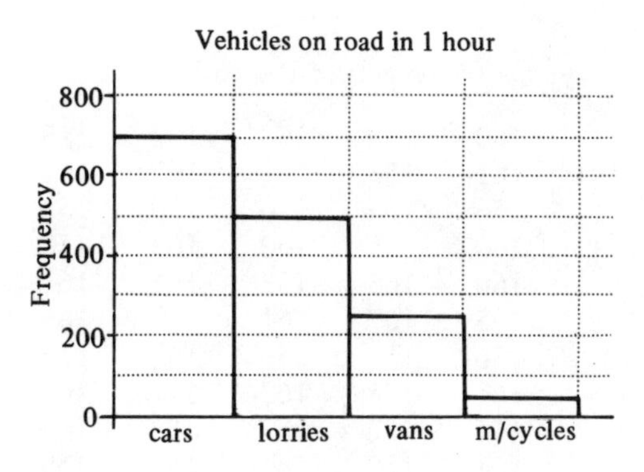

2. The following table gives the sizes of secondary schools in England and Wales in 1971.

School size (pupils)	under 200	201–400	401–600	601–800	801–1000	1001–1500
Number of schools	151	2000	1662	1160	568	466

Draw a histogram to represent this information using a scale of 1 cm per 100 schools on the vertical axis and a width of 2 cm per column for the first five columns. On this scale, what is the width and height of the last column?

3. The areas of the continents in 1 000 000 km^2 are approximately as shown:

Continent	Africa	Antarctic	Asia	Australia	Europe	N. America	S. America
Area (1 000 000 km^2)	12	14	44	8	10	24	18

   (a) What area does this give for the total land mass of the Earth?
   (b) Draw a pie chart to show the area of the Earth covered by each continent, working out each angle to the nearest 1°.

4. In a certain school there are 96 fifth-formers. For part of the week they divide into four option groups with the following numbers in each group: Physics 24, Biology 32, General Science 12, Rural Science 28. Illustrate this information with a pie chart showing clearly the size of the angle of each sector.

5. A Local Authority declares that its rates are spent in the following proportions given as pence per £: Education 72p, Health 12p, Roads 9p, Police and Fire Services 7p. Illustrate this information using:
   (a) a bar chart (b) a pie chart showing the angle of each sector.

6. The lengths of 161 crane fly larvae found in a badly infested garden were as shown below.

Length (mm)	22–24	24–26	26–28	28–30	30–32	32–34	34–36	36–38	38–40
Number found	6	9	14	20	28	30	26	24	4

Draw a histogram of this information using:
(a) intervals of 2 mm (b) intervals of 4 mm, starting at 20 mm.

9.2 Collection and representation of data

Here are the heights of 100 children, measured in cm to the nearest cm.

~~160~~	~~158~~	~~168~~	~~164~~	~~153~~	~~169~~	~~162~~	~~159~~	~~159~~	~~154~~
161	156	161	167	162	163	158	159	169	165
156	166	150	156	156	158	155	166	160	160
160	158	160	162	158	166	166	157	165	162
160	166	155	161	162	162	159	164	154	159
155	167	161	159	166	151	159	164	160	165
156	158	156	163	156	161	156	155	165	154
161	155	163	162	164	160	160	170	160	159
157	152	157	152	167	162	157	168	165	160
161	158	164	159	159	162	168	160	162	165

Information like this has to be sorted before any sense can be made of it.

1. Draw up a *frequency table*, dividing the *range* into *class intervals.* Usually 7 to 10 class intervals give a good idea of the shape of the distribution.
   In this example the smallest value is 150 cm and the largest is 170 cm. The range is therefore 21 cm.

Frequency table using class intervals of width 2 cm.

Height (cm)		Frequency
150–151		
152–153	/	
154–155	/	
156–157		
158–159	///	
160–161	/	
162–163	/	
164–165	/	
166–167		
168–169	//	
170–171		
	Total	

2. Put a tally mark in the frequency table for every number you cross off. The first row has been done. Grouping the tally marks in fives makes it easier to add up the frequency for each class interval.
3. Check that the total of the frequency column gives the overall total of the original information.

Exercise 9.2

1. Copy and complete the frequency table for the heights of the 100 children in the example on the opposite page. Draw a block graph of the distribution.
2. Draw up a frequency table for the heights of the 100 children in the example on the opposite page using class intervals of 3 cm and starting with 150–152 cm. Draw a block graph of the distribution.
3. The following scores were obtained from 100 throws of a die:

2	4	3	1	6	3	2	6	3	3	1	3	4	2	1
4	5	6	1	2	3	6	1	5	4	4	3	4	5	4
3	4	6	3	6	5	4	6	4	5	6	4	3	6	1
6	5	4	2	5	3	3	1	4	3	4	6	2	6	4
3	1	5	4	4	6	3	4	3	6	4	4	6	2	6
1	1	6	4	6	1	2	6	3	6	2	1	2	3	1
5	3	3	5	3	2	6	4	6	4					

(a) Draw up a table to show the frequency of each score
(b) Draw a block graph of the results
(c) Do you consider the die to have been biased–if so, to which corner?

4. Metal rods of 10 cm length are produced by a machine. Accurate measurements with a micrometer show that they lie between 99 mm and 101 mm. A sample of 50 gave the following readings in mm:

99·7	99·3	99·8	100·2	100·8	100·1	100·0	99·9	99·7	100·4
99·1	100·3	99·4	100·0	100·3	100·3	100·2	100·4	100·8	100·5
100·5	99·4	100·2	100·6	100·0	100·6	100·7	100·9	99·8	101·0
100·1	100·4	100·5	99·8	100·2	99·5	100·1	99·8	100·3	100·2
99·9	100·1	99·7	100·3	100·0	100·4	100·7	99·6	100·2	100·9

Draw a frequency table for these measurements using:
(a) a class interval width of 0·3 mm starting at 99·0–99·2
(b) a class interval width of 0·4 mm starting at 99·0–99·3.
Draw a block graph for each frequency table.

5. Here are the test scores for a group of 30 children in Arithmetic and in English.

Arithmetic

2	3	5	7	9	10
10	11	11	12	12	13
13	14	14	14	15	15
15	16	16	16	17	17
17	18	18	19	20	20

English

6	7	9	10	11	11
12	12	12	13	13	13
14	14	14	14	14	15
15	15	15	16	16	16
16	17	17	17	18	18

Draw up a frequency table for each set of marks using class intervals 0–2, 3–5, 6–8 etc. Draw a block graph for each set of results. Comment on the difference between the distributions.

9.3 Averages: mode, median, mean

An *average* is a representive number.

Seventeen measurements are given in order in the column on the right.

	16 cm
	15 cm
	15 cm
	14 cm
	13 cm
	12 cm
	12 cm
	11 cm
median →	11 cm
	11 cm
	10 cm
mode →	10 cm
	10 cm
	10 cm
	9 cm
	9 cm
	8 cm

1. The *median* is the middle measurement in order. If there are n measurements in order, the median is the value of the $\frac{1}{2}(n+1)$ th measurement. There are seventeen measurements here, so the median is the value of the ninth.

   The median length is 11 cm.

2. The *mode* is the most frequently occurring measurement. Here there are four.

   The mode is 10 cm.

3. The *mean* is found by dividing the sum of all the measurements by the number of measurements.

   Here, the mean $= \dfrac{8+9+9+10+\ \dots\ +15+16}{17}$

   The mean $\approx 11{\cdot}5$ cm.

Example 1

A sample of 120 bees is taken from a hive and the age of each bee is estimated and recorded as follows:

Age (weeks)	under 1	1	2	3	4	5	6
Frequency	20	30	24	19	15	9	3

Draw a block graph to illustrate this information.

(a) What is the modal class? (b) What is the median age of the bees?

The graph is shown on the right. The modal class is the class containing the greatest number of bees. The modal class is the one-week old bees.

The median is the age of the 60th bee in order of age (strictly speaking, the age of the $60\frac{1}{2}$ th bee!). The median age is 2 weeks.

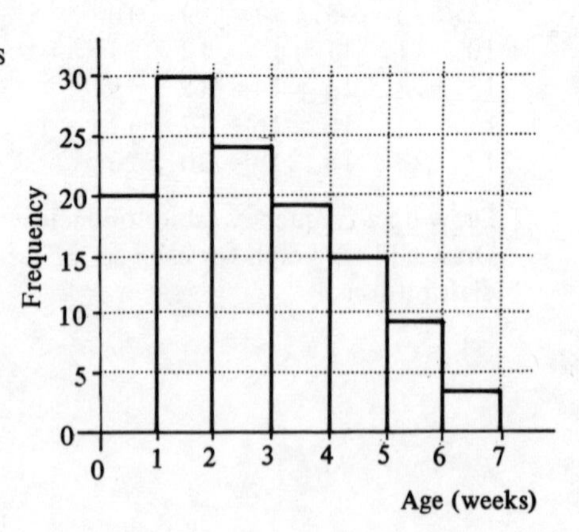

Exercise 9.3

1. Find the mode and the median score of the set of scores of a die recorded in question 3 of exercise 9.2 (page 161).
2. Find the mode and the median score of the set of micrometer readings recorded in question 4 of exercise 9.2 (page 161).
3. Find the mode and median test score for each set of test scores for 30 children given in question 5 of exercise 9.2 (page 161).
4. Find the mode and median shoe size of 30 children whose shoe sizes are:

   4 2 3 4 4 5 2 5 3 2 3 3 2 3 1
   2 3 2 4 6 7 4 2 4 3 2 1 3 1 3

5. Calculate the mean shoe size of the 30 children in question 4 above.
6. The examination marks of 20 pupils are: 72, 41, 87, 56, 64, 71, 45, 91, 38, 69, 54, 56, 58, 61, 83, 92, 62, 73, 57, 43. From the definition of median, there should be as many marks above the median as below. What value is the median mark?
7. The block graph shows the family sizes of 35 children.
   (a) What is the mode?
   (b) What is the median?
8. Find the mode and median heights of a group of 10 children whose heights in cm are: 119, 120, 121, 121, 121, 123, 124, 124, 125, 128.

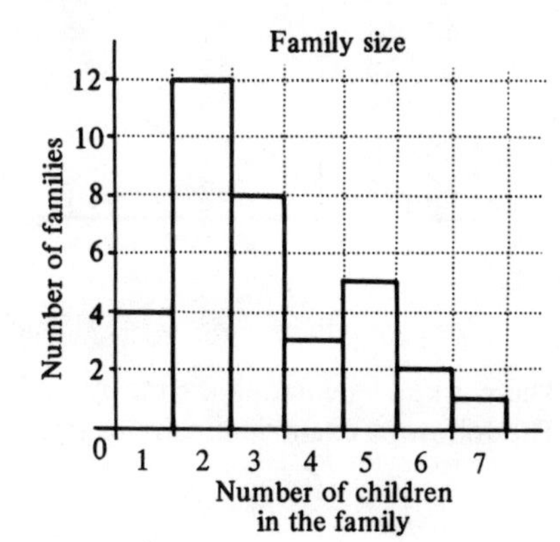

Example 2

Calculation of median. In question 6 of exercise 9.1 (page 159) the lengths of 161 crane fly larvae are given.

The median is the length of the 81st in order; this occurs in the class interval 32–34 cm. If we assume that the 30 larvae found in this interval are equally distributed, we can calculate the median length.

Number of larvae < 32 cm $= 6 + 9 + 14 + 20 + 28 = 77$
Number in 32–34 cm required to reach 81st $= 81 - 77 = 4$
Number of larvae in 32–34 cm class $= 30$

$$\text{Median} = 32 + \frac{4}{30} \times 2 = 32 + 0{\cdot}27 = 32{\cdot}27 \text{ cm.}$$

9. The table below gives the distribution of 100 metal rods according to length:

Length (cm)	150–155	155–160	160–165	165–170	170–175	175–180
Frequency	8	13	24	25	19	11

   Calculate the median length, taking it to be the 50th rod.

10. The lengths of 100 mice were carefully measured to the nearest mm.

Length (mm)	25–30	30–35	35–40	40–45	45–50	50–55	55–60	60–65
Frequency	3	12	20	34	25	4	1	1

   (a) In which class interval does the median lie? (b) Calculate the median length of the mice.

9.4 Calculating the mean

When data has been collected into a frequency table, the mean can be calculated direct from the table. The example below illustrates the method.

Example 1

Calculate the mean length of the 100 mice whose measurements are given in question 10 of exercise 9.3 on page 163.

Length (mm)	Mid-interval value x	Frequency f	fx
25–30	27·5	3	82·5
30–35	32·5	12	390
35–40	37·5	20	750
40–45	42·5	34	1445
45–50	47·5	25	1187·5
50–55	52·5	4	210
55–60	57·5	1	57·5
60–65	62·5	1	62·5
		Total fx =	4185

Mean length $= \frac{4185}{100} = 41{\cdot}85$ mm.

The work of calculating the mean can be greatly simplified by using an *assumed* or *working mean.* The following example illustrates the method.

Example 2

Calculate the mean length of the crane fly larvae from the data given in question 6 of exercise 9.1 on page 159, using a working mean of 33 mm.

Class interval (mm)	Mid-interval value x	Difference from working mean d	Frequency f	fd
22–24	23	−10	6	−60
24–26	25	−8	9	−72
26–28	27	−6	14	−84
28–30	29	−4	20	−80
30–32	31	−2	28	−56
32–34	33	0	30	0
34–36	35	2	26	52
36–38	37	4	24	96
38–40	39	6	4	24
			Total fd =	−180

Mean length $= 33 - \frac{180}{161} = 33 - 1{\cdot}12 \approx 31{\cdot}9$ cm.

Exercise 9.4

1. Calculate the mean of the set of numbers:
(a) 804, 882, 817, 853, 824
(b) 106 421, 105 987, 106 204, 106 311, 105 870
(c) 8·62, 8·58, 8·57, 8·65, 8·64.
2. Calculate the mean of the set of numbers:
(a) 18 047, 18 053, 18 021, 18 061, 18 018
(b) 0·0017, 0·0009, 0·0081, 0·0054, 0·005, 0·0005
(c) $9{\cdot}2 \times 10^{-3}$, $9{\cdot}9 \times 10^{-3}$, $9{\cdot}7 \times 10^{-3}$, $9{\cdot}4 \times 10^{-3}$, $1{\cdot}1 \times 10^{-2}$, $1{\cdot}2 \times 10^{-2}$, $9{\cdot}8 \times 10^{-3}$, $9{\cdot}6 \times 10^{-3}$, $9{\cdot}9 \times 10^{-3}$, $1{\cdot}3 \times 10^{-2}$.
3. The frequency table gives the scores of a pair of dice obtained in 100 throws. Calculate the mean score using a working mean of 8.

Score	2	3	4	5	6	7	8	9	10	11	12
Frequency	2	4	6	8	10	16	15	17	11	6	5

4. Calculate the mean of the set of numbers: 6, 2, 1, 5, 3, 1, 2, 4, 4, 3, 4, 5, 6, 2, 1, 1, 3, 2, 2, 5.
5. A record is made of the number of occupants of 50 cars and produces the following results: 1, 1, 1, 1, 2, 2, 1, 2, 2, 1, 1, 1, 1, 1, 4, 2, 1, 1, 3, 1, 1, 1, 1, 1, 2, 2, 1, 1, 3, 1, 2, 1, 1, 1, 1, 2, 1, 1, 3, 1, 1, 2, 4, 3, 1, 1, 2, 1, 1, 2.
Construct a frequency table of these results and use it to find the mean number of occupants per car.
6. The ages, in years and months of 40 children on 1 September are given in this table:

Age	12.0	12.1	12.2	12.3	12.4	12.5	12.6	12.7	12.8	12.9	12.10	12.11
Frequency	1	3	2	4	5	7	3	4	1	3	4	3

(Note that 12.10 means 12 years 10 months).
(a) Calculate the mean age of the 40 children.
(b) What will be their mean age on 1 December?
7. The lengths of a sample of 100 worms are as shown in the table:

Length (mm)	70–80	80–90	90–100	100–110	110–120	120–130	130–140
Frequency	5	8	25	30	20	10	2

Calculate the mean length using a working mean of 105 mm.
8. Calculate the mean length of the worms in question 7 using a working mean of 95 mm.
9. The following marks were obtained by a group of 80 candidates in a mathematics examination:

71	63	24	64	76	20	64	20	55	71	29	74	82	33	41	45
77	43	32	47	38	59	26	36	73	65	71	36	39	82	62	82
97	4	24	60	63	54	83	2	24	72	51	64	49	91	61	18
39	16	90	72	63	54	7	64	58	94	53	21	31	51	19	13
72	44	59	34	38	28	45	83	64	71	19	59	73	84	46	26

(a) Draw up a frequency table using intervals 0–9, 10–19, 20–29, etc.
(b) Calculate the mean mark from the frequency table taking the mid-interval values at 4·5, 14·5, 24·5, etc. and using a working mean of 44·5.
10. Calculate the mean mark of the candidates in question 9 above using a working mean of 54·5.

9.5 Cumulative frequency

In the example on page 160, the heights of 100 children are given to the nearest 1 cm. The frequency table uses class widths of 2 cm. The interval 152–153 cm of these children will contain all children whose heights range from 151·5 to 153·5 cm.

A *cumulative frequency table* gives the total up to each class boundary. Here is the cumulative frequency table for the heights of the 100 children:

Height (cm)	Frequency	Cumulative frequency	Heights of children represented by cumulative frequency
150–151	2	2	not more than 151·5 cm
152–153	3	5	not more than 153·5 cm
154–155	8	13	not more than 155·5 cm
156–157	12	25	not more than 157·5 cm
158–159	17	42	not more than 159·5 cm
160–161	19	61	not more than 161·5 cm
162–163	13	74	not more than 163·5 cm
164–165	11	85	not more than 165·5 cm
166–167	9	94	not more than 167·5 cm
168–169	5	99	not more than 169·5 cm
170–171	1	100	not more than 171·5 cm

A graph showing cumulative frequency is called an *ogive.* The cumulative frequency is plotted at the upper end of each interval. The graph can be used to find measures of *dispersion.*

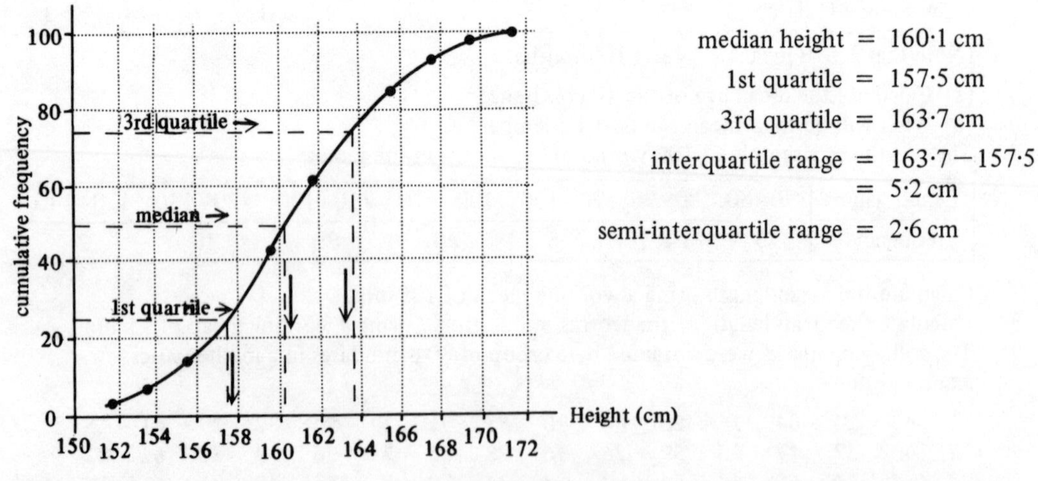

When measurements have been arranged in order, from the least to the greatest, the values of the measurements that divide the distribution into quarters are the *first quartile*, the *median*, and the *third quartile.* These are called the 25th, the 50th, and the 75th percentile.

The *interquartile range* measures the difference between the two quartiles. The semi-interquartile range is half the interquartile range.

Exercise 9.5

1. Draw up a cumulative frequency table for the size of schools given in question 2 of exercise 9.1 (page 159). What is the total number of schools? Draw a cumulative frequency graph and use it to find the value of the median and the interquartile range.
2. Draw up a cumulative frequency table for the length of crane fly larvae given in question 6 of exercise 9.1 (page 159). Use the graph to estimate the median length.
3. Draw up a cumulative frequency table for the age of bees given in example 1 on page 162. Use the intervals 'under 1 week', 'under 2 weeks', etc. From the table plot a cumulative frequency graph and hence estimate the value of the 25th, the 50th and the 75th percentile, and the semi-interquartile range.
4. Draw up a cumulative frequency table for the family size of the 35 children of question 7 in exercise 9.3 (page 163). What is the family size of the median, of the first quartile, and of the third quartile? (Answers are whole numbers.)
5. Draw a cumulative frequency graph for the length of the 100 rods in question 9 of exercise 9.3 (page 163).
   (a) Use the graph to find the value of the first and of the third quartile.
   (b) Find the semi-interquartile range as a percentage of the range.
6. Draw a cumulative frequency graph for the length of the 100 mice given in question 10 of exercise 9.3 (page 163).
   (a) Find the value of the 10th and of the 90th percentile.
   (b) Show that approximately 80% of the mice span $\frac{2}{5}$ of the range.
7. Draw up a cumulative frequency table for the age of the 40 children given in question 6 of exercise 9.4 (page 165). Hence estimate the median age.
8. Draw a cumulative frequency graph for the length of the 100 worms in question 7 of exercise 9.4 (page 165). Use it to find the approximate value of k where
   range $= k \times$ semi-interquartile range.
9. The following three graphs are each a cumulative frequency graph for a certain distribution. From the graphs, sketch the histogram (block graph) of the distribution using the same intervals.

(a)

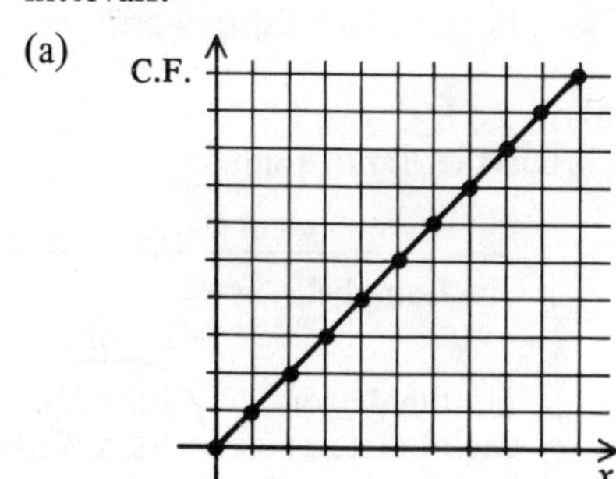

(b)

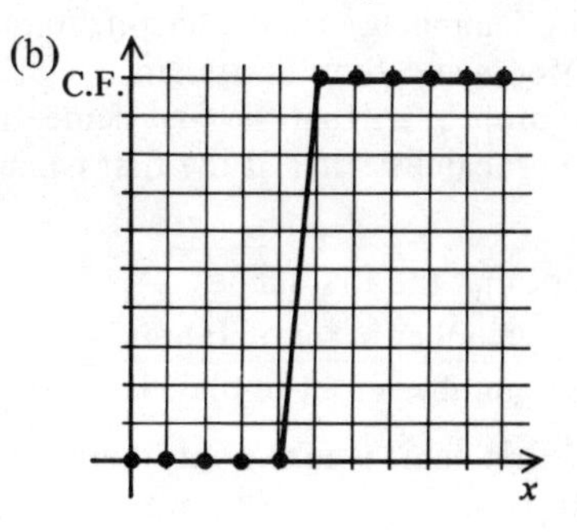

(c)

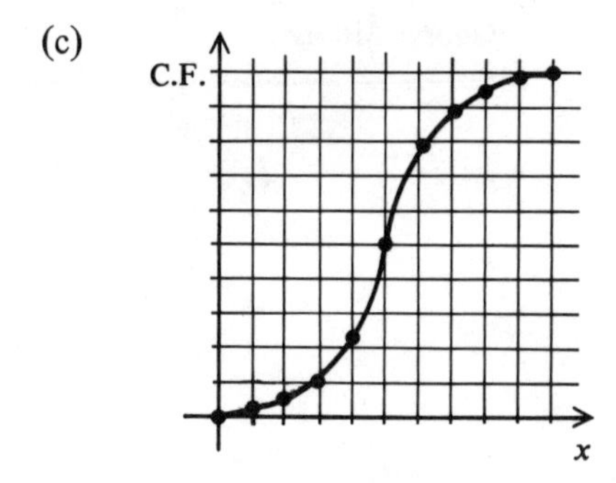

9.6 Simple probability

All simple probability theory depends on the idea of equally probable events. The probability p of an event occurring out of n equally probable events is $\frac{1}{n}$.

The probability of a certainty is 1.
The probability of an impossibility is 0.

If the probability of an event is p and the probability that it does not take place is q, then $p+q=1$.

Example 1

An ordinary pack of 52 playing cards is shuffled and dealt. Find the probability that:
(a) the first card is an Ace; (b) the second card is a Heart.

(a) There are 4 Aces in 52 cards.
Probability that the first card is an Ace $= \frac{4}{52} = \frac{1}{13}$.

(b) There are 13 Hearts in 52 cards.
Probability that the second card is a Heart $= \frac{13}{52} = \frac{1}{4}$.

Note: It makes no difference whether we are talking about the first card, the second card, or the last card. Each deal of a card is an event for which each of the 52 cards is equally probable.

Probabilities can be considered as numbers of elements in sets.

Example 2

Eight men's names were drawn in turn from a hat to find partners in a tennis tournament. Their names were: Jim Campbell, Martin Merson, John Walker, Jack Forster, Jack Butterworth, John Mills, George Moore and John Berryman.

Find: (a) the probability that the first name drawn is John;
(b) the probability that, if the first name is John, his partner is not John.

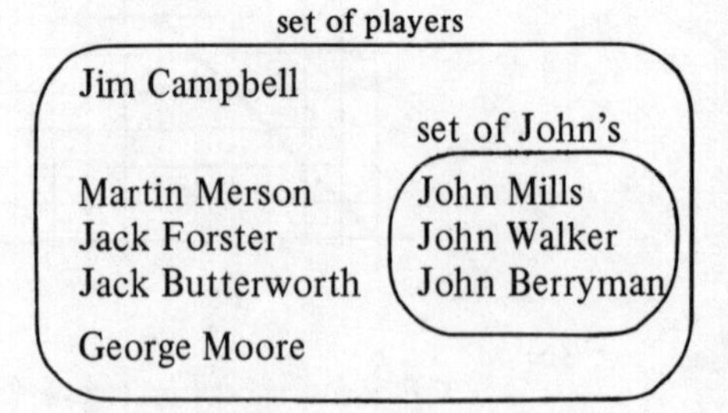

(a) probability that the first name is John $= \dfrac{\text{number in set of John's}}{\text{number in set of players}} = \dfrac{3}{8}$.

(b) After the first name is taken out, there are seven left of whom two are John's.

probability that the second name is John $= \dfrac{2}{7}$.

probability that the second name is not John $= 1 - \frac{2}{7} = \frac{5}{7}$.

Probabilities can be displayed on a graph.

Example 3

Two dice, one red, the other blue are thrown together. Find the probability that the total score is less than seven.

All possible scores are shown by the dots on the graph.
The subset of scores where the total is less than 7 is also shown.
From the graph the probability of a score less than 7 is $\frac{15}{36} = \frac{5}{12}$

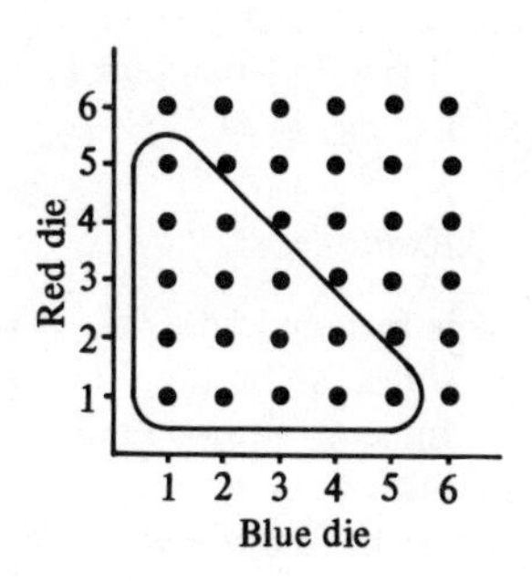

Exercise 9.6

An ordinary pack of 52 playing cards is shuffled and placed face down. The cards are turned over one by one. Find the probability that:

1. the first card is:
   (a) a Spade (b) not a Spade (c) the Ace of Spades.
2. the last card is:
   (a) the King of Hearts (b) a King (c) not a King.
3. if the first card is the Queen of Diamonds, the second card is:
   (a) a Queen (b) a Diamond (c) a Club.
4. if the first card is the Jack of Clubs, the second card is:
   (a) not a Jack (b) not a Club (c) a Heart.
5. if the first four cards are Kings, the next is:
   (a) a Heart (b) a King (c) the King of Hearts (d) not the King of Hearts.
6. if the first thirteen cards are Hearts, the next is:
   (a) not a Heart (b) a Club (c) a King (d) the King of Clubs.

7. A bag contains 26 cards each bearing a different letter of the alphabet. One card is withdrawn. State the probability that the letter on it is:
   (a) a vowel (b) not a vowel (c) B.
   The first card withdrawn is not replaced. If it bears the letter B, what is the probability that the second card is a vowel?
8. A bag contains 3 red and 5 green balls identical in all but colour. They are removed one by one and not replaced. Find the probability that:
   (a) the first ball is red (b) if the first ball is green, the second is red
   (c) if the first two balls are green, the third is also green.
9. Draw a graph like the one in example 3 to show all possible scores of a pair of red and blue dice. On your graph show those scores:
   (a) which total 5 (b) which total 9
   (c) in which only one of the dice shows a 3.
   Write down the probabilities of each of these events.
10. Two dice are thrown. Find the probability that:
    (a) the total score is greater than 6 (b) both dice display odd numbers.

9.7 Probability of independent events

Events are independent if the outcome of one does not affect the other. For example, if two out of ten coloured balls in a bag are red, and one is withdrawn and then replaced, the probability that it is a red ball is $\frac{1}{5}$, no matter how many times the experiment is repeated and no matter what the previous results may have been. Each selection is an independent event.

If two independent events are E_1 and E_2, the probability that both occur is $p(E_1 \textit{ and } E_2) = p(E_1) \times p(E_2)$.

Example

A die is thrown and a coin is tossed. Find the probability that:
(a) the score on the die is less than 5
(b) the coin shows a tail
(c) both these events occur.
Illustrate the answer with a graph.

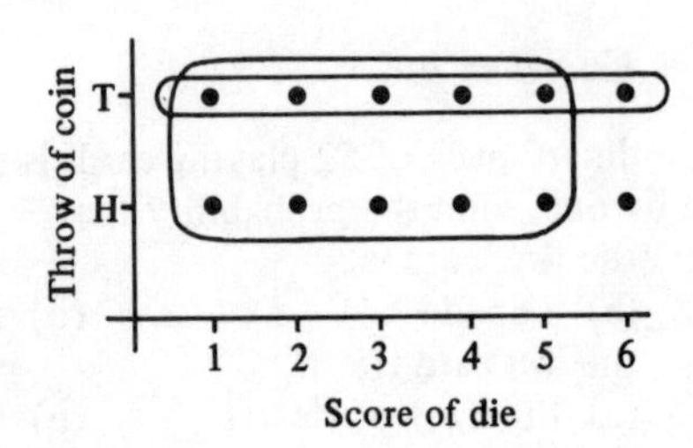

The graph shows the set of all possible results (tails T and heads H), together with the subset of tails and the subset of scores less than 6.

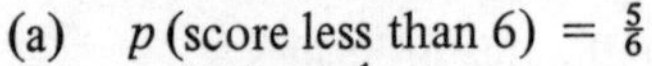

(a) $p(\text{score less than } 6) = \frac{5}{6}$
(b) $p(\text{tail}) = \frac{1}{2}$
(c) $p(\text{both events}) = \frac{5}{6} \times \frac{1}{2} = \frac{5}{12}$

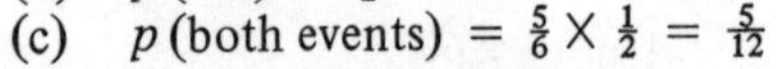

Exercise 9.7

1. Say whether or not these are independent events:
   (a) throwing a six with a die, picking an Ace from a pack of cards
   (b) picking an Ace from a pack of cards, picking an Ace from the same pack after returning the first card and shuffling
   (c) picking an Ace from a pack of cards, picking an Ace from the same pack without returning the first card.
2. Say whether or not these are independent events:
   (a) tossing a head, tossing a head again with the same coin
   (b) picking a Club from a pack of cards, picking a Club from a different pack
   (c) picking a Club from a pack of cards, picking a Club from the same pack after returning the first card and shuffling.
3. Find the probability of both events in question 1 (a) occurring.
4. Find the probability of both events in question 2 (b) occurring.
5. Two packs of cards are shuffled. One card is drawn from each pack. Find the probability that they will be in this order:
   (a) Heart, Heart (b) Heart, Club (c) King, King
   (d) King, Diamond (e) Diamond, King (f) Picture card, Ace.
6. One die is thrown twice. Find the probability that the scores will be, in order:
   (a) 6, 6 (b) even number, 3 (c) even number, odd number
   (d) greater than 4, not 1 (e) not 6, less than 3.
7. If x is selected at random from the set $X = \{1, 2, 3, 4, 5\}$ and y is selected at random from the set $Y = \{1, 2, 3, 4, 5\}$, find the probability that:
   (a) $x > 2$ *and* $y > 2$ (b) $x > 2$ *and* $y < 5$
   (c) x is odd *and* y is even (d) x and y are *both* even.

8. Two letters are written down. It is known that each is one of the first five letters of the alphabet. What is the probability that:
(a) both are consonants? (b) both are vowels?

9. One boy tosses a coin, a second cuts a pack of playing cards and a third throws a die. Find the probability of all three of the following results occurring:
(a) Head, King, 6 (b) Head, Hearts, even number
(c) Tail, King of Hearts, 3 (d) Head, Picture card, odd number.

10. The graph shows the set of all possible scores of a red and blue die and the subset of results for which the blue die score is even.
(a) What is the probability that the blue die score is even?
(b) Draw a copy of the graph and show the subset of scores for which the red die score is 1 or 5
(c) What is the probability that the red die score is 1 or 5?
(d) Find the probability that the blue die score is even *and* the red die score is 1 or 5.

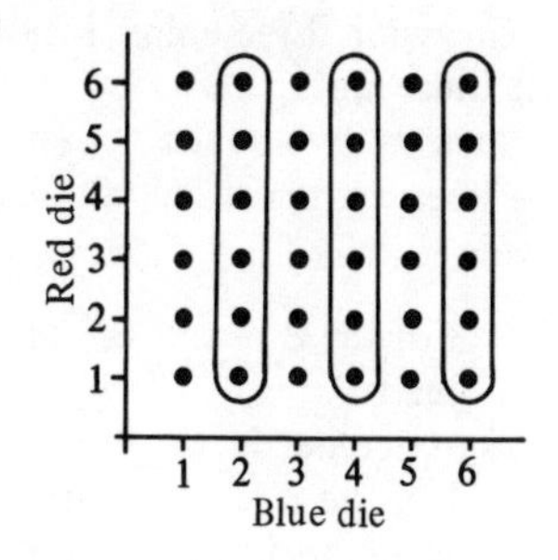

11. The probability of a man taking a packed lunch to work is $\frac{1}{5}$.
The probability of his going to work by car rather than bus is $\frac{3}{5}$.
The probability of his being phoned by a particular client is $\frac{1}{10}$.
Find the probability of all three events taking place and also the probability of none of the events taking place.

12. The probability of a particular football player scoring a goal is $\frac{1}{12}$.
The probability that he gets booked in any match is $\frac{2}{7}$.
The probability of his team winning is $\frac{2}{3}$.
Find the probability of all three events taking place and also the probability of none of the events taking place.

9.8 Use of tree diagrams

Probabilities of two or more events can be shown on a tree diagram.

Example 1

Use a tree diagram to show all possible results for throwing three coins. Find the probability that there are:
(a) exactly two heads (b) at least two heads.
The tree diagram shows all possible results (heads H, tails T).

(a) 3 of the 8 results contain exactly two heads.
$p(\text{two heads}) = \frac{3}{8}$.
(b) 4 of the 8 results contain at least two heads (including HHH).
$p(\text{at least two heads}) = \frac{4}{8} = \frac{1}{2}$.

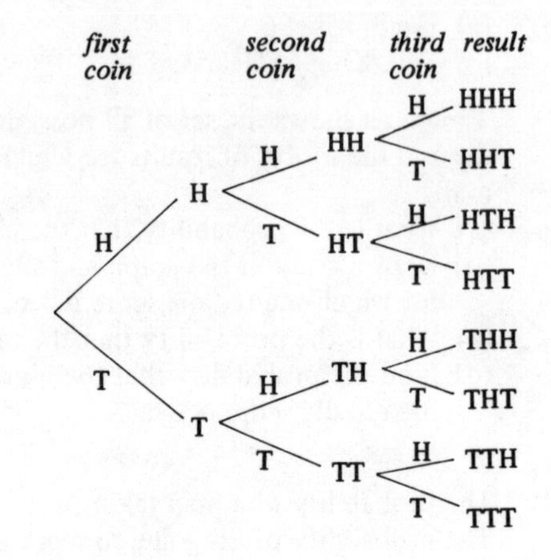

Example 2

A box contains two yellow beads and one blue bead. The beads are removed one at a time. Use a tree diagram to show all possible ways of withdrawing the beads. What is the probability that:
(a) the first bead is yellow? (b) the first two beads are yellow?

The possible results are yellow Y and blue B.
(a) 2 of the 3 possible results of the first selection are yellow.
$p(\text{yellow}) = \frac{2}{3}$.
(b) 2 of the 6 possible results after two selections are yellow.
$p(\text{yellow and yellow}) = \frac{2}{6} = \frac{1}{3}$.

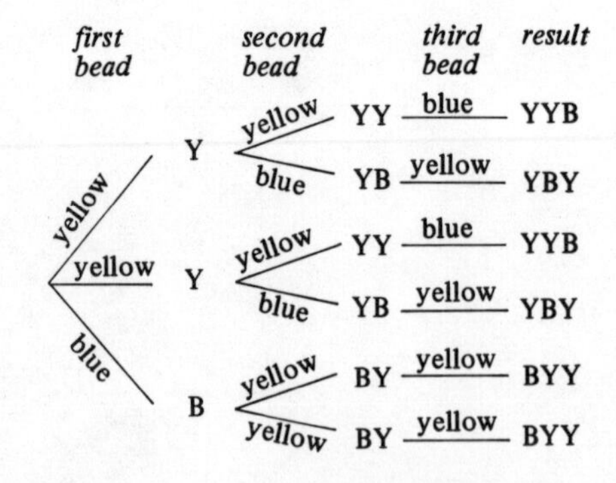

Tree diagrams can be used to compute probabilities directly. Each possible branch is given its probability value.

A *probability tree diagram* for Example 2 is shown here. Note that the sum of all probabilities at each stage is 1.

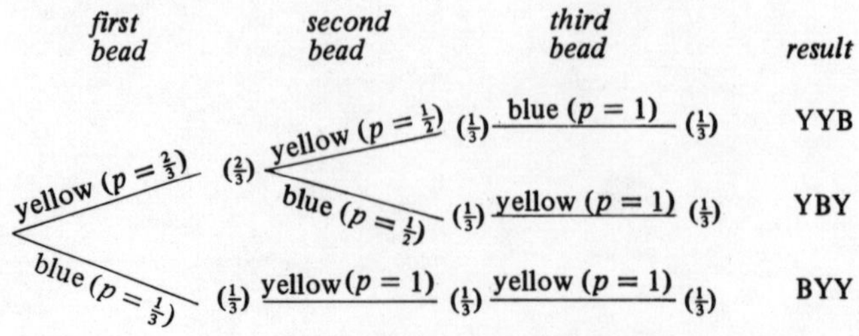

Exercise 9.8

1. Extend the diagram of Example 1 on page 172 to show all possible results of throwing 4 coins. Use the diagram to find the probability of:
(a) 2 heads (b) 3 heads (c) at least 3 heads.
2. A bag contains 3 beads, one black, one green and one red. The beads are removed one by one. Draw a tree diagram to show all possible ways in which they may be removed. What is the probability that the last one is red?
3. A bag contains 3 yellow beads and 2 blue beads. One bead is taken out and not put back, then a second bead is taken out. Draw a tree diagram to show all possible results. In how many of them are both the beads the same colour? What is the probability of this occurring?
4. Draw a tree diagram to show all possible results if a boy throws a die and a girl tosses a coin. What is the probability that the die will show a score of more than 4 and the coin will show a tail?
5. A boy picks a Heart from a pack of cards while a girl throws a die hoping for a 6.
Copy the probability tree diagram, completing the probability values.

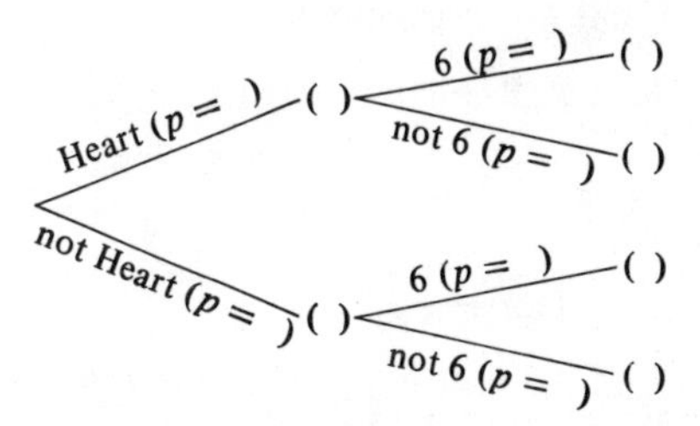

6. Draw a probability tree diagram similar to the one in question 5 to find the probability of drawing a King and throwing an even number.
7. A bag contains 5 yellow balls, 4 blue balls and 1 red ball. One ball is removed, its colour noted and replaced. This experiment is repeated. Draw a probability tree diagram for these two events. Write down the probability of removing:
(a) two yellow balls (b) two balls of the same colour
(c) one yellow and one red ball.
8. A picture card is drawn by chance from a pack of cards and replaced.
(a) What is the probability that a second card drawn from the pack is a picture card?
(b) Draw a tree diagram to show the probability of drawing two picture cards in succession.
(c) What is the probability of drawing two cards in succession which are not picture cards?
9. A bag contains ten coloured marbles of which just two are blue. They are withdrawn one by one and not replaced. What is the probability that the second marble is blue if the first is not blue?
Draw a tree diagram to show the probabilities of blue marbles occurring in the first two selections.
10. A bag contains 4 red counters, 3 green counters and 3 white counters. Find, by using a probability tree diagram, the probability that two counters taken from the bag are the same colour.

9.9 Probability laws

A pack of 52 playing cards contains the subsets of Aces (A), Picture cards (P) and Hearts (H). The probability that a card withdrawn at random belongs to one of these subsets is:

$$p\,(\mathrm{A}) = \frac{4}{52} = \frac{1}{13};\quad p\,(\mathrm{P}) = \frac{12}{52} = \frac{3}{13};\quad p\,(\mathrm{H}) = \frac{13}{52} = \frac{1}{4}.$$

The pack of cards is well shuffled and one card is withdrawn.

Example 1

Find the probability that it is a Picture card *or* an Ace.
This is the probability that the card lies in P or A.
P and A are disjoint sets.

So $p\,(\mathrm{P}\ \textit{or}\ \mathrm{A}) = p\,(\mathrm{P}) + p\,(\mathrm{A})$
$= \frac{3}{13} + \frac{1}{13} = \frac{4}{13}.$

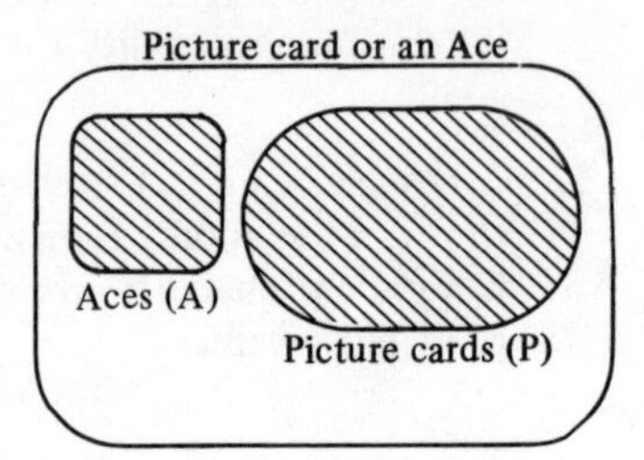

Example 2

Find the probability that it is a Picture card *and* a Heart.
This is the probability that the card lies in P ∩ H.

So $p\,(\mathrm{P}\ \textit{and}\ \mathrm{H}) = p\,(\mathrm{P}) \times p\,(\mathrm{H})$
$= \frac{3}{13} \times \frac{1}{4} = \frac{3}{52}.$

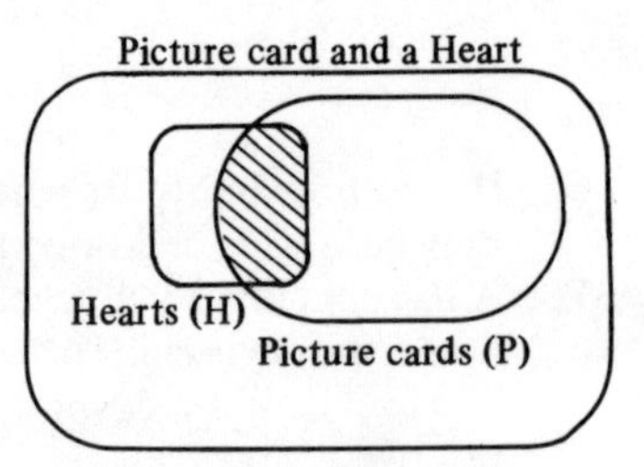

Example 3

Find the probability that it is a Picture card *or* a Heart.
This is the probability that the card lies in P ∪ H.

So $p\,(\mathrm{P}\ \textit{or}\ \mathrm{H}) = p\,(\mathrm{P}) + p\,(\mathrm{H}) - p\,(\mathrm{P}\ \textit{and}\ \mathrm{H})$
$= p\,(\mathrm{P}) + p\,(\mathrm{H}) - p\,(\mathrm{P}) \times p\,(\mathrm{H})$
$= \frac{3}{13} + \frac{1}{4} - \frac{3}{52} = \frac{11}{26}$

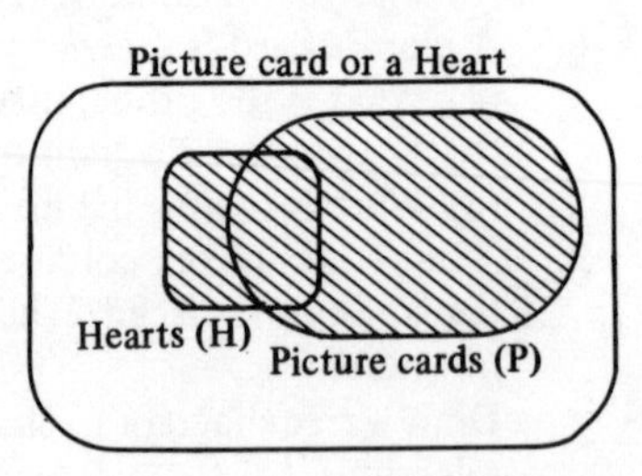

Note: If X and Y are independent events, the probability that:

both occur $= p\,(\mathrm{X}\ \textit{and}\ \mathrm{Y}) = p\,(\mathrm{X}) \times p\,(\mathrm{Y})$

either occurs $= p\,(\mathrm{X}\ \textit{or}\ \mathrm{Y}) = p\,(\mathrm{X}) + p\,(\mathrm{Y}) - p\,(\mathrm{X}) \times p\,(\mathrm{Y})$

The use of *or* includes *both*. For example, if a prize is won by picking a Heart *or* a King from a pack of cards, you certainly win if you pick the King of Hearts.

Exercise 9.9

1. X and Y are two independent events. If $p(X) = \frac{1}{3}$ and $p(Y) = \frac{3}{5}$, calculate the probability that:
   (a) X and Y both occur (b) either X or Y occurs.
2. A coin is tossed and a die is rolled. Calculate the probability that the result is:
   (a) a tail and a 3 (b) either a tail or a 3.
3. Two packs of cards are shuffled and the top card taken from each pack. Calculate the probability that:
   (a) they are both Aces (b) at least one of the them is an Ace.
4. For the two packs of cards in question 3, calculate the probability that at least one of the cards is a Heart.
5. Calculate the probability that a card selected at random from a pack will be:
   (a) a Club (b) a card less than 7 (c) a Club less than 7
   (d) either a Club or a card less than 7.
6. Calculate the probability that a card selected at random from a pack will be:
   (a) an Ace (b) the Ace of Diamonds (c) an Ace or a Diamond.
7. Two dice are rolled together. Calculate the probability that:
   (a) both are sixes (b) at least one is a 6.
8. Two dice are rolled together. Calculate the probability that:
   (a) both are greater than 4 (b) at least one is greater than 4.
9. In a class of 28 pupils, 15 like pop music, 12 like classical music and 3 like neither. (Obviously some like both.) One pupil is selected at random. What is the probability that:
   (a) he likes pop music? (b) he likes classical music?
   (c) he likes both types of music?
   Show how your answers can be used to calculate the probability that he likes either type of music.
10. A set of cards is made up as follows: $\frac{2}{5}$ are blue and $\frac{3}{5}$ are yellow; $\frac{1}{4}$ of the blue cards and $\frac{1}{4}$ of the yellow cards are marked with a circle and the rest with a square. If a prize is won by drawing a card that is either blue or is one that has a circle on it, what is the probability of winning a prize?

Part 10: Sets and structure

10.1 Modulo arithmetic

In Mod 6 arithmetic we only use 0, 1, 2, 3, 4, and 5. Addition and multiplication tables are shown on on the right.
The entries in the tables are the remainders when the results are divided by 6,
i.e. $3 + 5 = 2$ (Mod 6)
since $(3 + 5) \div 6 = 1$ remainder 2

+ Mod 6	0	1	2	3	4	5
0	0	1	2	3	4	5
1	1	2	3	4	5	0
2	2	3	4	5	0	1
3	3	4	5	0	1	2
4	4	5	0	1	2	3
5	5	0	1	2	3	4

× Mod 6	0	1	2	3	4	5
0	0	0	0	0	0	0
1	0	1	2	3	4	5
2	0	2	4	0	2	4
3	0	3	0	3	0	3
4	0	4	2	0	4	2
5	0	5	4	3	2	1

Exercise 10.1a

1. Find in Mod 6:
(a) $2 + 3$ (b) $3 + 4$ (c) $4 + 5$ (d) $5 - 2$ (e) $3 - 4$ (f) $2 - 5$
2. Working in Mod 6, find the number in the box:
(a) $2 + \square = 5$ (b) $1 + \square = 0$ (c) $4 + \square = 3$ (d) $4 - \square = 1$ (e) $1 - \square = 3$
3. Complete the addition table for Mod 5 which is started on the right. Use your results to write down:
(a) $3 + 1$ (b) $2 + 4$ (c) $4 + 3$
(d) $4 - 2$ (e) $3 - 4$ (f) $1 - 3$
Find the number in the $\square$:
(g) $3 + \square = 1$ (h) $2 - \square = 3$

+ Mod 5	0	1	2	3	4
0					4
1			3		
2					
3				1	
4		0			3

4. Make up a table for multiplication in Mod 5. Use your results to write down:
(a) 2×3 (b) 2×4 (c) 3×4 (d) 4×0
5. Working in Mod 5, find the number in the box:
(a) $2 \times \square = 4$ (b) $3 \times \square = 2$ (c) $\square \times 4 = 1$ (d) $\square \times 3 = 4$ (e) $1 \times \square = 0$
6. Find in Mod 7:
(a) $2 + 4$ (b) $3 + 5$ (c) $6 + 4$ (d) $5 - 3$ (e) $2 - 5$ (f) $4 - 6$
7. Find in Mod 7:
(a) 2×3 (b) 3×5 (c) 6×4 (d) 5×3 (e) 2×5 (f) 4×6
8. Working in Mod 7, find the number in the box:
(a) $2 \times \square = 6$ (b) $4 \times \square = 3$ (c) $5 \times \square = 1$ (d) $3 \times \square = 2$ (e) $\square \times 5 = 4$
9. Which result is the same as $(6 + 7)$ in Mod 9?
(a) $(3 + 4)$ in Mod 6 (b) $(5 + 6)$ in Mod 7 (c) $(5 + 8)$ in Mod 9
(d) $(10 + 6)$ in Mod 12
10. Which result is the same as (6×2) in Mod 9?
(a) (3×4) in Mod 6 (b) (5×2) in Mod 7 (c) (3×7) in Mod 9
(d) (4×6) in Mod 12
11. Find in Mod 8:
(a) $(5 + 4) + 7$ (b) $5 + (4 + 7)$ (c) $(5 + 4) \times 7$ (d) $5 + (4 \times 7)$
12. Find in Mod 6:
(a) $(2 + 4) + 3$ (b) $2 + (4 + 3)$ (c) $(2 + 4) \times 3$ (d) $2 + (4 \times 3)$
13. Find the number that can be written in the $\square$ to make the statement true:
(a) $2 + \square + 4 = 3$, in Mod 7 (b) $\square + \square + 5 = 2$, in Mod 9
(c) $2 \times \square \times 4 = 3$, in Mod 5 (d) $3\square + 4 = 1$, in Mod 9.

The Mod 7 multiplication table is shown on the right.

× Mod 7	1	2	3	4	5	6
1	1	2	3	4	5	6
2	2	4	6	1	3	5
3	3	6	2	5	1	4
4	4	1	5	2	6	3
5	5	3	1	6	4	2
6	6	5	4	3	2	1

Note:
The *identity* element is 1, since each number is unchanged when multiplied by 1. Each number has an *inverse.* For example the inverse of 3 is 5 since $3 \times 5 = 1$.

In Mod 7:

the equation $3x = 4$ has only one solution : 6
$(3 \times 6 = 4)$
the equation $x^2 = 2$ has two solutions : 3 and 4
$(3 \times 3 = 2,\ 4 \times 4 = 2)$

Exercise 10.1b

1. Find, in Mod 7, the solutions of:
(a) $x + 5 = 2$ (b) $x \times 3 = 5$ (c) $2x+3=4$ (d) $2x+4=3$ (e) $x^2 = 4$
2. Find, in Mod 5, the solutions of:
(a) $x + 4 = 1$ (b) $x \times 3 = 2$ (c) $3x+2=4$ (d) $3x+4=2$ (e) $x^2 = 4$
3. Using the Mod 6 multiplication table on the opposite page, write down:
(a) the identity element (b) the inverse of 5 (c) the numbers which have no inverses.
4. Write down the number of solutions, in Mod 6, of each equation:
(a) $x + 4 = 3$ (b) $5x = 3$ (c) $4x = 2$ (d) $3x = 0$ (e) $2x + 3 = 1$
(f) $x^2 = 3$
5. Complete the table on the right for multiplication in Mod 10, using {1, 3, 7, 9}:
(a) What is the identity element?
(b) Write down the inverse of:
(i) 1 (ii) 3 (iii) 7 (iv) 9
(c) Solve the equation $3x = 7$, in Mod 10.

× Mod 10	1	3	7	9
1				
3		9		
7	7			
9				1

6. Complete the table on the right for multiplication in Mod 10, using {2, 4, 6, 8}:
(a) What is the identity element?
(b) Write down the inverse of:
(i) 2 (ii) 4 (iii) 6 (iv) 8
(c) Solve the equation $8x = 2$, in Mod 10.

× Mod 10	2	4	6	8
2				2
4			6	
6				8
8	6			

7. Make up a table to show multiplication in Mod 8. Use your table to find the solutions of:
(a) $x \times 7 = 4$ (b) $x \times 3 = 7$ (c) $6x = 2$ (d) $4x = 0$ (e) $2x + 7 = 3$
(f) $x^2 = 1$
Do you agree that the equations $4x = 5$ and $2x + 7 = 6$ have no solutions in Mod 8?
8. What pairs of numbers x and y, $x \neq 0, y \neq 0$, will make $xy = 0$, when working in:
(a) Mod 6? (b) Mod 8? (c) Mod 10? (d) Mod 12? (e) Mod 24? (f) Mod 60?
9. A two-digit number p is such that $p = 0$, in Mod 3, and $p = 1$, in Mod 4. Find all the possible values of p.
10. Show that all numbers of the form $4(5n + 3)$ are equal to 0 in Mod 4, but are equal to 2 in Mod 5. Write down the numbers in this set between 0 and 100.
11. Show that all numbers of the form $6(8n + 3)$ are equal to 0 in Mod 6, but are equal to 2 in Mod 8. What are these numbers equal to in Mod 12?
12. Solve the equation $2(3x + 5) + 4 = 1$ in: (a) Mod 7 (b) Mod 11 (c) Mod 9

10.2 Set language and notation

A set is a collection of distinguishable elements.

A set can be defined (a) by listing the elements e.g. $\{2, 4, 6, 8, 10\}$
(b) by describing the elements precisely e.g. {the first five even numbers}.

$\in$ belongs to, is a member of, is an element of $3 \in \{2, 3, 4, 5, 6\}$
$\notin$ does not belong to $9 \notin \{2, 3, 4, 5, 6\}$
$\subset$ is a subset of, is contained within $\{3, 5\} \subset \{2, 3, 4, 5, 6\}$
$\supset$ contains $\{2, 3, 4, 5, 6\} \supset \{3, 6\}$
$\not\subset$ is not a subset of $\{3, 7\} \not\subset \{2, 3, 4, 5, 6\}$

$\cap$ intersection (the elements which are common) $\{3, 5, 7, 9\} \cap \{6, 7, 8, 9, 10\} = \{7, 9\}$
$\cup$ union (all the elements taken together) $\{3, 5, 7, 9\} \cup \{6, 7, 8, 9, 10\} = \{3, 5, 6, 7, 8, 9, 10\}$

&, or $\mathscr{U}$ the universal set (all the elements under consideration)

A′ the complement of A (all the elements of & which are not in A)
If & $= \{2, 3, 4, 5, 6, 7, 8\}$ and $A = \{3, 5, 7\}$ then $A' = \{2, 4, 6, 8\}$

{ }, or ϕ the empty set, the set with no elements $\{2, 4, 6, 8\} \cap \{1, 3, 5, 7\} = \phi$

$n(A)$ the number of elements in A $n\{2, 3, 4, 5\} = 4$ $n\{\ \} = 0$

Exercise 10.2

1. Describe the elements of the set:
(a) $\{1, 3, 5, 7, 9\}$ (b) $\{1, 4, 9, 16, 25\}$ (c) $\{1, 2, 4, 8, 16\}$ (d) $\{10, 20, 30, 40\}$
(e) {a, b, c, d, e} (f) {a, e, i, o, u} (g) $\{2, 3, 5, 7, 11\}$ (h) {b, c, d, f, g}
2. Describe the elements of the set:
(a) $\{3, 6, 9, 12, 15\}$ (b) $\{13, 14, 15, 16\}$ (c) {v, w, x, y, z} (d) {H, I, O, X}
(e) {a, c, e, g, i} (f) $\{21, 23, 25, 27, 29\}$
(g) $\{1, 8, 27, 64, 125\}$ (h) $\{80, 84, 88, 92, 96\}$
3. List the elements of the set:
(a) {the first four multiples of 7} (b) {the names of the suits in a pack of cards}
(c) {the factors of 12} (d) {the names of the countries in the E.E.C.}
4. List the elements of the set:
(a) {the first five multiples of 4} (b) {the letters in the alphabet between m and r}
(c) {the first six prime numbers} (d) {the months which have 31 days}
5. Copy and complete with the correct symbol:
(a) 6 ___ $\{2, 4, 6, 8, 10\}$ (b) $\{5, 9\}$ ___ $\{5, 6, 7, 8, 9\}$ (c) 11 ___ {the multiples of 3}
(d) {the factors of 24} ___ $\{3, 12\}$ (e) $\{3, 6, 9, 12, 15\}$ ___ $\{6, 12, 18, 24\} = \{6, 12\}$
6. Copy and complete with the correct symbol:
(a) $\{36, 64\}$ ___ {the square numbers} (b) 9 ___ {the prime numbers}
(c) $\{1, 4, 9, 16, 25\}$ ___ $\{4, 8, 12, 16\} = \{1, 4, 8, 9, 12, 16, 25\}$
(d) $\{3, 7\}$ ___ {the factors of 15}
7. Write *true* or *false* for each statement:
(a) $2 \in$ {the even numbers} (b) $\{7, 12, 19, 73\} \cap \{7, 21, 91, 73\} = \{12, 91\}$
(c) $\{7, 11\} \subset$ {the factors of 77} (d) $15 \notin$ {the multiples of 5}
8. Write *true* or *false* for each statement:
(a) $\{3, 5, 7, 9\} \subset$ {the prime numbers} (b) $\{a, c, e, g\} \cup \{a, e, i\} = \{a, c, e, g, i\}$
(c) $5 \subset$ {the odd numbers} (d) $\{0, 1, 2, 3\} \cap \{0, 5, 10, 15\} = \phi$

9. Complete the statement:
(a) $\{a, c, o, s\} \cap \{c, o, b, u\} =$ (b) $\{1, 4, 16, 64\} \cup \{1, 16, 256\} =$
(c) $\{1, 3, \ \ , 10\} \cap \{4, 6, 8, \ \ \} = \{6, 10\}$ (d) $\{j, l, \ \ , p\} \cup \{a, p\} = \{a, j, l, n, p\}$

10. Complete the statement:
(a) $\{1, 4, 7\} \cap \{2, 5, 8\} =$ (b) $\{1, 3, 5, 7, 9\} \cup \{1, 5, 9\} =$
(c) $\{2, 6, \ \ , \ \ \} \ \ \{ \ \ , \ \ , 10\} = \{8, 9\}$ (d) $\{a, m, r, z\} \cup \{a, \ \ , \ \ \} = \{a, m, r, y, z\}$

11. If A = {5, 10, 15, 20, 25}, B = {10, 20, 30}, and C = {1, 2, 3, 5, 20, 30} find:
(a) $A \cap B$ (b) $B \cap C$ (c) $A \cap C$ (d) $A \cup B$ (e) $B \cup C$ (f) $A \cup C$

12. If P = {3, 6, 9, 12, 15}, Q = {4, 8, 12, 16, 20}, and R = {1, 2, 3, 4, 6, 12} find:
(a) $P \cap Q$ (b) $Q \cap R$ (c) $P \cap R$ (d) $P \cup Q$ (e) $Q \cup R$ (f) $P \cup R$

13. Describe the elements of each set in questions 11 and 12.

14. Find for question 11:
(a) $A \cap B \cap C$ (b) $A \cup B \cup C$ (c) $(A \cap B) \cup C$

15. Find for question 12:
(a) $P \cap Q \cap R$ (b) $P \cup Q \cup R$ (c) $(P \cup Q) \cap R$

16. If & = {the whole numbers from 1 to 10}, A = {2, 4, 6, 8}, and B = {1, 3, 6, 10} find:
(a) A' (b) B' (c) $A \cap B$ (d) $(A \cap B)'$ (e) $A \cup B$ (f) $(A \cup B)'$

17. If & = {the square numbers less than 100}, P = {1, 9, 25, 49, 81}, and Q = {1, 4, 16, 64} find:
(a) P' (b) Q' (c) $P \cap Q$ (d) $(P \cap Q)'$ (e) $P \cup Q$ (f) $(P \cup Q)'$

18. Complete the statement:
(a) $n\{1, 5, 25, 125\} =$ (b) n {the square numbers less than 64}=
(c) n {the factors of 30} = (d) n {the prime numbers between 23 and 29} =

19. If A = {10, 13, 16, 19}, B = {5, 10, 15, 20}, and C = {11, 13, 15, 17, 19} find:
(a) $n(A)$ (b) $n(A \cap B)$ (c) $n(A \cap C)$ (d) $n(B \cup C)$ (e) $n(A \cup C)$

20. If P = {the factors of 12}, Q = {the whole numbers between 5 and 15}, and R = {the multiples of 7 less than 35} find:
(a) $n(P)$ (b) $n(Q)$ (c) $n(R)$ (d) $n(P \cap Q)$ (e) $n(Q \cap R)$
(f) $n(P \cap R)$ (g) $n(P \cup Q)$ (h) $n(Q \cup R)$ (i) $n(P \cup R)$ (j) $n(P \cap Q \cap R)$

21. If n(&) = 20, $n(A) = 7$ and $n(B) = 12$ find:
(a) $n(A')$ (b) $n(B')$ (c) greatest $n(A \cap B)$ (d) least $n(A \cup B)$

22. If n(&) = 16, $n(P) = 9$ and $n(Q) = 13$ find:
(a) $n(P')$ (b) $n(Q')$ (c) least $n(P \cap Q)$ (d) greatest $n(P \cup Q)$

23. If & = {the whole numbers from 1 to 20}, A = {the multiples of 3}, and B = {the even numbers}, write *true* or *false* for each statement:
(a) $A \cap B = \{3, 9, 15\}$ (b) $n(A) = 6$ (c) B' = {the odd numbers}
(d) $n(B') = 10$ (e) $n(A \cup B) = 12$ (f) $A' \cap B = \{2, 4, 8, 10, 14, 16\}$

24. If & = {the odd numbers less than 30}, A = {the square numbers}, B = {the multiples of 5} and C = {the factors of 225} find:
(a) $A \cap C$ (b) $n(B)$ (c) $B \cup C$ (d) C' (e) $n(A \cup B)$
(f) $(B \cap C)'$ (g) $n(A')$ (h) $(A \cup B) \cap C$ (i) $n(A \cap (B \cup C))$

10.3 Set algebra

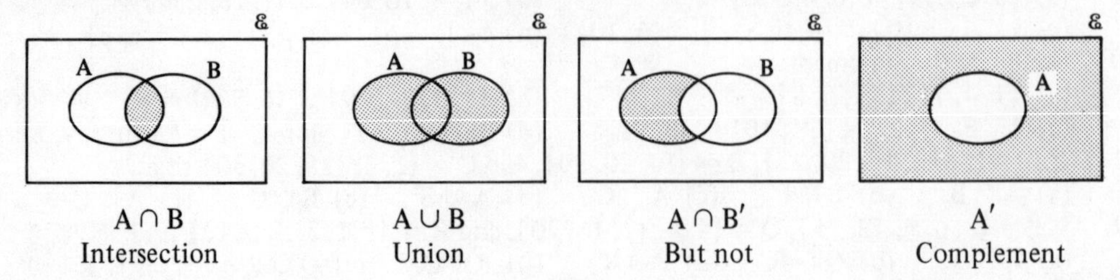

$A \cap B$ Intersection $A \cup B$ Union $A \cap B'$ But not A' Complement

Example 1

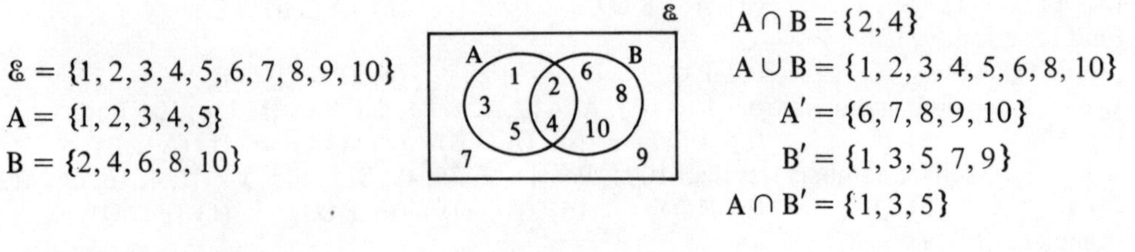

$\mathscr{E} = \{1, 2, 3, 4, 5, 6, 7, 8, 9, 10\}$
$A = \{1, 2, 3, 4, 5\}$
$B = \{2, 4, 6, 8, 10\}$

$A \cap B = \{2, 4\}$
$A \cup B = \{1, 2, 3, 4, 5, 6, 8, 10\}$
$A' = \{6, 7, 8, 9, 10\}$
$B' = \{1, 3, 5, 7, 9\}$
$A \cap B' = \{1, 3, 5\}$

Exercise 10.3a

1. Show the information as a Venn diagram.
   $\mathscr{E} = \{1, 2, 3, 4, 5, 6, 7, 8, 9, 10\}$, $A = \{1, 3, 6, 10\}$ and $B = \{1, 2, 4, 8\}$.
   Find (a) $A \cap B$ (b) $A \cup B$ (c) A' (d) B' (e) $(A \cap B)'$ (f) $(A \cup B)'$.
2. Show the information on a Venn diagram.
   $\mathscr{E} = \{31, 33, 35, 37, 39, 41, 43, 45, 47, 49\}$, $P = \{33, 35, 39, 45, 49\}$,
   $Q = \{31, 37, 41, 43, 47\}$.
   Find: (a) $P \cap Q$ (b) $P \cup Q$ (c) P' (d) Q' (e) $(P \cap Q)'$ (f) $(P \cup Q)'$.

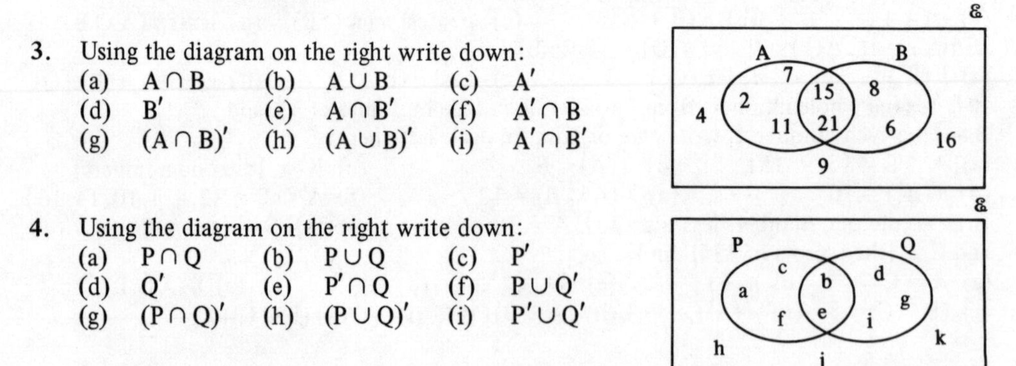

3. Using the diagram on the right write down:
   (a) $A \cap B$ (b) $A \cup B$ (c) A'
   (d) B' (e) $A \cap B'$ (f) $A' \cap B$
   (g) $(A \cap B)'$ (h) $(A \cup B)'$ (i) $A' \cap B'$
4. Using the diagram on the right write down:
   (a) $P \cap Q$ (b) $P \cup Q$ (c) P'
   (d) Q' (e) $P' \cap Q$ (f) $P \cup Q'$
   (g) $(P \cap Q)'$ (h) $(P \cup Q)'$ (i) $P' \cup Q'$
5. Describe, using symbols, the shaded region:

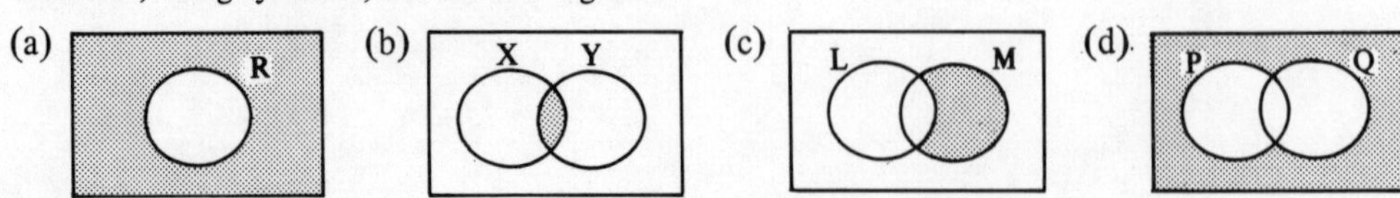

6. $\mathscr{E} = \{2, 4, 6, 8, 10, 12, 14, 16, 18, 20\}$, $A = \{4, 8, 12, 16, 20\}$ and $B = \{2, 8, 14, 20\}$.
   Find: (a) A' (b) B' (c) $A' \cap B'$ (d) $A \cup B$
   Show that $A' \cap B' = (A \cup B)'$.

7. $\mathscr{E} = \{a, b, c, d, e, f, g, h, i, j, k, l\}$, $P = \{a, c, e, g, j, h\}$ and $Q = \{b, d, f, h, j, l\}$.
Find: (a) P' (b) Q' (c) $P' \cup Q'$ (d) $P \cap Q$
Show that $P' \cup Q' = (P \cap Q)'$.
8. By shading a Venn diagram, or otherwise, show that:
(a) $(X \cap Y)' = X' \cup Y'$ (b) $(R \cup S)' = R' \cap S'$ (c) $(A' \cap B')' = A \cup B$
9. Use a diagram to simplify: (a) $A \cap A'$ (b) $A \cup A'$ (c) $A \cap \mathscr{E}$ (d) $A \cup \mathscr{E}$ (e) $\mathscr{E}'$
(f) $A \cap \phi$ (g) $A \cup \phi$ (h) $\mathscr{E} \cap \phi$ (i) $\mathscr{E} \cup \phi$ (j) ϕ'
10. Using a Venn diagram, or otherwise, simplify:
(a) $A \cap (A \cup B)$ (b) $P \cup (P \cap Q)$ (c) $(X \cup Y) \cup Y$ (d) $(R \cap S) \cap R$
11. Using a Venn diagram, or otherwise, simplify:
(a) $A \cap (A' \cup B)$ (b) $P' \cap (P \cup Q)$ (c) $(P' \cup Q) \cup Q'$ (d) $(R \cap S') \cup S$
12. Write *true* or *false* for each statement:
(a) $P \cap P' = \mathscr{E}$ (b) $A \cup B = B \cup A$ (c) $X \cup X' = \mathscr{E}$ (d) $R \cap S' = R' \cap S$

Example 2

$\mathscr{E} = \{1, 2, 3, 4, 5, 6, 7, 8, 9, 10\}$

$A = \{2, 4, 6, 8, 10\}$

$B = \{3, 6, 9, 10\}$

$C = \{2, 3, 4, 6\}$

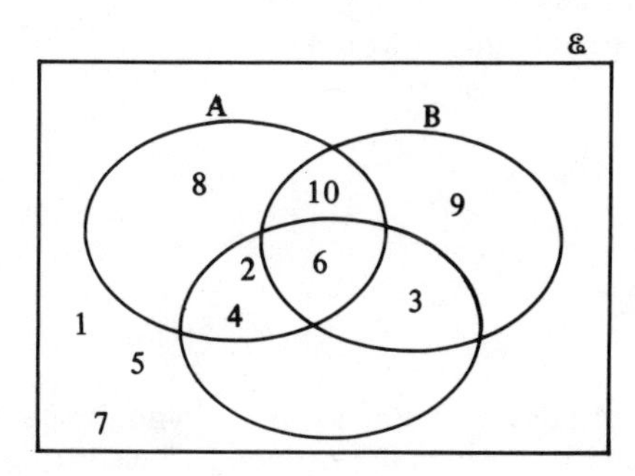

$A \cap B = \{6, 10\}$

$A \cap C = \{2, 4, 6\}$

$B \cup C = \{2, 3, 4, 6, 9, 10\}$

$A \cap B \cap C = \{6\}$

$A \cap C \cap B' = \{2, 4\}$

$(A \cup B \cup C)' = \{1, 5, 7\}$

$C \cap (A \cup B)' = \phi$

Exercise 10.3b

1. Show the information on a Venn diagram:
$\mathscr{E} = \{a, b, c, d, e, f, g, h, i, j, k, l\}$, $A = \{a, d, f, g, i, j, l\}$, $B = \{a, e, i, j\}$ and $C = \{b, c, f, g, i, j\}$.
Find: (a) $A \cap B$ (b) $A \cap C$ (c) $B \cup C$ (d) $A \cap B \cap C$ (e) $A \cap B \cap C'$
(f) $B \cap C \cap A'$
2. Show the information on a Venn diagram:
$\mathscr{E} = \{1, 2, 3, 4, 5, 6, 7, 8, 9, 10, 11, 12\}$, $P = \{3, 6, 9, 12\}$, $Q = \{2, 4, 6, 8, 10, 12\}$ and $R = \{1, 3, 6, 10\}$.
Find: (a) $P \cap Q$ (b) $P \cup R$ (c) $P \cap Q \cap R$ (d) $P \cup Q \cup R$ (e) $P \cap Q \cap R'$
(f) $P \cap (Q \cup R)'$
3. Using the diagram on the right write down:
(a) $R \cap S$ (b) $S \cup T$
(c) R' (d) $(R \cup T)'$
(e) $(S \cap T)'$ (f) $R \cap S \cap T$
(g) $R \cup S \cup T$ (h) $R \cap S \cap T'$
(i) $R \cap (S \cup T)'$ (j) $R \cup (S \cap T)'$
(k) $(R \cup S \cup T)'$ (l) $R' \cup (S \cap T)$

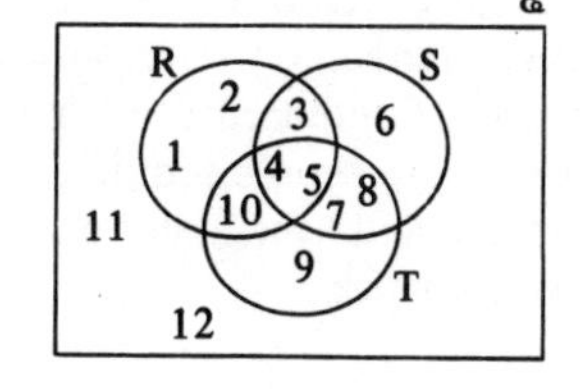

4. By using a Venn diagram, or otherwise, write *true* or *false* for each statement.
(a) $(A \cap B) \cup (A \cap C) = A \cap (B \cup C)$ (b) $(P \cup Q) \cap (R \cup Q) = (P \cap R) \cup Q$
(c) $(A \cap B \cap C)' = A' \cap B' \cap C'$ (d) $(X \cup Y \cup Z)' = X' \cap Y' \cap Z'$
5. Shade on a Venn diagram the regions which represent:
(a) $(A \cup B) \cap (C \cup B)'$ (b) $(P \cap Q) \cup (P \cup Q)'$ (c) $(X' \cup Y) \cap (Z \cup X)$

10.4 Problems using Venn diagrams

Example 1

There are 43 members in a youth club. 25 play badminton, 16 play chess and 20 play tennis. 5 play all three games, 2 play chess and tennis but not badminton, 12 play tennis and badminton, and 10 play only badminton. How many do not play any of these games?

Number in youth club (43)

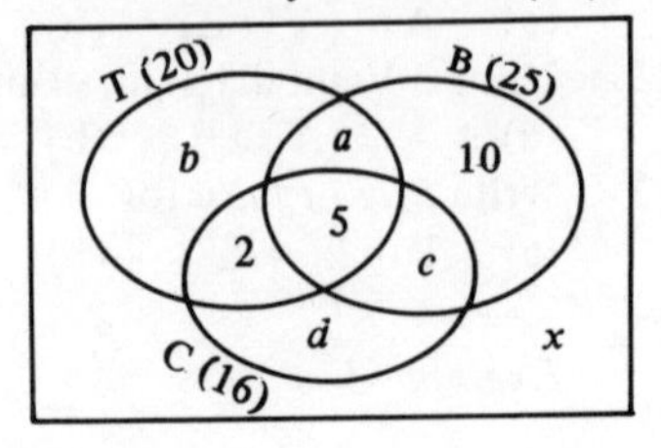

Certain information in this problem can be entered directly onto the diagram namely the 2, 5, and 10.
The number playing tennis and badminton only, denoted by a, can be found by subtracting the 5 who play all three games from 12. Since the total number playing each of tennis and badminton is known b and c can now be found, similarly d. The number not playing any game, x, is finally found by subtracting from 43 the total numbers within the three loops.

$$a = 12 - 5 = 7$$
$$b = 20 - 2 - 5 - 7 = 6$$
$$c = 25 - 5 - 10 - 7 = 3$$
$$d = 16 - 2 - 5 - 3 = 6$$
$$x = 43 - 39 = 4$$

Exercise 10.4a

1. During the summer holiday 24 boys were able to go to a wild life park and a car museum; 8 chose to go to both, 12 went to the wild life park and 17 went to the car museum. How many boys did not go to either?
2. 30 fifth-formers had the opportunity of doing a shorthand course and a first-aid course. 3 decided to do both, 18 chose the first-aid course and 7 did neither. How many chose: (a) shorthand? (b) shorthand only?
3. Of 53 anglers, 17 caught only carp, 25 caught only bream, and 9 were unlucky and caught neither. Find how many caught both.
4. There are 56 sixth-formers who are taking English, History or Geography; 32 take English, 24 take History and 29 take Geography; 7 take all three subjects. If 12 take History and Geography only, and 8 take Geography only, show that 22 must be taking only English.
5. When 92 housewives shop in a local dairy, 41 buy butter, cheese and yoghurt, 14 buy butter and cheese only, 9 buy yoghurt only. If 58 buy butter and yoghurt, 75 buy some butter and 68 buy some yoghurt find how many buy only cheese. (Assume all buy at least one.)
6. On booking their holiday at a travel agency, 72 people required French currency, 53 required German currency and 29 required Swiss currency. 19 required French and Swiss currency, 14 required French and German, whilst 7 required German and Swiss currency. If 44 of these people required French currency only, find how many required all three currencies, and how many people booking at the travel agency required at least one of these three currencies.

Example 2

There are 86 students taking languages; 25 take French and German, 17 take French and Spanish, and 5 take German and Spanish; 57 take French, 31 take Spanish and 42 take German. Find how many students take all three languages.

In this problem it is necessary to denote the number in the required region by a letter, say x. The numbers in the other regions may be found in terms of x. For example 25 take French and German, so if x take all three languages, $(25 - x)$ take French and Germany only. Similarly $(17 - x)$ take French and Spanish only, and $(5 - x)$ take German and Spanish only. The numbers taking one language only may then be found as is shown.

Number of students (86)

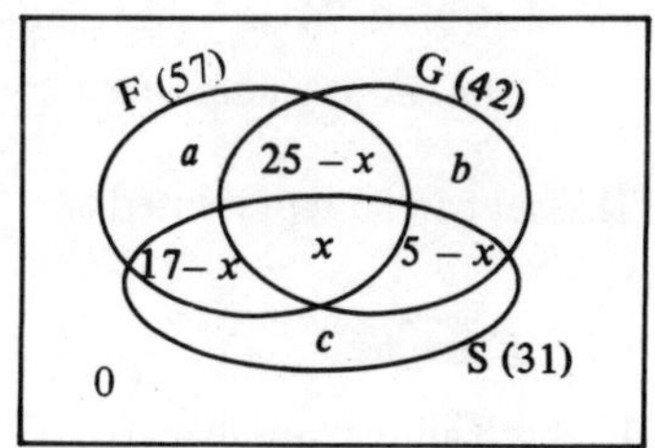

$$a = 57 - (25 - x) - x - (17 - x) = 15 + x$$
$$b = 42 - (25 - x) - x - (5 - x) = 12 + x$$
$$c = 31 - (17 - x) - x - (5 - x) = 9 + x$$

Since the total number of students within the three loops is 86 and equation may be formed. i.e. $86 = 57 + (5 - x) + (12 + x) + (9 + x)$.

Exercise 10.4b

1. Solve the equation above and find the numbers of students taking all three languages. Check that the total numbers within the 7 regions of the diagram come to 86 as is required.
2. There are 89 students taking sciences; 18 take Physics and Chemistry, 9 take Physics and Biology and 26 take Chemistry and Biology; 43 take Biology, 56 take Chemistry and 35 take Physics. How many students take all three subjects?
3. 47 gardeners grow roses, dahlias or chrysanthemums. At the local flower show, 25 show roses, 18 show dahlias and 21 show chrysanthemums; 12 show only roses, 3 show only dahlias and 4 show only chrysanthemums; 5 show all three flowers. How many show roses and dahlias but not chrysanthemums? How many do not send any entries to the flower show?
4. In a popularity poll three T.V. programmes A, B, and C were being compared. Of 100 viewers, 13 liked none of the three programmes; 56 liked A, 47 liked B, and 45 liked C. 18 liked A and B only, 12 liked B and C only, and 15 liked A and C only. Find how many of the 100 viewers liked all three programmes.
5. A newsagent sold Sunday papers to 79 people. 27 bought the *Sunday Times*, 58 the *Sunday Express*, and 35 the *News of the World.* 16 bought the *Times* and *Express* only, 18 bought the *Express* and *News of the World* only, whilst 1 person bought the *News of the World* and *Times* only. If all 79 people bought one of these papers, find how many bought the *Times* only.

10.5 Solution sets

The solution set for $x + 3 \leqslant 7$ is $\{0, 1, 2, 3, 4\}$ if x is a positive whole number,
or $\{x : x \leqslant 4\}$ if x is any real number.

$\{x : 2 \leqslant x < 7\}$ is a shorthand way of writing:

'the set of values of x between 2 and 7, including 2 but excluding 7'.

This set can be represented as 0 1 2 3 4 5 6 7 8 9 on the number line.

Example 1

If $\mathscr{E}$ = {all real numbers from 0 to 10 inclusive}, $A = \{x : 1 < x < 5\}$, and $B = \{x : 3 \leqslant x \leqslant 7\}$ find $A \cap B$ and $A \cup B$.

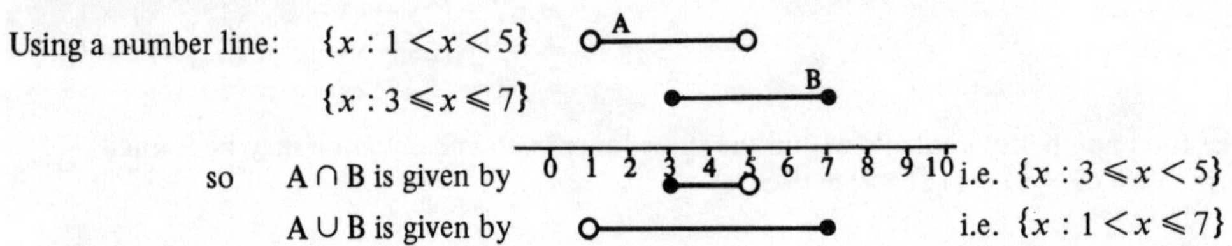

Exercise 10.5a

1. Assuming that you can use any whole number from 0 to 10 write down the solution set for:
(a) $x < 5$ (b) $x > 3$ (c) $x + 5 < 9$ (d) $x - 3 \geqslant 4$
2. Assuming that you can use any whole number from 0 to 20 find the solution set for:
(a) $x + 8 < 19$ (b) $x - 12 \geqslant 3$ (c) $2x + 5 \leqslant 17$ (d) $3x - 5 > 16$
3. Writing your answer as $\{x : x \quad\}$ find the solution set for:
(a) $x + 9 > 15$ (b) $x - 5 < 17$ (c) $5x + 4 > 39$ (d) $4x - 13 < 19$ (e) $14 - x > 9$
4. Writing your answer as $\{x : x \quad\}$ find the solution set for:
(a) $3x + 7 < 22$ (b) $6x - 5 > 19$ (c) $18 - 3x < 6$ (d) $2(3x + 5) > 52$
(e) $5 - x < 68 - 8x$
5. Using a number line, find the solution set for:
(a) $\{x : 1 < x < 9\} \cap \{x : 5 \leqslant x < 13\}$ (b) $\{x : 5 \leqslant x < 12\} \cup \{x : 2 < x \leqslant 11\}$
(c) $\{x : 4 < x < 17\} \cap \{x : 6 \leqslant x \leqslant 14\}$ (d) $\{x : 2 \leqslant x < 19\} \cup \{x : 7 < x \leqslant 19\}$
6. Describe each of A, B, C and D below as: $\{x : \quad x \quad\}$

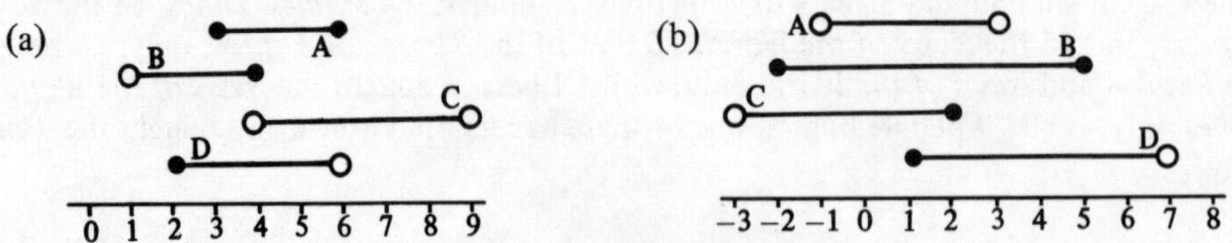

7. Using the information in question 6, write down for each diagram:
(a) $A \cap B$ (b) $B \cap C$ (c) $C \cap D$ (d) $A \cup B$ (e) $B \cup C$ (f) $C \cup D$ (g) $C \cap D'$
8. Writing your answer as $\{x : \quad\}$ find the solution set for:
(a) $\{x : 2x + 3 > 15\} \cap \{x : 4x - 7 < 41\}$ (b) $\{x : 5x - 1 < 19\} \cup \{x : 2x - 9 < 11\}$
(c) $\{x : 2 \leqslant x < 13\} \cap \{x : 4x - 3 > 21\}$ (d) $\{x : 4 < x \leqslant 9\} \cup \{x : 3x + 5 < 23\}$
(e) $\{x : 12 < x < 15\} \cap \{x : 17 - x > 8\}$ (f) $\{x : 6 - x > 1\} \cup \{x : 4 - x \leqslant 1\}$
(g) $\{x : 2x + 1 \leqslant 5x - 8\} \cap \{x : 4x - 1 > 11 - 2x\}$
(h) $\{x : 4/x > 2\} \cup \{x : 1 \leqslant x < 5\}$

Example 2

Find the solution set for $(x-1)(x-2)>0$.

The solution set may be found either by drawing the graph of $y=(x-1)(x-2)$ and finding when $y>0$ as is shown on the right, or by considering the sign of each of $(x-1)$ and $(x-2)$ as below.

Both $(x-1)$ and $(x-2)$ are positive so $x>2$, *or* both $(x-1)$ and $(x-2)$ are negative so $x<1$.

The solution set is $\{x : x<1\} \cup \{x : x>2\}$

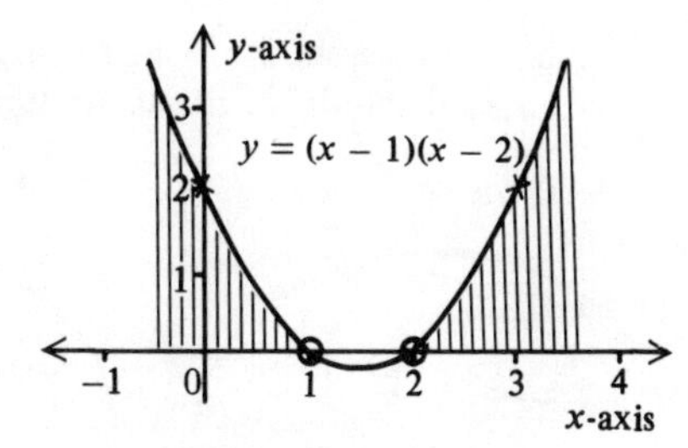

$y>0$ when $x>2$ or when $x<1$

Exercise 10.5b

1. By drawing a graph of $y=(x-2)(x-3)$ find the solution set for $(x-2)(x-3)>0$.
2. By drawing a graph of $y=(x-1)(x-4)$ find the solution set for $(x-1)(x-4)<0$.
3. By drawing a graph of $y=(x+2)(x-3)$ find the solution set for $(x+2)(x-3)<0$.
4. By drawing a graph of $y=(x+4)(x+3)$ find the solution set for $(x+4)(x+3)>0$.
5. For what values of x is:
   (a) $(x-2)>0$? (b) $(x-5)>0$? (c) $(x-2)>0$ *and* $(x-5)>0$?
   (d) $(x-2)<0$? (e) $(x-5)<0$? (f) $(x-2)<0$ *and* $(x-5)<0$?
   Use your results for (c) and (f) to write down the solution set for $(x-2)(x-5)>0$.
6. For what values of x is:
   (a) $(x-3)>0$? (b) $(x-4)<0$? (c) $(x-3)>0$ *and* $(x-4)<0$?
   (d) $(x-3)<0$? (e) $(x-4)>0$? (f) $(x-3)<0$ *and* $(x-4)>0$?
   Use your results for (c) and (f) to write down the solution set for $(x-3)(x-4)<0$.
7. Find the solution set for $(x+3)(x-4)>0$ by using the fact that *either* the value of each bracket is positive, *or* the value of each bracket is negative.
8. Find the solution set for $(x+4)(x+5)<0$ by using the fact that *either* the value of the first bracket is positive whilst the second is negative, *or* the value of the first bracket is negative whilst the second is positive.
9. Look at the diagram on the right.
   When is $(x-5)>0$ *and* $(x-3)>0$?
   When is $(x-5)<0$ *and* $(x-3)<0$?
   Write down the solution set for
   $$(x-5)(x-3)>0$$

$(x-5)<0$ $(x-5)>0$
$(x-3)<0$ $(x-3)>0$
0 1 2 3 4 5 6 7 8 9 10

10. Look at the diagram on the right.
   When is $(x+4)>0$ *and* $(1-x)>0$?
   When is $(x+4)<0$ *and* $(1-x)<0$?
   Write down the solution set for
   $$(x+4)(1-x)>0$$

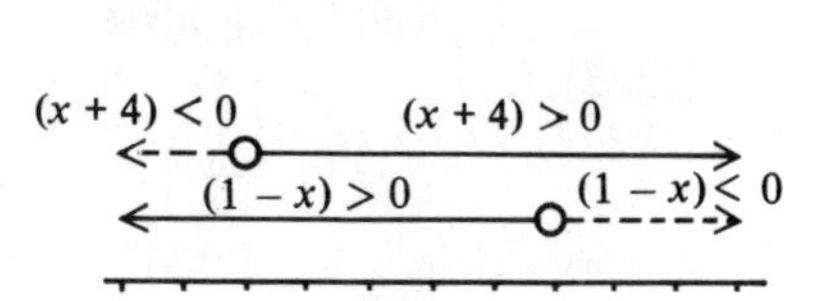

11. Use a number line as in question 10 to show when $(x-4)$ and $(x+2)$ are each positive or negative. Hence find the solution set for $(x-4)(x+2)>0$.
12. Use a number line as in question 11 to show when $(3-x)$ and $(3+x)$ are each positive or negative. Hence find the solution set for $(3-x)(3+x)>0$.
13. Find the solution set for:
   (a) $(x-1)(x-6)>0$ (b) $(x-3)(x+2)<0$ (c) $(x+5)(4-x)>0$
   (d) $(x+1)(x+2)<0$ (e) $(3-x)(2-x)>0$ (f) $(2x-1)(x-4)>0$
   (g) $(x+3)(4x-1)<0$ (h) $(2x+1)(1-5x)>0$

10.6 Arrow graphs and mapping diagrams

Information can be shown as an *arrow graph.* It is important to remember that each time the arrow is used it stands for the same thing. Three examples are shown below.

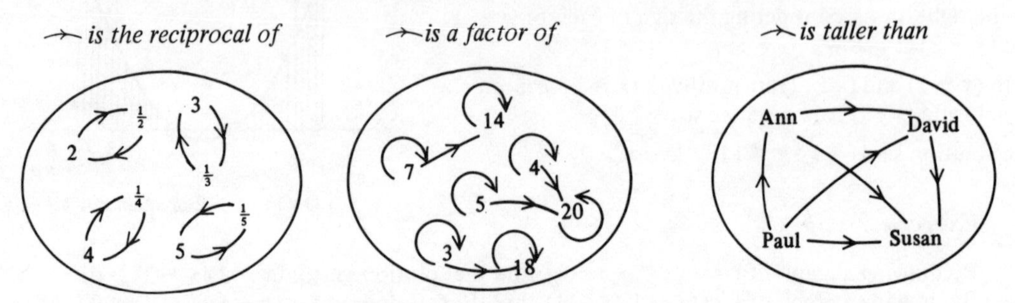

Each pair of elements linked by an arrow can also be represented by an ordered pair. There will be one ordered pair for each arrow.
In the first arrow graph above, the relationship *is the reciprocal of* works in both directions, so the arrows go both ways. Two of the eight ordered pairs are $(2, \frac{1}{2})$ and $(\frac{1}{2}, 2)$.
In the second arrow graph above, each number *is a factor of* itself so each number has an arrow going to itself. Three of the eleven ordered pairs are (7, 7), (14, 14) and (7, 14). (14, 7) is not included since 14 is not a factor of 7 and there is no arrow from 14 to 7.

Exercise 10.6a

1. Copy and complete the arrow graph. What does the arrow stand for?

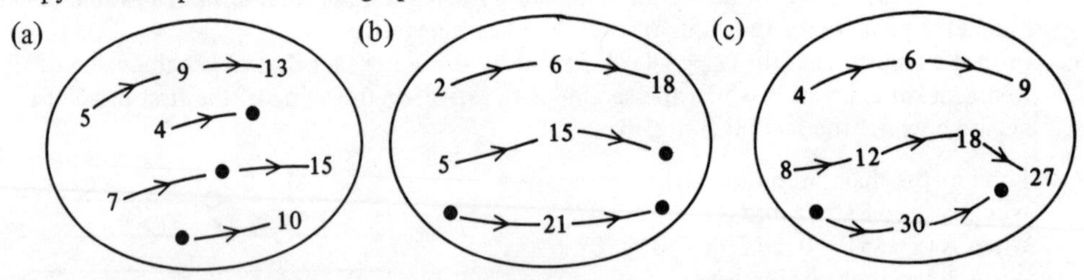

   Write down for each arrow graph the corresponding set of ordered pairs.
2. Draw an arrow graph for each set of ordered pairs:
   (a) {(2, 2), (2, 6), (6, 18), (2, 18), (6, 6), (18, 18)}
   (b) {(12, 9), (9, 6), (6, 3), (7, 4), (4, 1), (8, 5)}
   (c) {(1, 2), (2, 4), (4, 8), (8, 16), (3, 6), (6, 12)}
   Write down for each arrow graph what the arrow stands for.
3. Draw an arrow graph for each set using the arrow as defined:
   (a) {1, 2, 3, 4, 5, 6, 7, 8} : → *is three more than*
   (b) {1, 3, 4, 12, 16, 48, 64} : → *is a quarter of*
   (c) {1, 2, 3, 4, 5, 9, 16, 25} : → *is the square root of*
   (d) $\{\frac{1}{2}, \frac{1}{3}, \frac{1}{4}, \frac{2}{4}, \frac{2}{6}, \frac{3}{6}, \frac{3}{9}, \frac{2}{8}, \frac{9}{12}\}$: → *is the same value as*
   Write down for each arrow graph the corresponding set of ordered pairs.

There are four types of *mapping diagram*. Examples of these are shown.

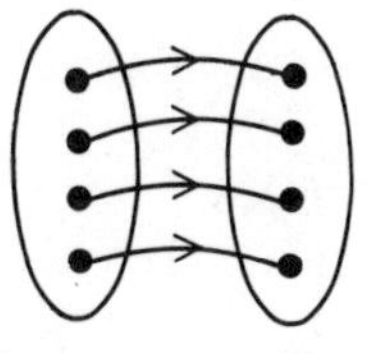
one to one (1–1)

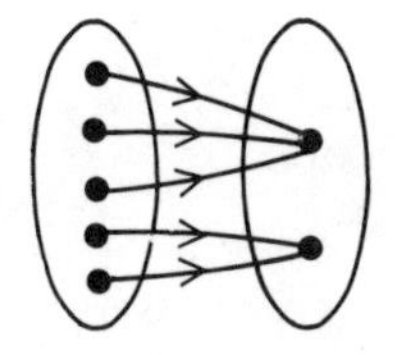
many to one (M–1)

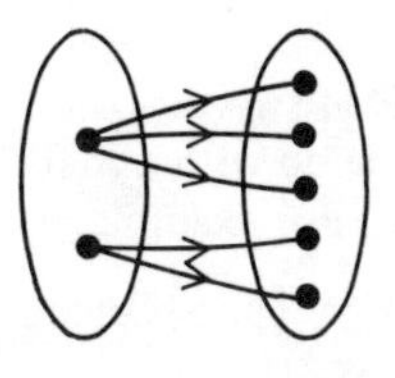
one to many (1–M)

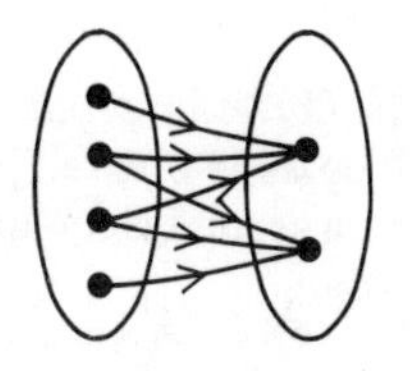
many to many (M–M)

Exercise 10.6b

1. Copy and complete the mapping diagram. What does the arrow mean? (Remember each time the arrow is used it stands for the same thing.)

(a)

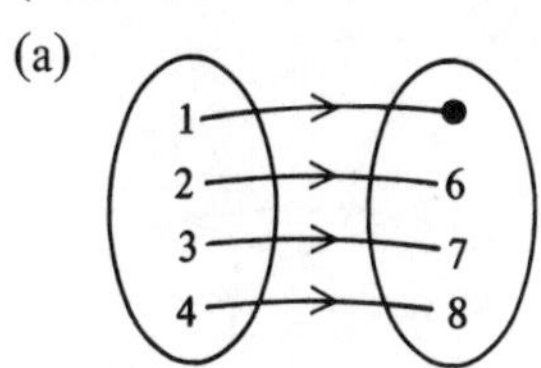

(b)

1 2 3 4 — 3 6 12

(c)

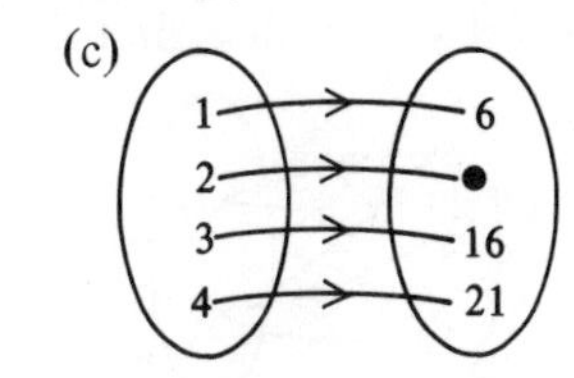

2. Draw a mapping diagram for each set of ordered pairs:
   (a) {(apple, A), (banana, B), (orange, O), (pear, P), (pineapple, P)}
   (b) {(31, January), (28, February), (31, March), (30, April), (31, May)}
   (c) {(□, square), (△ , triangle), (○ , circle), (⬠ , pentagon)}
   (d) {(2, 6), (2, 8), (2, 10), (3, 6), (4, 8), (5, 10)}
   In each case say what the arrow stands for, and whether the mapping diagram is (1–1), (M–1), (1–M), or (M–M).
3. Draw a mapping diagram for the two sets using the arrow as defined:
   (a) {1, 4, 9, 16, 25}, {−3, −2, 1, 2, 3, 4, 5} : ⟶ *is the square of*
   (b) {8, 10, 12, 16, 18}, {2, 3, 4, 5, 6} : ⟶ *is a multiple of*
   (c) {1, 2, 3, 4}, {10, 100, 1000, 10000} : ⟶ *is the log of*
   (d) {Ann, June, Jack, Jill, Andrew}, {A, B, J} : ⟶ *has as its first letter*
   In each case say whether the mapping diagram is (1–1), (M–1), (1–M), or (M–M), and write down the corresponding set of ordered pairs.

Remember: A mapping diagram which is (1–1) or (M–1) represents a *function*, since there is only one arrow going from each element in the left hand set.

Exercise 10.6c

1. Which of the diagrams in questions 2 and 3 above represent a function?
2. Draw a mapping diagram for each set of ordered pairs. Say whether it represents a function, or if not, what type it is.
   (a) {(1, 5), (2, 8), (3, 11), (4, 14), (5, 17)}
   (b) {(2, 2), (2, 4), (2, 8), (3, 3), (3, 9), (3, 27)}
   (c) {(3, 21), (4, 20), (5, 20), (7, 21), (11, 22)}
   (d) {(4, 0), (5, 1), (6, 2), (7, 3), (8, 4), (9, 5)}
   (e) {(0, 0), (0·5, 30), (1, 90), (0·5, 150)}
   (f) {(0, 0), (45, 1), (135, −1), (180, 0), (225, 1)}

10.7 Cartesian products

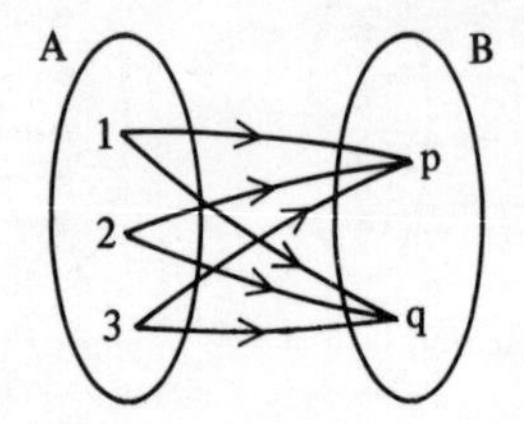

The *Cartesian product* A × B of two sets A and B is the set of all possible ordered pairs made up with a first element from A and a second element from B. For example:

If $\quad A = \{1, 2, 3\}$ and $B = \{p, q\}$ then

$A \times B = \{(1, p), (1, q), (2, p), (2, q), (3, p), (3, q)\}$

Note: Each ordered pair in A × B corresponds to one arrow in the arrow diagram.

Exercise 10.7a

1. Using the arrow diagram to help, write down the ordered pairs in A × B.

(a)

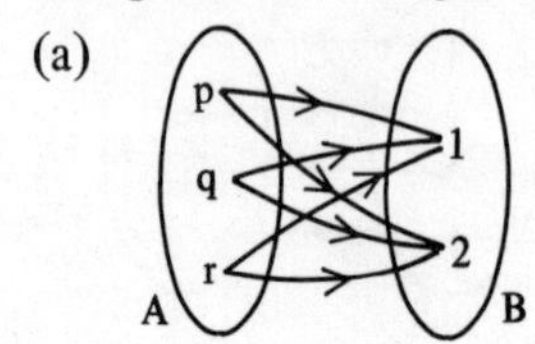

(b)

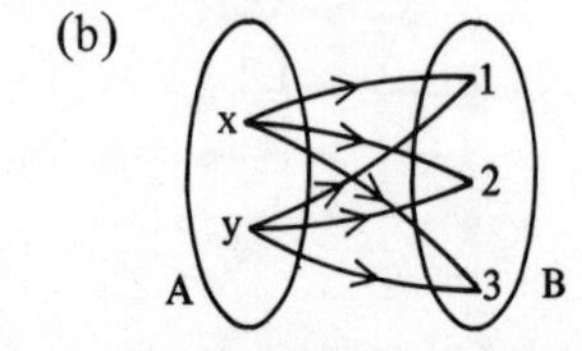

(c)

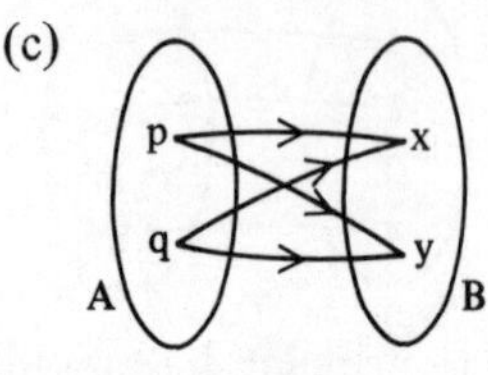

2. For each diagram in question 1, write down the ordered pairs in B × A.
3. Draw an arrow diagram to show the ordered pairs in P × Q where:
   (a) $P = \{1, 2, 3, 4\}$ and $Q = \{p, q\}$ (b) $P = \{p, q, r, s\}$ and $Q = \{1, 2\}$
   (c) $P = \{x, y, z\}$ and $Q = \{1, 2, 3\}$ (d) $P = \{a, b, c, d\}$ and $Q = \{5\}$
   For each arrow diagram write down the ordered pairs in P × Q.
4. For each pair of sets in question 3 write down the ordered pairs in Q × P.
5. Write down the number of ordered pairs in U × V where:
   (a) $U = \{1, 3, 5, 7\}$ and $V = \{x, y\}$ (b) $U = \{p, q, r\}$ and $V = \{a, b, c\}$
   (c) $U = \{1, 2\}$ and $V = \{p, q\}$ (d) $U = \{x, y, z\}$ and $V = \{4\}$
6. Draw an arrow diagram to show the ordered pairs in A × B where:
   (a) $A = \{1, 2, 3\}$ and $B = \{1, 2\}$ (b) $A = \{1, 2, 3\}$ and $B = \{1, 2, 3\}$
   (c) $A = \{1, 2, 3\}$ and $B = \{1, 2, 3, 4\}$
   In which case above is $A \times B = B \times A$? List the ordered pairs.

Example

Show that if $A \subset P$ and $B \subset Q$ then $(A \times B) \subset (P \times Q)$.

This result can be shown using an arrow diagram as on the right. $A \times B$ is shown by the heavy arrows, each of which also belongs to $P \times Q$.

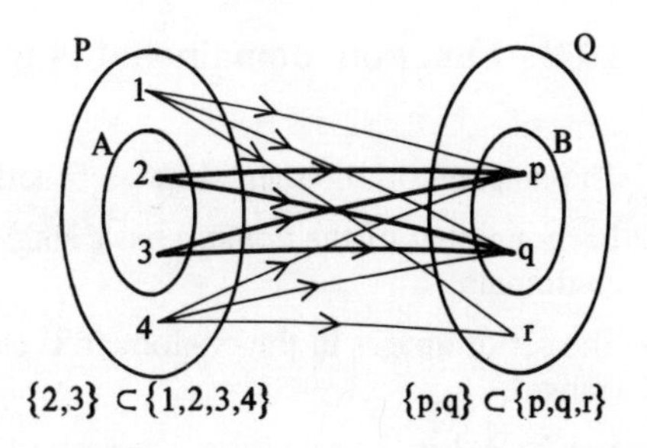

Exercise 10.7b

1. If $P = \{a, b, c\}$, $Q = \{1, 2, 3, 4\}$, $A = \{b, c\}$ and $B = \{1, 2\}$ write down the ordered pairs in:
   (a) $P \times Q$ (b) $A \times B$
   Hence show that $(A \times B) \subset (P \times Q)$.
2. If $P = \{p, q, r, s\}$, $Q = \{q, r, s, t\}$, $S = \{s, t, u\}$ and $T = \{t, u, v\}$ find:
   (a) $P \cap Q$ (b) $S \cap T$
   Hence write down the ordered pairs in $(P \cap Q) \times (S \cap T)$.
3. If $P = \{1, 2, 3, 4, 5\}$, $Q = \{3, 4, 5, 6\}$, $S = \{7, 8, 9\}$ and $T = \{8, 9, 10\}$, show that $(P \cap Q) \times (S \cap T) = (P \times S) \cap (Q \times T)$.
4. If $A = \{1, 2, 3, 4, 5\}$ write down the ordered pairs in $A \times A$.

   Copy the diagram on the right and show the ordered pairs as points.

   If $B = \{2, 3, 4\}$ use the diagram to show that $(B \times B) \subset (A \times A)$
5. If $P = \{a, b, c\}$, $Q = \{c, d, e\}$, $R = \{1, 2, 3\}$ and $S = \{3, 4, 5\}$ find:
   (a) $P \cup Q$ (b) $R \cup S$
   Hence describe the ordered pairs in $(P \cup Q) \times (R \cup S)$.
6. If $P = \{1, 2, 3, 4\}$, $Q = \{3, 4, 5\}$, $R = \{a, b, c\}$ and $S = \{b, c, d\}$ find whether $(P \cup Q) \times (R \cup S) = (P \times R) \cup (Q \times S)$ or not.
7. Explain the significance of the Cartesian product with reference to:
   (a) two dice each numbered from 1 to 6.
   (b) choosing a mixed doubles tennis pair from a team of 3 boys and 3 girls.
   (c) a table tennis match with two teams each of 4 players.
8. If $A = \{x : 1 \leqslant x \leqslant 6\}$, $B = \{y : 2 \leqslant y \leqslant 5\}$, $C = \{x : 3 \leqslant x \leqslant 4\}$ and $D = \{y : 1 \leqslant y \leqslant 6\}$ shade a region on an (x, y) graph to show:
   (a) $A \times B$ (b) $C \times D$ (c) $(A \times B) \cap (C \times D)$
   (Hint: each Cartesian product represents a rectangular region.)

10.8 Function, domain and range

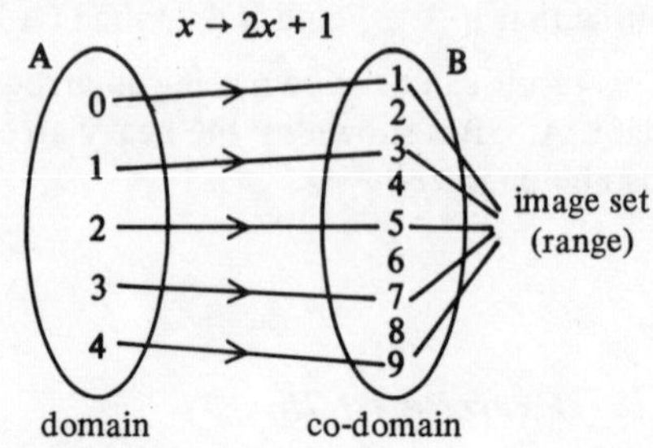

The diagram on the right shows a function which is (1–1).

Each member of the *domain* has a single image in the co-domain.

The set of images in the co-domain is called the *image set* (range).

$x \to 2x + 1$ is the *rule* for the function.
$2x + 1$ is the image of x.

Note: a function is specified when both the rule *and* domain are given.

Exercise 10.8a

1. Using the rule, $x \to 3x$, write down the image of: (a) 2 (b) 5 (c) 10
2. Using the rule, $x \to 4x - 3$, write down the image of: (a) 1 (b) 4 (c) 8
3. Using the domain $\{1, 2, 3, 4\}$ find the image set for the rule:
   (a) $x \to 2x$ (b) $x \to x + 2$ (c) $x \to 5x - 3$ (d) $x \to 10 - x$
4. Using the domain $\{1, 3, 7, 8\}$ find the image set for the rule:
   (a) $x \to 4x$ (b) $x \to 3x + 5$ (c) $x \to 8x - 7$ (d) $x \to 16 - 2x$
5. Using the domain $\{-3, -1, 2, 4\}$ find the image set for the rule:
   (a) $x \to 5x$ (b) $x \to x - 3$ (c) $x \to 3x + 2$ (d) $x \to 7 - x$
6. Using the domain $\{1, 2, 3, 4\}$ find the image set of the rule:
   (a) $x \to \frac{1}{2}x$ (b) $x \to \frac{1}{2}x + \frac{3}{2}$ (c) $x \to 1/x$ (d) $x \to x^2$
7. For each diagram write down: 1. the domain 2. the co-domain 3. the image set.

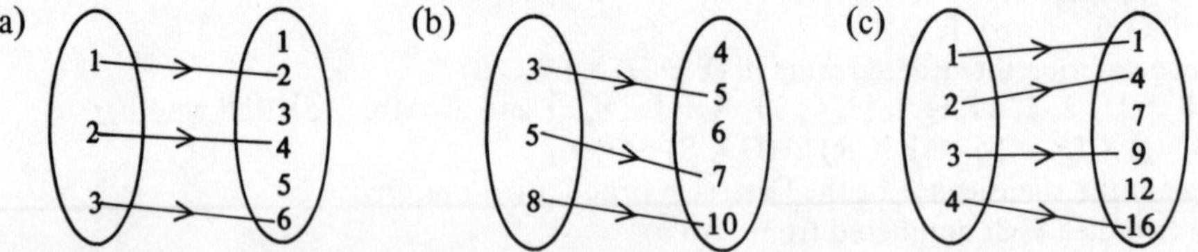

8. For each diagram in question 7 write down the rule as $x \to$.
9. Copy and complete each diagram using the given rule and the left hand set as the domain.

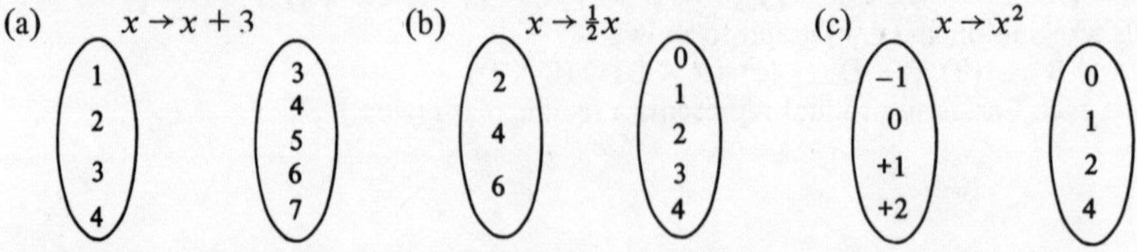

10. For each diagram in question 9 write down: 1. the co-domain 2. the image set.
11. Draw an arrow diagram for the given domain, co-domain and rule:
    (a) Domain $\{0, 1, 2\}$, co-domain $\{1, 2, 3, 4, 5, 6\}$, rule $x \to x + 4$
    (b) Domain $\{-2, 1, 5\}$, co-domain $\{-11, -7, -2, 4, 10\}$, rule $x \to 3x - 5$
    (c) Domain $\{5, 12, 31\}$, co-domain $\{7, 17, 64, 81, 85\}$, rule $x \to 100 - 3x$
12. For each diagram in question 11 write down the image set.

f is the name, or label, of a function.

$f(x)$ is the image of x, using *f*.

$f: x \to x^2 - 4$ is read as '*f* is a function such that x maps onto $x^2 - 4$'.

In the diagram on the right, $f(x) = x^2 - 4$.
$f(3)$, the image of 3, is 5.

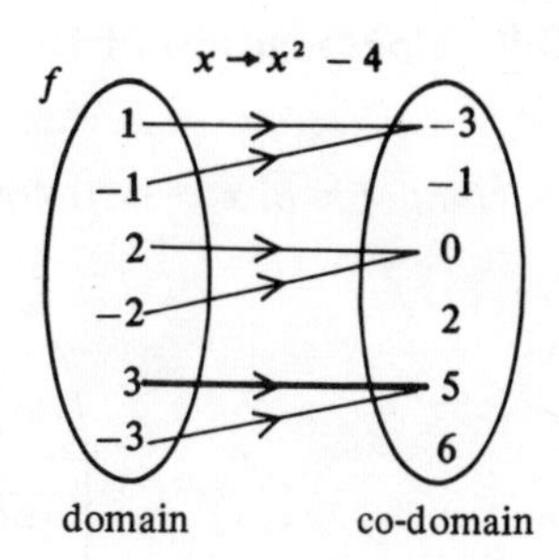

Exercise 10.8b

1. Using the diagram above, write down:
   (a) the co-domain (b) the image set (c) the image of 2 (d) the image of -1
   (e) $f(1)$ (g) $f(-2)$ (h) $f(-3)$
2. For each function write down 1. $f(2)$; 2. $f(7)$; 3. $f(0)$; 4. $f(-3)$:
   (a) $f(x) = 7x - 4$ (b) $f(x) = 4x + 19$ (c) $f(x) = 1 - 5x$ (d) $f(x) = 4 - 3x$
3. For each function write down 1. $g(1)$; 2. $g(2)$; 3. $g(4)$; 4. $g(\frac{1}{2})$:
   (a) $g(x) = x^2 + 5$ (b) $g(x) = x^2 - x$ (c) $g(x) = 1 - x^3$ (d) $g(x) = x + 1/x$
4. Copy and complete each diagram using the given rule and the left-hand set as the domain:
   (a) $f(x) = x + 5$ (b) $f(x) = 9x - 4$ (c) $f(x) = 6 - 5x$

   (a) {2, 3, 5, 7} {4, 7, 8, 10, 12}
   (b) {2, 9, 13} {14, 33, 77, 113, 117}
   (c) {4, 2, −2, −4} {26, 16, −4, −14}

5. For each diagram in question 4 write down:
   (a) the co-domain (b) the image set (c) $f(2)$
6. Draw an arrow graph for the given rule and domain:
   (a) $g(x) = 7x + 11$, $\{0, 1, 4, 9\}$
   (b) $g(x) = 1 - 6x$, $\{1, 2, 3, 4, 5\}$
   (c) $g(x) = \frac{1}{4}x - 5$, $\{4, 8, 12, 16\}$
   (d) $g(x) = x^2 + x$, $\{1, 2, 3, 4, 5\}$
7. For each diagram in question 6, write down $g(4)$.
8. For each function write down: 1. $f(2)$; 2. $f(10)$; 3. $f(-3)$; 4. $f(0)$:
   (a) $f: x \to 12x + 5$ (b) $f: x \to 12 - 5x$ (c) $f: x \to x^2 - x$
9. If $k: x \to 1/x - 3$, find:
   (a) $k(1)$ (b) $k(2)$ (c) $k(\frac{1}{2})$ (d) $k(\frac{1}{3})$
10. Find x, if $f(x) = 20$ and:
    (a) $f: x \to 4x$ (b) $f: x \to x + 7$ (c) $f: x \to 3x + 2$
11. Find x, if $g(x) = 15$ and $g: x \to 2(x - 1) + 3$.
12. If $f: x \to x + 3$ and $g: x \to 5x$ find:
    (a) $f(2)$ and the image of $f(2)$ using g
    (b) $g(4)$ and the image of $g(4)$ using f.

10.9 Composition of functions

The composite of $x \to 3x$ followed by $x \to x + 2$ is $x \to 3x + 2$

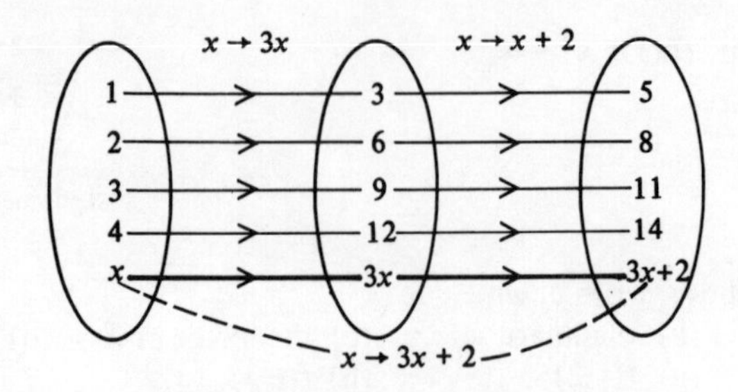

Exercise 10.9a

1. Draw an arrow graph, using $\{1, 2, 3, 4, 5\}$ as domain, to show:
   (a) $x \to x + 2$ followed by $x \to 3x$
   (b) $x \to 5x$ followed by $x \to x - 3$
   (c) $x \to x - 4$ followed by $x \to 7x$
   For each part write down the composite as $x \to$
2. Draw an arrow graph, using $\{0, 1, 2, 3, 4\}$ as domain, to show:
   (a) $x \to x^2$ followed by $x \to x + 4$
   (b) $x \to x + 4$ followed by $x \to x^2$
   (c) $x \to x + 1$ followed by $x \to 1/x$
   For each part write down the composite as $x \to$.
3. Write down the composite of:
   (a) $x \to x - 3$ followed by $x \to 5x$
   (b) $x \to 2x$ followed by $x \to x + 5$
   (c) $x \to x - 2$ followed by $x \to x^2$
   (d) $x \to x + 3$ followed by $x \to 1/x$
4. Using $\{-2, -1, 0, 1, 2\}$ as domain, find the image set for each composite function in question 3.
5. Write down the composite of:
   (a) $x \to 3x + 4$ followed by $x \to 2x + 1$
   (b) $x \to 2x - 7$ followed by $x \to 3x + 5$
   (c) $x \to 5x + 1$ followed by $x \to 4x - 3$
   (d) $x \to 4x - 1$ followed by $x \to x^2 + 2$
6. Using $\{0, 1, 2, 3, 4, 5\}$ as domain, find the image set for each composite function in question 5.
7. Which is the composite of $x \to 3x + 1$ followed by $x \to 2x + 5$?
   (a) $x \to 6x + 6$ (b) $x \to 6x + 5$ (c) $x \to 6x + 16$ (d) $x \to 6x + 7$
8. Is the composite of $x \to 2x + 5$ followed by $x \to 3x + 1$ given in question 7?
9. Which is the composite of $x \to 5x - 2$ followed by $x \to 3x - 1$?
   (a) $x \to 15x - 3$ (b) $x \to 15x + 7$ (c) $x \to 15x - 7$ (d) $x \to 15x + 2$
10. Is the composite of $x \to 3x - 1$ followed by $x \to 5x - 2$ given in question 9?
11. Is $x \to 4x + 3$ followed by $x \to 3x + 2$ the same as $x \to 3x + 2$ followed by $x \to 4x + 3$? For each composite find the image of $\{1, 2, 3, 4, 5\}$.

$gf(x)$ is the image of x using f first and then g.

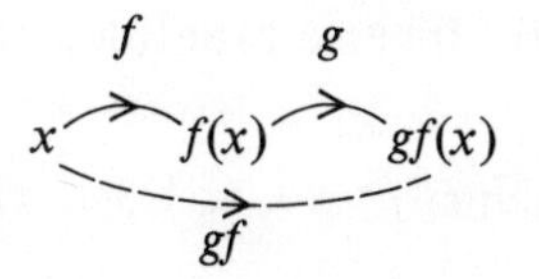

For example: if $f : x \to 3x + 1$
and $g : x \to x^2$
then $gf(x) = (3x + 1)^2$

$fg(x)$ is the image of x using g first and then f.

So in the above example:

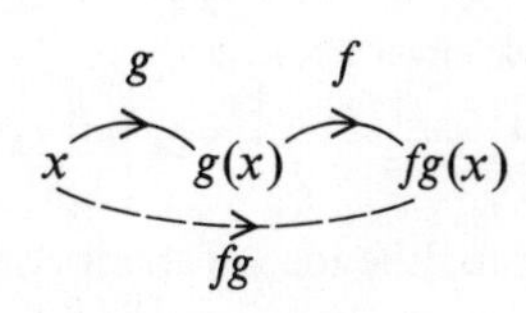

$$fg(x) = 3x^2 + 1$$

gf (or $g \circ f$) is called the composite of g on f.
fg (or $f \circ g$) is the composite of f on g.

Exercise 10.9b

1. If $f : x \to 2x + 1$ and $g : x \to x^2$, find:
   (a) $f(3)$ (b) $g(7)$ (c) $gf(3)$ (d) $fg(7)$ (e) $gf(7)$ (f) $fg(3)$
   Write down $gf(x)$ and $fg(x)$.
2. If $f : x \to 5x - 2$ and $g : x \to x^2$, find:
   (a) $f(2)$ (b) $g(8)$ (c) $gf(2)$ (d) $fg(8)$ (e) $gf(8)$ (f) $fg(2)$
   Write down $gf(x)$ and $fg(x)$.
3. If $f : x \to 3x - 7$ and $g : x \to 2x + 5$, find gf and fg. Hence write down:
   (a) $gf(2)$ (b) $fg(2)$ (c) $gf(-3)$ (d) $fg(-3)$
4. If $f : x \to 2x - 7$ and $g : x \to 3x + 5$, find gf and fg. Hence write down:
   (a) $gf(4)$ (b) $fg(1)$ (c) $gf(0)$ (d) $fg(-2)$
5. If $f : x \to 3x - 1$ and $g : x \to x^2$, find:
   (a) $gf(x)$ (b) $fg(x)$ (c) $gg(x)$ (d) $ff(x)$
6. If $f : x \to x^2$ and $g : x \to 4x + 1$, find:
   (a) $fg(x)$ (b) $gf(x)$ (c) $ff(x)$ (d) $gg(x)$
7. If $h : x \to 2x - 1$ and $k : x \to x^3$, find:
   (a) kh (b) hk (c) kk (d) hh
8. If $h : x \to x^3$ and $k : x \to 1 - x$, find:
   (a) hk (b) kh (c) hh (d) kk
9. If $f : x \to x + 1$ and $g : x \to x^2$, express in terms of f and g:
   (a) $x \to (x + 1)^2$ (b) $x \to x^2 + 1$ (c) $x \to x^4$ (d) $x \to x + 2$
10. If $h : x \to x^2$ and $k : x \to x - 1$, express in terms of h and k:
    (a) $x \to (x - 1)^2$ (b) $x \to x^2 - 1$ (c) $x \to x - 3$ (d) $x \to (x^2 - 1)^2$
11. For which pair is $fg = gf$?
    (a) $f : x \to 3x$, $g : x \to 2x$ (b) $f : x \to x + 5$, $g : x \to x - 2$ (c) $f : x \to 3x$, $g : x \to x - 2$ (d) $f : x \to 3x - 2$, $g : x \to 4x - 3$
    (e) $f : x \to 5x + 3$, $g : x \to 2x - 1$
12. Show in each case that ff is the identity, i.e. $x \to x$.
    (a) $f : x \to -x$ (b) $f : x \to 4 - x$ (c) $f : x \to 1/x$
13. Show in each case that gf is the identity. Show also that $fg = gf$.
    (a) $f : x \to x + 7$, $g : x \to x - 7$ (b) $f : x \to 5x$, $g : x \to \frac{1}{5}x$ (c) $f : x \to 2x + 5$, $g : x \to \frac{1}{2}(x - 5)$

10.10 Inverse functions and equations

A function $f : x \to 2x + 3$ can be shown as

$x \to$ [×2] $\to$ [+3] $\to 2x + 3$

Its *inverse*, denoted by f^{-1}, can be found by using the flow chart backwards.

$\frac{(x-3)}{2} \leftarrow$ [÷2] $\leftarrow$ [−3] $\leftarrow x$

i.e. $f^{-1} : x \to \frac{(x-3)}{2}$, or $f^{-1}(x) = \frac{(x-3)}{2}$

We can use the idea of the inverse process to solve the equation $2x + 3 = 13$. By using 13 as the right-hand side, the solution is $x = 5$.

$5 \leftarrow$ [÷2] $\leftarrow$ [−3] $\leftarrow 13$

Exercise 10.10a

1. Describe each flow chart as $x \to$.
   (a) $\to$ [×3] $\to$ [+2] $\to$ (b) $\to$ [×4] $\to$ [−3] $\to$ (c) $\to$ [+3] $\to$ [×5] $\to$
   Draw the inverse of each flow chart and write down the rule for the inverse.
2. Describe each flow chart as $x \to$.
   (a) $\to$ [−5] $\to$ [×4] $\to$ (b) $\to$ [+2] $\to$ [÷3] $\to$ (c) $\to$ [÷4] $\to$ [−3] $\to$
   Draw the inverse of each flow chart and write down the rule for the inverse.
3. Draw a flow chart to show:
   (a) $x \to 4x + 7$ (b) $x \to 3x - 5$ (c) $x \to 5(x + 2)$ (d) $x \to 7(x - 5)$
   Use your flow charts to help you write down the inverse of each rule.
4. Draw a flow chart to show:
   (a) $x \to 6x - 1$ (b) $x \to (x + 3) \div 2$ (c) $x \to x^2 + 3$ (d) $x \to (x + 1)^2$
   Use your flow charts to help you write down the inverse of each rule.
5. Write down the inverse of each rule:
   (a) $x \to 3x - 7$ (b) $x \to 2(x + 5)$ (c) $x \to \frac{1}{3}x + \frac{2}{3}$ (d) $x \to (x - 4) \div 3$
6. $f : x \to 4x + 1$. Find: (a) f^{-1} (b) $f(2)$ (c) $f^{-1}(9)$ (d) $f^{-1}(2)$
7. $g : x \to 2(x + 5)$. Find: (a) g^{-1} (b) $g(1)$ (c) $g^{-1}(12)$ (d) $g^{-1}(1)$
8. $h : x \to (x - 3) \div 2$. Find: (a) h^{-1} (b) $h(3)$ (c) $h^{-1}(0)$ (d) $h^{-1}(3)$
9. $k : x \to x^2 - 4$. Find: (a) k^{-1} (b) $k(5)$ (c) $k^{-1}(21)$ (d) $k^{-1}(5)$
10. $f : x \to 3(2x + 1) + 4$. Find: (a) f^{-1} (b) $f(3)$ (c) $f^{-1}(25)$ (d) $f^{-1}(4)$
11. Draw a flow chart to show $x \to 3x + 4$ and hence solve $3x + 4 = 19$.
12. Draw a flow chart to show $x \to 5x - 2$ and hence solve $5x - 2 = 13$.
13. Use a flow chart and its inverse to solve:
    (a) $5x + 3 = 38$ (b) $7x - 2 = 47$ (c) $4x + 7 = 3$ (d) $6x - 15 = 6$
14. Use a flow chart and its inverse to solve:
    (a) $3(2x + 5) = 57$ (b) $4(5x - 9) = 184$ (c) $2(x + 3) + 5 = 9$ (d) $5(x - 2) + 4 = -21$
15. If $f(x) = 4x - 7$, find x when: (a) $f(x) = 9$ (b) $f(x) = 21$ (c) $f(x) = -15$
16. If $g(x) = 6(x + 2) - 5$, find x when: (a) $g(x) = 25$ (b) $g(x) = 73$ (c) $g(x) = -17$
17. Solve:
    (a) $7(2x + 3) + 5 = 33$ (b) $2(9x - 4) - 3 = -47$ (c) $\frac{1}{4}(x - 5) + \frac{3}{4} = 15$
18. Draw a flow chart to show $f : x \to [7(2x - 3) + 5] \div 2$. Find f^{-1}. Hence find x when $f(x) = 20$. Find also $f^{-1}(6)$.

Example

Solve $x^2 + 4x - 5 = 0$.

The left-hand side can be written as $(x + 2)^2 - 9$.

Hence $x \to$ [+2] $\to$ [square] $\to$ [−9] $\to (x + 2)^2 - 9$

Working backwards using 0 as the right-hand side we get

1 or −5 $\leftarrow$ [−2] $\leftarrow$ [$\pm\sqrt{}$] $\leftarrow$ [+9] $\leftarrow$ 0

so the solution of $x^2 + 4x - 5 = 0$ is $x = 1$ or $x = -5$

Exercise 10.10b

1. Copy and complete:
   (a) $x^2 + 4x - 12 = (x + \quad)^2 -$ (b) $x^2 + 6x - 7 = (x + \quad)^2 -$
   (c) $x^2 + 8x - 20 = (x + \quad)^2 -$ (d) $x^2 - 4x - 5 = (x - \quad)^2 -$
2. Copy and complete:
   (a) $x^2 + 4x + 3 = (x + \quad)^2 -$ (b) $x^2 - 10x + 21 = (x - \quad)^2 -$
   (c) $x^2 - 12x + 11 = (x - \quad)^2 -$ (d) $x^2 + 16x + 64 = (x + \quad)^2 -$
3. By using your results from question 1 and a flow chart, solve:
   (a) $x^2 + 4x - 12 = 0$ (b) $x^2 + 6x - 7 = 0$ (c) $x^2 + 8x - 20 = 0$
   (d) $x^2 + 4x - 12 = 9$ (e) $x^2 + 6x - 7 = 20$ (f) $x^2 + 8x - 20 = -11$
4. Solve: (a) $x^2 - 4x - 5 = 0$ (b) $x^2 - 4x - 5 = 55$ (c) $x^2 - 4x - 5 = -8$
5. By using your results from question 2 and a flow chart, solve:
   (a) $x^2 + 4x + 3 = 0$ (b) $x^2 - 10x + 21 = 0$ (c) $x^2 - 12x + 11 = 0$
   (d) $x^2 + 4x + 3 = 8$ (e) $x^2 - 10x + 21 = 12$ (f) $x^2 - 12x + 11 = 39$
6. Solve: (a) $x^2 + 16x + 64 = 0$ (b) $x^2 + 16x + 64 = 25$ (c) $x^2 + 16x + 64 = 81$

Inverse of a composite function

$gf : x \to 2x + 3$

$(gf)^{-1} : x \to (x - 3) \div 2$

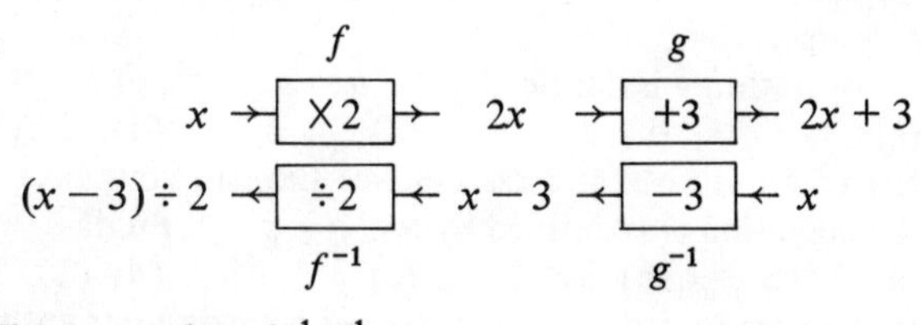

The inverse of gf is written as $(gf)^{-1}$. This is equal to $f^{-1}g^{-1}$.

Exercise 10.10c

1. $f : x \to 3x, g : x \to x + 2$. Find: (a) gf (b) f^{-1} (c) g^{-1} (d) $(gf)^{-1}$
   Find also fg and show that $(fg)^{-1} = g^{-1}f^{-1}$.
2. $f : x \to 4x, g : x \to x - 3$. Find: (a) fg (b) f^{-1} (c) g^{-1} (d) $(fg)^{-1}$
   Find also gf and show that $(gf)^{-1} = f^{-1}g^{-1}$.
3. $f : x \to 3x + 2, g : x \to 4x$. Find: (a) gf (b) $(gf)^{-1}$ (c) fg (d) $(fg)^{-1}$
4. $h : x \to x + 5, k : x \to 2x - 1$. Find: (a) hk (b) $(hk)^{-1}$ (c) kh (d) $(kh)^{-1}$
5. $f(x) = 5x, g(x) = x - 2$. Find: (a) $gf(3)$ (b) $g^{-1}(13)$ (c) $f^{-1}(15)$
   Show also that $(fg)^{-1}(20) = 6$.
6. $f(x) = 3x + 1, g(x) = x + 2$. Find: (a) $fg(4)$ (b) $f^{-1}(19)$ (c) $g^{-1}(6)$
   Show also that $(gf)^{-1}(15) = 4$.
7. $f(x) = 2x + 1, g(x) = 3x - 2$. Find x when:
   (a) $gf(x) = 19$ (b) $fg(x) = 21$

10.11 Properties of operations

Example 1

An operation $*$ is defined by $\boxed{a * b = 2a + b}$

$3 * 5 = 2 \times 3 + 5 = 11$ but $5 * 3 = 2 \times 5 + 3 = 13$, so $*$ is *not commutative.*

$(3 * 5) * 6 = 11 * 6 = 2 \times 11 + 6 = 28$
$3 * (5 * 6) = 3 * 16 = 2 \times 3 + 16 = 22$ so $*$ is *not associative.*

Example 2

An operation $\square$ is defined by $\boxed{a \square b = 2ab}$

$3 \square 5 = 2 \times 3 \times 5 = 30$ and $5 \square 3 = 2 \times 5 \times 3 = 30$

also $a \square b = b \square a = 2ab$, for all a and b, so $\square$ is *commutative.*

$(3 \square 5) \square 6 = 30 \square 6 = 2 \times 30 \times 6 = 360$ and
$3 \square (5 \square 6) = 3 \square 60 = 2 \times 3 \times 60 = 360$

also $(a \square b) \square c = a \square (b \square c) = 4abc$, for all a, b and c, so $\square$ is *associative.*

Exercise 10.11a

1. An operation $*$ is defined by $a * b = 3a + b$. Find:
   (a) $4 * 6$ (b) $6 * 4$ (c) $6 * 2$ (d) $(4 * 6) * 2$ (e) $4 * (6 * 2)$
   Is $*$ commutative? Is $*$ associative?
2. An operation $\square$ is defined by $a \square b = 3ab$. Find:
   (a) $4 \square 6$ (b) $6 \square 4$ (c) $6 \square 2$ (d) $(4 \square 6) \square 2$ (e) $4 \square (6 \square 2)$
   Explain why $\square$ is commutative. Find $(a \square b) \square c$ and $a \square (b \square c)$ and hence show that $\square$ is associative.
3. An operation $*$ is defined by $p * q = p^2 + q^2$. Find:
   (a) $2 * 3$ (b) $3 * 2$ (c) $3 * 4$ (d) $(2 * 3) * 4$ (e) $2 * (3 * 4)$
   Is $*$ commutative? Is $*$ associative? Explain your answers.
4. An operation $\square$ is defined by $p \square q = p^2 q^2$. Find:
   (a) $2 \square 3$ (b) $3 \square 2$ (c) $3 \square 4$ (d) $(2 \square 3) \square 4$ (e) $2 \square (3 \square 4)$
   Is $\square$ commutative? Is $\square$ associative? Explain your answers.
5. An operation $*$ is defined by $x * y = x$.
   (a) By finding $3 * 4$ and $4 * 3$ show that $*$ is not commutative.
   (b) By finding $(x * y) * z$ and $x * (y * z)$ show that $*$ is associative.
6. The operation $\circ$ is defined by $a \circ b = \sqrt{a^2 + b^2}$.
   (a) Find $5 \circ 12$ and $12 \circ 5$ and explain why $\circ$ is commutative.
   (b) Find $(3 \circ 4) \circ 12$ and $3 \circ (4 \circ 12)$. Do you think $\circ$ is associative? Justify your answer by finding $(a \circ b) \circ c$ and $a \circ (b \circ c)$.
7. Repeat question 6 if $\circ$ is defined by $(a + b)^{-1}$.
8. The operation $*$ is defined by $p * q = pq + p + q$.
   (a) Find $2 * 3$ and $3 * 2$ and explain why $*$ is commutative.
   (b) Find $(2 * 3) * 4$ and $2 * (3 * 4)$. Do you think $*$ is associative?
   Justify your answer by finding $(p * q) * r$ and $p * (q * r)$.

If $a \circ (b * c) = (a \circ b) * (a \circ c)$ and $(b * c) \circ a = (b \circ a) * (c \circ a)$ for all a, b, and c then $\circ$ is *distributive* over $*$ (i.e. distributive from both left and right).

Example 3

An operation $*$ is defined by $x * y = \dfrac{x+y}{2}$

An operation $\circ$ is defined by $x \circ y = 2xy$

$3 \circ (5 * 7) = 3 \circ \left(\dfrac{5+7}{2}\right) = 3 \circ 6 = 2 \times 3 \times 6 = 36$, and also

$(3 \circ 5) * (3 \circ 7) = (2 \times 3 \times 5) * (2 \times 3 \times 7) = 30 * 42 = \dfrac{30+42}{2} = 36.$

The operation $\circ$ is *distributive from the left* over the operation $*$

$(5 * 7) \circ 3 = \dfrac{5+7}{2} \circ 3 = 6 \circ 3 = 2 \times 6 \times 3 = 36$, and also

$(5 \circ 3) * (7 \circ 3) = (2 \times 5 \times 3) * (2 \times 7 \times 3) = 30 * 42 = \dfrac{30+42}{2} = 36.$

The operation $\circ$ is *distributive from the right* over the operation $*$.

Exercise 10.11b

1. An operation $*$ is defined by $x * y = \dfrac{x-y}{2}$
   An operation $\circ$ is defined by $x \circ y = 2xy$.
   Find: (a) $8 * 4$ (b) $3 \circ 8$ (c) $3 \circ 4$ (d) $3 \circ (8 * 4)$ (e) $(3 \circ 8) * (3 \circ 4)$
   Is $\circ$ distributive from the left over $*$?
   Show that $\circ$ is also distributive from the right over $*$.
2. An operation $*$ is defined by $p * q = 3(p+q)$.
   An operation $\circ$ is defined by $p \circ q = pq/2$.
   Find: (a) $7 * 5$ (b) $7 \circ 4$ (c) $5 \circ 4$ (d) $(7 * 5) \circ 4$ (e) $(7 \circ 4) * (5 \circ 4)$
   Is $\circ$ distributive from the right over $*$?
   Show that $\circ$ is also distributive from the left over $*$.
3. Find, in question 1, whether $*$ is: (a) commutative (b) associative.
4. Find, in question 2, whether $*$ is: (a) commutative (b) associative.
5. An operation $\square$ is defined by $a \square b = a + b + 1$.
   An operation $*$ is defined by $a * b = 3ab$.
   Find: (a) $4 \square 7$ (b) $5 * 4$ (c) $5 * 7$ (d) $5 * (4 \square 7)$ (e) $(5 * 4) \square (5 * 7)$
   Is $\square$ distributive from the left over $*$?
   Is $\square$ distributive from the right over $*$? Justify your answer.
6. Find, in question 5, whether $\square$ is: (a) commutative (b) associative.
7. An operation $\oplus$ is defined by $p \oplus q = (p+q)$ Mod 7.
   Find: (a) $2 \oplus 3$ (b) $3 \oplus 2$ (c) $3 \oplus 4$ (d) $4 \oplus 3$
   Do you think $\oplus$ is commutative? Try for other pairs of numbers.
   Find: (e) $(2 \oplus 3) \oplus 4$ (f) $2 \oplus (3 \oplus 4)$. Do you think $\oplus$ is associative?
   If $*$ is multiplication in Mod 7, find: (g) $(2 \oplus 3) * 4$ (h) $(2 * 4) \oplus (3 * 4)$
   Is $*$ distributive from the right over $\oplus$?
   Show that $*$ is distributive from the left over $\oplus$.

10.12 Properties and laws of number

Reminder

Natural (or counting) numbers: 1, 2, 3, 4, 5, 6, ...

Integers (both positive and negative): ... −5, −4, −3, −2, −1, 0, 1, 2, 3, 4, 5, ...

Rational numbers include:
- common fractions i.e. $\frac{2}{5}, \frac{3}{7}, \frac{91}{100}, \ldots$
- vulgar fractions i.e. $\frac{5}{2}, \frac{7}{3}, \frac{100}{91}, \ldots$
- terminating decimals i.e. 0·25, 0·125, 0·0001, ...
- recurring decimals i.e. $0{\cdot}333\dot{3}, 0{\cdot}1212\dot{1}\dot{2}, \ldots$

Irrational numbers include:
- non-recurring decimals i.e. 0·101001000100001 ...
- numbers such as $\pi, \sqrt{2}, \sqrt{3}$, etc.

Exercise 10.12a

1. Express as a recurring decimal:
   (a) $\frac{1}{3}$ (b) $\frac{1}{6}$ (c) $\frac{1}{9}$ (d) $\frac{1}{7}$ (e) $\frac{2}{11}$
2. Express as a common fraction:
   (a) 0·4444 ... (b) 0·363636 ... (c) 0·571428 ...
3. Write down a rational number which is between:
   (a) $\frac{2}{5}$ and $\frac{3}{7}$ (b) 0·261 and 0·352 (c) 0·197 and $\frac{3}{16}$
4. Write down an irrational number which is between:
   (a) 1 and 2 (b) 3 and 4 (c) $\sqrt{2}$ and 2 (d) π and 4 (e) $\sqrt{2}$ and π

Commutative

Addition and multiplication are commutative.

For example: $3 + 4 = 4 + 3$
$3 \times 4 = 4 \times 3$

Commutative law $a * b = b * a$

Associative

Addition and multiplication are associative.

For example: $(3 + 4) + 5 = 3 + (4 + 5)$
$(3 \times 4) \times 5 = 3 \times (4 \times 5)$

Assocative law $(a * b) * c = a * (b * c)$

Exercise 10.12b

1. Give two examples, using numbers to show why subtraction and division are not commutative.
2. Write *true* or *false* for each statement:
   (a) $7 + 13 = 13 + 7$ (b) $16 \div 4 = 4 \div 16$ (c) $7 - 3 = 3 - 7$ (d) $14 \times 5 = 5 \times 14$
3. Give two examples, using numbers to show why subtraction and division are not associative.
4. Write *true* or *false* for each statement:
   (a) $(7 + 13) + 4 = 7 + (13 + 4)$
   (b) $(16 \div 4) \div 2 = 16 \div (4 \div 2)$
   (c) $(7 - 3) - 1 = 7 - (3 - 1)$

Distributive

Multiplication is distributive from the *left* over addition

$$3 \times (4+5) = 3 \times 4 + 3 \times 5$$

Multiplication is distributive from the *right* over addition

$$(7+8) \times 9 = 7 \times 9 + 8 \times 9$$

Distributive laws

$$p * (q+r) = p * q + p * r$$
$$(x+y) * z = x * z + y * z$$

Exercise 10.12c

1. Write *true* or *false* for each statement:
   (a) $9 \times (8+3) = 9 \times 8 + 9 \times 3$ (b) $(5-3) \times 7 = 5 \times 7 - 3 \times 7$
   (c) $4 + (5 \times 7) = (4+5) \times (4+7)$ (d) $(20+15) \div 5 = 20 \div 5 + 15 \div 5$
   (e) $(12 \times 5) - 3 = (12-3) \times (5-3)$ (f) $2 \div (6+8) = 2 \div 6 + 2 \div 8$
2. Give two examples, using numbers, to show why multiplication is distributive over subtraction: (a) from the left (b) from the right.
3. Write down the value of:
   (a) $(14+6) \div 10$ (b) $(14 \div 10) + (6 \div 10)$ (c) $(27-9) \div 3$ (d) $(27 \div 3) - (9 \div 3)$
   Hence find, in two different ways, the value of $(35+25) \div 5$.
4. Give two examples, using numbers, to show why division is distributive from the right over addition, but *not* from the left.

Identity 0 is the identity for addition $n + 0 = 0 + n = n$
1 is the identity for multiplication $m \times 1 = 1 \times m = m$

Inverse ^-n is the inverse of n for addition $n + {}^-n = 0$
$\frac{1}{m}$ is the inverse of m for multiplication $m \times \frac{1}{m} = 1$

Note: The commutative, associative, and distributive properties are often used in simplifying calculations.

For example: $\frac{1}{4} \times 20 \times 4$
$= \frac{1}{4} \times (4 \times 20)$ multiplication is commutative
$= (\frac{1}{4} \times 4) \times 20$ multiplication is associative
$= 20$ 1 is the identity for multiplication

Exercise 10.12d

1. For each number write down the additive inverse:
   (a) 7 (b) $^-5$ (c) $\frac{1}{3}$ (d) $\frac{2}{5}$ (e) $^-\frac{1}{8}$ (f) $^-\frac{5}{7}$
2. For each number write down the multiplicative inverse:
   (a) 4 (b) $\frac{1}{5}$ (c) $\frac{2}{7}$ (d) $^-3$ (e) $^-\frac{1}{6}$ (f) $^-\frac{3}{8}$
3. State the property of numbers used at each step in the calculation:
   (a) $\frac{1}{5} \times 12 \times 5 = \frac{1}{5} \times (5 \times 12) = (\frac{1}{5} \times 5) \times 12 = 1 \times 12 = 12$
   (b) $3 \times (4+5) + {}^-15 = (3 \times 4 + 3 \times 5) + {}^-15 = 12 + (15 + {}^-15) = 12 + 0 = 12$
   (c) $(5 + 4 + {}^-5) \times 3 = ((4+5) + {}^-5) \times 3 = (4 + (5 + {}^-5)) \times 3 = (4+0) \times 3 = 12$

10.13 Group structure

Look at the Mod 10 multiplication table shown on the right for the set $\{2, 4, 6, 8\}$.

× Mod 10	2	4	6	8
2	4	8	2	6
4	8	6	4	2
6	2	4	6	8
8	6	2	8	4

Note:
Each element appears once in each row and each column.

The result of combining any pair of elements is also a member of the set. We say the system is *closed.*

When 6 is combined with any element the result is the same element. 6 is the *identity* element.

Each element can be combined with another element to give the identity. So each element has an *inverse.*

The operation 'multiplication in Mod 10' is *associative*. i.e. $(2 \times 4) \times 8 = 2 \times (4 \times 8)$.

A system like that above is called a *group* because:
1. it is closed
2. it has an identity
3. it has inverses
4. it is associative

Exercise 10.13a

1. Look at the composition table on the right.
   Write down:
   (a) the identity element
   (b) each element and its inverse.

+ Mod 4	0	1	2	3
0	0	1	2	3
1	1	2	3	0
2	2	3	0	1
3	3	0	1	2

2. Make up a composition table for multiplication in Mod 5 using $\{1, 2, 3, 4\}$. Write down:
   (a) the identity element (b) each element and its inverse.
3. $\mathbf{R}_{90}$, $\mathbf{R}_{180}$, $\mathbf{R}_{270}$, and $\mathbf{R}_{360}$ represent rotations of $\frac{1}{4}$, $\frac{1}{2}$, $\frac{3}{4}$ and a whole turn respectively.
   Make up a table to show how they combine. Write down:
   (a) the identity element (b) each element and its inverse.
4. Make up a composition table for matrix multiplication using $\{\mathbf{A}, \mathbf{B}, \mathbf{C}, \mathbf{D}\}$ where
   $\mathbf{A} = \begin{pmatrix} 1 & 0 \\ 0 & 1 \end{pmatrix}$; $\mathbf{B} = \begin{pmatrix} -1 & 0 \\ 0 & -1 \end{pmatrix}$; $\mathbf{C} = \begin{pmatrix} 1 & 0 \\ 0 & -1 \end{pmatrix}$; $\mathbf{D} = \begin{pmatrix} -1 & 0 \\ 0 & 1 \end{pmatrix}$.
   Write down:
   (a) the identity element (b) each element and its inverse.
5. Which composition table below represents a group? Explain your answer.

(a)

×	O	E
O	O	E
E	E	E

(b)

×	1	−1
1	1	−1
−1	−1	1

(c)

+	0	1	2
0	0	1	2
1	1	2	3
2	2	3	0

(d)

+	2	4	6
2	4	6	0
4	6	0	2
6	0	2	4

(e)

*	*a*	*b*	*c*
a	*c*	*a*	*b*
b	*a*	*b*	*c*
c	*b*	*c*	*a*

6. For each group in question 5 write down:
   (a) the identity element (b) each element and its inverse.

For a *group* there has to be a set of elements S, and an operation $*$ combining them with the following four conditions.

1. *Closure*; that is $p * q$ also belongs to the set, for all p and q in S.
2. There is an *identity* element e; that is $e * p = p * e = p$, for all p in S.
3. Each element p has an *inverse* p^{-1}; that is $p * p^{-1} = p^{-1} * p = e$, for all p in S.
4. $*$ is *associative*; that is $(p * q) * r = p * (q * r)$ for all p, q and r in S.

Example

Show that all four properties of a group are used when solving the equation $p * x = q$.

$p * x = q$	
$p^{-1} * (p * x) = p^{-1} * q$	p has an inverse p^{-1}
$(p^{-1} * p) * x = p^{-1} * q$	$*$ is associative
$e * x = p^{-1} * q$	e is the identity
$x = (p^{-1} * q)$	$(p^{-1} * q)$ is in S, closure.

Exercise 10.13b

1. The set $S = \{p, q, r, s, t, u\}$ under the operation $*$ forms a group. The composition table for this group is shown.
   Write down:
   (a) the identity element
   (b) each element and its inverse.
   Simplify the expression:
   (c) $q * s * u$
   (d) $s * q * t * p$

*	p	q	r	s	t	u
p	r	u	p	t	s	q
q	u	t	q	r	p	s
r	p	q	r	s	t	u
s	t	r	s	u	q	p
t	s	p	t	q	u	r
u	q	s	u	p	r	t

2. Using the composition table in question 1, solve the equation:
   (a) $q * x = t$ (b) $x * t = s$ (c) $p * x * q = t$ (d) $q * x = x * q$
3. Which composition tables below represent a group? Explain your answer.

(a)

*	a	b	c	d
a	a	b	c	d
b	b	a	d	c
c	c	d	a	b
d	d	c	b	a

(b)

*	a	b	c	d
a	b	c	d	a
b	c	d	a	b
c	d	a	b	c
d	a	b	c	d

(c)

*	a	b	c	d
a	d	c	b	a
b	c	a	d	b
c	b	d	a	c
d	a	b	c	d

(d)

*	a	b	c	d
a	b	d	a	c
b	c	a	d	b
c	d	b	c	a
d	a	c	b	d

4. For each composition table in question 3, simplify the expression:
   (a) $c * b$ (b) $(a * b) * c$ (c) $a * (b * c)$ (d) $(a * c) * (b * d)$
5. For each composition table in question 3 solve the equation:
   (a) $b * x = c$ (b) $x * c = d$ (c) $x * x = a$ (d) $b * x * c = d$
6. Write down the multiplicative inverse of $\begin{pmatrix} 2 & 5 \\ 1 & 3 \end{pmatrix}$ and $\begin{pmatrix} 5 & 8 \\ 4 & 7 \end{pmatrix}$.
   Use the inverses you have just found to solve the equations:
   (a) $\begin{pmatrix} 2 & 5 \\ 1 & 3 \end{pmatrix} \times \begin{pmatrix} a & b \\ c & d \end{pmatrix} = \begin{pmatrix} 1 & 1 \\ 1 & 1 \end{pmatrix}$ (b) $\begin{pmatrix} 5 & 8 \\ 4 & 7 \end{pmatrix} \times \begin{pmatrix} p & q \\ r & s \end{pmatrix} = \begin{pmatrix} 0 & 1 \\ 1 & 0 \end{pmatrix}$
7. Find the multiplicative inverse of $\mathbf{A} = \begin{pmatrix} 8 & 5 \\ 3 & 2 \end{pmatrix}$ and $\mathbf{B} = \begin{pmatrix} 3 & 4 \\ 1 & 2 \end{pmatrix}$.
   Hence, find the matrix $\mathbf{X}$ where:
   (a) $\mathbf{A} \times \mathbf{X} = \mathbf{B}$ (b) $\mathbf{B} \times \mathbf{A} \times \mathbf{X} = \mathbf{A}$.

Part 11: Calculus

11.1 Scale factors

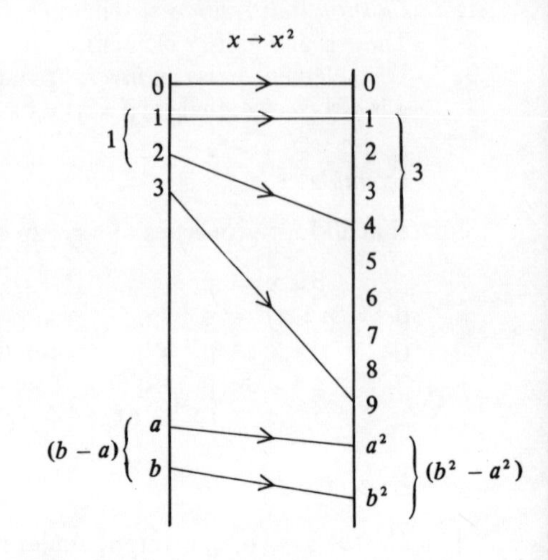

The *average* scale factor for the interval [1, 2] is

$$\frac{4-1}{2-1} = \frac{3}{1} = 3.$$

The average scale factor for the interval $[a, b]$ is

$$\frac{(b^2-a^2)}{(b-a)} = \frac{(b-a)(b+a)}{(b-a)} = (b+a)$$

The *local* scale factor associated with $x = a$ is the value of $(b+a)$ as b approaches a i.e. $2a$.

Exercise 11.1a

1. Show that the average scale factor for $x \to 7x$ for any interval is 7.
2. Find the average scale factor for $x \to x^2$ for the interval:
   (a) [0, 1] (b) [2, 3] (c) [3, 4] (d) [4, 5] (e) [1, 3]
3. Find the average scale factor for $x \to 3x^2$ for the interval:
   (a) [0, 1] (b) [1, 2] (c) [2, 3] (d) [3, 4] (e) [1, 3]
   How do your results compare with those for question 2?
4. Find the average scale factor for $x \to 3x^2$ for the interval $[a, b]$. Find the value of this as b approaches a.
5. Find the average scale factor for $x \to x^3$ for the interval:
   (a) [0, 1] (b) [1, 2] (c) [2, 3] (d) [0, 3] (e) $[a, b]$
6. Repeat question 5 for $x \to 2x^3$. How do your results compare with those for question 5?
7. Find the average scale factor for $x \to 1/x$ for the interval:
   (a) [1, 2] (b) [2, 3] (c) [3, 4] (d) [1, 4] (e) [2, 7]
8. Repeat question 7 for $x \to 5/x$. How do your results compare with those for question 7?
9. Find the average scale factor for $x \to 1/x$ for the interval $[a, b]$. Find the value of this as b approaches a.
10. Find the average scale factor for $x \to x^2 + 4x$ for the interval:
    (a) [0, 1] (b) [1, 2] (c) [2, 3] (d) $[a, b]$
    Use your result from (d) to find the local scale factor associated with $x = a$.
11. Find the average scale factor for the interval $[a, b]$ for:
    (a) $x \to 3x + 2$ (b) $x \to x^2 + 5$ (c) $x \to 2x^2 + x$ (d) $x \to 1/x^2$
12. Use your results from question 11 to find for each the local scale factor associated with:
    (a) $x = a$ (b) $x = 10$

Scale factors and gradients

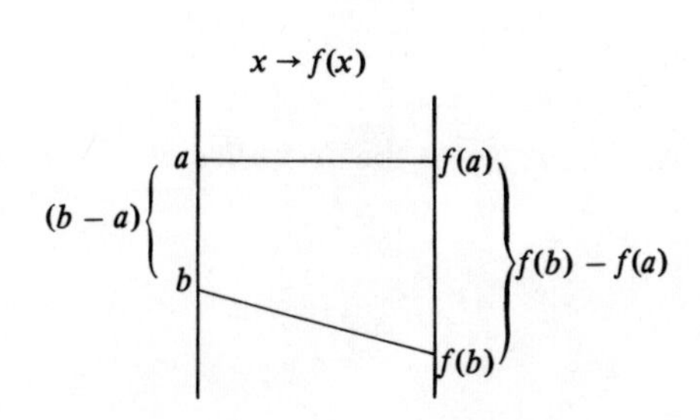

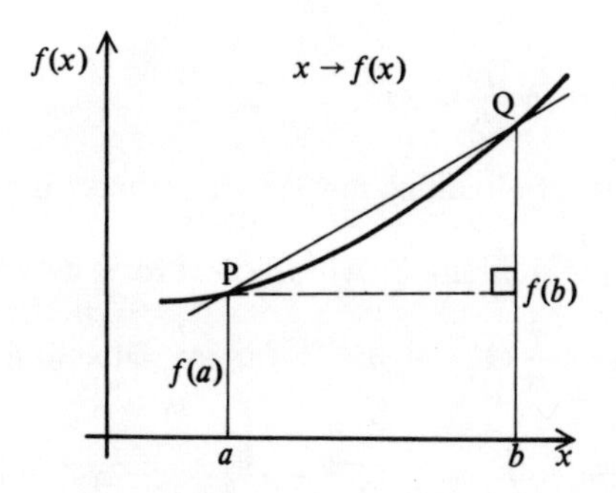

The average scale factor for the interval $[a, b]$ is
$\dfrac{f(b)-f(a)}{(b-a)}$

The local scale factor associated with $x = a$ is the value of $\dfrac{f(b)-f(a)}{(b-a)}$ as b approaches a.

The gradient of the chord PQ is
$\dfrac{f(b)-f(a)}{(b-a)}$

The gradient of the tangent at P is the value of $\dfrac{f(b)-f(a)}{(b-a)}$ as b approaches a.

Remember: The value of the local scale factor at $x = a$ gives the *gradient* of the tangent to $x \to f(x)$ where $x = a$, or the value of the *derivative* of $f(x)$ when $x = a$.

Exercise 11.1b

1. Find the average scale factor for $x \to x^2$ for the interval $[2, 5]$. Hence find the gradient of the chord joining the points $(2, 4)$ and $(5, 25)$.
2. Find the average scale factor for $x \to 4x^2 - 3x$ for the interval $[1, 3]$. Hence find the gradient of the chord joining the points $(1, 1)$ and $(3, 27)$.
3. Find the average scale factor for $x \to x^2 - 3$ for the interval $[a, b]$. By letting b approach a, find the local scale factor associated with $x = a$. Hence write down the gradient of the tangent to $x \to x^2 - 3$ at the point where:
(a) $x = a$ (b) $x = 5$.
4. Find the average scale factor for $x \to 3/x$ for the interval $[a, b]$. By letting b approach a, find the local scale factor associated with $x = a$. Hence write down the gradient of the tangent to $x \to 3/x$ at the point where:
(a) $x = a$ (b) $x = 5$.
5. Find the gradient of the chord joining the points where $x = p$ and $x = q$ on $x \to 4 + 1/x$. By letting q approach p, find the gradient of the tangent where $x = p$.
6. Repeat question 5 for:
(a) $x \to x^2 + 1/x$ (b) $x + 1/x^2$

11.2 Differentiation

If $y = x^3$ then $\frac{dy}{dx} = 3x^2$.

The process of obtaining the $3x^2$ from the x^3 is called *differentiation*: $3x^2$ is the *derivative* of x^3.

When we *differentiate* x^3 with respect to x, we write it as $\frac{d}{dx}(x^3)$.

Remember: $\frac{d}{dx}(x^n) = nx^{n-1}$ for any value of n except 0.

Example 1 $\quad \frac{d}{dx}[x^{-3} + x^{\frac{3}{2}}] = -3x^{-4} + \frac{3}{2}x^{\frac{1}{2}}$

Example 2 $\quad \frac{d}{dx}(\sqrt{x}) = \frac{d}{dx}(x^{\frac{1}{2}}) = \frac{1}{2}x^{-\frac{1}{2}}$

Exercise 11.2a

1. Write down the expression for $\frac{dy}{dx}$ if y is:
   (a) x^3 (b) x^4 (c) x^{10} (d) x^{-7} (e) x^{-2}
2. Differentiate with respect to x:
   (a) x^5 (b) x^{-4} (c) $x^{\frac{4}{3}}$ (d) $x^{-\frac{3}{2}}$ (e) $x^{\frac{1}{2}}$
3. Find:
   (a) $\frac{d}{dx}(x^7)$ (b) $\frac{d}{dx}(x^{-5})$ (c) $\frac{d}{dy}(y^{-\frac{1}{3}})$ (d) $\frac{d}{dz}(z^{\frac{2}{3}})$
4. What expression has as its derivative:
   (a) $9x^8$? (b) x^9? (c) $\frac{5}{2}x^{\frac{3}{2}}$? (d) x^{-10}? (e) $x^{-\frac{4}{3}}$?
5. Write down the expression for $\frac{ds}{dt}$ if s is:
   (a) $t^2 + t^3$ (b) $t^5 - 10t$ (c) $t^{-3} + t^{-4}$ (d) $2t^{\frac{3}{2}} + t^{-\frac{2}{3}} + 6$
6. Differentiate with respect to r:
   (a) $2r^5 + 3r^8 + 1$ (b) $3r^{-4} + 7r^{-6} - 5$ (c) $r^{-\frac{5}{2}} + r^{-\frac{1}{4}} - 0{\cdot}5$
7. Find:
   (a) $\frac{d}{dx}\left(\frac{x^5}{5} + \frac{x^7}{7}\right)$ (b) $\frac{d}{dt}\left(\frac{3}{2}t^{\frac{2}{3}} - 2t^{-\frac{1}{2}}\right)$ (c) $\frac{d}{dx}(\sqrt{x} + \sqrt[3]{x})$
8. Taking π and h as constants, differentiate with respect to r:
   (a) $2\pi r$ (b) πr^2 (c) $\frac{4}{3}\pi r^3$ (d) $\pi r^2 h$ (e) $2\pi r^2 + \pi rh$
9. Find the gradient of the tangent to $y = x^2$ at the point (3, 9).
10. Find the gradient of the tangent to $y = x^3 + x$ at the point (2, 10).
11. The distance travelled by a car at time t is given by the expression $3t^2 - t$. Find the speed (ds/dt) of the car when $t = 4$.
12. Differentiate the expression $5x^4 - 3x^{-4}$ twice with respect to x.

Example 1

Find the gradients of the tangents to the curve $y = (x-1)(x-2)$ at the points where $x = 1$ and where $x = 2$. Find also the equations of these tangents.

$$y = (x-1)(x-2) = x^2 - 3x + 2$$

$$\frac{dy}{dx} = 2x - 3$$

$\frac{dy}{dx} = -1$ when $x = 1$. $\frac{dy}{dx} = 1$ when $x = 2$.

Equation of tangent at (1, 0) is $(y-0) = -1\,(x-1)$
i.e. $y = 1 - x$

Equation of tangent at (2, 0) is $(y-0) = 1\,(x-2)$
i.e. $y = x - 2$

Example 2

If $y = x(x^2-1)$ find $\left(\frac{dy}{dx}\right)^2$ and $\frac{d^2y}{dx^2}$

$$y = x(x^2-1) = x^3 - x$$

$$\frac{dy}{dx} = 3x^2 - 1 \quad \text{so} \quad \left(\frac{dy}{dx}\right)^2 = (3x^2-1)^2$$

$$\frac{d^2y}{dx^2} = \frac{d}{dx}\left(\frac{dy}{dx}\right) = \frac{d}{dx}(3x^2-1) = 6x$$

Exercise 11.2b

1. Find $\frac{dy}{dx}$ and $\frac{d^2y}{dx^2}$ if y is:

   (a) $x^5 - x^2$ (b) $x^{-7} + x$ (c) $x^{\frac{5}{2}} + x^{-\frac{1}{2}}$ (d) $x^{\frac{4}{3}} - x^{-\frac{3}{4}} + 6$

2. Find $\frac{ds}{dt}$ and $\frac{d^2s}{dt^2}$ if s is:

   (a) $3t^4 - 2t^{-1}$ (b) $t^2(t-7)$ (c) $(t-2)(t-3)$ (d) $\sqrt{t} + 16t$

3. Find $\frac{dy}{dx}$ and $\frac{d^2y}{dx^2}$ when $x = 4$, if y is:

   (a) $x^4 - x^{-3}$ (b) $x^{\frac{1}{2}} + x^{-\frac{1}{2}}$ (c) $x^3\left(x + \frac{1}{x}\right)$ (d) $\frac{x^5 - x^2 + 1}{x}$

4. Find the gradients of the tangents to the curve $y = x(x-3)$ at the points where $x = 0$ and $x = 3$. Sketch the curve and explain why these gradients have the same magnitude but opposite signs.

5. Find the gradients of the tangents to the curve $y = 4 - x^2$ at the points where $x = -2$, $x = 0$ and $x = 2$. Find also the equations of these tangents. Explain the significance of the gradient at $x = 0$.

6. Show that if $y = x^{\frac{3}{2}} + x^{\frac{1}{2}}$ then $\left[2x\frac{d^2y}{dx^2} - \frac{dy}{dx}\right]^2 = \frac{1}{x}$.

11.3 Rates of change

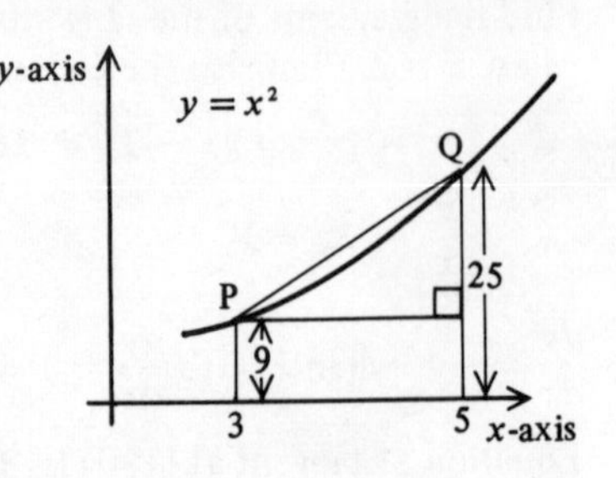

The gradient of a chord joining two points on a curve is the *average rate of change* of y with respect to x over the particular interval. In the diagram this is $\dfrac{25-9}{5-3} = \dfrac{16}{2} = 8$.

Example 1

When a car travels 75 m in 3 s, the *average rate of change* of distance with respect to time (i.e. speed) is $\frac{75}{3} = 25$ m/s.

Example 2

When the speed of a car increases from 60 km/h to 80 km/h in 1 minute, the *average rate of change* of speed with respect to time (i.e. acceleration) is $\dfrac{80-60}{\frac{1}{60}} = 1200$ km per hour per hour.

Example 3

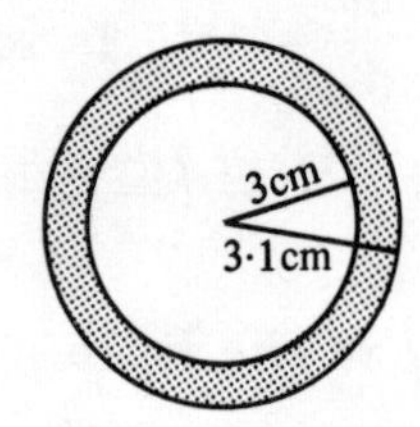

When the radius of a circle changes from 3 cm to 3·1 cm, the area of the circle increases from 9π cm^2 to $9{\cdot}61\pi$ cm^2.
The *average rate of change* of area with respect to radius is therefore

$$\frac{9{\cdot}61\pi - 9\pi}{3{\cdot}1 - 3} = \frac{0{\cdot}61\pi}{0{\cdot}1} = 6{\cdot}1\pi \text{ cm}^2/\text{cm}.$$

Exercise 11.3a

1. Find the gradient of the chord joining the points:
(a) (0, 0) and (3, 12) (b) (1, 2) and (5, 10) (c) (−3, 1) and (1, 3)
2. Find the average rate of change of y with respect to x, on the curve $y = x^3$ for the interval $x = 1$ to $x = 4$.
3. Find the gradient of the chord joining the points on $y = x^2 + x$ where:
(a) $x = 0$ and $x = 3$ (b) $x = 6$ and $x = 7$ (c) $x = 3$ and $x = 5$.
4. Find the average speed in km/h of a car which travels:
(a) 150 km in $2\frac{1}{2}$ hours (b) 12 km in 15 minutes (c) 126 m in 3 seconds.
5. Find the average acceleration in km/h^2 of a car whose speed changes from:
(a) 72 km/h to 83 km/h in $\frac{1}{4}$ of an hour (b) 56 km/h to 59 km/h in 20 seconds.
6. Find the average rate of change of the area of a circle with respect to its radius when the radius changes from:
(a) 5 cm to 7 cm (b) 1·2 cm to 1·5 cm (c) 1·9 cm to 2·1 cm.
7. Find the average rate of change of the volume of a cube with respect to the length of its edge, if the edge length changes from:
(a) 2 cm to 6 cm (b) 1·1 cm to 1·3 cm (c) 0·7 cm to 1·2 cm.
8. Find the average rate of change of the volume of a spherical bubble with respect to its radius when the radius changes from 3 cm to 6 cm. Find also the average rate of change of the volume with respect to time if the radius changes from 3 cm to 6 cm over half a second.

The gradient of the tangent to a curve at a particular point is a measure of *the rate of change* of y with respect to x at that particular point. In the diagram this is $2 \times 5 = 10$. (i.e. $2x$ when $x = 5$).
The rate of change of y with respect to x is $\frac{dy}{dx}$.

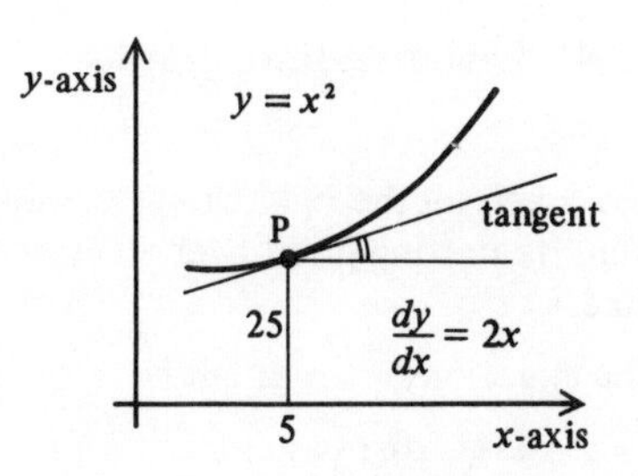

The *rate of change* of the area of a circle with respect to its radius is obtained from the formula $A = \pi r^2$ by calculating $\frac{dA}{dr}$ for the given radius.

Example 4

The population of bacteria in a colony is given at time t by the formula $p = 3t^2 - 2t + 1$. Find the average rate of increase in the population during the third second and the rate of increase after 3 seconds.

The average rate of increase during the third second is $22 - 9 = 13$.

The rate of increase when $t = 3$ is $\frac{dp}{dt} = 6t - 2 = 18 - 2 = 16$.

Exercise 11.3b

1. Find the gradient of the tangent at the point where $x = 3$ for:
   (a) $y = x^2$ (b) $y = 5x^2 - 2x$ (c) $y = x^3$ (d) $y = x + x^{-1}$
2. Find the rate of change of y with respect to x, when $x = 4$ for:
   (a) $y = x^4$ (b) $y = x^3 - x^2$ (c) $y = x^{-4}$ (d) $y = x^{\frac{1}{2}} + x^{\frac{3}{2}}$
3. Find the rate of change of s with respect to t, when $t = 2$ for:
   (a) $s = 3t - 6$ (b) $s = 7 - t^2$ (c) $s = t - t^3$ (d) $s = t^4 - 3t^2 + 2$
4. Using the formula $A = \pi r^2$ find the rate of change of the area of a circle with respect to its radius when:
   (a) $r = 3$ (b) $r = 5$ (c) $r = 1{\cdot}2$
5. Using the formula $V = \frac{4}{3}\pi r^3$ find the rate of change of the volume of a sphere with respect to its radius when:
   (a) $r = 2$ (b) $r = 4$ (c) $r = 10$
6. Find the rate of change of distance s with respect to time t, when $t = 2$ for:
   (a) $s = t^4 + 2$ (b) $s = 9t^2 - 3$ (c) $s = 2t^3 + 7$
   In which of these is the speed greatest when $t = 2$?
7. Find the rate of change of speed v with respect to time t, when $t = 9$ for:
   (a) $v = t^2 - t$ (b) $v = 36t^{\frac{1}{2}} + 2$ (c) $v = 81t^{-1} + 6$
   In which of these is the acceleration smallest when $t = 9$?
8. The population of bacteria in a colony is given at time t by the formula $p = 2t^3 - t + 1$. Find the average rate of increase in the population during the fourth second and the rate of increase after 4 seconds.
9. The amount of fuel left in a rocket is given at time t by the formula $f = 2000 - t^2 - t^3$. Find the average rate of decrease of fuel during the tenth second and the rate of decrease at the beginning and end of the tenth second.

11.4 Distance-time graphs

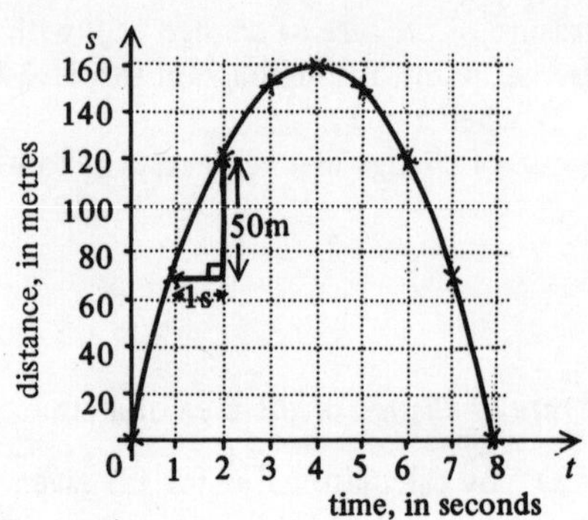

The graph on the right shows the distance of a particle from its starting point over an eight-second interval of time.

The equation of the graph is

$$s = 10t(8-t)$$

$$\frac{ds}{dt} = 80 - 20t$$

Note: (i) The *average* speed during the 2nd second is $\dfrac{(120-70)}{1}$ m/s.

(ii) The speed after two seconds is given by $\dfrac{ds}{dt}$ when $t = 2$.

Exercise 11.4a

1. Find the average speed of the particle above during:
   (a) each one-second time interval (b) each two-second time interval.
   Describe how the average speed changes over the eight seconds.
2. Find the speed of the particle above, after each whole second. Describe how the speed changes over the eight seconds. Explain why $\dfrac{ds}{dt}$ is negative for $4 < t \leqslant 8$.
3. At what time during the eight seconds is the particle above:
   (a) at rest? (b) at its starting point?
4. The distance s of a particle from its starting point at time t seconds is given in metres by the equation $s = t^3 - 6t^2 + 12t$. Find:
   (a) the average speed of the particle during each of the first four seconds.
   (b) the speed of the particle after each of the first four seconds.
   (c) at what time the particle is at rest.
5. The distance s moved by a particle in t seconds is given in metres by $s = t^2 + 4t + 3$. Find:
   (a) the distance moved during the first second.
   (b) the average speed during the first second.
   (c) the initial speed and the speed after one second.
6. A car is moving at a constant speed when the brakes are applied. The distance, s metres, travelled by the car in t seconds, is given by $s = 27t - t^3$. Find:
   (a) the distance travelled and time taken before coming to rest.
   (b) the initial speed and the average speed over the first two seconds.

Acceleration

Speed is the rate of change of distance with respect to time i.e. $\frac{ds}{dt}$.

Acceleration is the rate of change of speed with respect to time i.e. $\frac{d^2s}{dt^2}$.

Note: If the speed is increasing the acceleration is positive.
If the speed is decreasing the acceleration is negative.
If the speed is constant the acceleration is zero.

Example

If $s = t^3 - 2t^2 - 5$, where s is the distance travelled in time t, find the speed and acceleration when $t = 3$.

$$\text{Speed} = \frac{ds}{dt} = 3t^2 - 4t = 15, \text{ when } t = 3$$

$$\text{Acceleration} = \frac{d^2s}{dt^2} = 6t - 4 = 14, \text{ when } t = 3.$$

Exercise 11.4b

1. The distance s moved by a particle in t seconds is given in metres by $s = 20t - t^4 + 1$. Find:
   (a) the initial speed and the initial acceleration.
   (b) the speed and the acceleration after 2 seconds.
2. The speed v of a particle after t seconds is given in metres per second by $v = 3t^2 + 5$. Find:
   (a) the initial acceleration and the acceleration after 4 seconds.
   (b) the distance travelled in the first second.
   (c) the average speed in the first second.
3. A body moves s metres in t seconds given by $s = t^4 - 4t^3$. Find the acceleration when the speed of the body is zero. Find also when the acceleration is zero.
   For what values of t is the acceleration:
   (a) positive? (b) negative?
4. A body moves s metres in t seconds given by $s = 15t - 2t^{\frac{3}{2}}$. Find the acceleration when the speed of the body is zero. Find the speed and acceleration after 9 seconds.
   For what values of t is the acceleration:
   (a) positive? (b) negative?
5. Readings were taken every second to record the distance in metres travelled by a particle. The results are shown below.

Time in seconds	1	2	3	4	5	6	7	8
Distance in metres	0·4	7·2	20·4	40	66	98·4	137·2	182·4

   (a) Draw a graph to show this information and find the average speed during each one-second time interval.
   (b) By drawing a tangent to the curve, find the speed after four seconds.
   (c) What can you say about the acceleration during the eight seconds?

11.5 Maximum and minimum values

At a *maximum*, the gradient of the tangent is zero; i.e. $\frac{dy}{dx} = 0$, *and* the gradient changes from positive to negative; i.e. $\frac{d^2y}{dx^2} < 0$.

At a *minimum*, the gradient of the tangent is zero; i.e. $\frac{dy}{dx} = 0$, *and* the gradient changes from negative to positive; i.e. $\frac{d^2y}{dx^2} > 0$.

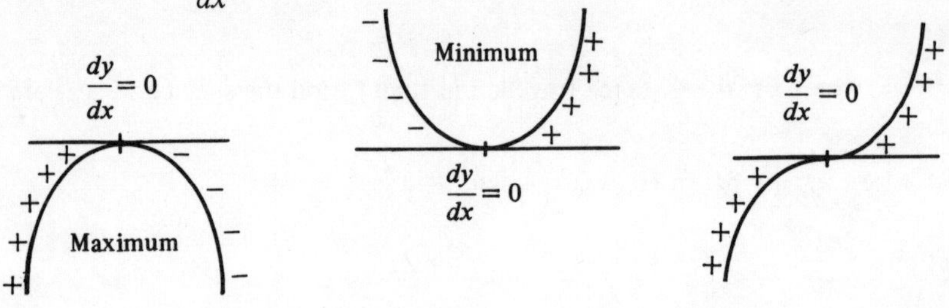

Remember: It is possible for $\frac{dy}{dx} = 0$ at a point which is neither a maximum nor a minimum as is shown on the right above.

Exercise 11.5a

1. If $y = x^2 - 4x + 3$ find where $\frac{dy}{dx} = 0$. By considering the gradient on either side of this point determine the nature of the turning value. Find the value of y at this point.
2. If $y = x^2 + 3x - 4$ find where $\frac{dy}{dx} = 0$. By considering the sign of $\frac{d^2y}{dx^2}$ determine the nature of the turning value.
3. Find the minimum value of $x^2 - 5x + 1$.
4. Find the maximum value of $4x - x^2$.
5. Find the co-ordinates of the points on the curve $y = x^3 - 12x$ where the tangent is parallel to the x-axis. Find in each case the nature of the turning value.
6. Find the co-ordinates of the points on the curve $y = 5 + 3x^2 - x^3$ where the tangent is parallel to the x-axis. Find in each case the nature of the turning value.
7. The derivative of $f(x)$ is $(x - 1)(x - 3)$. For what values of x is $\frac{df}{dx} = 0$? Which of these gives a minimum value for $f(x)$?
8. The derivative of $g(x)$ is $(x - 1)(x - 3)(x - 5)$. For what values of x is $\frac{dg}{dx} = 0$? Find in each case the nature of the turning value.

Example

A cylinder is to be made so that the height plus the radius is 3. Find the size of the radius which would give the greatest volume.

Now $V = \pi r^2 h$ and $h + r = 3$ so $h = 3 - r$

$$V = \pi r^2(3 - r);\ \frac{dV}{dr} = 6\pi r - 3\pi r^2 = 3\pi r(2 - r)$$

$$\frac{dV}{dr} = 0 \text{ when } r = 0 \text{ or } r = 2$$

$$\frac{d^2V}{dr^2} = -6\pi \text{ when } r = 2.$$

Since $\frac{d^2V}{dr^2}$ is negative, $r = 2$ will give a maximum value for V.

The greatest volume possible is 4π.

Exercise 11.5b

1. The perimeter of a rectangle is 20 cm. Find the area of the rectangle in terms of the length of one edge. Hence find the greatest possible area.
2. A cardboard box has a square base. If the length of the three edges of the box must not exceed 72 cm, find the greatest possible volume.
3. When a ball is thrown into the air, its height above the ground is given by the equation $h = 9{\cdot}8t - 4{\cdot}9t^2$ where h is the height in metres and t the time in seconds. Find the greatest height of the ball above the ground.
4. A particle moves so that its distance s from the starting point is given by $s = 2t^3 - 9t^2 + 12t$ where t is the time taken. Find the two times when the particle is stationary. Hence find the *maximum* and *minimum* distances from the starting point for $t \leqslant 3$. Find also the *greatest* and *least* distances for $0 \leqslant t \leqslant 3$.
5. An open box is to be made from a square piece of card 20 cm × 20 cm as shown on the right. Find the value of x which would give a box of greatest volume. What is this volume?

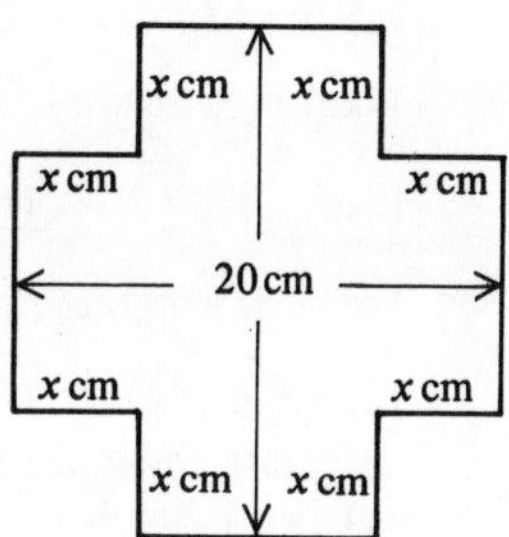

6. Find the maximum and minimum values of $x^3 - 2x^2 + x$. Hence sketch the curve $y = x^3 - 2x^2 + x$.
7. The surface area of a closed cylinder of height h cm and radius r cm is 96π cm^2. Write down the volume of the cylinder in terms of r only. Find the value of r which gives the greatest volume. Find also this greatest volume and the corresponding value of h.

11.6 Integration

If $\frac{dy}{dx} = x^2$ then $y = \frac{x^3}{3} + c$ where c is a constant.

The process of finding an expression (y) when its derivative $\left(\frac{dy}{dx}\right)$ is known is called *integration*; $\left(\frac{x^3}{3} + c\right)$ is the *integral* of x^2.

When we *integrate* x^2 with respect to x, we write it as $\int x^2\, dx$.

Remember: $\int x^n\, dx = \frac{x^{n+1}}{n+1} + c$ for any value of n except -1.

Example 1

$$\int (x^{\frac{1}{2}} + x^{-5})\, dx = \frac{x^{\frac{3}{2}}}{\frac{3}{2}} - \frac{x^{-4}}{4} + c$$

Example 2

$$\int \frac{1}{\sqrt{x}}\, dx = \int x^{-\frac{1}{2}}\, dx = 2x^{\frac{1}{2}} + c$$

Exercise 11.6a

1. Write down an expression for y if $\frac{dy}{dx}$ is:
   (a) x^3 (b) x^4 (c) x^{10} (d) x^{-7} (e) x^{-2}
2. Integrate with respect to x:
   (a) x^5 (b) x^{-3} (c) $x^{\frac{4}{3}}$ (d) $x^{-\frac{5}{4}}$ (e) $x^{-\frac{1}{2}}$
3. Find:
   (a) $\int x^7 dx$ (b) $\int x^{-4} dx$ (c) $\int y^{\frac{2}{3}} dy$ (d) $\int z^{-\frac{1}{3}} dz$
4. What expression has as its integral:
   (a) $\frac{x^7}{7} + c$? (b) $x^9 + 6$? (c) $3x^{\frac{1}{3}} - 5$? (d) $x^{-\frac{3}{4}} + 4$?
5. Write down an expression for s if $\frac{ds}{dt}$ is:
   (a) $(t^2 + t^3)$ (b) $(5t^4 - 10)$ (c) $(t^{-3} + t^{-4})$ (d) $(2t^{\frac{3}{2}} + t^{-\frac{2}{3}})$
6. Integrate the expression with respect to r.
   (a) $2r^5 + 3r^8 + 1$ (b) $3r^{-4} + 7r^{-6} - 5$ (c) $r^{-\frac{5}{2}} + r^{-\frac{1}{4}} - 0{\cdot}5$
7. Find:
   (a) $\int (5x^4 + 6x^7) dx$ (b) $\int (\frac{5}{3}t^{\frac{2}{3}} - \frac{1}{2}t^{-\frac{1}{2}})\, dt$ (c) $\int (\sqrt{x} + \sqrt[3]{x})\, dx$
8. What expression has as its derivative:
   (a) $s^{-2} + s^{-3} - 4$? (b) $t^{\frac{1}{4}} + t^{\frac{1}{5}} - t$? (c) $y^{20} - y^{-20} + y^{\frac{1}{20}}$?
9. A curve passes through the point (0, 3) and its gradient at any point is $3x^2$. Find the equation of the curve.
10. A curve passes through the point (2, 8) and its gradient at any point is $2x + x^3$. Find the equation of the curve.
11. A car starts from rest and its speed at any time on its journey is given by the expression $3t^2 - t$. Find how far the car has travelled when $t = 4$.
12. Find an expression which when differentiated twice is $5x^4 - 3x^{-4}$.

Definite integrals

$\int_2^5 x^2\,dx = \left[\frac{x^3}{3}\right]_2^5 = \frac{5^3}{3} - \frac{2^3}{3} = \frac{117}{3}$ is a *definite* integral of x^2.

Remember: There is no need to write the constant c in [ ] since

$\left[\frac{x^3}{3} + c\right]_2^5 = \left(\frac{5^3}{3} + c\right) - \left(\frac{2^3}{3} + c\right) = \frac{5^3}{3} - \frac{2^3}{3} = \frac{117}{3}$ as above.

$\int x^2\,dx = \frac{x^3}{3} + c$ is the *indefinite* integral of x^2 since the limits of integration are not specified.

Example 3

Find the value of $\int_1^2 (x^3 + x^{-2} + 1)\,dx$

$$\int_1^2 (x^3 + x^{-2} + 1)\,dx = \left[\frac{x^4}{4} - x^{-1} + x\right]_1^2 = \left[\frac{16}{4} - \frac{1}{2} + 2\right] - \left[\frac{1}{4} - 1 + 1\right] = 5\tfrac{1}{4}$$

Exercise 11.6b

1. Find the value of:

   (a) $\int_1^2 x^2\,dx$ (b) $\int_2^3 x^4\,dx$ (c) $\int_1^2 x^{-3}\,dx$ (d) $\int_{-1}^{+3} x^3\,dx$

2. Find the value of:

   (a) $\int_0^1 x^{\frac{1}{3}}\,dx$ (b) $\int_2^5 t^{-2}\,dt$ (c) $\int_{-2}^1 y\,dy$ (d) $\int_4^9 s^{-\frac{1}{2}}\,ds$

3. Find the value of:

   (a) $\int_1^2 (x^2 + x + 1)\,dx$ (b) $\int_{-1}^1 (y^5 + y^3 + y)\,dy$ (c) $\int_1^4 (t^{-\frac{3}{2}} + t^{-\frac{1}{2}} + t^{\frac{1}{2}})\,dt$

4. Find the value of:

   (a) $\int_0^5 (3x + 2)\,dx$ (b) $\int_{-2}^{+2} (t^3 + 2)\,dt$ (c) $\int_{-2}^{-1} \left(x^2 - \frac{1}{x^2}\right)dx$

5. Find the value of:

   (a) $\int_0^{\frac{1}{2}} (6x^2 - 4x + 3)\,dx$ (b) $\int_4^9 \left(\sqrt{x} + \frac{1}{\sqrt{x}}\right)dx$ (c) $\int_{-2}^{-1} \left(x^4 - \frac{1}{x^4}\right)dx$

6. Find the value of:

   (a) $\int_0^1 x(x + 1)\,dx$ (b) $\int_{-1}^1 x^2(x^2 + 1)\,dx$ (c) $\int_1^2 x(x^2 - 3)\,dx$

7. Find the value of:

   (a) $\int_0^1 (x + 1)(x + 2)\,dx$ (b) $\int_{-2}^{+2} (x - 1)(x - 3)\,dx$ (c) $\int_{-2}^1 (1 - x)(x + 2)\,dx$

8. Find the value of:

   (a) $\int_p^q x^2\,dx$ (b) $\int_0^n (x^3 + x)\,dx$ (c) $\int_{-r}^{+r} x(x^3 - 1)\,dx$

11.7 Areas under curves

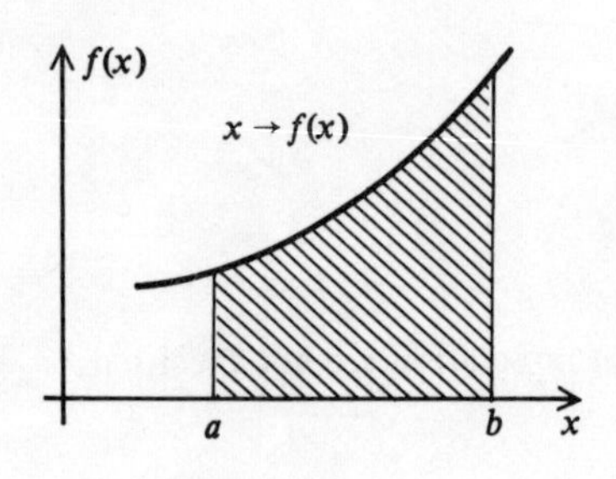

The shaded area is

$$\int_a^b f(x)\,dx$$

The shaded area is

$$\int_2^3 (x^2 - 2x)\,dx = \left[\frac{x^3}{3} - x^2\right]_2^3$$

$$= (\tfrac{27}{3} - 9) - (\tfrac{8}{3} - 4) = \tfrac{4}{3}$$

Note: In the right hand diagram above $\int_0^2 (x^2 - 2x)\,dx = -\tfrac{4}{3}$.

This integral is negative since it represents an area *below* the x-axis.

Exercise 11.7a

1. Find the area under $y = x^2 - 2x$ between:
   (a) $x = 3$ and $x = 4$ (b) $x = 2$ and $x = 4$ (c) $x = -1$ and $x = 0$
2. Find the area under $y = x^2 + x$ between:
   (a) $x = 0$ and $x = 1$ (b) $x = 2$ and $x = 3$ (c) $x = 1$ and $x = 4$
3. Find the area under $y = x^3$ between:
   (a) $x = 0$ and $x = 1$ (b) $x = 1$ and $x = 3$ (c) $x = 0$ and $x = 3$
   Do you agree that the sum of the first two areas is equal to the third area?
4. Find the area under $y = x^3 - x$ between:
   (a) $x = 1$ and $x = 2$ (b) $x = 2$ and $x = 3$ (c) $x = 1$ and $x = 3$
   Write down $\int_1^3 (x^3 - x)\,dx$ as the sum of the two other integrals.
5. Sketch the graph of $y = x^2 - 4$.
   Find the values of the integrals $\int_2^3 (x^2 - 4)\,dx$ and $\int_0^2 (x^2 - 4)\,dx$.
   Explain why one of these values is positive and the other negative.
6. For each of the areas A, B, and C in the diagram, write down an appropriate integral. Hence find the three areas.
   Find also $\int_{-2}^{2} (x^3 - 4x)\,dx$.
   Explain why the value of this integral is zero.

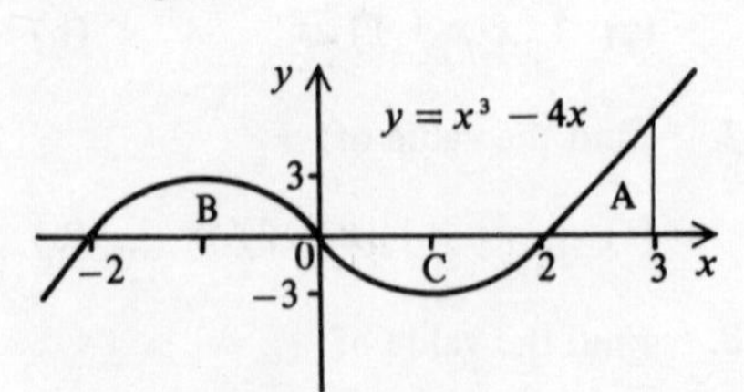

Areas between curves

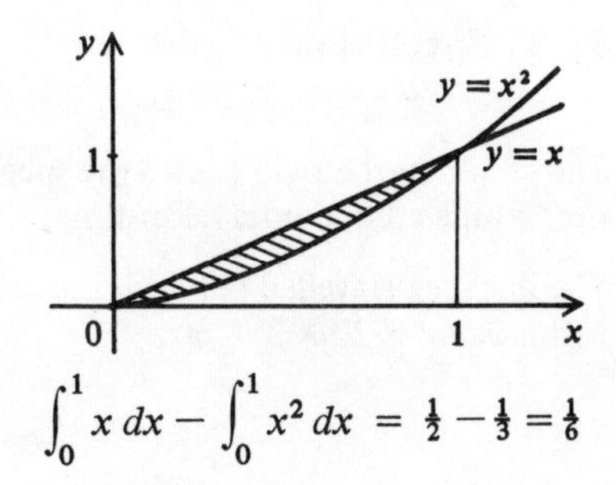

The shaded area on the right is the area between the curves $y = x$ and $y = x^2$.
To find this area it is necessary to find the area under each curve between $x = 0$ and $x = 1$ and then find the difference between them.

$$\int_0^1 x\,dx - \int_0^1 x^2\,dx = \tfrac{1}{2} - \tfrac{1}{3} = \tfrac{1}{6}$$

Example

Find the area between $y = x^2 + 2$ and $y = 4x - 1$.

The two curves meet where $x^2 + 2 = 4x - 1$

i.e. where $x^2 - 4x + 3 = 0$
$(x - 3)(x - 1) = 0$

i.e. where $x = 3$ or $x = 1$

The area between the curves is $\int_1^3 (4x - 1)\,dx - \int_1^3 (x^2 + 2)\,dx.$

Exercise 11.7b

1. Evaluate the two integrals given above and find the area between the curves $y = x^2 + 2$ and $y = 4x - 1$.
2. Show that the curves $y = x^2 - 2$ and $y = 3x - 4$ meet at the points $(1, -1)$ and $(2, 2)$.
   Evaluate the integrals $\int_1^2 (x^2 - 2)\,dx$ and $\int_1^2 (3x - 4)\,dx$ and hence find the area between the two curves.
3. Find the two points of intersection of the curves $y = x^2 + 1$ and $y = 2x + 4$. Hence find the area between the two curves.
4. For each of the areas A, B, and C in the diagram, write down the two integrals from which you can find the area. Find these areas.

   Find also $\int_0^2 (x^3 - x)\,dx$

   and explain why $\int_{-2}^2 (x^3 - x)\,dx = 0$.

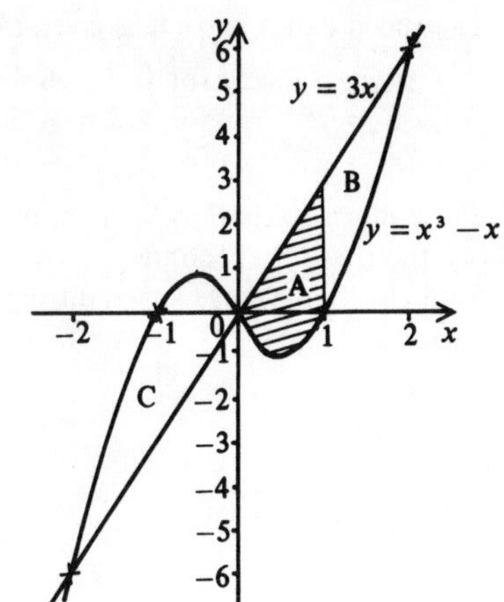

11.8 Speed-time graphs

The graph on the right shows the speed of a particle over a nine-second interval of time.

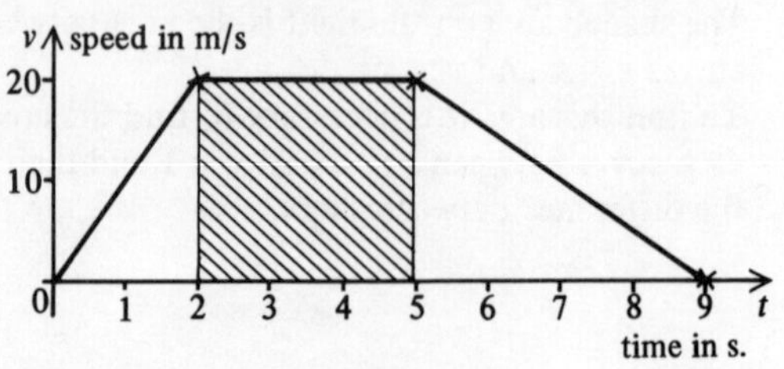

The distance travelled between the second and the fifth second is $20 \times 3 = 60$ m.

The distance travelled during the first two seconds is the area under the graph between $t = 0$ and $t = 2$, i.e. $\frac{1}{2} \times 20 \times 2 = 20$ m.

The distance travelled during a given interval of time can be found from the area under the speed-time graph.

Note: If $v = 10t$ then $\int_0^2 10t\,dt$ is the distance travelled between $t = 0$ and $t = 2$.

Exercise 11.8a

1. Using the graph above, find the distance travelled during the last four seconds.

   Explain why this distance is $\int_5^9 (45 - 5t)\,dt$.

2. The speed of a car increases from 0 km/h to 60 km/h in twenty seconds. It then increases from 60 km/h to 90 km/h in the next fifteen seconds and finally reduces to 0 km/h over another eighteen seconds. Show this information on a graph and find the distances travelled during each stage.
3. The speed v of a particle is given in m/s by $v = 3t^2 + 5$ where t is the time in seconds. Find the distance travelled in the first three seconds and also during the fourth second.
4. The speed v of a particle is given in m/s by $v = 2 + 3t + t^2$ where t is the time in seconds. Find the distance travelled in the first two seconds and also during the fourth second.
5. The speed v of a particle is given in m/s as follows:

   $v = 3 + 2t$ for $0 \leqslant t \leqslant 4$
   $v = t^2 - 5$ for $4 < t \leqslant 5$
   $v = 40 - 4t$ for $5 < t \leqslant 10$

   Show this information on a graph and find the distance travelled during:
   (a) the first four seconds (b) the fifth second (c) the last five seconds.
   Find also the average speed during each of these intervals, and where the acceleration is greatest.

The speed of a particle is given by

$$v = 8 + 7t - t^2$$

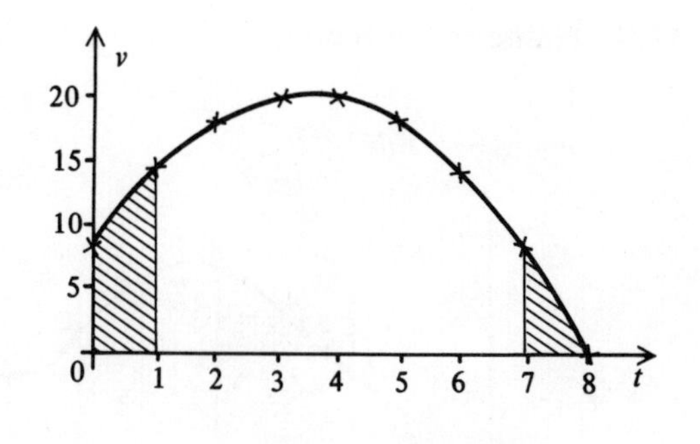

The distance travelled during the first second is

$$\int_0^1 (8 + 7t - t^2)\, dt.$$

The distance travelled during the eighth second is

$$\int_7^8 (8 + 7t - t^2)\, dt.$$

The acceleration is given by $\dfrac{dv}{dt} = 7 - 2t$.

The maximum speed is when $\dfrac{dv}{dt} = 0$ i.e. when $t = 3{\cdot}5$.

Exercise 11.8b

1. Using the graph above find:
   (a) the distance travelled during the first second
   (b) the distance travelled during the fourth second
   (c) the acceleration when: (i) $t = 2$ (ii) $t = 5$.
2. The speed of a particle is given by $v = t^2 - 5t + 7$. Find:
   (a) the distance travelled during the first second
   (b) the distance travelled in the next three seconds
   (c) the acceleration when: (i) $t = 1$ (ii) $t = 3$
   (d) the minimum speed
   (e) the greatest speed during the first six seconds.
3. The speed of a particle is given by $v = t^3 - 8t^2 + 13t + 20$. Find:
   (a) the distance travelled during the first two seconds
   (b) the acceleration when: (i) $t = 0$ (ii) $t = 2$ (iii) $t = 5$
   (c) the maximum and minimum speeds during the first ten seconds
   (d) the greatest and least speeds during the first ten seconds
   (e) the minimum acceleration during the first ten seconds.
4. The acceleration of a moving particle is $10\,\text{m/s}^2$. If the initial speed is 5 m/s, find the speed and also the distance travelled after 6 seconds.
5. The acceleration of a moving particle is $(3 + 2t)\,\text{m/s}^2$. If the initial speed is 10 m/s, find the speed and also the distance travelled after 2 seconds.
6. A bullet fired from a rifle leaves the barrel with a speed of 800 m/s. It decelerates at a rate of $10\,\text{m/s}^2$. How far will the bullet travel in 10 seconds?
7. Repeat question 6 if the deceleration is $6t\,\text{m/s}^2$.
8. A car decelerates from a speed of 100 m/s at a rate of $5\,\text{m/s}^2$. Find how long the car takes to stop and how far it travels.

11.9 Numerical integration

Mid-ordinate rule

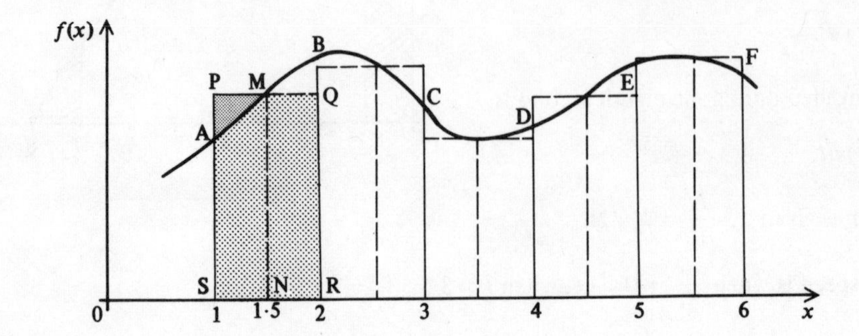

The area under the above curve between A and B is approximately equal to the area of the rectangle PQRS. The rectangle is formed by drawing PQ through M, where the x-co-ordinate of M is 1·5. MN is the *mid-ordinate* in the interval [1, 2]. The area of this rectangle is $1 \times f(1{\cdot}5)$.

The area under the curve between A and F is the sum of the areas of all the rectangles like PQRS, formed by using the mid-ordinates of each interval.

i.e. $$\int_1^6 f(x)\,dx \approx 1 \times f(1{\cdot}5) + 1 \times f(2{\cdot}5) + 1 \times f(3{\cdot}5) + 1 \times f(4{\cdot}5) + 1 \times f(5{\cdot}5)$$

Note: The area of each rectangle is obtained by multiplying the mid-ordinate by the width of the interval.

Exercise 11.9a

1. Use the mid-ordinate rule with four intervals to find an approximation to the area under $f(x) = x^2$ between $x = 0$ and $x = 4$. Compare your result with the actual area found by integration.

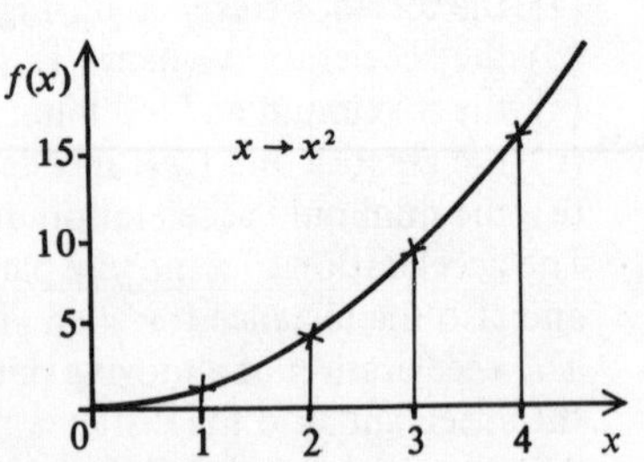

2. Repeat question 1 but using eight intervals each of width 0·5.
3. Use the mid-ordinate rule with four intervals to find an approximation to the area under $f(x) = x^2 - x$ between $x = 1$ and $x = 9$. Compare your result with the actual area.
4. Repeat question 3 but using eight intervals each of width 1.
5. Use the mid-ordinate rule with six intervals to find an approximation to the area under $f(x) = x^3$ between $x = 0$ and $x = 3$.
6. Use the mid-ordinate rule with ten intervals to find an approximation to the area under $f(x) = 4 - x^2$ between $x = -1$ and $x = 1$. Compare your result with the actual area.

Trapezium rule

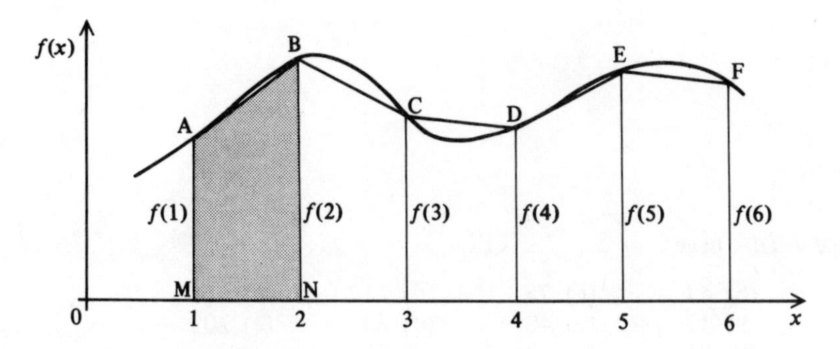

The area under the above curve between A and B is approximately equal to the area of the trapezium ABNM.

The area under the curve between A and F is the sum of the areas of all the trapeziums like ABNM.

i.e. $$\int_1^6 f(x)\,dx \approx \tfrac{1}{2}(f(1)+f(2))\times 1 + \tfrac{1}{2}(f(2)+f(3))\times 1 + \tfrac{1}{2}(f(3)+f(4))\times 1 + \tfrac{1}{2}(f(4)+f(5))\times 1 + \tfrac{1}{2}(f(5)+f(6))\times 1$$

Note: The area of each trapezium is obtained by finding half the sum of the ordinates at each end and then multiplying by the width of the interval.

Exercise 11.9b

1. Use the trapezium rule with four intervals to find an approximation to the area under $f(x) = x^2$ between $x = 0$ and $x = 4$. Compare your results with exercise 11.9a question 1.
2. Repeat question 1 but using:
   (a) two intervals of width 2 (b) eight intervals of width 0·5
3. Use the trapezium rule with four intervals to find an approximation to the area under $f(x) = x^2 - x$ between $x = 1$ and $x = 9$. Compare your results with exercise 11.9a question 3.
4. Repeat question 3 but using:
   (a) two intervals of width 4 (b) eight intervals of width 1.
5. Use the trapezium rule with six intervals to find an approximation to the area under $f(x) = x^3$ between $x = 0$ and $x = 3$. Compare your results with exercise 11.9a question 5.
6. Use the trapezium rule with ten intervals to find an approximation to the area under $f(x) = 4 - x^2$ between $x = -1$ and $x = 1$. Compare your results with exercise 11.9a question 6.
7. Sketch the curve $f(x) = x^3 + 8$ for $-2 \leqslant x \leqslant 2$. Use the trapezium rule with two intervals to find an approximation to the area under the curve between:
   (a) $x = -2$ and $x = 0$ (b) $x = 0$ and $x = 2$.
   For which of (a) or (b) is the area more than the actual area? Explain your answer in relation to the shape of the graph over the interval.

Answers

Part 1

Exercise 1.1a page 2

1.	(a) 20	(b) 43	(c) 75	(d) 217	(e) 551	
2.	(a) 9	(b) 17	(c) 40	(d) 83	(e) 201	
3.	(a) 10	(b) 91	(c) 13	(d) 57		
4.	(a) 13	(b) 271	(c) 40	(d) 514		
5.	(a) 41	(b) 111	(c) 200	(d) 343	(e) 2045	
6.	(a) 32	(b) 121	(c) 1001	(d) 1312	(e) 10002	
7.	(a) 45	(b) 242	(c) 100	(d) 173	(e) 22221	(f) 1100100
8.	(a) 205	(b) 522	(c) 3003	(d) 2130		
9.	(a) 111	(b) 120	(c) 330	(d) 1100		
10.	(a) 62	(b) 203	(c) 650	(d) 1221		

Exercise 1.1b page 3

1.	(a) 5	(b) 13	(c) 9	(d) 21	(e) 27
2.	(a) 110	(b) 1101	(c) 11001	(d) 100001	(e) 1001011
3.	(a) 1000	(b) 10011	(c) 10100	(d) 100010	
4.	(a) 10	(b) 111	(c) 1010	(d) 1100	
5.	(a) $5+3$, $5-3$	(b) $13+6$, $13-6$	(c) $15+5$, $15-5$	(d) $23+11$, $23-11$	
6.	(a) 1000001	(b) 101010	(c) 1000110	(d) 1111110	
7.	(a) 151	(b) 335	(c) 124	(d) 242	
8.	(a) 32	(b) 302	(c) 23	(d) 223	
9.	(a) 178	(b) 356	(c) 222	(d) 6261	
10.	(a) 112	(b) 340	(c) 1105	(d) 3123	
11.	(a) 1101, 1111	(b) 1001, 1012	(c) 144, 224	(d) 10, 2	(e) 20, 2
12.	(a) 10122	(b) 174	(c) 121021		

Exercise 1.2a page 4

1. (a) 3, 5, 7, 9, ... odd numbers. (b) 2, 3, 4, 5, ... natural numbers.
(c) 1, 2, 4, 8, ... powers of two. (d) 2, 6, 18, 54, ... twice the powers of three.
(e) 0, 1, 1, 2, 3, 5, ... Fibonacci numbers.

2. (a) $4n, 4n+1, 4n+3$. (b) $2^{n-1}, 2^{n-1}-1, 2^{n-1}+2$. (c) n^2, n^2+2, n^2+5.

3. (a) $22, 25, 3n+1$ (b) $50, 65, n^2+1$ (c) $29, 37, \dfrac{n(n+1)}{2}+1$
(d) $1458, 4374, 2.3^{n-1}$

4. $n^2, (n+1)^2, (n+1)^2 - n^2 = 2n+1$

5. $\dfrac{n(n+1)}{2}, \dfrac{(n+1)(n+2)}{2}, \dfrac{(n+1)(n+2)}{2} + \dfrac{n(n+1)}{2} = \dfrac{(n+1)(2n+2)}{2} = (n+1)^2$

6. For example: $5, 8, 13,\ 5 \times 13 = 65 = 8^2 + 1;\ 8, 13, 21,\ 8 \times 21 = 168 = 13^2 - 1$

Exercise 1.2b page 5

1. 1, 2, 4, 8, 16, i.e. 2^{n-1}
2. 1, 1, 2, 3, 5, 8, 13, 21, the Fibonacci numbers
3. $1 + 8x + 28x^2 + 56x^3 + 70x^4 + 56x^5 + 28x^6 + 8x^7 + x^8$

Exercise 1.2c page 5

1. (a) A.P (b) G.P. (c) G.P (d) A.P
2. (a) $4n - 1$ (b) 4.3^{n-1} (c) $(0{\cdot}1)^{n-1}$ (d) $77 - 6n$ or $71 - 6(n-1)$
3. (a) 820 (b) 600 (c) $2(3^{20} - 1)$ (d) $8(1 - (\frac{1}{2})^{20})$
4. $7 + 18 + 29 + 40 + 51 + 62$
5. $36 + 12 + 4 + \frac{4}{3} + \frac{4}{9} + \frac{4}{27}$
6. 565; $54(1 - (\frac{1}{3})^{10})$

Exercise 1.3a page 6

1. (a) 1, 2, 3, 4, 6, 12 (b) 1, 2, 4, 5, 10, 20
   (c) 1, 2, 4, 8, 16, 32 (d) 1, 2, 3, 4, 6, 7, 12, 14, 21, 28, 42, 84
2. (a) 1, 2, 4, 5, 7, 10, 14, 20, 28, 35, 70, 140
   (b) 1, 2, 3, 4, 6, 12, 19, 38, 57, 76, 114, 228
   (c) 1, 2, 4, 5, 10, 20, 25, 50, 100, 125, 250, 500
   (d) 1, 2, 3, 4, 6, 8, 9, 12, 18, 24, 27, 36, 54, 72, 108, 216
3. (a)

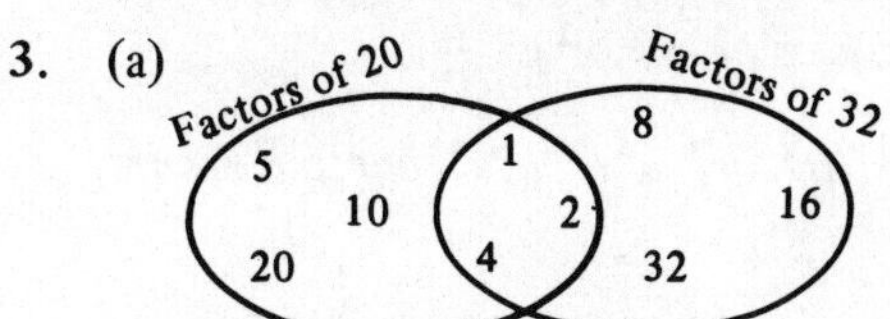

H.C.F. 4

(b)

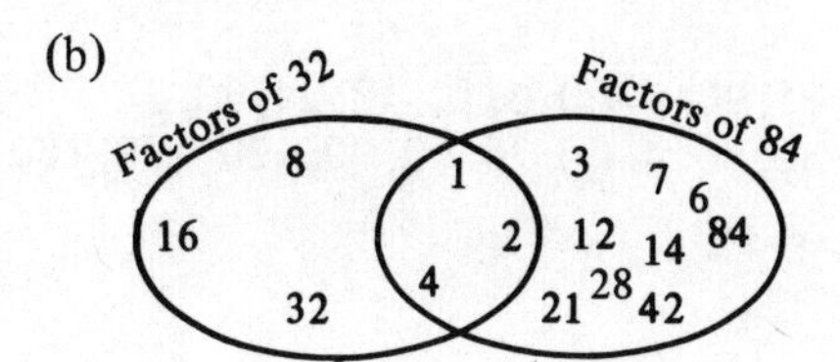

H.C.F. 4

(c)

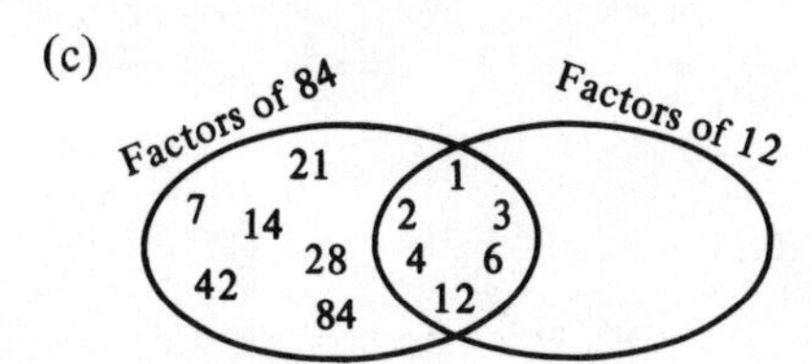

H.C.F. 12

4. (a) 4 (b) 20 (c) 12
5. (a) $2^2 \times 3^2$ (b) $2 \times 3 \times 7$ (c) $2^3 \times 7$ (d) $2^4 \times 5$ (e) $2^5 \times 3$
6. (a) 6 (b) 8 (c) 12
7. 31, 59, and 97 are prime.

8.

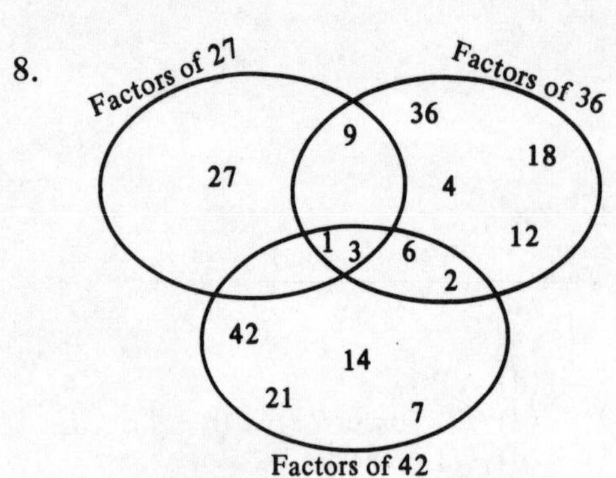

(a) 9 (b) 6 (c) 3

H.C.F. of 27, 36, and 42 is 3

9.

Factors of 64
Factors of 72
Factors of 96
64
9
18
36
72
1
4
2
32
3
6
8
16
12
24
48
96

Factors of 64 and 72 but not 96; none.

Factors of 72 and 96 but not 64; 3, 6, 12, and 24

Factors of 64 and 96 but not 72; 16 and 32

10. (a) bc (b) ac (c) qr (d) p^2qr (e) xy^2z^2 (f) $(x-y)$

11. (a) $1, p, q, r, pq, qr, pr, pqr$ (b) $1, x, x^2, y, xy, x^2y$

(c) $1, u, u^2, u^3, v, v^2, uv, u^2v, u^3v, uv^2, u^2v^2, u^3v^2$

(d) $1, p, q, r, q^2, r^2, r^3, pq, pq^2, pr, pr^2, pr^3, qr, q^2r, qr^2, qr^3, q^2r^2, q^2r^3, pqr, pq^2r, pqr^2, pq^2r^2, pqr^3, pq^2r^3$

Exercise 1.3b page 7

1. (a) 7, 14, 21, 28, 35, 42 (b) 15, 30, 45, 60, 75, 90 (c) 6, 12, 18, 24, 30, 36

2. (a) 24, 48, 72, 96, 120, 144 (b) 16, 32, 48, 64, 80, 96 (c) 30, 60, 90, 120, 150, 180

3. (a)

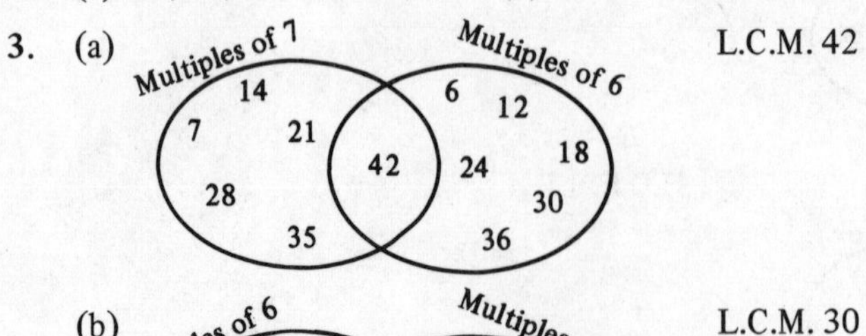

L.C.M. 42

(b)

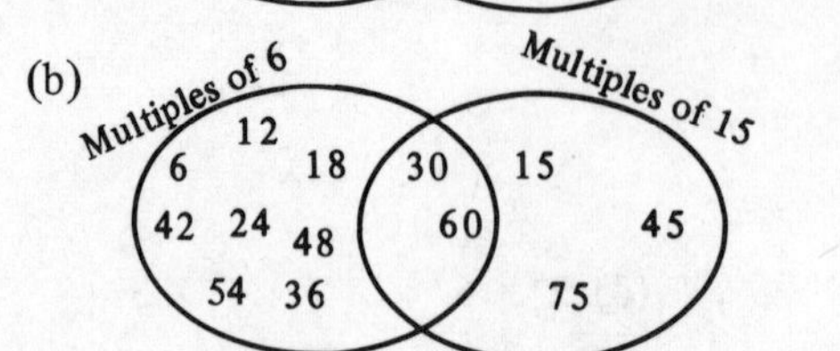

L.C.M. 30

(c)

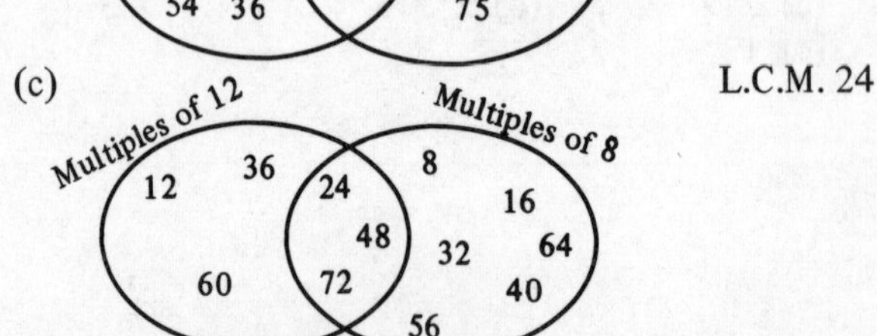

L.C.M. 24

4. (a) 48 (b) 240 (c) 120

5. (a) $2^3 \times 3$ (b) $2 \times 3 \times 5$ (c) $3^2 \times 5$ (d) 2^6 (e) $2^2 \times 3 \times 7$ (f) $2^5 \times 3$

6. (a) 120 (b) 90 (c) 1440 (d) 168 (e) 192 (f) 2880

7. 48, 96, 144, 264, and 456 are multiples of 24.

8. (a) $2^6 \times 7 \times 11, 2^3 \times 7 \times 11$: Yes (b) $2^3 \times 3 \times 5 \times 7, 2^2 \times 3^2$: No
(c) $2 \times 3 \times 5 \times 7 \times 13, 2 \times 3 \times 5 \times 7$: Yes

9. (a) 216 (b) 108 (c) 72

10. (a) every 36 and every 48 days (b) every 144 days

11. (a) $pqrs$ (b) p^2qr (c) x^2y^2z (d) $u^2v^2w^3$ (e) $72\,abc$
(f) $(x-y)(x+y)(2x+1)$

Exercise 1.4a page 8

1. (a) 11 (b) $^-3$ (c) $^-3$ (d) 3 (e) $^-11$

2. (a) 7 (b) 7 (c) $^-7$ (d) $^-7$ (e) $^-17$

3. (a) 4 (b) $^-8$ (c) 4 (d) $^-8$ (e) $^-4$

4. (a) 6 (b) $^-6$ (c) $^-6$ (d) 6 (e) $^-20$

5. (a) $^-1$ (b) 7 (c) 7 (d) 1 (e) $^-1$

6. (a) 6 (b) $^-24$ (c) 24 (d) 24 (e) $^-6$

7. (a) 10 (b) $^-6$ (c) $^-13$ (d) $^-9$

8. (a) 3 (b) 2 (c) $^-20$ (d) $^-15$

9. (a) 10 (b) 4 (c) 0 (d) 15

10. (a) $^-15$ (b) 45 (c) $^-25$

11. (a) 9 (b) $^-9$ (c) $^-42$

12. (a) 3 (b) $^-2$ (c) 7 (d) 7 (e) 7 (f) $^-3$ (g) $^-5$ (h) $^-2$

13. 12°C; 3°C per hour

14. (a) 7 (b) 6 (c) 4

Exercise 1.4b page 9

1. (a) 21 (b) $^-21$ (c) $^-21$ (d) 21 (e) $^-21$

2. (a) 24 (b) 24 (c) $^-24$ (d) $^-24$ (e) $^-24$

3. (a) 4 (b) $^-4$ (c) $^-4$ (d) 4 (e) $^-\frac{1}{4}$

4. (a) 3 (b) $^-3$ (c) 3 (d) $^-\frac{1}{3}$ (e) $\frac{1}{3}$

5. (a) $^-10$ (b) $^-16$ (c) 6 (d) $^-9$

6. (a) 18 (b) $^-66$ (c) 8 (d) 32

7. (a) $^-1$ (b) $^-3$ (c) 2 (d) 1

8. (a) $^-3$ (b) 2 (c) 1 (d) $^-2$

9. (a) 4 (b) $^-2$ (c) 2 (d) $^-3$ (e) 4 (f) $^-5$ (g) 5 (h) $^-3$

10. (a) 4 (b) $^-8$ (c) $^-24$ (d) 12 (e) 2 (f) $^-4$ (g) 36 (h) $^-10$

11. (a) 36 (b) $^-40$ (c) $^-120$ (d) $^-264$

12. (a) $^-41$ (b) $^-15$ (c) 22 (d) $^-24$

Exercise 1.5a page 10

1. (a) 4^3 (b) 7^6 (c) 2^2 (d) 5^4

2. (a) $3 \times 3 \times 3 \times 3$ (b) $4 \times 4 \times 4 \times 4 \times 4$ (c) $2 \times 2 \times 2$ (d) $\frac{1}{6 \times 6}$
(e) $10 \times 10 \times 10 \times 10 \times 10 \times 10$ (f) $\frac{1}{7 \times 7 \times 7}$

3. (a) 4^8 (b) 7^5 (c) 5^{10} (d) 2^{10} (e) 7^6
4. (a) 5^2 (b) 4^3 (c) 2 (d) 1 (e) 3^5
5. (a) 3^{-3} (b) 4^3 (c) 8^{-5} (d) 7^{-1} (e) 2^{-6}
6. (a) 6^{-5} (b) 7^{-7} (c) 5^5 (d) 2^{-1} (e) 1
7. (a) 16 (b) 27 (c) $\frac{1}{8}$ (d) 0·0001 or $\frac{1}{10000}$ (e) 1 (f) 0·2 or $\frac{1}{5}$
8. (a) 5 (b) 4 (c) 2 (d) 5 (e) 2 (f) −3 (g) −1 (h) 2

Exercise 1.5b page 11

1. (a) 2^5 (b) 3^{-1} (c) 1 (d) 5^{-4}
2. (a) $4^{2/3}$ (b) 2^4 (c) $5^{3/2}$ (d) 6
3. (a) $2^6 = 64$ (b) $3^6 = 729$ (c) 2 (d) 3 (e) 4 (f) 0·01 or $\frac{1}{100}$
4. (a) $2 \times 3^3 = 54$ (b) $7^5 \times 5^{-2}$ (c) $5^4 \times 6^{-1}$ (d) $2^{12} \times 3^{-3}$
5. (a) p^{12} (b) x^8 (c) w^{-3} (d) $l^{3/2}$
6. (a) y^{12} (b) x^{2p} (c) w^{n^2} (d) z^4 (e) l^{-1} (f) p^2
7. (a) 4 (b) −7 (c) −3
8. (a) 6 (b) 6 (c) 3 (d) 4
9. (a) 6×10^5 (b) 6×10^2 (c) 2×10^{-7}
10. (a) $1{\cdot}2 \times 10^6$ (b) 5×10^9 (c) 4×10^{-2}
11. (a) $\frac{4}{9}$ (b) $\frac{27}{64}$ (c) $\frac{5}{2}$ (d) 49 (e) $\frac{27}{8}$ (f) $\frac{8}{27}$
12. (a) 4 (b) $\frac{2}{3}$ (c) 5 (d) $\frac{4}{5}$ (e) 3 (f) $\frac{2}{3}$
13. (a) $\frac{1}{4}$ (b) $\frac{2}{7}$ (c) $\frac{1}{5}$ (d) $\frac{5}{4}$ (e) 1000 (f) 25
14. (a) 2 (b) 2 (c) 1 (d) 81

Exercise 1.6a page 12

1. (a) $\frac{4}{5}$ (b) $\frac{3}{7}$ (c) $\frac{14}{11}$ (d) $\frac{21}{53}$ (e) $\frac{24}{30} = \frac{4}{5}$
2. (a) $\frac{10}{15} = \frac{2}{3}$ (b) $\frac{13}{21}$ (c) $\frac{32}{33}$ (d) $\frac{5}{106}$ (e) $\frac{58}{90} = \frac{29}{45}$
3. (a) $\frac{13}{15}$ (b) $-\frac{1}{28}$ (c) $\frac{68}{77}$ (d) $\frac{21}{106}$ (e) $\frac{19}{48}$

Exercise 1.6b page 12

1. (a) 4 (b) 5 (c) 12 (d) 20 (e) $\frac{33}{4}$
2. (a) $\frac{3}{10}$ (b) $\frac{4}{21}$ (c) $\frac{9}{40}$ (d) $\frac{7}{10}$ (e) $\frac{9}{56}$
3. (a) 16 (b) 45 (c) $\frac{16}{3}$ (d) $\frac{45}{4}$ (e) 36
4. (a) $\frac{6}{5}$ (b) $\frac{12}{7}$ (c) $\frac{33}{70}$ (d) $\frac{35}{32}$ (e) $\frac{45}{56}$

Exercise 1.6c page 13

1. (a) 0·75 (b) 0·375 (c) 0·625 (d) 0·6 (e) 0·5555 (f) 0·8333
2. (a) $0{\cdot}\dot{7}1428\dot{5}$ (b) $0{\cdot}36\dot{3}\dot{6}$ (c) $0{\cdot}4166\dot{6}$ (d) $0{\cdot}2083\dot{3}$ (e) 0·1875 (f) 0·16
3. (a) $\frac{7}{10}$ (b) $\frac{3}{5}$ (c) $\frac{11}{100}$ (d) $\frac{3}{4}$ (e) $\frac{9}{20}$ (f) $\frac{2}{25}$

Exercise 1.6d page 13

1. (a) $\frac{3}{5}$ (b) $\frac{1}{9}$ (c) $\frac{7}{20}$
2. (a) $\begin{pmatrix} 5 \\ \frac{3}{4} \end{pmatrix}$ (b) $\begin{pmatrix} \frac{4}{15} \\ -2 \end{pmatrix}$ (c) $\begin{pmatrix} \frac{29}{36} \\ \frac{13}{21} \end{pmatrix}$
3. (a) $\frac{3}{2}$ (b) $\frac{5}{14}$ (c) $\frac{5}{8}$ (d) $\frac{13}{36}$ (e) $\frac{5}{14}$

4. (a) $\frac{5}{3}$ kg (b) $\frac{15}{8}$ kg (c) $\frac{1}{2}$ litre
5. $\frac{4}{15}$ 6. $\frac{11}{84}$ 7. $\frac{11}{56}$
8. (a) $\frac{133}{24}$ (b) $\frac{7}{8}$ (c) $\frac{259}{16}$

Exercise 1.7a page 14

1. (a) 3·8 (b) 6·2 (c) 45·1 (d) 104·1 (e) 1·2 (f) 3·0 (g) 10·5 (h) 19·8
2. (a) 5·59 (b) 7·00 (c) 16·21 (d) 30·25 (e) 5·12 (f) 1·62 (g) 6·15 (h) 19·76
3. (a) 5·38 (b) 14·31 (c) 10·48 (d) 1·763 (e) 8·872 (f) 0·141

Exercise 1.7b page 14

1. (a) 2·1 (b) 4·8 (c) 13·5 (d) 50·4 (e) 111·6
2. (a) 0·63 (b) 5·36 (c) 22·08 (d) 34·65 (e) 155·34
3. (a) 1·4 (b) 8·4 (c) 14·5 (d) 45·5 (e) 137·7
   (f) 0·16 (g) 0·42 (h) 1·16 (i) 3·9 (j) 3·06
4. (a) 12·6 (b) 30·36 (c) 170·25 (d) 331·08 (e) 69·58
5. (a) 0·08 (b) 0·42 (c) 0·95 (d) 1·61 (e) 7·02
   (f) 0·006 (g) 0·048 (h) 0·084 (i) 1·974 (j) 2·18
6. (a) 7·52 (b) 9·8 (c) 2·892 (d) 14·076 (e) 16·1448

Exercise 1.7c page 15

1. (a) 0·2 (b) 0·8 (c) 0·4 (d) 1·5 (e) 1·2
2. (a) 0·16 (b) 0·18 (c) 0·94 (d) 0·009 (e) 0·312
3. (a) 70 (b) 140 (c) 800 (d) 9000 (e) 2300
   (f) 40 (g) 50 (h) 200 (i) 600 (j) 26 000
4. (a) 0·126 (b) 0·21 (c) 0·27 (d) 0·004 (e) 0·005
5. (a) 4 (b) 2 (c) 8 (d) 8 (e) 12
   (f) 40 (g) 30 (h) 0·4 (i) 70 (j) 270
6. (a) 4 (b) 9 (c) 3·2 (d) 0·4 (e) 4·25

Exercise 1.7d page 15

1. 96 : (a) 9·6 (b) 9·6 (c) 0·96 (d) 0·096 (e) 0·0096
2. 276 : (a) 27·6 (b) 276 (c) 2·76 (d) 2·76 (e) 0·0276
3. 17 : (a) 1·7 (b) 170 (c) 17 (d) 1700 (e) 0·17
4. 6 : (a) 0·6 (b) 600 (c) 0·6 (d) 60 (e) 60
5. (a) 2 (b) 0·2 (c) 0·3 (d) 0·2 (e) 0·4 (f) 0·8
6. (a) 3 (b) 0·3 (c) 0·01 (d) 2·1 (e) 0·125 (f) 0·0032
7. 5·457 cm^2
8. 4·7 cm

Exercise 1.8a page 16

1. (a) 13 700 (b) 13 714·26 (c) 14 000 (d) 13 714·3 (e) 10 000
2. (a) 0·0134 (b) 0·013 426 9 (c) 0·013 43 (d) 0·013 427 (e) 0·01
3. (a) 1·67 (b) 1·672 35 (c) 1·6724 (d) 1·672 352 (e) 1·7
4. (a) 0·03 (b) 0·0326 (c) 0·032 618 2 (d) 0·033 (e) 0·0

5. (a) 12·63 (b) 0·12 (c) 0·02 (d) 135·63
6. (a) 9·763 (b) 0·483 (c) 14·236 (d) 0·002
7. (a) 12·6 (b) 56 400 (c) 0·001 24 (d) 1400
8. (a) 56 000 (b) 0·036 (c) 0·0032 (d) 50
9. (a) 2 (b) 1 (c) 5 (d) 4 (e) 0
10. (a) 3 (b) 3 (c) 2 (d) 2 (e) 4
11. (a) 2·52 (b) 0·11 (c) 0·05 (d) 1·96
12. (a) 114 (b) 1050 (c) 22·6 (d) 2·80

Exercise 1.8b page 17

1. (a) $4{\cdot}62 \times 10^2$ (b) $5{\cdot}106 \times 10^4$ (c) $1{\cdot}25 \times 10^{-1}$ (d) $2{\cdot}61 \times 10^{-3}$ (e) $9{\cdot}65 \times 10^1$
2. (a) $5{\cdot}6 \times 10^3$ (b) $5{\cdot}172 \times 10^2$ (c) $6{\cdot}7 \times 10^{-6}$ (d) $2{\cdot}4 \times 10^4$ (e) $1{\cdot}27 \times 10^1$ (f) $1{\cdot}786 \times 10^1$ (g) $7{\cdot}12 \times 10^{-5}$ (h) $1{\cdot}25 \times 10^4$
3. (a) $3{\cdot}044 \times 10^3$ (b) $4{\cdot}305 \times 10^{-2}$ (c) $2{\cdot}1012 \times 10^1$ (d) $1{\cdot}22 \times 10^{-4}$
4. (a) $1{\cdot}2 \times 10^6$ (b) $1{\cdot}02 \times 10^{-3}$ (c) $1{\cdot}4 \times 10^5$
5. (a) $6{\cdot}3 \times 10^2$ (b) $1{\cdot}842 \times 10^{-1}$ (c) $6{\cdot}229 \times 10^4$ (d) $6{\cdot}216 \times 10^7$ (e) $1{\cdot}2 \times 10^3$ (f) $4{\cdot}2 \times 10^{-7}$
6. (a) $3000 \times 80 = 240\,000$ (b) $0{\cdot}007 \times 30 = 0{\cdot}21$ (c) $(600)^2 \times 0{\cdot}08 = 28\,800$ (d) $\dfrac{40 \times 100}{80 \times 6} \approx 8$ (e) $\dfrac{0{\cdot}007 \times 700}{50 \times 0{\cdot}1} \approx 1$ (f) $\dfrac{30000 \times 40}{200 \times 0{\cdot}08} = 75\,000$
7. (a) $500 \times 80 = 40\,000$ (b) $0{\cdot}08 \times 0{\cdot}5 = 0{\cdot}04$ (c) $2000 \times 0{\cdot}009 = 18$ (d) $\dfrac{30 \times 5000}{800 \times 2} \approx 100$ (e) $\dfrac{2000 \times 0{\cdot}2}{40 \times 0{\cdot}01} = 1000$ (f) $\dfrac{(70)^2 \times 0{\cdot}0002}{6000 \times 40} \approx 0{\cdot}000\,004$

Exercise 1.9a page 18

1. (a) $\frac{1}{2}$ (b) $\frac{4}{5}$ (c) $\frac{3}{20}$ (d) $\frac{16}{25}$ (e) $\frac{1}{20}$ (f) $\frac{6}{5}$
2. (a) 75% (b) 40% (c) 12% (d) 35% (e) 12·5% (f) 125%
3. (a) 8 (b) 3 (c) 12 (d) 66 (e) 42 (f) 3 (g) 1·26 (h) 2
4. (a) 1·2 kg (b) 22·5 m (c) £97·50 (d) £16 600
5. (a) 42p (b) 301p (c) £74·60
6. (a) 77 (b) 175 (c) £1684·80 (d) 76·076
7. (a) 38 (b) 198 (c) £1942·50 (d) 26·946
8. (a) 25% (b) $33\frac{1}{3}$% (c) 12·5%
9. (a) 80% (b) 62·5% (c) 86%
10. (a) 16% (b) 25% (c) $33\frac{1}{3}$% (d) 20% (e) 12·5%
11. £3568; 11·5%
12. 1386; 15·5%

Exercise 1.9b page 19

1. £300; £75
2. £2·52; 36p
3. (a) £252 (b) £264 (c) £270 (d) £276 (e) £320
4. (a) 25% (b) 5% (c) 12%
5. (a) 20% (b) 8% (c) 12·5%
6. £1040; £1081·60; £1124·86; £1169·85
7. £72; the result is the same; 64%

8. £15 000
9. £800
10. £1200; 4·5%

Exercise 1.10a page 20

1. (a) 12 (b) 20 (c) 42 (d) 72 (e) 90
2. (a) 4·5 (b) 6·3 (c) 7·1 (d) 8·4 (e) 8·9
3. (a) 90 000 (b) 900 (c) 0·09 (d) 0·0009 (e) 0·000 009
4. (a) 30 (b) 95 (c) 0·3 (d) 0·95 (e) 0·095
5. (a) 4×10^2; 20 (b) 4×10^4; 200 (c) 4×10^{-2}; 0·2 (d) 4×10^{-4}; 0·02
6. (a) 9×10^4; 300 (b) 9×10^{-2}; 0·3 (c) 9×10^2; 30 (d) 9×10^{-4}; 0·03
7. (a) 40 (b) 500 (c) 0·06 (d) 0·7 (e) 9000
8. (a) 300 (b) 0·002 (c) 200 (d) 0·3 (e) 900
9. (a) 22·36 (b) 70·71 (c) 0·7071 (d) 0·2236 (e) 0·07071
10. (a) 54·77 (b) 17·32 (c) 547·7 (d) 0·1732 (e) 0·01732
11. (a) 10·95 (b) 34·64 (c) 109·5 (d) 0·3464 (e) 0·1095
12. (a) 26·65 (b) 0·8426 (c) 84·26 (d) 0·2665 (e) 266·5

Exercise 1.10b page 21

1. (a) 1·46 (b) 1·42 (c) 4·51 (d) 4·71 (e) 4·47
2. (a) 14·6 (b) 14·2 (c) 45·1 (d) 47·1 (e) 44·7
(f) 46·2 (g) 142 (h) 0·142 (i) 0·471 (j) 0·0447
3. (a) 23·7 (b) 75·0 (c) 0·75 (d) 0·237 (e) 237
4. (a) 93·3 (b) 29·5 (c) 0·933 (d) 0·0933 (e) 0·295
5. 2·04; 6·45 (a) 204 (b) 0·645 (c) 20·4 (d) 0·204 (e) 0·0204
6. 1·33; 4·21 (a) 13·3 (b) 0·133 (c) 133 (d) 0·421 (e) 0·0421
7. (a) 5·92 (b) 2·66 (c) 9·80 (d) 2·12 (e) 1·26
(f) 4·86 (g) 9·05 (h) 2·38 (i) 1·16 (j) 7·76
8. (a) 18·7 (b) 0·843 (c) 98·0 (d) 0·212 (e) 12·6
(f) 154 (g) 0·286 (h) 0·754 (i) 36·6 (j) 0·0776
9. (a) 70 (b) 20 (c) 0·9 (d) 0·3 (e) 0·02

Exercise 1.11a page 22

1. (a) 3·2 (b) 32
2. $10^{0\cdot5} \times 10^{1\cdot5} = 10^2 = 100$
3. (a) 1·30 (b) 1·48 (c) 1·60 (d) 1·78
4. (a) 3·08 (b) 3·08 (c) 3·08
5. (a) 3 (b) 5
6. (a) $64 = 4^3$ (b) $25 = 5^2$
7. (a) 4 (b) 6 (c) −3
8. (a) 6 (b) 4 (c) 3

Exercise 1.11b page 23

1. (a) 0·778 (b) 1·079 (c) 1·255 (d) 0·602 (e) 0·954 (f) 0·903 (g) 1·431 (h) 1·857
2. (a) 1·544 (b) 1·398 (c) 1·690 (d) 2·097 (e) 2·535 (f) 2·243 (g) 2·796 (h) 2·389
3. 1·041; 1·114 (a) 2·082 (b) 2·228 (c) 2·155 (d) 3·123
4. 0·398; 0·491 (a) 0·796 (b) 0·889 (c) 1·473 (d) 1·778
5. (a) $x + y$ (b) $2x$ (c) $3y$ (d) $3x + 2y$

Exercise 1.11c page 23

1. (a) 0·332 (b) 0·575 (c) 0·701 (d) 0·875 (e) 0·954 (f) 0·962 (g) 1·332 (h) 2·575 (i) 3·701 (j) $\bar{2}$·875 (k) $\bar{1}$·954 (l) 4·962
2. (a) 1·60 (b) 3·10 (c) 4·93 (d) 7·31 (e) 6·76 (f) 1·05 (g) 16·0 (h) 3100 (i) 493 (j) 0·731 (k) 0·006 76 (l) 0·000 105
3. (a) 4·42 (b) 5·16 (c) 35·5 (d) 82·5 (e) 4·41
4. (a) 4·47 (b) 7·65 (c) 59·6 (d) 2410 (e) 2·50
5. (a) 11·9 (b) 66·7 (c) 0·0181 (d) 0·000 28 (e) 0·0272

Exercise 1.11d page 24

1. (a) 2 (b) $\bar{2}$ (c) 3 (d) $\bar{2}$
2. (a) 0·145 (b) 3·035 (c) 1·792
3. (a) 100; 0·01; 1000; 0·01
   (b) 1·40; 1080; 61·9
4. (a) 4·57 (b) 0·0450 (c) 0·0453 (d) 0·000 529
5. (a) 6·73 (b) 2640 (c) 125 (d) 1860 (e) 638
6. (a) 0·0410 (b) 77·9 (c) 0·0137 (d) 4·29
7. 162 m^2
8. 41 500 mm^3 or 41·5 cm^3
9. 5·19 cm
10. £540
11. (a) 400; 384 (b) 110; 114 (c) 7000; 4570
12. (a) 140; 185 (b) 0·002; 0·002 52 (c) 0·0003; 0·000 479

Exercise 1.11e page 25

1. (a) 40·3 (b) 468 (c) 1520 (d) 0·326 (e) 0·000 017 6
2. (a) 2·09 (b) 16·2 (c) 19·4 (d) 2·36 (e) 3·16
3. (a) 0·195 (b) 0·210 (c) 0·155 (d) 0·769 (e) 0·3
4. (a) 31·6; $(3{\cdot}16)^3$ (b) 0·849; $(0{\cdot}721)^{1/2}$
5. (a) 7·19 (b) 3·98 (c) 172 (d) 45·7
6. (a) 0·765 (b) 1·50 (c) $\bar{2}$·863 or −1·137 (d) $\bar{1}$·204 or −0·796
7. (a) 5740 (b) 3 (c) 15·9
8. (a) 2·32 (b) 1·77 (c) 3·06 (d) −2·32

Part 2

Exercise 2.1 page 27

1. (a) 10 (b) 26 (c) 12·5 (d) 25
2. (a) 11 (b) 0·9 (c) 1 (d) 2
3. (a) 1·41 cm (b) 14·1 cm (c) 141 cm
4. 10·6 cm
5. 21·4 cm, 53·5 cm^2
6. 10·9 cm, 54·5 cm^2
7. 8·67 m, 17·3 m
8. 7·07 m, 14·1 m
9. $AB = \sqrt{10} \approx 3{\cdot}16$, $BC = \sqrt{34} \approx 5{\cdot}83$
$CD = \sqrt{20} \approx 4{\cdot}47$, $DA = \sqrt{8} \approx 2{\cdot}82$
diagonal $BD = \sqrt{26} \approx 5{\cdot}10$
10. Side $= \sqrt{20} \approx 4{\cdot}47$, diagonal $= \sqrt{40} \approx 6{\cdot}32$
11. (a) $XY = \sqrt{149} \approx 12{\cdot}2$ (b) $XZ = \sqrt{185} \approx 13{\cdot}6$ (c) $WZ = \sqrt{73{\cdot}25} \approx 8{\cdot}56$
12. (a) true (b) false (c) true (d) false (e) false

Exercise 2.2a page 28

1. 15·5 cm^2, 7·75 cm
2. 5 cm, 3·75 cm
3. 2·5 cm, 1·67 cm
4. 10·5 cm, 6 cm
5. 0·63 m^2
6. 1·35 m^2
7. 2 cm
8. 54 cm^2
9. 1·08 m

Exercise 2.2b page 29

1. (a) 160 cm^3 (b) 125
2. (a) 30 cm^2 (b) 6 cm
3. (a) 3·21 m^2 (b) 8·988 m^3
4. (a) 5·25 mm^2 (b) 31·5 mm^2 (c) 5·0 cm^3
5. 805 litres
6. 2·27 m^3, 0·19 m^3

Exercise 2.3a page 30

1. 154 cm^2, 44 cm
2. 24·4 cm
3. 150 cm
4. 7·16 cm
5. 4 cm
6. 7·98 cm
7. 31·8 cm^2
8. 35·4 cm
9. (a) (2) 50 cm (b) (3) 21 cm^2
10. (a) true (b) false (c) true (d) true (e) false
11. 55·7 m
12. (a) 5 cm (b) 7·07 cm (c) 5·64 cm

Exercise 2.3b page 31

1. Volume = 15·4 m^3, curved surface area = 44 m^2.
2. Volume = $\frac{99}{875}$ = 0·113 m^3, total surface area = $\frac{33}{25}$ = 1·32 m^2.
3. 1·59 cm, 95·4 cm^3

4. (a) 396 cm^2 (b) 198 g
5. 880 cm^3 **6.** 528 cm^3
7. (a) 1260 cm^3 (b) 75·4 litres per minute **8.** 943 litres
9. (a) 32 cm^3, 48 cm^2 (b) 108 mm^3, 108 mm^2 (c) 43 cm^3, 58 cm^2
10. (a) 1437 cm^3, 616 cm^2 (b) 38·8 cm^3, 55·4 cm^2 (c) 4·2 m^3, 12·6 m^2
11. 0·10 m^3 **12.** 1·41 m^2 **13.** 170 cm^3
14. 0·634 m^3

Exercise 2.4 page 32

1. (a) 6 (b) a^3 (c) $\frac{1}{6}a^3$ (d) height of pyramid = $\frac{1}{2}a$
2. (a) 36 cm^2 (b) 4 cm (c) 48 cm^3 (d) 15 cm^2 (e) 96 cm^2
3. 0·427 m^3 **4.** 2 560 000 m^3
5. (a) 5 cm (b) 12 cm (c) 192 cm^3 **6.** (c) 5·48 cm
7. VM = 20 mm, VN = 16 mm, total volume = 51840 mm^3
8. (a) $21\frac{1}{3}h$ cm^3 (b) 256 cm^3
9. (a) 49·5 cm^3, 29·5 cm^2 (b) $1\frac{1}{8}$ cm^3 (c) $229\frac{1}{2}$ cm^3
10. AM = 1·73 cm, VG = 1·63 cm
(a) 1·73 cm^2 (b) 0·94 cm^3

Exercise 2.5a page 34

1. (a) 2·44 cm, 6·11 cm^2 (b) 12·2 cm, 61·1 cm^2 (c) 12·8 cm, 44·9 cm^2
2. (a) 5·86 cm, 23·5 cm^2 (b) 40·3 cm, 302 cm^2 (c) 2·44 cm, 17·1 cm^2
3. curved surface area = 96 cm^2, circumference of base = 19·2 cm
4. 62·8 cm, 171·4°

Exercise 2.5b page 35

1. 167 cm^3 **2.** 56·5 cm^3
3. (a) 5 cm (b) 50·3 cm^3 (c) 62·8 cm^2
4. (a) 5 cm (b) 314 cm^3 (c) 204 cm^2
5. radius 3 cm, height 2 cm
6. 3 cm **7.** 67 cm^3 **8.** 90·9 cm^2 **9.** 8 cm
10. (c) $h = 2r$
11. (b) $x = 5$ cm, $y = 4$ cm (c) 8 cm
12. (a) $\frac{1}{2}r$ (b) $\frac{3}{8}\times$ volume of the cone

Exercise 2.6 page 37

1. (a) 1:3 (b) 4:5 (c) 1:8 (d) 1:250
2. (a) 1:2·5 (b) 1:$333\frac{1}{3}$ (c) 1:36000 (d) 1:2000
3. 12·5 cm **4.** 10·5 m **5.** 60 mm **6.** 6·25 km
7. 25 mm **8.** 750 m **9.** 1·6 cm **10.** 3·75 m
11. (a) 250 m (b) 575 m (c) 75 m
12. (a) 2·5 cm (b) 1 cm (c) 1·5 cm
13. 1/50 000 **14.** 1/200 000 **15.** 312 500 m^2 **16.** 82 000 m^2
17. 0·5 cm^2 **18.** 12 cm^2 **19.** 1:12 500 **20.** 1:100
21. (a) 50 m (b) 180 m (c) 187 m
22. (a) 150 m (b) 581 m (c) 1·16 cm

Exercise 2.7 page 39

1. 6·4 cm **2.** 7·3 cm **3.** AX = 5·1 cm **4.** 5·6 cm
5. 1 cm **6.** 3·1 cm
7. (a) 35 cm, 65 cm (b) 35 cm, 45 cm
(c) 20 cm, 40 cm, 60 cm (d) 24 cm, 18 cm, 36 cm
8. (a) 40p, 60p (b) 36p, 64p (c) 30p, 20p, 50p
(d) 11p, 22p, 67p (e) 20p, 27p, 53p
9. £7000, £8333·33, £9666·67
10. £375, £675
11. (a) BZ = 8 cm, 2 : 5 (b) 1 **12.** (a) BS = 3·4 cm, 2 : 5 (b) 1

Part 3

Exercise 3.1 page 41

1. (a) 0·500 (b) 0·707 (c) 0·988 (d) 0·182 (e) 0·964 (f) 0·900
2. (a) 0·500 (b) 0·707 (c) 0·588 (d) 0·745 (e) 0·988 (f) 0·958
3. (a) 30° (b) 60° (c) 50° (d) 0·5° (e) 62·5° (f) 25·7°
4. (a) 60° (b) 59° (c) 72·6° (d) 24·5° (e) 25·2° (f) 77·4°
5. (a) 3·42 cm, 9·40 cm (b) 6·30 cm, 4·93 cm (c) 5·16 cm, 9·71 cm
6. (a) 23·6°, 66·4° (b) 48·2°, 41·8° (c) 56·4°, 33·6°
7. (a) 2·68 m (b) 16·53 m (c) 32·86 m
8. (a) 0° (b) 90° above horizontal
(c) 30° below horizontal (d) 19·9° above horizontal
9. 47·7° **10.** 65·4° to horizontal, 2·18 m
11. 6·68 cm **12.** 2·46 m

Exercise 3.2 page 43

1. (a) true (b) true (c) false (d) false (e) false
2. AM = $\sqrt{3}$, (a) $\sqrt{3}$ (b) $\frac{1}{\sqrt{3}}$
3. (a) 0·839 (b) 0·306 (c) 2·14 (d) 5·85 (e) 5·00
4. (a) 22° (b) 61° (c) 78·9° (d) 20·7° (e) 82·3°
5. (a) 31°, 59° (b) 35·2°, 54·8° (c) 37·8°, 52·2°
6. (a) 2·33 cm (b) 5·64 cm (c) 36·0 cm
7. (a) 32·5 m (b) 33°
8. (a) 7·97° (b) 197 m

Exercise 3.3 page 45

1. (a) 0·574 (b) −0·574 (c) −0·574
2. 72·5°, 287·5°
3. (a) 0·743 (b) −0·906 (c) −0·122 (d) −0·921
4. (a) 0·643 (b) −0·438 (c) −0·829 (d) −0·515
5. between 128° and 129° **6.** between 55° and 56°
7. 0°, 14°, 293° **8.** 0°, 47°, 180°, 313° **9.** 30°

Exercise 3.4 page 47

1. 28·8°
2. 59·2°
3. 38·2°, 123·8°
4. 63·3°, 33·7°
5. 21·5 cm
6. 16·9 cm
7. 26·2 cm
8. 6·01 cm
9. 6·10 cm
10. 6·52 cm
11. $\hat{C} = 86°$, $b = 34·4$ cm, $c = 34·5$ cm
12. $\hat{B} = 81°$, $b = 3·08$ cm, $c = 1·32$ cm
13. $\hat{A} = 26·2°$, $\hat{C} = 91·8°$, $c = 9·06$ cm
14. impossible
15. $\hat{C} = 92·2°$, $a = 5·40$ cm, $b = 2·85$ cm
16. $\hat{A} = 126·6°$, $b = 3·39$ cm, $c = 6·87$ cm
17. $\hat{A} = 38°$, $a = 5·05$ cm, $c = 6·72$ cm
18. $\hat{C} = 125°$, $a = 3·20$ cm, $b = 1·40$ cm
19. impossible
20. $\hat{C} = 22·9°$, $\hat{A} = 149·1°$, $a = 55·3$ cm *or* $\hat{C} = 157·1°$, $\hat{A} = 14·9$, $a = 27·7$ cm
21. 1·22 m
22. 3770 m
23. 9 paces
24. 319 m, 71·8 m
25. AB = 93·3 m, BC = 116 m, CD = 125 m, AD = 71·2 m

Exercise 3.5 page 49

1. 8·1 cm
2. 3·80 cm
3. 24·1 cm
4. 13·0 cm
5. 1·79 cm
6. 1·54 cm
7. 2·39 cm
8. 8·80 cm
9. 7·04 cm
10. 56·4 cm
11. 78·5°
12. 80·2°
13. 80·8°
14. 86·6°
15. 104·5°
16. 90°
17. 102·1°
18. 104·9°
19. 130·5°
20. 99·7°
21. 3·12 km
22. 18·2°, 51·3°, 110·5°
23. 6·08 cm, 9·64 cm
24. 44·3 m
25. 82·8°, 69·3°, 27·9°

Exercise 3.6 page 51

1. (a) 207° (b) 333° (c) 153° (d) 270°
2. $201\frac{1}{2}°$
3. 10·07 km south, 9·73 km east
4. 10·9 km south, 16·8 km west
5. first ship 17·7 km north, second ship 17·4 km north; first ship attains the more northerly position.
6. the aeroplane that flies 160 km on a bearing of 262°.
7. 30·8 km
8. 145 km
9. 16·9 km on a bearing of 175·1°
10. 11·1 km

Exercise 3.7 page 53

1. (a) 29·2° (b) 67·0° (c) 51·3°
2. (a) 63·4° (b) 53·1°
3. (a) 39°, 70·5°, 70·5° (b) 9·43 cm (c) 97·2° (d) 69·3°
4. (a) 78·2° (b) 78·2°
5. (a) 18·7 cm (b) 21·6 cm (c) 33·8°
6. 14·1°
7. 66 m

Exercise 3.8 page 54

1. (a) $17{\cdot}2\ \text{cm}^2$ (b) $11{\cdot}3\ \text{cm}^2$ (c) $7{\cdot}94\ \text{cm}^2$
2. (a) $7{\cdot}04\ \text{cm}^2$ (b) $39{\cdot}4\ \text{cm}^2$ (c) $536\ \text{cm}^2$
3. $32{\cdot}0\ \text{cm}^2$
4. $77{\cdot}2\ \text{cm}^2$
5. $6{\cdot}64\ \text{cm}^2$
6. $339\ \text{cm}^2$
7. $43{\cdot}3\ \text{cm}^2$
8. $85{\cdot}6\ \text{cm}^2$
9. square $9\ \text{cm}^2$, equilateral triangle $6{\cdot}93\ \text{cm}^2$, hexagon $10{\cdot}4\ \text{cm}^2$
10. (a) $1570\ \text{cm}^2$ (b) $2410\ \text{cm}^2$ (c) $1570\ \text{cm}^2$ (d) $838\ \text{cm}^2$; $90°$
11. (a) $0{\cdot}225\ \text{cm}^2$ (b) $39{\cdot}3\ \text{cm}^2$
12. (a) $2{\cdot}65\ \text{cm}^2$ (b) $169\ \text{cm}^2$
13. $11{\cdot}2\ \text{cm}^2$
14. $54{\cdot}4\ \text{cm}^2$
15. $65{\cdot}7\ \text{cm}^2$

Exercise 3.9 page 57

1. (a) 4 m (b) $4\sqrt{2}$ m (c) $4\sqrt{3}$ m
2. (a) $\sqrt{3} - \dfrac{3}{2}$ (b) 2 (c) $\dfrac{5}{\sqrt{2}}$
3. (a) $\dfrac{\sqrt{3}}{2}$ (b) $-\dfrac{1}{2}$ (c) $-\sqrt{3}$ (d) $\dfrac{1}{\sqrt{2}}$
4. (a) $\dfrac{1}{2}$ (b) $-\dfrac{\sqrt{3}}{2}$ (c) $-\dfrac{1}{\sqrt{3}}$ (d) $-\dfrac{1}{\sqrt{2}}$
5. (a) $\frac{4}{5}$ (b) $\frac{3}{4}$
6. (a) $\frac{12}{13}$ (b) $\frac{5}{13}$
7. (a) $\frac{4}{5}$ (b) $-\frac{3}{5}$
8. (a) $-\dfrac{\sqrt{21}}{5}$ (b) $-\dfrac{2}{\sqrt{21}}$
9. (a) $5\ \text{cm}^2$ (b) $\sqrt{(41 - 20\sqrt{3})} \approx 2{\cdot}52$ cm
10. (a) $20\sqrt{2} \approx 28{\cdot}3\ \text{cm}^2$ (b) $\sqrt{(164 + 80\sqrt{2})} \approx 16{\cdot}6$ cm
11. ± 1
12. $\pm \frac{1}{2}$
13. 1
14. $-\frac{1}{4}, \frac{1}{3}$
15. $-\frac{1}{2}$
16. $0{\cdot}82$

Exercise 3.10 page 58

1. (a) $180°$ (b) $30°$ (c) $60°$ (d) $120°$ (e) $540°$ (f) $135°$
2. (a) $\dfrac{3\pi}{2}$ (b) 4π (c) $\dfrac{\pi}{4}$ (d) $\dfrac{2\pi}{3}$ (e) $\dfrac{5\pi}{4}$ (f) $\dfrac{5\pi}{6}$
3. (a) $\frac{1}{2}$ (b) $\dfrac{1}{\sqrt{2}}$ (c) -1 (d) 0 (e) 1
4. (a) $\dfrac{\sqrt{3}}{2}$ (b) $\dfrac{\sqrt{3}}{2}$ (c) 0 (d) 0 (e) -1
5. (a) $\dfrac{\pi}{2}$ (b) $\dfrac{\pi}{3}, \dfrac{5\pi}{3}$ (c) $\dfrac{2\pi}{3}, \dfrac{4\pi}{3}$ (d) $\dfrac{\pi}{4}, \dfrac{5\pi}{4}$ (e) $\dfrac{3\pi}{4}, \dfrac{7\pi}{4}$
6. (a) π (b) $\dfrac{\pi}{3}, \dfrac{2\pi}{3}$ (c) $\dfrac{3\pi}{4}, \dfrac{5\pi}{4}$ (d) $\dfrac{3\pi}{2}$ (e) $\dfrac{7\pi}{6}, \dfrac{11\pi}{6}$
7. (a) $0{\cdot}84$ (b) $0{\cdot}62$ (c) $1{\cdot}03$ (d) $0{\cdot}91$ (e) $-0{\cdot}59$ (f) $-0{\cdot}14$
8. (a) $1{\cdot}16^c$ (b) $0{\cdot}78^c$ or $2{\cdot}37^c$ (c) $2{\cdot}03^c$ (d) $1{\cdot}95^c$ (e) $0{\cdot}18^c$ or $2{\cdot}96^c$
9. (a) 10 cm (b) 20 cm (c) 24 cm (d) $10\pi = 31{\cdot}4$ cm
   (e) $5\pi = 15{\cdot}7$ cm
10. (a) $50\ \text{cm}^2$, $100\ \text{cm}^2$, $120\ \text{cm}^2$, $50\pi = 157\ \text{cm}^2$, $78{\cdot}5\ \text{cm}^2$
    (b) $7{\cdot}93\ \text{cm}^2$, $54{\cdot}5\ \text{cm}^2$, $86{\cdot}2\ \text{cm}^2$, $50\pi = 157\ \text{cm}^2$, $28{\cdot}5\ \text{cm}^2$

11. (a) 5 cm (b) $1{\cdot}85^c$ (c) $11{\cdot}2\text{ cm}^2$
12. $26{\cdot}6\text{ cm}^2$
13. (a) $0{\cdot}000\,29^c$ (b) 1·86 km
14. $0{\cdot}0093^c \approx 0{\cdot}534°$
15. angle $\approx 0{\cdot}07^c \approx 4{\cdot}01°$

Part 4

Exercise 4.1a page 60

1. (a) $8x$ (b) $10a$ (c) $12b$ (d) $-3y$ (e) $9a$ (f) $3b$ (g) $-8p$ (h) $17ab$
(i) $9a^2$ (j) $7b^2$ (k) $11x^2$ (l) $2y^3$
2. (a) $6a$ (b) $20b$ (c) $-21y$ (d) $30p$ (e) $12ab$ (f) $14a^2$ (g) $-12pq$ (h) $15y^2$
(i) $24bc$ (j) $30abc$ (k) $6a^3$ (l) $40x^3$
3. (a) $3a$ (b) 2 (c) $\frac{5}{x}$ (d) $\frac{1}{2y}$ (e) $\frac{5a}{2b}$ (f) $2q$ (g) $\frac{3}{y}$ (h) 3
(i) a (j) ab (k) $\frac{7b^3}{3a^2}$ (l) $\frac{4x}{y}$
4. (a) $11x^2$ (b) $4mn$ (c) $-3l^2m^2$ (d) $-16pqr$ (e) $-15m$ (f) $66ab$
(g) $21a^5$ (h) $8p^3q^4$ (i) $5m^2$ (j) $\frac{7ab^2}{2}$ (k) $\frac{4}{7x^3}$ (l) $\frac{2n^2}{3l^2}$
5. (a) $7a + 12b$ (b) $11a + 5b$ (c) $2a + 6b$ (d) $4x + 4y$ (e) $-7x - 10y$
(f) $2xy - 3z$ (g) $2ab + 6pq$ (h) $7a^2 - 2b^2$ (i) $3a^3 + 4a$
6. (a) $2a + 2b$ (b) $3x + 6y$ (c) $12p - 20q$ (d) $5ab + 5ac$
(e) $3pq - 6pr$ (f) $5a - 10b + 15c$ (g) $2xy - 4x + 6xw$ (h) $-6l - 3m$
(i) $-2lm + 10ln$ (j) $p^2 + 3pq - 5pr$ (k) $-12xy + 8y$ (l) $4x^3 - 4x^2 + 4x$
7. (a) $5x + y$ (b) $10x + 19y$ (c) $4x + 5y$ (d) $2a - 5b$
(e) $11x - 8$ (f) $-7ab$ (g) $3x^3 + 11x$ (h) $-5a - 4ab$
(i) $xy + 8xz$
8. (a) $-18x - 3y - 2z$ (b) $5ab + 6ac + bc$
(c) $6m + 5n$ (d) $2x^3y + xy^3 + x^4$

Exercise 4.1b page 61

1. (a) $pr + ps + qr + qs$ (b) $ln - lp + mn - mp$ (c) $ac - ad - bc + bd$
(d) $6pr - 2ps + 3qr - qs$ (e) $3ln + 15lp + mn + 5mp$ (f) $24ac - 16ad - 3bc + 2bd$
(g) $l^2 + 2lm + m^2$ (h) $p^2 - q^2$ (i) $a^2 - 2ab + b^2$
(j) $4p^2 + 12pq + 9q^2$ (k) $9a^2 - b^2$ (l) $16l^2 - 24lm + 9m^2$
2. (a) $x^2 + 3x + 2$ (b) $x^2 + 8x + 15$ (c) $x^2 + 3x - 4$
(d) $x^2 - 7x + 12$ (e) $2 - 3x + x^2$ (f) $12 - x - x^2$
(g) $6x^2 + 13x + 5$ (h) $12x^2 + 11x + 2$ (i) $14x^2 + 33x - 5$
(j) $8x^2 - 18x + 9$ (k) $6p^2 + 7pq + 2q^2$ (l) $35l^2 + 6lm - 8m^2$
3. (a) $3ac + 12ad + 6bc + 24bd$ (b) $5x^2 + 30x + 40$
(c) $12a^2 + 34ab + 10b^2$ (d) $p^2 + 2pq + q^2 + pr + qr$
(e) $x^2 + 3xy + 2y^2 - 3x - 6y$ (f) $2l^2 - 5lm + 3m^2 + 4ln - 4mn$
4. (a) $2x^2 + 2x + 8$ (b) $2x^2 - 5x - 1$ (c) $-12x + 6$ (d) $-2x + 13$
5. (a) $2ac + 2bd$ (b) $2pr - 2qs$ (c) $4lm$ (d) 0
6. (a) $12x^2 - 8x - 5$ (b) $-x^2 - 2x + 1$ (c) $16ac - 13ad + 7bc + 40bd$
(d) $-2p^2 + 12pq + 25q^2$

Exercise 4.1c page 61

1. (a) $b+a$ (b) $2q-3p$ (c) $z+2x+3y$
   (d) lm^2+l^3 (e) $\frac{3x^4}{y}-2x^2y^4$ (f) $a^4b-2ab^3-\frac{5a^2c^2}{b}$
2. (a) $2+\frac{x}{y}+\frac{y}{x}$ (b) $\frac{4b}{a}-\frac{9a}{b}$ (c) $18-15p+\frac{5p}{q}-\frac{8q}{p}+6q$
   (d) $b+\frac{a^3}{b^2}+\frac{b^2}{a^2}+\frac{a}{b}$ (e) $\frac{x^4}{y^3}-xy-\frac{x}{y}+\frac{y^3}{x^2}$ (f) $-\frac{1}{6}-\frac{q}{6}+\frac{q}{2p}-\frac{p}{3q}+\frac{p}{6}$

Exercise 4.2a page 62

1. (a) $3(x+y)$ (b) $5(a-b)$ (c) $4(x+y+z)$ (d) $6(a-b+c)$
   (e) $2(x+3y)$ (f) $4(2a-b)$ (g) $3(x+2y+3z)$ (h) $5(5a-2b-c)$
2. (a) $a(x+y)$ (b) $p(a-b)$ (c) $p(x+y+z)$ (d) $r(a-b+c)$
   (e) $q(x+3y)$ (f) $s(5a-b)$ (g) $t(2x+5y+z)$ (h) $l(7a-4b-c)$
3. (a) $(p+q)x$ (b) $(a-b)s$ (c) $(p+q+r)x$ (d) $(r-s+t)a$
   (e) $(3l+2m)y$ (f) $(6f-5g)h$ (g) $(4x+9y+z)t$ (h) $(2l-7m-3n)g$
4. (a) $2(a+b)$ (b) $3(a-b)$ (c) $4(x+3y)$ (d) $3(3p-2q)$
   (e) $p(x+y)$ (f) $r(a-b)$ (g) $s(7x+4y)$ (h) $t(2a-7b)$
   (i) $x(a+b+c)$ (j) $l(a-b-c)$ (k) $r(4x+5y+z)$ (l) $p(a-6b+8c)$
5. (a) $(l+m)x$ (b) $(a-b)n$ (c) $(7p+2q)y$ (d) $(r-5s)t$
   (e) $(p+q+r)t$ (f) $(a+b-c)n$ (g) $(5l+m+2n)x$ (h) $(4k-2l-m)g$
6. (a) $x(x+3)$ (b) $y(y-5)$ (c) $z(2z+3)$ (d) $m(4m-1)$
   (e) $x(x^2+2y)$ (f) $y^2(4z-y)$ (g) $ab(b+a)$ (h) $xyz^2(x-z)$
   (i) $\pi r(r+2h)$ (j) $2lm(m+4l)$ (k) $x^2(x^2+x+1)$ (l) $8y(4+2y^2+y^4)$
7. (a) $ab(c^2+b+a)$ (b) $pq^2r(p^2+pr+r^2)$ (c) $7xy(a+2b+3c)$
   (d) $8x^3(x^3+2x+6)$ (e) $lm(2p-1+5m)$ (f) $fg^2(f^3-6fg+2g^2)$
   (g) $5bcd(a+7e)$ (h) $8klm^2n(3k-4ln^2)$ (l) $4cx(4ab-7bd-5de)$
8. (a) $\frac{1}{4}\left(p-\frac{q}{2}+\frac{r}{3}\right)$ (b) $\frac{6}{x}(2a-b-4c)$ (c) $\frac{lm}{n}\left(l+\frac{m}{n}+\frac{lm}{n^2}\right)$

Exercise 4.2b page 63

1. (a) $(a-b)(x+y)$ (b) $(p+q)(x-y)$ (c) $(u+v)(t^2+s)$
   (d) $(u+v)(t^2-s)$ (e) $(2l+m)(x-y)$ (f) $(p-q)(r+3s)$
2. (a) $(s+5t)(x+y)$ (b) $(a+b)(m-2l)$ (c) $(s+t)(u^2+4v)$
   (d) $(a^2+b^2)(l-m)$ (e) $(x+y)(2b+3c)$ (f) $(3z-y^2)(w^2+t)$
3. (a) $(a+b-c)(x+y)$ (b) $(p+q)(x^2-y-z^3)$ (c) $(l+m+n)(s+t)$
   (d) $(f+g)(2x+y-3z)$
4. (a) $(x-y^2)(3m^2+4n^3)$ (b) $(x-y)(4a-5b)$ (c) $(2a+b)(c-2d)$
   (d) $(2q-r)(5p-s)$

Exercise 4.2c page 63

1. (a) $(x+4)(x+1)$ (b) $(x+2)(x+7)$ (c) $(x-1)(x+13)$ (d) $(x-8)(x-2)$
   (e) $(x+8)(x+8)$ (f) $(x-5)(x-5)$
2. (a) $(x+2)(x+5)$ (b) $(x-3)(x-6)$ (c) $(x+6)(x-3)$ (d) $(x+7)(x-2)$
   (e) $(x+9)(x+9)$ (f) $(x-4)(x-4)$ (g) $(x+8)(x-3)$ (h) $(x-7)(x+3)$

3. (a) $(3x+1)(3x+1)$ (b) $(3x+1)(x+1)$ (c) $(5x-1)(x-1)$ (d) $(4x+1)(2x-1)$
(e) $(3x+2)(x+1)$ (f) $(5x-1)(x-2)$ (g) $(7x+4)(x-1)$ (h) $(4x+1)(2x-3)$

Exercise 4.3a page 64

1. (a) 40 (b) 160 (c) 7800 (d) 460 (e) 1720 (f) 3420
2. (a) 72 (b) 18 (c) 24 (d) 14·8 (e) 19·6 (f) 44
3. (a) 60 (b) 80 (c) 600 (d) 3240 (e) 10 800 (f) 63 600
(g) 131 000 (h) 25 100
4. (a) 14 (b) 62 (c) 4·8 (d) 18·8 (e) 28·6 (f) 54
(g) 3·2 (h) 65·8
5. (a) $(p-q)(p+q)$ (b) $(l-m)(l+m)$ (c) $(p-2q)(p+2q)$
(d) $(3l-m)(3l+m)$ (e) $(3p-2q)(3p+2q)$ (f) $(4l-3m)(4l+3m)$
6. (a) $(x-y)(x+y)$ (b) $(s-t)(s+t)$ (c) $(2a-b)(2a+b)$
(d) $(l-3m)(l+3m)$ (e) $(4a-b)(4a+b)$ (f) $(p-5q)(p+5q)$
(g) $(4a-5b)(4a+5b)$ (h) $(7p-4q)(7p+4q)$
7. (a) $(b-c)(b+c)$ (b) $(9x-y)(9x+y)$ (c) $(p-10q)(p+10q)$
(d) $(10a-9b)(10a+9b)$ (e) $(xy-z)(xy+z)$ (f) $(a-bc)(a+bc)$
(g) $(2p-qr)(2p+qr)$ (h) $(3lm-n^2)(3lm+n^2)$
8. (a) $(ab-cd)(ab+cd)$ (b) $(xyz-w)(xyz+w)$ (c) $(4pq-9r)(4pq+9r)$
(d) $(l^2-n^2)(l^2+n^2)$ (e) $(a+b-c)(a+b+c)$ (f) $(p-q-r)(p+q+r)$
(g) $(x-z)(x+2y+z)$ (h) $(2m)(2l)$
9. (a) BC = 5 cm (b) BC = 24 cm (c) AB = 40 cm
10. (a) 60 cm (b) 84 cm (c) 3·6 cm (d) 2·97 cm

Exercise 4.3b page 65

1. (a) 14 cm² (b) 128 cm² (c) 40·32 cm²
2. (a) 600π cm² (b) 272π cm² (c) $21{\cdot}12\pi$ cm²
3. (a) 95·16 cm³ (b) 10·075 m³ (c) 283·05 cm³
4. (a) 28·94 m³ (b) 1·21 m³ (c) 3010 cm³
5. (a) 2170 cm³ (b) 0·205 m³
6. (a) BC = 9·74 cm (b) DC = 7·25 cm

Exercise 4.4a page 66

1. (a) 144 (b) 121 (c) 4 (d) 9
2. (a) $7^2+2.7.8+8^2$ (b) $5^2+2.5.9+9^2$ (c) $8^2-2.8.3+3^2$
(d) $11^2-2.11.7+7^2$
3. (a) 441 (b) 2704 (c) 841 (d) 5929
4. (a) 82·81 (b) 151·29 (c) 24·01 (d) 392·04
5. (a) $l^2+2lm+m^2$ (b) $x^2-2xy+y^2$ (c) $4p^2+4pq+q^2$
(d) $a^2-6ab+9b^2$ (e) $16m^2+24mn+9n^2$
6. (a) $x^2+12x+36$ (b) $x^2+22x+121$ (c) $x^2-10x+25$
(d) $x^2-26x+169$ (e) $4x^2+4x+1$ (f) $25x^2+10x+1$
(g) $9x^2-6x+1$ (h) $49x^2-14x+1$ (i) $9x^2+30x+25$
(j) $16x^2-56x+49$
7. (a) $(p+q)^2$ (b) $(l-m)^2$ (c) $(x+6)^2$ (d) $(y-9)^2$
8. (a) $(x+4)^2$ (b) $(x+10)^2$ (c) $(x-7)^2$ (d) $(x-\frac{1}{2})^2$
(e) $(3x+1)^2$ (f) $(5x+1)^2$ (g) $(4x-1)^2$ (h) $(6x-1)^2$

9. (a) $(x+3y)^2$ (b) $(2l+m)^2$ (c) $(a+5b)^2$ (d) $(p-4q)^2$
(e) $(8x-y)^2$ (f) $(10l-m)^2$ (g) $(2a+3b)^2$ (h) $(4p-3q)^2$

10. (a) $(m^2-2mn+n^2)+(m^2+2mn+n^2) = 2(m^2+n^2)$
(b) $(m^2-2mn+n^2)+4mn = (m+n)^2$

11. $(m^4-2m^2n^2+n^4)+4m^2n^2 = (m^2+n^2)^2$; $(m^2-n^2), 2mn, (m^2+n^2)$; 3, 4, 5

12. (a) $3x^2+12x+14$ (b) $4x^2+20$ (c) $2x^2+2y^2+2z^2+2xy+2yz+2zx$
(d) y^2-4x^2+20xy

Exercise 4.4b page 67

1. (a) $(x+2)^2-4$ (b) $(x+4)^2-16$ (c) $(x+6)^2-36$ (d) $(x+10)^2-100$
(e) $(x-3)^2-9$ (f) $(x-5)^2-25$ (g) $(x-2)^2-4$ (h) $(x-7)^2-49$

2. (a) $(x+1)^2+4$ (b) $(x+3)^2+3$ (c) $(x+8)^2+36$ (d) $(x+9)^2+19$
(e) $(x+5)^2-22$ (f) $(x-4)^2+9$ (g) $(x-6)^2+4$ (h) $(x-8)^2-54$

3. (a) $(x+7)^2-49$ (b) $(x+11)^2-121$ (c) $(x-9)^2-81$ (d) $(x-10)^2-100$
(e) $(x+\frac{3}{2})^2-\frac{9}{4}$ (f) $(x+\frac{1}{2})^2-\frac{1}{4}$ (g) $(x-\frac{3}{2})^2-\frac{9}{4}$ (h) $(x-\frac{5}{2})^2-\frac{25}{4}$

4. (a) $(x+3)^2+8$ (b) $(x+5)^2-5$ (c) $(x-4)^2+5$ (d) $(x-12)^2-244$
(e) $(x+\frac{3}{2})^2+\frac{11}{4}$ (f) $(x+\frac{5}{2})^2-\frac{13}{4}$ (g) $(x-\frac{3}{2})^2+\frac{19}{4}$ (h) $(x-\frac{1}{2})^2+\frac{15}{4}$

Exercise 4.4c page 67

1. (a) $(x+2)^2+8$; 8 (b) $(x+5)^2+5$; 5 (c) $(x+8)^2+6$; 6 (d) $(x+1)^2-1$; -1
(e) $(x-1)^2+4$; 4 (f) $(x-3)^2+6$; 6 (g) $(x-5)^2-7$; -7 (h) $(x-\frac{3}{2})^2-\frac{9}{4}$; $-\frac{9}{4}$

2. (a) $(2x+3)^2+2$; 2 (b) $(3x+2)^2+1$; 1 (c) $(4x+1)^2$; 0
(d) $(5x+2)^2-4$; -4 (e) $(2x-4)^2+4$; 4 (f) $(7x-4)^2+3$; 3
(g) $(6x-1)^2$; 0 (h) $(9x-1)^2-1$; -1

3. (a) 13 (b) 20 (c) 36 (d) 16

Exercise 4.5a page 68

1. (a) 14 (b) 8 (c) 22 (d) -4 (e) -27
2. (a) 24 (b) 16 (c) 149 (d) -5 (e) 1
3. (a) 4 (b) 12 (c) 9 (d) $\frac{1}{2}$ (e) $\frac{1}{5}$
4. (a) 10 (b) 42 (c) 38 (d) -12 (e) -66
5. (a) 3 (b) 7 (c) 11 (d) 8 (e) 4
6. (a) 12 (b) 63 (c) 108 (d) 160 (e) 121
7. (a) 7 (b) 9 (c) 11 (d) 1 (e) -2 (f) -1 (g) $\frac{5}{2}$ (h) $\frac{6}{5}$
8. (a) 6 (b) 2 (c) 9 (d) 5 (e) -6 (f) -7 (g) $\frac{3}{5}$ (h) -3
9. (a) 1 (b) 5 (c) 4 (d) -3 (e) 7 (f) 4·8 (g) 22 (h) $-\frac{36}{7}$
10. (a) 20 (b) 54 (c) 70 (d) $\frac{84}{11}$
11. (a) -1 (b) 29
12. (a) 8 (b) 10

Exercise 4.5b page 69

1. $2(x+x+7)=98$; 21 cm, 28 cm
2. $2(x+3x)=72$; 9 cm, 27 cm
3. 25, 26, 27
4. 34, 36, 38
5. $2(x+x+2{\cdot}3)=17$; 3·1 m, 5·4 m
6. $2(x+x+3{\cdot}1)=25{\cdot}5+1{\cdot}1$; 5·1 m, 8·2 m; no
7. $x-60=2(330-x+60)$; £2·80, £0·50
8. $n-11=\frac{1}{2}(72-n+11)$; 24, 48

Exercise 4.6 page 71

1. (a) $x<4$ (b) $x\geqslant 17$ (c) $x<-5$ (d) $x>13$ (e) $x<6$ (f) $x>5$
(g) $x\leqslant 4$ (h) $x>7$

2. (a) 1, 2, 3 (b) none (c) none (d) none (e) 1, 2, 3, 4, 5
(f) 6, 7, 8, 9 (g) 1, 2, 3, 4 (h) 8, 9

3. (a) $x>4$ (b) $x\leqslant 6$ (c) $x<3$ (d) $x>4$ (e) $x\geqslant 4{\cdot}5$ (f) $x<-2$
(g) $x\leqslant 4$ (h) $x>-2$

4. (a) 5 (b) $-5, -4, -3, -2, -1, 0, 1, 2, 3, 4, 5$ (c) $-5, -4, -3, -2, -1, 0, 1, 2$
(d) 5 (e) 5 (f) $-5, -4, -3$
(g) $-5, -4, -3, -2, -1, 0, 1, 2, 3, 4$ (h) $-1, 0, 1, 2, 3, 4, 5$

5. (a) $x\geqslant 5$ (b) $x<2$ (c) $x>12$ (d) $x<4$ (e) $x\leqslant -3$ (f) $x>-5$

7. (a) $x\geqslant 3$ (b) $x\leqslant 1$ (c) $x>-4$ (d) $x<5$ (e) $x>3$ (f) $x\leqslant -2$

9. (a) $x\geqslant 6$ (b) $x\leqslant \frac{4}{3}$ (c) $x>4$ (d) $x\leqslant 6$ (e) $x>2$ (f) $x\leqslant -2$
(g) $x>2$ (h) $x\leqslant -\frac{1}{3}$ (i) $x<5$

10. (a) $x>4$ (b) $x\geqslant -4$ (c) $x<5$ (d) $x\leqslant -5$

11. (a) $x>4$ (b) $x\geqslant -4$ (c) $x<5$ (d) $x<-5$

12. (a) $x<1$ (b) $x\geqslant -6$ (c) $x\leqslant 3$ (d) $x>5$

13. (a) $-4, -3, -2, -1, 0$ (b) $-4, -3, -2, -1, 0, 1, 2, 3, 4$
(c) $-4, -3, -2, -1, 0, 1, 2, 3$ (d) none

14. (a) $x<1$ (b) $x>1$ (c) $x\leqslant 6$ (d) $x\geqslant -4$ (e) $x>3$ (f) $x\geqslant -5$

15. (a) $x>8$ (b) $x\geqslant \frac{3}{2}$ (c) $x\geqslant -2$ (d) $x>7$ (e) $x\geqslant \frac{23}{5}$ (f) $x>2$

16. (a) $0<x<5$ (b) $x>\frac{1}{2}$ or $x<0$ (c) $0<x\leqslant \frac{1}{4}$ (d) $x>6$ or $x<0$

Exercise 4.7a page 72

1. (a) 12 000 (b) 140 (c) 251·2 (d) 65 (e) 25·55 (f) 0·628

2. (a) $l \to$ [× b] $\to$ [× h] $\to V$ (b) $r \to$ [× 2] $\to$ [× π] $\to$ [× h] $\to S$
(c) $t \to$ [× a] $\to$ [+ u] $\to v$ (d) $r \to$ [square] $\to$ [× π] $\to$ [× h] $\to V$
(e) $h \to$ [× 2] $\to$ [+ h] $\to$ [× r] $\to$ [× π] $\to S$ (f) $F \to$ [−32] $\to$ [× 5] $\to$ [÷ 9] $\to C$
(g) $C \to$ [× 9] $\to$ [÷ 5] $\to$ [+32] $\to F$ (h) $a \to$ [× t] $\to$ [÷ 2] $\to$ [+ u] $\to$ [× t] $\to s$

3. (a) $l=\dfrac{V}{b\times h}$ (b) $r=\dfrac{S}{2\pi h}$ (c) $t=\dfrac{(v-u)}{a}$ (d) $r=\sqrt{\dfrac{V}{\pi h}}$
(e) $h=\dfrac{1}{2}\left(\dfrac{S}{\pi r}-r\right)$ (f) $F=\dfrac{9}{5}C+32$ (g) $C=\dfrac{5}{9}(F-32)$ (h) $a=\dfrac{2}{t}\left(\dfrac{s}{t}-u\right)$

4. (a) 314 (b) $R \to$ [square] $\to$ [$-r^2$] $\to$ [× h] $\to$ [× π] $\to V$
(c) $R=\sqrt{\dfrac{V}{\pi h}+r^2}$

Exercise 4.7b page 73

1. (a) $l = \frac{A}{b}$ (b) $r = \frac{C}{2\pi}$ (c) $I = \frac{V}{R}$ (d) $L = \frac{S}{\pi r}$

   (e) $f = \frac{(v-u)}{t}$ (f) $x = \frac{(y-c)}{m}$ (g) $T = \frac{PV}{R}$ (h) $R = \frac{E}{I}$

   (i) $m = \frac{Pt}{(v-u)}$ (j) $r = \sqrt{\frac{V}{\pi h}}$ (k) $u = \sqrt{v^2 - 2as}$ (l) $x = \sqrt{r^2 - y^2} + a$

   (m) $g = \frac{4\pi^2 l}{T^2}$ (n) $v = \frac{uf}{(u-f)}$ (o) $x = \frac{c}{y} - a$

2. (a) $r = \sqrt{\frac{S}{4\pi}}$ (b) $r = \sqrt[3]{\frac{3V}{4\pi}}$ (c) $r = \frac{2A}{h} - R$ (d) $r = \sqrt{R^2 - \frac{V}{\pi h}}$

3. (a) $t = \frac{2s}{(u+v)}$ (b) $t = \frac{(v-u)}{f}$ (c) $t = \frac{PV}{mR}$ (d) $t = \frac{1}{\alpha}\left(\frac{T}{k} - \beta\right)$

4. (a) $x = \frac{y}{m} + a$ (b) $x = a\left(1 - \frac{y}{b}\right)$ (c) $x = \frac{c}{y} + a$ (d) $x = \frac{1}{1-y}$

5. (a) $R = 100\left(\frac{A}{P} - 1\right)$ (b) $R = \frac{t}{(t-2)}$ (c) $R = \frac{Vr}{(V-1)}$ (d) $R = \frac{1-2K}{K-1}$

Exercise 4.7c page 73

1. b/a 2. $\sqrt{\frac{2gP}{m}}$ 3. $h = \frac{A}{2\pi r} - r$ *or* $\frac{A - 2\pi r^2}{2\pi r}$ 4. $\frac{(2y-1)}{(3-y)}$

Exercise 4.8a page 74

1. (a) $x = -2, y = 16$ (b) $x = 4, y = -1$ (c) $x = -1, y = \frac{17}{3}$
2. (a) $x = 6, y = 4$ (b) $x = -2, y = 4$ (c) $x = 5, y = -3$
3. (a) $x = 3, y = 4$ (b) $x = 2, y = -3$ (c) $x = 5, y = 1$
4. (a) $x = 2, y = 5$ (b) $x = 5, y = 6$ (c) $x = 3, y = -4$
   (d) $x = 2, y = 3$ (e) $a = 7, b = -2$ (f) $p = \frac{11}{3}, q = \frac{4}{3}$

Exercise 4.8b page 75

1. (a) $x = 1, y = 5$ (b) $x = 3, y = 7$ (c) $x = 1, y = 2$
2. (a) $x = 2, y = 3$ (b) $x = 7, y = 1$ (c) $x = -2, y = 3$
4. (a) $x = 4, y = 3$ (b) $x = 3, y = -2$ (c) $x = 3, y = -4$ (d) $x = 4, y = 3$
   (e) $x = \frac{3}{2}, y = 2$ (f) $x = -\frac{29}{5}, y = -\frac{31}{5}$ (g) $x = 2, y = 1$ (h) $x = 2, y = -1$

Exercise 4.8c page 75

1. 51, 26 2. A pencil costs 8p; a rubber costs 12p.
3. An adult's ticket costs 75p; a child's ticket costs 40p
4. 90 km; 200 km

Exercise 4.9a page 76

1. (a) 3,4 (b) 5, −2 (c) −3, 7 (d) −2, −9 (e) $\frac{1}{2}$, 3 (f) $\frac{1}{3}$, −4
(g) $\frac{1}{4}, \frac{1}{5}$ (h) 0, 3 (i) 0, $-\frac{1}{3}$ (j) $-\frac{1}{7}, -\frac{1}{4}$ (k) $\frac{3}{2}, \frac{2}{3}$ (l) $\frac{2}{7}, -\frac{4}{3}$
2. (a) 5, 2 (b) −4, 3 (c) −2, $\frac{1}{3}$
3. (a) 5, 1 (b) 3, 2 (c) 5, 2 (d) −6, −2 (e) −3, −1 (f) −8, −2
(g) −7, 1 (h) −7, 2 (i) −7, 4 (j) 6, −1 (k) −2, 8 (l) −3, 13
4. (a) $(2x-1)(x-1)$; $\frac{1}{2}$, 1 (b) $(3x-1)(2x-1)$; $\frac{1}{3}, \frac{1}{2}$ (c) $(4x-1)(2x-1)$; $\frac{1}{4}, \frac{1}{2}$
(d) $(7x+1)(x+1)$; $-\frac{1}{7}$, −1 (e) $(5x-1)(2x-1)$; $\frac{1}{5}, \frac{1}{2}$ (f) $(4x-1)(3x-1)$; $\frac{1}{4}, \frac{1}{3}$
(g) $(2x-1)(x-4)$; $\frac{1}{2}$, 4 (h) $(3x-1)(x-5)$; $\frac{1}{3}$, 5 (i) $(4x-3)(x-1)$; $\frac{3}{4}$, 1
(j) $(2x+1)(x-3)$; $-\frac{1}{2}$, 3 (k) $(3x+2)(x-1)$; $-\frac{2}{3}$, 1 (l) $(4x-3)(x+2)$; $\frac{3}{4}$, −2
5. (a) $(4-x)(3-x)$; 4, 3 (b) $(4+x)(3-x)$; −4, 3 (c) $(5+x)(2-x)$; −5, 2
(d) $(2+x)(1-3x)$; −2, $\frac{1}{3}$ (e) $(3-x)(1+2x)$; 3, $-\frac{1}{2}$ (f) $(3-2x)(1+4x)$; $\frac{3}{2}, -\frac{1}{4}$
6. $(x+8)^2 = (x+7)^2 + x^2$; 13, 12, 5
7. $x^2 + (x+1)^2 + (x+2)^2 = 110$; 5, 6, 7
8. $x(x+10) = 56$; 4 cm and 14 cm

Exercise 4.9b page 77

1. (a) 7, −3 (b) 13, −3 (c) 7, −1 (d) −3, −5 (e) 1, −13 (f) 0, −4
2. (a) 4, 2 (b) 11, 1 (c) 2, 12 (d) −6, −2 (e) 3, −13 (f) 7, −9
3. (a) $1+\sqrt{5}, 1-\sqrt{5}$ (b) $4+\sqrt{7}, 4-\sqrt{7}$ (c) $6+\sqrt{11}, 6-\sqrt{11}$
(d) $-2+\sqrt{3}, -2-\sqrt{3}$ (e) $-5+\sqrt{2}, -5-\sqrt{2}$ (f) $-1+\sqrt{13}, -1-\sqrt{13}$
4. (a) $3+\sqrt{8}, 3-\sqrt{8}$ (b) $8+\sqrt{61}, 8-\sqrt{61}$ (c) $2+\sqrt{3}, 2-\sqrt{3}$
(d) $-1+\sqrt{8}, -1-\sqrt{8}$ (e) $-3+\sqrt{21}, -3-\sqrt{21}$ (f) $-7+\sqrt{52}, -7-\sqrt{52}$

Exercise 4.9c page 77

1. (a) $\frac{1}{3}$, 2 (b) $\frac{3}{5}$, 1 (c) $-\frac{2}{3}+\frac{\sqrt{10}}{3}, -\frac{2}{3}-\frac{\sqrt{10}}{3}$
(d) $-\frac{11}{14}+\frac{\sqrt{65}}{14}, -\frac{11}{14}-\frac{\sqrt{65}}{14}$
2. (a) 3, 1 (b) 7, 2 (c) $\frac{1}{2}$, 3 (d) −2, −5 (e) −10, 2 (f) $\frac{1}{4}$, −3
3. (a) $\frac{5}{2}+\frac{\sqrt{21}}{2}, \frac{5}{2}-\frac{\sqrt{21}}{2}$ (b) $\frac{7}{2}+\frac{\sqrt{37}}{2}, \frac{7}{2}-\frac{\sqrt{37}}{2}$ (c) $-2+\sqrt{13}, -2-\sqrt{13}$
(d) $-3+\sqrt{7}, -3-\sqrt{7}$ (e) $\frac{1}{3}+\frac{\sqrt{19}}{3}, \frac{1}{3}-\frac{\sqrt{19}}{3}$ (f) $\frac{9}{10}+\frac{\sqrt{61}}{10}, \frac{9}{10}-\frac{\sqrt{61}}{10}$
4. 16·36 m, 20·36 m

Exercise 4.10a page 78

1. 65
2. 11
3. 74·5p per gallon
4. 115 units; the first method is cheaper.
5. £1475
6. £2·20, £3·70, £1·85

Exercise 4.10b page 79

1. 360 cars and 8 coaches.
2. 121 7p stamps, and 59 9p stamps; 112 7p stamps, and 66 9p stamps.
3. 8 cm and 4 cm.
4. £18.
5. The boys are paid 84p per hour; the men are paid £1.20 per hour; £162, £126.77.

Exercise 4.10c page 79

1. 13 cm, 12 cm and 5 cm
2. 11 cm
3. 90 km/h

Part 5

Exercise 5.1a page 80

2. (a) $y = x + 2$ (b) $y = 2x$ (c) $y = 3x - 1$ (d) $x + y = 6$

3. (a)

	x	−2	−1	0	1	2	3	4	5	6
(a)	y	1	2	3	4	5	6	7	8	9
(b)	y	−6	−3	0	3	6	9	12	15	18
(c)	y	−1	1	3	5	7	9	11	13	15
(d)	y	9	8	7	6	5	4	3	2	1

4. $y = x + 2, y = x + 3$ are parallel
   $y = 2x, y = 3x$ both go through (0, 0)
   $x + y = 6, y = 7 - x$ are parallel
   $y = 3x, y = 3x - 1$ are parallel
   $y = 2x, y = 2x + 3$ are parallel
5. (a) $y = x + 5$ (b) $y = 4x$ (c) $x + y = 5$ (d) $y = 3x - 2$
   (e) $y = 10 - 2x$ (f) $y = \frac{1}{2}x + 3$

6.

	x	−2	−1	0	1	2	3	4	5	6
(a)	y	−13	−8	−3	2	7	12	17	22	27
(b)	y	10	7	4	1	−2	−5	−8	−11	−14
(c)	y	−0·5	0	0·5	1·0	1·5	2·0	2·5	3·0	3·5
(d)	y	−12	−9	−6	−3	0	3	6	9	12

7. $y = 3x - 2, y = 3(x - 2)$ are parallel
   $y = \frac{1}{2}x + 3, y = \frac{1}{2}x + \frac{1}{2}$ are parallel
   $y = 10 - 2x, y = 4 - 3x$ both slope downwards, as does $x + y = 5$

Exercise 5.1b page 81

1. ① : (−8, −2), (−6, 0), (−4, 2), (−2, 4), (0, 6), (2, 8), (4, 10), (6, 12), (8, 14): $y = x + 6$
   ② : (2, −4), (4, 0), (8, 8), (10, 12); $y = 2x - 8$
   ③ : (−2, 1), (2, 3), (4, 4), (8, 6), (10, 7), (12, 8); $y = \frac{1}{2}x + 2$
2. (a) 6; −8; 2 (b) 1; 2; $\frac{1}{2}$
5. (a) $y = 2x$ (b) $y = 2x + 1$ (c) $y = 2x + 5$ (d) $y = 2x - 3$ (e) $y = 2x + \frac{3}{2}$
6. (a) $y = 3x + 2$ (b) $y = 5x + 2$ (c) $y = \frac{1}{2}x + 2$ (d) $y = 2 - 2x$ (e) $y = 2$
7. (a) $y = 3x + 5$ (b) $y = 4x - 2$ (c) $y = 1 - x$
8. (a) 3; $y = 3x + 2$ (b) 2; $y = 2x + 4$ (c) 4; $y = 4x - 1$ (d) 4; $y = 4x - 3$

Exercise 5.2a page 82

1. (a) Yes (b) Yes (c) No (d) No
2. (a) $y = 75x$ (b) $y = 12x + 50$
3.

	0	1	2	3	4	5	6	7
(a)	0	83	166	249	332	415	498	581
(b)	5	22·5	40	57·50	75	92·50	110	127·50
(c)	200	172	144	116	88	60	32	4

4. (a) $y = 83x$ (b) $y = \frac{35}{2}x + 5$ (c) $y = 200 - 28x$
5. (a) is a linear relation (b) is not a linear relation

Exercise 5.2b page 83

1. $E = 20, R = \frac{50}{6}$
2. (a) $s = 13{\cdot}6t$ (b) $s = 3{\cdot}5t$
3. $I = 10,\ P = 5{\cdot}3$
4. 19 → 11·1 should be 19 → 12·7
5. $a = -3{\cdot}5$; $b = 9{\cdot}5$ so $E = 9{\cdot}5 - 3{\cdot}5W$
   1·1 → 6·0 should be 1·1 → 5·6; 2·7 → 0·6 should be 2·7 → 0

Exercise 5.3a page 84

1. (a) $y = 2x, x + y = 6$; (2, 4) (b) $y = 2x + 3, y = 3x - 2$; (5, 13)
2. (a) $x + y = 7, y = 3x + 1$; $(\frac{3}{2}, \frac{11}{2})$ (b) $y = 2x + 1, y = 4x - 4$; $(\frac{5}{2}, 6)$
3. (a) $(\frac{8}{3}, \frac{16}{3})$ (b) (5, 6) (c) (3, 7) (d) $(\frac{5}{2}, 7)$
4. (a) (2, 3) (b) (2, 7) (c) $(\frac{3}{2}, \frac{7}{2})$ (d) $(\frac{5}{2}, \frac{9}{2})$
5. $y = 6 - x$; $y = 3x + 2$; (2, 4), (2, 8), (1, 5) and (3, 7)

Exercise 5.3b page 85

1. 4 hours 40 minutes; $5\frac{1}{3}$ km from A
   (a) 2 hours (b) 1 hour
   (c) First hiker: 4 km/h, 4 km/h; second hiker: 1·33 km/h, 2 km/h
   (d) First hiker: 4 km/h; second hiker: 1·6 km/h
2. (a) 9.20 a.m.; 40 m.p.h. (b) 60 m.p.h.; 11.00 a.m. (c) 60 miles; 10.50 a.m.
3. The first method is cheapest if less than 50 litres are used.
   The second method is cheapest if between 50 litres and 150 litres are used.
   The third method is cheapest if more than 150 litres are used.

Exercise 5.4a page 86

1. (a) Yes (b) No (c) No (d) Yes
2. (2, 5) is in the region. (6, 1) is not in the region.
3. (3, 4) is in the region. (4, 6) is not in the region although on the line $y = x + 2$.
5. (2, 7) is not in the region. (3, 6) is not in the region although on the line $y = x + 3$.

Exercise 5.4b page 87

1. (3, 4); $x + y = 7$
2. (6, 3) and (5, 4); (6, 4) is on the line $x + 3y = 18$ and therefore not in the region.
3. (a) region ⑦ (b) region ② (c) region ⑤
4. (a) $x + y \leqslant 6, y \geqslant 2x + 3, x + 3y \geqslant 12$ (b) $x + 3y \leqslant 12, y \leqslant 2x + 3, x + y \leqslant 6$
   (c) $x + 3y \leqslant 12, x + y \geqslant 6$

Exercise 5.4c page 88

1. $12N + 9L \leqslant 1080$; $5N + 8L \leqslant 800$;
   To make his maximum profit he should manufacture 27 of type N and 83 of type L.
2. $8L + 3S \geqslant 72$; $L + S \leqslant 18$; $8L + 2S$ as small as possible;
   For minimum cost, the camp should have 4 large tents and 14 small tents; £60.
3. $33\frac{1}{3}$ hectares for wheat, $66\frac{2}{3}$ hectares for potatoes; £5200; 150 hectares.

Exercise 5.4d page 89

1. (2, 16), (3, 14) or (4, 12)
2. v vans, l lorries: (a) $v + l \leqslant 10$ (b) $6v + 15l \leqslant 90$ (c) $600v + 1100l \geqslant 6600$
   $5v$ and $4l$, $6v$ and $3l$, $7v$ and $3l$, or $8v$ and $2l$.
3. 1 part of Feed 1 to 3 parts of Feed 2.

Exercise 5.5a page 90

1. (a) translation of 3 units upwards (b) translation of 2 units upwards
   (c) translation of 5 units upwards.
2. (a) $f(x) = x^2 - 4$ (b) $f(x) = x^3 - 4$ (c) $f(x) = \frac{1}{x} - 4$
3. (a) translation of 3 units to the right (b) translation of 2 units to the right
   (c) translation of 5 units to the right.
4. (a) $f(x) = (x + 4)^2$ (b) $f(x) = (x + 4)^3$ (c) $f(x) = \frac{1}{(x + 4)}$
5. (a) translation of 1 unit upwards (b) translation of 1 unit to the left
   (c) translation of 1 unit downwards (d) translation of 1 unit to the right
6. (a) (10, 10) (b) (16, 4) (c) (8, 8) (d) (4, 16)
   $x = 0$ is a line of symmetry for $f(x) = x^2 + 1$ and $f(x) = x^2 - 1$
   $x = -1$ is a line of symmetry for $f(x) = (x + 1)^2$; $x = 1$ is a line of symmetry for $f(x) = (x - 1)^2$
7. (12, 3), (9, 4), (6, 6); e.g. (2, 18), (1, 36), (18, 2), (36, 1)

Exercise 5.5b page 91

1. $f(n) = 2^n - 1$; a translation of $f(n) = 2^n$, 1 unit downwards.
2. 3, 9, 27, 81, 243; $f(n) = 3^n$.
3. $f(x) = \dfrac{100}{x}$, domain $\{x: 20 \leqslant x \leqslant 100\}$, image set $\{f(x): 1 \leqslant f(x) \leqslant 5\}$; $f(x) = \dfrac{100}{x} + \dfrac{1}{2}$
4. $f(x) = \dfrac{10x}{100} + 10$, domain $\{x: 0 \leqslant x \leqslant 100\}$; $f(x) = \dfrac{9x}{100} + 11$, domain $\{x: 100 \leqslant x \leqslant 200\}$;

   $f(x) = \dfrac{8x}{100} + 13$, domain $\{x: 200 \leqslant x \leqslant 300\}$

Exercise 5.6a page 92

1. 48 2. $\frac{10}{7}$ 3. 125 km 4. 150 5. 4
6. $S = kl^2$; $k = 6, 96\ \text{cm}^2$ 7. 2·7 8. 12 9. 2 seconds
10. (a) 6 (b) 2·4 11. (a) 0·6 (b) 20 12. 5 amps

Exercise 5.6b page 93

1. 3·2; $S = 3{\cdot}2\,T^2$
2. Yes; $k = 5{\cdot}4, c = 3$: $y = \dfrac{5{\cdot}4}{x^2} + 3$
3. $n = 1{\cdot}5 : k = 4 : v = 4p^{1\cdot5}$: 11·31

Exercise 5.7a page 94

1. $-3 \leqslant f(x) \leqslant 6, -\frac{3}{4}, 3\frac{1}{4}, -\frac{3}{4}$
2. (a) 0, 4 (b) 1, 3 (c) -1 (d) 0·3, 3·7
4. (a) $-4 \leqslant f(x) \leqslant 5, f(\frac{1}{2}) = -1\frac{3}{4}$ (b) $0 \leqslant f(x) \leqslant 4, f(\frac{1}{2}) = 2\frac{1}{4}$
   (c) $-\frac{1}{4} \leqslant f(x) \leqslant 20, f(\frac{1}{2}) = 3\frac{3}{4}$ (d) $-2\frac{1}{4} \leqslant f(x) \leqslant 10, f(\frac{1}{2}) = 1\frac{3}{4}$
5. (a) $x = -2$ or -1, $x = 1$ or 4, $x = -3$ or 1, $x = 2$
   (b) $x = -3$ or 0, $x = 4{\cdot}6$ or 0·4, $x = -3{\cdot}4$ or 1·4, $x = 3{\cdot}4$ or 0·6
   (c) $x = -3{\cdot}6$ or 0·6, $x = 0$ or 5, $x = -3{\cdot}8$ or 1·8, $x = 0$ or 4
   (d) none, none, $x = 0$ or -2, none
7. (a) $-5 \leqslant f(x) \leqslant 4, 3\frac{3}{4}$ (b) $-10 \leqslant f(x) \leqslant 2\frac{1}{4}, 1\frac{1}{4}$
   (c) $-5 \leqslant f(x) \leqslant 4, 3\frac{3}{4}$ (d) $-5 \leqslant f(x) \leqslant 4, 1\frac{3}{4}$
8. (a) $x = -2$ or 2, $x = 0$ or 3, $x = -1$ or 3, $x = -3$ or 1
   (b) $x = -1{\cdot}7$ or 1·7, $x = 2{\cdot}6$ or 0·4, $x = -0{\cdot}7$ or 2·7, $x = -2{\cdot}7$ or 0·7
   (c) $x = -2{\cdot}4$ or 2·4, $x = -0{\cdot}6$ or 3·6, $x = -1{\cdot}4$ or 3·4, $x = -3{\cdot}4$ or 1·4
   (d) $x = -1$ or 1, none, $x = 0$ or 2, $x = -2$ or 0
9. graphs (a), (c) and (d) have their vertex at the same height from the x-axis; they are exactly the same. A translation of 1 unit to the right would map $f(x) = 4 - x^2$ onto $f(x) = 3 + 2x - x^2$. A translation of 1 unit to the left would map $f(x) = 4 - x^2$ onto $f(x) = 3 - 2x - x^2$.
10. (a) 2, 3 (b) -4, 2 (c) 1·4, 3·6 (d) $-3{\cdot}6$, 1·6

Exercise 5.7b page 95

1. (a) $x = 0, x = 5, x = 2$ or $-2, x = -8$ or -4
(b) $x = -1$ or $1, x = 4$ or $6, x = -2{\cdot}2$ or $2{\cdot}2, x = -8{\cdot}2$ or $-3{\cdot}8$
(c) $x = -2$ or $2, x = 3$ or $7, x = -2{\cdot}8$ or $2{\cdot}8, x = -8{\cdot}8$ or $-3{\cdot}2$
(d) $x = -2{\cdot}2$ or $2{\cdot}2, x = 2{\cdot}8$ or $7{\cdot}2, x = -3$ or $3, x = -9$ or -3
2. ① $-3 \leqslant x \leqslant 3, 0 \leqslant f(x) \leqslant 9$; ② $2 \leqslant x \leqslant 8, 0 \leqslant f(x) \leqslant 9$; ③ $-3 \leqslant x \leqslant 3, -4 \leqslant f(x) \leqslant 5$;
④ $-9 \leqslant x \leqslant -3, -4 \leqslant f(x) \leqslant 5$
3. ① 6 units to the right, 4 units upwards; ② 11 units to the right, 4 units upwards;
③ 6 units to the right
4. (a) $f(x) = x^2 + 3$ (b) $f(x) = x^2 - 2$ (c) $f(x) = (x-4)^2$ (d) $f(x) = (x+1)^2$
(e) $f(x) = (x-2)^2 + 5$ (f) $f(x) = (x+3)^2 - 1$
5. $f(x) = (x-3)^2 + 2$
7. (a) $f(x) = (x+2)^2 - 4$ (b) $f(x) = (x-3)^2 - 9$ (c) $f(x) = (x+2)^2 + 3$
(d) $f(x) = (x-3)^2 - 4$ (e) $f(x) = (x+3)^2 - 5$ (f) $f(x) = (x-2)^2 - 5$
8. $f(x) = (x-3)^2 + 2$

Exercise 5.8a page 96

1. (a) $0 < x < 2$ (b) $x > 4$ or $x < -2$ (c) all x except $x = 1$ (d) -1 or 3
$-1{\cdot}4 < x < 3{\cdot}4$
2. (a) $x > 2$ or $x < -3$ (b) $-4 < x < 3$ (c) $x > 0$ or $x < -1$ (d) -5 or 4
$x > 1$ or $x < -2$
3. (a) $-2 < x < 4$ (b) $x > 2$ or $x < 0$ (c) $x \geqslant 5$ or $x \leqslant -3$ (d) $-1 \leqslant x \leqslant 3$
$x^2 - 2x - 8 \geqslant -9$ for all values of x.
4. (a) $-3 < x < 3$ (b) $x > 1$ or $x < -1$ (c) $-2 \leqslant x \leqslant 2$ (d) $x \geqslant 4$ or $x \leqslant -4$
$\sqrt{2} = 1{\cdot}4$ or $-1{\cdot}4$
5. (a) $x > 3$ or $x < -1$ (b) 4 or -2 (c) $0 \leqslant x \leqslant 2$ (d) $x = 1$ only
$3 + 2x - x^2 \geqslant \frac{1}{2}$ for $-0{\cdot}9 \leqslant x \leqslant 2{\cdot}9$
6. (a) $2 < x < 3$ (b) $x \geqslant 1$ or $x \leqslant -5$ (c) $x > 5$ or $x < -1$ (d) $x > 4$ or $x < 1$
(e) $-3 \leqslant x \leqslant -1$ (f) $-0{\cdot}6 \leqslant x \leqslant 4{\cdot}6$
7. (a) $0{\cdot}3 < x < 3{\cdot}7$ (b) $-0{\cdot}3 \leqslant x \leqslant 3{\cdot}3$ (c) $0{\cdot}4 < x < 4{\cdot}6$
8. 2 or 4; $2 \leqslant x \leqslant 4$
9. 3 or -4; $x > 3$ or $x < -4$

Exercise 5.8b page 97

5. $y \leqslant \dfrac{6}{x}$ and $x + y \leqslant 7$
6. $y \leqslant (x-1)^2$ and $y < 4 - x^2$

Exercise 5.9a page 98

1. (a) 2 or -1 (b) $-2{\cdot}4$ or $2{\cdot}4$ (c) $-1{\cdot}4$ or $1{\cdot}4$ (d) 1
2. (a) 0 or 1 (b) -3 or 2 (c) -2 or 1 (d) 2
3. (a) -2 or 4 (b) 0 or 3 (c) -1 or 4
4. (a) -1 or 5 (b) 1 or 4 (c) -1 or 4
5. (a) -2 or 5 (b) -1 or 5 (c) -2 or 4
6. (a) -2 or 2 (b) -4 or 3 (c) -1 or 2
7. (a) $8 + 2x - x^2 = x + 6$ (b) $8 + 2x - x^2 = 8 - x$ (c) $8 + 2x - x^2 = 2$
8. -1 or 4

Exercise 5.9b page 99

1. (a) $\{3\}$ (b) $\{-3, 3\}$ (c) $\{-1{\cdot}7, +1{\cdot}7\}$
2. (a) $\{-5, 2\}$ (b) $\{-2, 5\}$
3. (a) $\frac{10}{x} = x + 3$ is $10 = x^2 + 3x$ i.e. $x^2 + 3x - 10 = 0$

   (b) $\frac{10}{x} = x - 3$ is $10 = x^2 - 3x$ i.e. $x^2 - 3x - 10 = 0$
4. $\{-0{\cdot}36, 1{\cdot}36\}$
5. (a) $\{0, 1, -1\}$ (b) $\{0, 2, -2\}$ (c) $\{0, 3, -3\}$
6. $\{-3, 1, 2\}$; $\{-3, 2\}$

Part 6

Exercise 6.1a page 100

1. (a) and (d), (b) and (c)
2. (a) and (c), (b) and (d)

		nodes	*arcs*	*regions*
3.	(a)	4	5	3
	(b)	2	2	2
	(c)	2	2	2
	(d)	4	5	3
4.	(a)	3	4	3
	(b)	3	5	4
	(c)	3	4	3
	(d)	3	5	4

5. 3 regions in each network
6. (a) 1 (b) 3 (c) 2
7. (a) yes (b) no
8. (a) 3 regions, 2 nodes (b) 4 nodes, 1 region

Exercise 6.1b page 101

3. 3 regions in each network
4. (a) 2 regions (b) 1 region
5. 6 faces, 12 edges, 8 vertices; $6 - 12 + 8 = 2$
6. (a) 5 faces, 9 edges, 6 vertices; $5 - 9 + 6 = 2$
   (b) no; 6 faces, 11 edges, 7 vertices
7. (b) and (d)
8. (a) 8 odd nodes (b) 6 even nodes
   cube not traversible because it has more than 2 odd nodes

Exercise 6.1c page 102

The matrices in questions **1** and **2** may be different if the nodes are taken in a different order.

1. (a) and (d) $\begin{pmatrix} 0 & 1 & 1 & 0 \\ 1 & 0 & 1 & 1 \\ 1 & 1 & 0 & 1 \\ 0 & 1 & 1 & 0 \end{pmatrix}$, (b) and (c) $\begin{pmatrix} 1 & 1 \\ 1 & 0 \end{pmatrix}$;

(a) and (c) $\begin{pmatrix} 0 & 1 & 1 \\ 1 & 0 & 2 \\ 1 & 2 & 0 \end{pmatrix}$, (b) and (d) $\begin{pmatrix} 0 & 2 & 1 \\ 2 & 0 & 2 \\ 1 & 2 & 0 \end{pmatrix}$.

2. (a) $\begin{pmatrix} 0 & 1 & 1 & 0 & 1 \\ 1 & 0 & 0 & 1 & 1 \\ 1 & 0 & 0 & 1 & 1 \\ 0 & 1 & 1 & 0 & 1 \\ 1 & 1 & 1 & 1 & 0 \end{pmatrix}$ (b) $\begin{pmatrix} 0 & 1 & 1 & 0 & 1 & 1 \\ 1 & 0 & 0 & 1 & 1 & 1 \\ 1 & 0 & 0 & 1 & 1 & 0 \\ 0 & 1 & 1 & 0 & 1 & 0 \\ 1 & 1 & 1 & 1 & 0 & 0 \\ 1 & 1 & 0 & 0 & 0 & 0 \end{pmatrix}$

(c) $\begin{pmatrix} 0 & 2 & 1 & 2 \\ 2 & 0 & 1 & 0 \\ 1 & 1 & 0 & 1 \\ 2 & 0 & 1 & 0 \end{pmatrix}$ (d) $\begin{pmatrix} 0 & 1 & 0 & 0 & 0 & 0 \\ 1 & 0 & 1 & 0 & 2 & 1 \\ 0 & 1 & 0 & 2 & 0 & 1 \\ 0 & 0 & 2 & 0 & 1 & 1 \\ 0 & 2 & 0 & 1 & 0 & 1 \\ 0 & 1 & 1 & 1 & 1 & 0 \end{pmatrix}$

The number of odd and even nodes can be found from adding the numbers in each row of the matrix.

3. (a) 2 odd nodes, 1 even node, traversible (b) 2 odd nodes, traversible
(c) 4 odd nodes, not traversible

5. sum of entries: 12, 6; number of arcs is half these numbers.

6. (a) odd number in leading diagonal (b) no symmetry about leading diagonal
(c) no symmetry about leading diagonal
Networks with one-way routes could be

(a)  (b)  (c) 

Exercise 6.1d page 103

2. $\begin{pmatrix} 0 & 0 & 1 \\ 1 & 0 & 1 \\ 1 & 1 & 0 \end{pmatrix}$

3. (a) $\begin{pmatrix} 0 & 2 \\ 1 & 0 \end{pmatrix}$ (b) $\begin{pmatrix} 0 & 2 & 0 \\ 1 & 0 & 1 \\ 0 & 0 & 0 \end{pmatrix}$ (c) $\begin{pmatrix} 0 & 1 & 0 & 0 \\ 0 & 0 & 1 & 1 \\ 0 & 0 & 0 & 1 \\ 1 & 1 & 0 & 0 \end{pmatrix}$

4. (a) $\begin{pmatrix} 1 & 0 & 1 \\ 1 & 0 & 0 \\ 0 & 1 & 0 \end{pmatrix}$ (b) $\begin{pmatrix} 1 & 1 \\ 1 & 1 \end{pmatrix}$ (c) $\begin{pmatrix} 0 & 1 & 0 \\ 0 & 2 & 1 \\ 0 & 0 & 0 \end{pmatrix}$

5. $\mathbf{M} = \begin{pmatrix} 0 & 0 & 0 \\ 1 & 0 & 0 \\ 1 & 1 & 0 \end{pmatrix}$, $\mathbf{M}' = \begin{pmatrix} 0 & 1 & 1 \\ 0 & 0 & 1 \\ 0 & 0 & 0 \end{pmatrix}$

$\mathbf{M}'$ represents *is less than*

6. $\mathbf{X} = \begin{array}{c} \\ 3 \\ 5 \\ 9 \\ 15 \\ 45 \end{array} \begin{array}{c} \begin{array}{ccccc} 3 & 5 & 9 & 15 & 45 \end{array} \\ \begin{pmatrix} 1 & 0 & 0 & 0 & 0 \\ 0 & 1 & 0 & 0 & 0 \\ 1 & 0 & 1 & 0 & 0 \\ 1 & 1 & 0 & 1 & 0 \\ 1 & 1 & 1 & 1 & 1 \end{pmatrix} \end{array}$ $\mathbf{X}'$ represents *is a factor of*

7. $\mathbf{A}' = \begin{pmatrix} 1 & 1 \\ 0 & 0 \end{pmatrix}$, $\mathbf{B}' = \begin{pmatrix} 3 & -1 \\ 2 & 0 \end{pmatrix}$, $\mathbf{C}' = \begin{pmatrix} 0 & 2 \\ -4 & 1 \end{pmatrix}$, $\mathbf{D}' = \begin{pmatrix} 5 & 3 \\ 2 & -1 \end{pmatrix}$

$\mathbf{A} + \mathbf{A}' = \begin{pmatrix} 2 & 1 \\ 1 & 0 \end{pmatrix}$, $\mathbf{B} + \mathbf{B}' = \begin{pmatrix} 6 & 1 \\ 1 & 0 \end{pmatrix}$, $\mathbf{C} + \mathbf{C}' = \begin{pmatrix} 0 & -2 \\ -2 & 2 \end{pmatrix}$, $\mathbf{D} + \mathbf{D}' = \begin{pmatrix} 10 & 5 \\ 5 & -2 \end{pmatrix}$

All the matrices are symmetrical; they are self-transpose.

Exercise 6.2 page 104

1. $\begin{pmatrix} -1 & -38 \\ 20 & -48 \end{pmatrix}$

2. $\begin{pmatrix} 11 & 1 & 30 \\ 12 & 13 & 3 \\ 33 & -3 & 29 \end{pmatrix}$

3. (a) $\begin{pmatrix} 4 & 11 \\ 7 & 18 \end{pmatrix}$ (b) $\begin{pmatrix} 32 & 4 \\ 20 & 7 \end{pmatrix}$ (c) $\begin{pmatrix} -15 & -5 \\ -6 & 0 \end{pmatrix}$ (d) $\begin{pmatrix} 1 & 0 \\ 0 & 1 \end{pmatrix}$

(e) $\begin{pmatrix} 4 & 2 \\ 1 & 3 \end{pmatrix}$

4. (a) $\begin{pmatrix} 9 & 12 \\ 15 & 9 \end{pmatrix}$ (b) $\begin{pmatrix} -9 & 12 \\ -33 & -12 \end{pmatrix}$ (c) $\begin{pmatrix} 0 & 20 \\ 20 & 0 \end{pmatrix}$ (d) $\begin{pmatrix} 3 & 2 \\ 4 & 1 \end{pmatrix}$

(e) $\begin{pmatrix} -1 & 0 \\ 0 & 1 \end{pmatrix}$

5. $a = 5\frac{1}{2}, b = -12\frac{1}{2}, c = -10, d = -61$
6. $x = -2, y = -1, z = 16, w = -8$

7. (a) $\begin{pmatrix} 6 & 0 & 8 \\ 0 & 4 & 0 \\ 0 & 0 & 24 \end{pmatrix}$ (b) $(6 \quad -69)$

8. (a) $\begin{pmatrix} 2 & 5 & 8 \\ -6 & -1 & 4 \end{pmatrix}$ (b) $\begin{pmatrix} 35 & 19 & -13 \\ 0 & 9 & -10 \\ 0 & 0 & -4 \end{pmatrix}$

9. (a) $\begin{pmatrix} 5 \\ 5 \end{pmatrix}$ (b) $\begin{pmatrix} -1 & -8 & 7 \\ 4 & 15 & 16 \end{pmatrix}$

10. (a) $\begin{pmatrix} 7 & 11 & -4 \\ -4 & 1 & -2 \\ 5 & 3 & 0 \end{pmatrix}$ (b) $\begin{pmatrix} 6 & -5 \\ 13 & 2 \end{pmatrix}$

11. (a) (18) (b) $\begin{pmatrix} 6 & 0 & 3 & 12 \\ -2 & 0 & -1 & -4 \\ 8 & 0 & 4 & 16 \\ 4 & 0 & 2 & 8 \end{pmatrix}$

12. (a) an $n \times n$ matrix, all of whose entries are 1. (b) (n)

13. (a) $\begin{pmatrix} 3 & 8 \\ 9 & 16 \end{pmatrix}$ (b) $\begin{pmatrix} 3 & 6 \\ 12 & 16 \end{pmatrix}$ (c) $\begin{pmatrix} 1 & 0 \\ 0 & 1 \end{pmatrix}$ (d) $\begin{pmatrix} 1 & 0 \\ 0 & 1 \end{pmatrix}$ (e) $\begin{pmatrix} -6 & 3 \\ 6 & -2 \end{pmatrix}$

14. (a) $\begin{pmatrix} 2 & 0 \\ 0 & 2 \end{pmatrix}$ (b) $\begin{pmatrix} 2 & 0 \\ 0 & 2 \end{pmatrix}$ (c) $\begin{pmatrix} 8 & 6 \\ 4 & 2 \end{pmatrix}$ (d) $\begin{pmatrix} -2 & 6 \\ 4 & -8 \end{pmatrix}$ (e) $\begin{pmatrix} -8 & 4 \\ 6 & -2 \end{pmatrix}$

15. $\begin{pmatrix} 6 & -1 \\ 6 & 1 \end{pmatrix}$, same result

16. $\begin{pmatrix} 10 & -14 \\ 6 & -10 \end{pmatrix}$, same result

17. $\mathbf{XY} = \begin{pmatrix} 12-5b & 3a-20 \\ -4+2b & -a+8 \end{pmatrix}$, $\mathbf{YX} = \begin{pmatrix} 12-a & -20+2a \\ 3b-4 & -5b+8 \end{pmatrix}$

$\mathbf{XY} = \mathbf{YX}$ if $a = b = 0$

20. $\mathbf{P}^2 = \begin{pmatrix} 4 & 0 \\ 0 & 4 \end{pmatrix}$, $\mathbf{P}^3 = \begin{pmatrix} 0 & 8 \\ 8 & 0 \end{pmatrix}$, $\mathbf{P}^4 = \begin{pmatrix} 16 & 0 \\ 0 & 16 \end{pmatrix}$

$\mathbf{P}^n = 2^n \mathbf{I}$, n even

$= 2^{n-1}\mathbf{P}$, n odd

Exercise 6.3 page 106

1. $\begin{array}{c} \\ A \\ B \end{array}\begin{array}{c} E \\ \begin{pmatrix} 3 \\ 2 \end{pmatrix} \end{array}$, 5 routes

2. $\begin{pmatrix} & E & F \\ A & 1 & 2 \\ B & 2 & 4 \\ C & 1 & 2 \end{pmatrix}$, 4 routes from B to F

3. (a) L M N

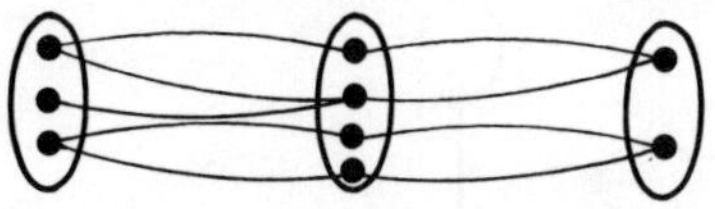

(b) $\begin{pmatrix} 1 & 1 & 0 & 0 \\ 0 & 1 & 0 & 0 \\ 0 & 0 & 1 & 1 \end{pmatrix}$, $\begin{pmatrix} 1 & 0 \\ 1 & 0 \\ 0 & 1 \\ 0 & 1 \end{pmatrix}$ (c) $\begin{pmatrix} 2 & 0 \\ 1 & 0 \\ 0 & 2 \end{pmatrix}$ (i) 2 ways by air (ii) 3 ways by sea

4. $\begin{array}{c} \\ a \\ b \\ c \end{array}\begin{array}{c} \begin{matrix} p & q & r & s \end{matrix} \\ \begin{pmatrix} 1 & 0 & 0 & 0 \\ 0 & 1 & 1 & 0 \\ 0 & 0 & 0 & 1 \end{pmatrix} \end{array}$, $\begin{array}{c} \\ p \\ q \\ r \\ s \end{array}\begin{array}{c} \begin{matrix} x & y \end{matrix} \\ \begin{pmatrix} 0 & 1 \\ 1 & 0 \\ 0 & 1 \\ 0 & 1 \end{pmatrix} \end{array}$, $\begin{array}{c} \\ a \\ b \\ c \end{array}\begin{array}{c} \begin{matrix} x & y \end{matrix} \\ \begin{pmatrix} 0 & 1 \\ 1 & 1 \\ 0 & 1 \end{pmatrix} \end{array}$

5. $\mathbf{Y} = \begin{array}{r} \\ \textit{Potatoes} \\ \textit{Carrots} \\ \textit{Onions} \end{array}\begin{array}{c} \begin{matrix} A & B \end{matrix} \\ \begin{pmatrix} 12 & 10 \\ 14 & 16 \\ 20 & 18 \end{pmatrix} \end{array}$ $\mathbf{XY} = \begin{array}{r} \\ \textit{week 1} \\ \textit{week 2} \end{array}\begin{array}{c} \begin{matrix} A & B \end{matrix} \\ \begin{pmatrix} 94 & 84 \\ 168 & 150 \end{pmatrix} \end{array}$

XY represents total bill for shopping at each shop; shop B gives better value.

6. $N = \begin{array}{c} \\ \textit{25 mm} \\ \textit{50 mm} \\ \textit{75 mm} \end{array} \overset{\text{Price (p)}}{\begin{pmatrix} 5 \\ 8 \\ 12 \end{pmatrix}}, \quad MN = \begin{array}{c} S_1 \\ S_2 \end{array} \overset{\text{Price (p)}}{\begin{pmatrix} 19800 \\ 36900 \end{pmatrix}}$

Total value of stock £567.

7. $\mathbf{X}^2 = \begin{pmatrix} 5 & 0 & 2 \\ 3 & 1 & 1 \\ 2 & 0 & 1 \end{pmatrix}, \mathbf{X}^3 = \begin{pmatrix} 12 & 0 & 5 \\ 8 & 1 & 3 \\ 5 & 0 & 2 \end{pmatrix}$

Networks show total number of routes between P, Q and R using 2 journeys and 3 journeys respectively.

8. $\mathbf{M} = \begin{array}{c} \\ A \\ B \\ C \end{array} \begin{array}{c} \begin{array}{ccc} A & B & C \end{array} \\ \begin{pmatrix} 1 & 1 & 1 \\ 0 & 0 & 1 \\ 1 & 0 & 0 \end{pmatrix} \end{array}$ (i) after 2 moves, $\mathbf{M}^2 = \begin{pmatrix} 2 & 1 & 2 \\ 1 & 0 & 0 \\ 1 & 1 & 1 \end{pmatrix}$

(ii) after 3 moves, $\mathbf{M}^3 = \begin{pmatrix} 4 & 2 & 3 \\ 1 & 1 & 1 \\ 2 & 1 & 2 \end{pmatrix}$

Fred, 4 ways; George, 1 way; Harry, 2 ways.

9. $\mathbf{T} = \begin{array}{c} \\ k \\ l \\ m \end{array} \begin{array}{c} \begin{array}{ccc} k & l & m \end{array} \\ \begin{pmatrix} 0 & 1 & 1 \\ 1 & 0 & 0 \\ 1 & 0 & 0 \end{pmatrix} \end{array}$

(a) $\mathbf{T}^2 = \begin{array}{c} \\ k \\ l \\ m \end{array} \begin{array}{c} \begin{array}{ccc} k & l & m \end{array} \\ \begin{pmatrix} 2 & 0 & 0 \\ 0 & 1 & 1 \\ 0 & 1 & 1 \end{pmatrix} \end{array}$ (b) [diagram: k, l, m]

(c) $\mathbf{T}^2$ describes *is parallel to*

10. $\mathbf{M} = \begin{array}{c} \\ A \\ B \\ C \end{array} \begin{array}{c} \begin{array}{ccc} A & B & C \end{array} \\ \begin{pmatrix} 0 & 1 & 0 \\ 0 & 0 & 1 \\ 1 & 0 & 0 \end{pmatrix} \end{array}$

$\mathbf{M}^2 = \begin{array}{c} \\ A \\ B \\ C \end{array} \begin{array}{c} \begin{array}{ccc} A & B & C \end{array} \\ \begin{pmatrix} 0 & 0 & 1 \\ 1 & 0 & 0 \\ 0 & 1 & 0 \end{pmatrix} \end{array}$ describes *is on the right of*

$\mathbf{M}^3 = \begin{array}{c} \\ A \\ B \\ C \end{array} \begin{array}{c} \begin{array}{ccc} A & B & C \end{array} \\ \begin{pmatrix} 1 & 0 & 0 \\ 0 & 1 & 0 \\ 0 & 0 & 1 \end{pmatrix} \end{array}$ describes *is in the same place as*

Exercise 6.4 page 108

1. (a) $\begin{pmatrix} 3 & -7 \\ -2 & 5 \end{pmatrix}$ (b) $\begin{pmatrix} \frac{3}{2} & -1 \\ -\frac{5}{2} & 2 \end{pmatrix}$ (c) $\begin{pmatrix} 4 & -3 \\ 3 & -2 \end{pmatrix}$ (d) $\begin{pmatrix} 0{\cdot}5 & -0{\cdot}2 \\ -1 & 0{\cdot}6 \end{pmatrix}$
(e) $\begin{pmatrix} \frac{1}{6} & \frac{1}{4} \\ \frac{1}{18} & -\frac{1}{12} \end{pmatrix}$

2. (a) $\begin{pmatrix} 3 & -4 \\ -5 & 7 \end{pmatrix}$ (b) $\begin{pmatrix} 2 & -5\frac{1}{2} \\ -1 & 3 \end{pmatrix}$ (c) no inverse (d) $\begin{pmatrix} -4 & 5 \\ 5 & -6 \end{pmatrix}$
(e) $\begin{pmatrix} 0 & -\frac{1}{2} \\ -1 & -1 \end{pmatrix}$

3. $\mathbf{PQ} = \begin{pmatrix} 4 & 0 \\ 0 & 4 \end{pmatrix}$, $\mathbf{P}^{-1} = \begin{pmatrix} 2 & -2\frac{1}{4} \\ 1 & -1\frac{1}{4} \end{pmatrix}$, $\mathbf{Q}^{-1} = \begin{pmatrix} 1\frac{1}{4} & -2\frac{1}{4} \\ 1 & -2 \end{pmatrix}$

4. $\mathbf{XY} = \begin{pmatrix} -1 & 0 \\ 0 & -1 \end{pmatrix}$, $\mathbf{X}^{-1} = \begin{pmatrix} -1 & 2 \\ 3 & -5 \end{pmatrix}$

5. $\mathbf{A}^{-1} = \begin{pmatrix} 4 & \frac{1}{2} \\ -1 & 0 \end{pmatrix}$, $\mathbf{B}^{-1} = \begin{pmatrix} -1 & 3 \\ -2 & 5 \end{pmatrix}$
(a) $\begin{pmatrix} -2 & 1 \\ 26 & -14 \end{pmatrix}$ (b) $\begin{pmatrix} -7 & -\frac{1}{2} \\ -13 & -1 \end{pmatrix}$ (c) $\begin{pmatrix} -7 & -\frac{1}{2} \\ -13 & -1 \end{pmatrix}$
(b), (c) are the same

6. $\mathbf{U}^{-1} = \begin{pmatrix} -2\frac{1}{2} & 2 \\ 1\frac{1}{2} & -1 \end{pmatrix}$, $\mathbf{V}^{-1} = \begin{pmatrix} 1 & 0 \\ -2\frac{1}{3} & \frac{1}{3} \end{pmatrix}$
(a) $\begin{pmatrix} 30 & 12 \\ 38 & 15 \end{pmatrix}$ (b) $\begin{pmatrix} -7\frac{1}{6} & \frac{2}{3} \\ 3\frac{5}{6} & -\frac{1}{3} \end{pmatrix}$ (c) $\begin{pmatrix} -2\frac{1}{2} & 2 \\ 6\frac{1}{3} & -5 \end{pmatrix}$
$(\mathbf{UV})^{-1} = \mathbf{V}^{-1}\mathbf{U}^{-1}$

7. (a) $\begin{pmatrix} 1 & 0 \\ 0 & 1 \end{pmatrix}, \begin{pmatrix} 4 & -5 \\ 3 & -4 \end{pmatrix}$ (b) $\begin{pmatrix} 4 & 0 \\ 0 & 4 \end{pmatrix}, \begin{pmatrix} -\frac{1}{2} & 0 \\ -1 & \frac{1}{2} \end{pmatrix}$
(c) $\begin{pmatrix} 9 & 0 \\ 0 & 9 \end{pmatrix}, \begin{pmatrix} \frac{10}{9} & -\frac{7}{9} \\ \frac{13}{9} & -\frac{10}{9} \end{pmatrix}$ (d) $\begin{pmatrix} x^2 & 0 \\ 0 & x^2 \end{pmatrix}, \begin{pmatrix} 0 & 1/x \\ 1/x & 0 \end{pmatrix}$

8. (a) $\begin{pmatrix} 5 & -8 \\ 3 & -5 \end{pmatrix}, \begin{pmatrix} 1 & 0 \\ 0 & 1 \end{pmatrix}, \begin{pmatrix} 5 & -8 \\ 3 & -5 \end{pmatrix}$ (b) $\begin{pmatrix} -7 & 16 \\ -3 & 7 \end{pmatrix}, \begin{pmatrix} 1 & 0 \\ 0 & 1 \end{pmatrix}, \begin{pmatrix} -7 & 16 \\ -3 & 7 \end{pmatrix}$
(c) $\frac{1}{4}\begin{pmatrix} -6 & 8 \\ -4 & 6 \end{pmatrix}, \begin{pmatrix} 4 & 0 \\ 0 & 4 \end{pmatrix}, 4\begin{pmatrix} -6 & 8 \\ -4 & 6 \end{pmatrix}$ (d) $\begin{pmatrix} k & 1-k \\ 1+k & -k \end{pmatrix}, \begin{pmatrix} 1 & 0 \\ 0 & 1 \end{pmatrix}, \begin{pmatrix} k & 1-k \\ 1+k & -k \end{pmatrix}$

9. $x = -10, y = 4, z = 8, w = -3$

10. inverse $= \begin{pmatrix} -2 & 5 \\ \frac{3}{2} & -\frac{7}{2} \end{pmatrix}$, $x = 11,\ y = -7, z = -7\frac{1}{2}, w = 5$

11. $\mathbf{P}^{-1} = \begin{pmatrix} 0 & 1 \\ 1 & 0 \end{pmatrix}$, $\mathbf{Q}^{-1} = \begin{pmatrix} -1 & 0 \\ 0 & -1 \end{pmatrix}$, $\mathbf{PQ} = \begin{pmatrix} 0 & -1 \\ -1 & 0 \end{pmatrix}$, $(\mathbf{PQ})^{-1} = \begin{pmatrix} 0 & -1 \\ -1 & 0 \end{pmatrix}$
The only other matrix produced by these products is $\begin{pmatrix} 1 & 0 \\ 0 & 1 \end{pmatrix}$

12. $\mathbf{A}^2 = \begin{pmatrix} -1 & 0 \\ 0 & -1 \end{pmatrix}$, $\mathbf{A}^3 = \begin{pmatrix} 0 & 1 \\ -1 & 0 \end{pmatrix}$, $\mathbf{A}^4 = \begin{pmatrix} 1 & 0 \\ 0 & 1 \end{pmatrix}$
$\mathbf{A}^{-1} = \mathbf{A}^3$, $(\mathbf{A}^2)^{-1} = \mathbf{A}^2, (\mathbf{A}^3)^{-1} = \mathbf{A}$

13. (a) $\mathbf{R}^{-1} = \begin{pmatrix} \frac{1}{2} & \frac{\sqrt{3}}{2} \\ -\frac{\sqrt{3}}{2} & \frac{1}{2} \end{pmatrix}$ (b) $\mathbf{R}^2 = \begin{pmatrix} -\frac{1}{2} & -\frac{\sqrt{3}}{2} \\ \frac{\sqrt{3}}{2} & -\frac{1}{2} \end{pmatrix}$ (c) $(\mathbf{R}^2)^{-1} = \begin{pmatrix} -\frac{1}{2} & \frac{\sqrt{3}}{2} \\ -\frac{\sqrt{3}}{2} & -\frac{1}{2} \end{pmatrix}$

(d) $\mathbf{R}^3 = \begin{pmatrix} -1 & 0 \\ 0 & -1 \end{pmatrix}$ (e) $\mathbf{R}^6 = \begin{pmatrix} 1 & 0 \\ 0 & 1 \end{pmatrix}$

Six distinct matrices can be produced from products of **R** and $\mathbf{R}^{-1}$

14. $k = 0, 1, -1$

15. $\mathbf{M}' = \begin{pmatrix} 6 & 1 \\ 5 & 2 \end{pmatrix}$, $\mathbf{M}^{-1} = \begin{pmatrix} \frac{2}{7} & -\frac{5}{7} \\ -\frac{1}{7} & \frac{6}{7} \end{pmatrix}$, yes

16. $\mathbf{A}^{-1} = \frac{1}{ad-bc} \begin{pmatrix} d & -b \\ -c & a \end{pmatrix}$ $\mathbf{A}' = \begin{pmatrix} a & c \\ b & d \end{pmatrix}$

$(\mathbf{A}')^{-1} = \frac{1}{ad-bc} \begin{pmatrix} d & -c \\ -b & a \end{pmatrix}$, yes

Exercise 6.5 page 110

1. $\left.\begin{array}{l} 2x + y = 7 \\ x + y = 3 \end{array}\right\}$ $x = 4, y = -1$

2. $\begin{pmatrix} 1 & 3 \\ 4 & -1 \end{pmatrix}\begin{pmatrix} x \\ y \end{pmatrix} = \begin{pmatrix} 1 \\ -2 \end{pmatrix}$, $x = -\frac{5}{13}$, $y = \frac{6}{13}$

3. (a) $x = 0$, $y = 1$ (b) $x = -9$, $y = 20$ (c) $x = -\frac{1}{5}$, $y = 1\frac{3}{5}$

4. (a) $x = 5$, $y = -8$ (b) $x = -0{\cdot}1$, $y = -1{\cdot}7$ (c) $x = 0, y = -1$

5. (a) $u = 1, v = -1$ (b) $u = \frac{7}{9}$, $v = \frac{13}{9}$ (c) $u = 0{\cdot}8$, $v = 0{\cdot}4$

6. (a) $r = -7$, $s = 11$ (b) $r = 1{\cdot}4$, $s = 0{\cdot}8$ (c) $r = \frac{18}{35}$, $s = \frac{11}{35}$

7. matrix has no inverse, lines are parallel

8. solutions are pairs (x, y) of the form $(3k - 1, k)$, both equations represent the same line.

9. $x = 1 + 0{\cdot}4\,k, y = -1 + 0{\cdot}6\,k$;
$k = 0$: $x = 1$, $y = -1$;
$k = 5$: $x = 3$, $y = 2$;
$k = 15$: $x = 7$, $y = 8$;

10. $a = \frac{3}{2}$ does not give a unique solution
$a = 1$ gives the solution $x = -\frac{1}{2}$, $y = \frac{5}{2}$

11. (a) $x = -1$, $y = 4$ (b) $x = 3, y = -2$ **12.** (a) $r = -4, s = 3$ (b) $u = -10, v = 9$

Exercise 6.6 page 113

1. (a) $\begin{pmatrix} 5 \\ 5 \end{pmatrix}$ (b) $\begin{pmatrix} 2 \\ -2 \end{pmatrix}$ (c) $\begin{pmatrix} 4 \\ 5 \end{pmatrix}$ (d) $\begin{pmatrix} 4 \\ 5 \end{pmatrix}$

2. $\mathbf{r} + \mathbf{s} = \begin{pmatrix} 4 \\ -5 \end{pmatrix}$, $\mathbf{s} + \mathbf{t} = \begin{pmatrix} 2 \\ -2 \end{pmatrix}$

3. (a) $\begin{pmatrix} 1 \\ 2{\cdot}5 \end{pmatrix}$ (b) $\begin{pmatrix} 3 \\ 4 \end{pmatrix}$ **4.** (a) $\begin{pmatrix} -4 \\ 2 \end{pmatrix}$ (b) $\begin{pmatrix} 1 \\ 1 \end{pmatrix}$

5. (a) $x = 5$, $y = -10$ (b) $x = -1$, $y = -5$

6. (a) $x = -2$, $y = 9$ (b) $x = -8$, $y = -4$

7. $-\mathbf{a} = \begin{pmatrix}-2\\-3\end{pmatrix}$, $-\mathbf{b} = \begin{pmatrix}-2\\1\end{pmatrix}$, $-(\mathbf{a}+\mathbf{b}) = \begin{pmatrix}-4\\-2\end{pmatrix}$

8. $-\mathbf{a} = \begin{pmatrix}-1\\-2\end{pmatrix}$, $-\mathbf{b} = \begin{pmatrix}3\\0\end{pmatrix}$

9. $\overrightarrow{OP} = \begin{pmatrix}\frac{1}{2}\\8\frac{1}{2}\end{pmatrix}$

10. $\overrightarrow{OA} = \begin{pmatrix}3\\-1\end{pmatrix}$, $\overrightarrow{OB} = \begin{pmatrix}-1\\6\end{pmatrix}$, $\overrightarrow{OP} = \begin{pmatrix}1\\2\frac{1}{2}\end{pmatrix}$

11. $\overrightarrow{OA} = \begin{pmatrix}-3\\2\end{pmatrix}$, $\overrightarrow{OB} = \begin{pmatrix}2\\6\end{pmatrix}$, $\overrightarrow{OX} = \begin{pmatrix}-\frac{1}{2}\\4\end{pmatrix}$, $2\,\overrightarrow{OX} = \begin{pmatrix}-1\\8\end{pmatrix}$, $(-1, 8)$

12. $\overrightarrow{AB} = \begin{pmatrix}4\\2\end{pmatrix}$, $\overrightarrow{AC} = \begin{pmatrix}2\\4\end{pmatrix}$, $\overrightarrow{AD} = \begin{pmatrix}6\\6\end{pmatrix}$, $AB = AC = \sqrt{20}$

13. The fourth vertex is (22, 44); sides are all of length 25.

14. (a) 5 (b) 13 (c) 18 (d) $8\sqrt{5}$

Exercise 6.7a page 114

2. (a) $2(\mathbf{x}+\mathbf{y})$ (b) $2(\mathbf{u}+\mathbf{v})$

3. $\overrightarrow{CM} = \frac{1}{2}(\mathbf{a}+\mathbf{b})$, $\overrightarrow{CG} = \frac{1}{3}(\mathbf{a}+\mathbf{b})$, $\overrightarrow{AN} = -\mathbf{a}+\frac{1}{2}\mathbf{b}$

   $\overrightarrow{AG} = -\mathbf{a}+\frac{1}{3}(\mathbf{a}+\mathbf{b}) = -\frac{2}{3}\mathbf{a}+\frac{1}{3}\mathbf{b}$

4. $\overrightarrow{CX} = \frac{2}{3}\mathbf{a}+\frac{1}{3}\mathbf{b}$, $\overrightarrow{AY} = -\frac{2}{3}\mathbf{a}+\frac{1}{6}\mathbf{b}$, $\overrightarrow{CY} = \frac{1}{3}\mathbf{a}+\frac{1}{6}\mathbf{b}$

   $\overrightarrow{YZ} = -\frac{1}{3}\mathbf{a}+\frac{1}{12}\mathbf{b}$; $AY = \frac{2}{3}AZ$

Exercise 6.7b page 115

1. $(k-1)\mathbf{b} = -\mathbf{a}+l(2\mathbf{a}-\mathbf{b})$; $k = l = \frac{1}{2}$

2. $\overrightarrow{PZ} = k(2\mathbf{a}+3\mathbf{b})$, $\overrightarrow{YZ} = l(\mathbf{a}-2b)$; $k = \frac{2}{7}, l = \frac{4}{7}$

4. 4 : 3

5. $\overrightarrow{BK} = (1-k)\mathbf{a}+\frac{1}{2}k\mathbf{c}$, $\overrightarrow{BN} = \frac{1}{3}\mathbf{a}+\frac{2}{3}\mathbf{c}$, $k = \frac{4}{5}$

6. (a) $\frac{1}{2}\mathbf{a}+\frac{1}{2}\mathbf{b}+\mathbf{c}$ (b) $\frac{1}{4}\mathbf{a}+\frac{1}{4}\mathbf{b}-\frac{1}{2}\mathbf{c}$

Part 7

Exercise 7.1a page 116

3. (a) 100° (b) 140° (c) 20°
4. (a) 138° (b) 35° (c) 36°
5. (a) 80 (b) 10 (c) 30
6. 40°, 100° *or* 70°, 70°
7. 30°
8. 15°

Exercise 7.1b page 117

1. 1440°
2. 1800°
3. (a) 108° (b) 144° (c) 156°
4. (a) 135° (b) 150° (c) 162°
5. 12
7. 36°
8. $136\frac{2}{3}$°

Exercise 7.1c page 119

3. Triangle DAY is isosceles
4. (a) true (b) false (c) false (d) true
6. (a) right-angled isosceles, XA = XC (b) isosceles, XA = XC
(c) isosceles, CA = CX
7. (a) kite (b) parallelogram (c) rectangle
8. (a) rectangle (b) rhombus (c) square
9. (a) B (b) A (c) C (d) B (e) C
10. (a) A (b) B (c) C (d) A (e) B

Exercise 7.1d page 120

2. n is odd
5. (a) T (b) F (c) T (d) F (e) T
6. (a) T (b) T (c) F (d) T (e) T

Exercise 7.1e page 121

1. (a) (i) (b) (i) (c) (iii) (d) (iii) (e) (iii)
2. (a), (b), (c), (d) are isosceles
3. 540° **4.** 1080° **5.** 54°, 108° **6.** 40°
7. 140°, 40° **8.** 66°, 48°, 90° **9.** 36°, 144° **10.** $66\frac{2}{3}$°

Exercise 7.3a page 125

1. (a) $a = 4$, $b = 6$ (b) $c = 13\frac{1}{3}$, $d = 14\frac{2}{3}$
(c) $e = 4$, $f = 3$ (d) $h = 3$, $g = 8$
(e) $i = 9$, $j = 6\frac{1}{4}$ (f) $k = 5\frac{1}{3}$, $l = 6\frac{2}{3}$
2. 15 cm, $17\frac{1}{2}$ cm
3. QR = 8·33 cm, ZW = 6 cm, WX = 2·4 cm
4. $6\frac{2}{3}$ m **5.** 2·5 cm **8.** 4·5 cm **9.** UX = $3\frac{3}{7}$cm, 40 : 21

Exercise 7.3b page 127

1. 16 : 49 **2.** 9 : 16 **3.** XY = 9 cm, WZ = 4·5 cm, 4 : 1
4. 3 : 1 **5.** 56 cm, 25: 49 **6.** 6·37 cm
7. (a) 35 cm (b) 12 m (c) 400 cm^2 (d) 20 (e) 150 cm^3
8. (a) 16 cm (b) 24 cm (c) 80 cm^2 (d) 2 (e) 32 cm^3

Exercise 7.4a page 129

1. (a) $a = 80°$, $b = 40°$ (b) $c = 56°$, $d = 41°$ (c) $e = 27°$, $f = 70°$
(d) $g = 49°$
2. (a) $a = 28°$ (b) $b = 170°$, $c = 95°$ (c) $d = 106°$
(d) $e = 42°$, $f = 138°$
3. all 41°
4. (a) 90° (b) 45° (c) 45°

Exercise 7.4b page 130

1. (a) $a = 57°$, $b = 81°$ (b) $c = 28°$, $d = 83°$ (c) $e = 60°$
   (d) $f = 116°$
2. (a) $a = 90°$ (b) $b = 70°$, $c = 42°$ (c) $d = 132°$
3. 27°, 72°, 81°
4. 99°, 17°
5. all are 36°

Exercise 7.5 page 133

1. (a) $a = 4$ (b) $b = 10$ (c) $c = 12$ (d) $d = 5$ (e) $e = 7{\cdot}5$ (f) $f = 8$, $g = 8$
2. (a) $13\frac{1}{3}$ (b) 9 (c) 1
3. (a) 8 (b) 14 (c) 6
4. (b) 4 cm

Exercise 7.6 page 135

3. 5·2 cm
4. 152 m
5. 8·7 cm
6. 10 cm
7. 2·9 cm
8. 1·3 cm
12. 7·2 cm, 5·9 cm

Exercise 7.7 page 137

1. perpendicular bisector of AB
2. pair of lines parallel to given line
3. bisector of angle ABC
4. pair of angle bisectors
5. pair of lines parallel to AB and 8 cm from AB
6. circle radius 5 cm, centre mid-point of AB
7. can satisfy (a) and (c), (a) and (b) but not all three
8. perpendicular bisector of XY
9. two concentric circles, radius 4 cm and 6 cm

Exercise 7.8 page 139

1. (a) 900 n.m. (b) 3240 n.m.
2. 1620 n.m.
3. (a) 21 600 n.m. (b) 5400 n.m.
4. (a) (0°, 67°W) (b) (43°N, 15°E) (c) (40°S, 15°E)
5. 2640 n.m.
6. (a) 5400 n.m. (b) 3600 n.m.
7. 12 knots
8. 10 knots
9. 5180 km, 1720 km
10. (a) 11 700 km (b) 12 200 km (c) 7040 km
11. 112 km, 55·9 km; 30 n.m.
12. 1050 km

Part 8

Exercise 8.1 page 140

1. (a), (b) and (d)
2. (a) kite (b) rhombus
8. (a) $(1, 2)$ (b) $(-1, -2)$ (c) $(1, -2)$ (d) $(1, -2)$ (e) $(-2, 1)$
9. (a) $(3, 1)$ (b) $(3, 1)$ (c) $(-3, 1)$ (d) $(3, -1)$ (e) $(1, 3)$
10. $(0, 3)$, gradient of mirror line $\frac{3}{4}$, equation $4y - 3x = 12$
11. (a) $y = 0$ (b) $x = 0$ (c) $y = x$ (d) $x + y = 0$ (e) $x = 1\frac{1}{2}$
16. $x = 4$
17. $y = 1$

Exercise 8.2 page 142

1. (a), (b) are true, (c), (d) are false
2. (a) O, 90° (b) C, 90° (c) O, 180°
4. single rotation of 180° about the point vertically below X and to the left of Y.
6. (c) centre $(0, 0)$, angle 270° or $-90°$ (d) centre $(4, 0)$, angle 270° or $-90°$
8. (b) and (d) are equivalent
9. (a) $(0, 1)$, 180° (b) $(-1, 3)$, 90° (c) $(1, -1)$, 270° or $-90°$
10. (a) $(2, 0)$, 270° or $-90°$ (b) $(3, 1)$ (c) $(0, 3)$

Exercise 8.3a page 144

3. $\begin{pmatrix} -2 \\ -2 \end{pmatrix}$
4. $\begin{pmatrix} -1 \\ 4 \end{pmatrix}$

Exercise 8.3b page 145

1. translation, 6 units parallel with x-axis
2. yes
3. translation, 6 units opposite direction
5. (a) true (b) false (c) true (d) true
6. (b) and (c) are true
7. translation $\sqrt{50}$ units parallel with $y = x$
8. (a) false (b) true (c) true (d) false
9. rotation centre $(0, 1\frac{1}{2})$, 180°
10. glide -4 units along $y = 2$ as axis

Exercise 8.4 page 147

3. centre $(2, 0)$ scale factor -4
4. centre $(2\frac{2}{3}, 0)$ scale factor -2
7. (a) true (b) false (c) false (d) true (e) true
8. $(4, 2), (-4, 2), (-4, -4)$, 24

Exercise 8.5 page 148

1. $YY' = 1{\cdot}33$ cm
3. (a) true (b) true (c) false (d) false (e) true
4. (a) true (b) false (c) false (d) false (e) true
5. (a) BD (b) ADE (c) 9·2 cm, 2·9 cm (d) 13·4 cm²
6. 14·9 cm
8. an ellipse

Exercise 8.6a page 150

1. (a) $(-1,-1), (2,0), (4,2), (3,3)$ (b) $(-2,0), (0,2), (4,2), (6,0)$
   (c) $(-2,-1), (0,0), (4,2), (6,3)$ (d) $(-1,0), (2,2), (4,2), (3,0)$
   (e) $(-2,0), (0,4), (4,4), (6,0)$
2. (a) $(0,0), (0,1), (-2,1)$ (b) $(0,0), (1,0), (1,2)$
   (c) $(0,0), (0,2), (4,2)$ (d) $(0,0), (1,0), (1,-2)$
   (e) $(0,0), (0,0), (4,2)$

Exercise 8.6b page 151

1. (a) $(2,2), (0,4), (4,6), (6,2)$ (b) $(3,3), (0,6), (6,9), (9,3)$
   (c) $(\frac{1}{2},\frac{1}{2}), (0,1), (1,1\frac{1}{2}), (1\frac{1}{2},\frac{1}{2})$ (d) $(-2,-2), (0,-4), (-4,-6), (-6,-2)$
   (e) $(-\frac{1}{2},-\frac{1}{2}), (0,-1), (-1,-1\frac{1}{2}), (-1\frac{1}{2},-\frac{1}{2})$
2. (a) $(2,2), (2,-4), (-2,-2), (-4,2)$ (b) $(3,3), (3,-6), (-3,-3), (-6,3)$
   (c) $(\frac{1}{2},\frac{1}{2}), (\frac{1}{2},-1), (-\frac{1}{2},-\frac{1}{2}), (-1,\frac{1}{2})$ (d) $(-2,-2), (-2,4), (2,2), (4,-2)$
   (e) $(-\frac{1}{2},-\frac{1}{2}), (-\frac{1}{2},1), (\frac{1}{2},\frac{1}{2}), (1,-\frac{1}{2})$
3. $(0,0), (k,0), (k,k), (0,k)$
   area increased in ratio $1:k^2$; k^2 is the determinant of the matrix
4. (a) $\begin{pmatrix}-\frac{1}{2} & 0\\ 0 & -\frac{1}{2}\end{pmatrix}$ (b) $\begin{pmatrix}-3 & 0\\ 0 & -3\end{pmatrix}$

Exercise 8.7a page 152

1. (a) $\begin{pmatrix}1 & 0\\ 0 & -1\end{pmatrix}$ (b) $\begin{pmatrix}0 & 1\\ 1 & 0\end{pmatrix}$ (c) $\begin{pmatrix}0 & 1\\ -1 & 0\end{pmatrix}$
2. (a) $\begin{pmatrix}0 & 1\\ 1 & 0\end{pmatrix}$ (b) $\begin{pmatrix}-1 & 0\\ 0 & -1\end{pmatrix}$ (c) $\begin{pmatrix}0 & -1\\ -1 & 0\end{pmatrix}$
3. $\begin{pmatrix}0 & 1\\ -1 & 0\end{pmatrix}$, rotation 270°
4. $\begin{pmatrix}0 & -1\\ 1 & 0\end{pmatrix}$, rotation 90°
5. $\mathbf{X} = \begin{pmatrix}1 & 0\\ 0 & -1\end{pmatrix}$, $\mathbf{R} = \begin{pmatrix}0 & -1\\ 1 & 0\end{pmatrix}$, $\mathbf{XR} = \begin{pmatrix}0 & -1\\ -1 & 0\end{pmatrix}$, $\mathbf{RX} = \begin{pmatrix}0 & 1\\ 1 & 0\end{pmatrix}$,
   $\mathbf{XR}(\mathrm{P}) = (-4,1)$, $\mathbf{RX}(\mathrm{P}) = (4,-1)$
6. $\mathbf{Y} = \begin{pmatrix}-1 & 0\\ 0 & 1\end{pmatrix}$, $\mathbf{H} = \begin{pmatrix}-1 & 0\\ 0 & -1\end{pmatrix}$, reflection in x-axis
7. (a) reflection in x-axis (b) $\begin{pmatrix}1 & 0\\ 0 & -1\end{pmatrix}$

Exercise 8.7b page 153

2. (a) $\begin{pmatrix}\frac{1}{\sqrt{2}} & -\frac{1}{\sqrt{2}}\\ \frac{1}{\sqrt{2}} & \frac{1}{\sqrt{2}}\end{pmatrix}$ (b) $\begin{pmatrix}\frac{1}{2} & -\frac{\sqrt{3}}{2}\\ \frac{\sqrt{3}}{2} & \frac{1}{2}\end{pmatrix}$
   $\left(\frac{1}{\sqrt{2}}, \frac{3}{\sqrt{2}}\right)$, $\left(1-\frac{\sqrt{3}}{2}, \sqrt{3}+\frac{1}{2}\right)$

3. (a) $\begin{pmatrix} -0{\cdot}5 & -0{\cdot}866 \\ 0{\cdot}866 & -0{\cdot}5 \end{pmatrix}$ (b) $\begin{pmatrix} -0{\cdot}707 & -0{\cdot}707 \\ 0{\cdot}707 & -0{\cdot}707 \end{pmatrix}$

(c) $\begin{pmatrix} -0{\cdot}866 & -0{\cdot}5 \\ 0{\cdot}5 & -0{\cdot}866 \end{pmatrix}$ (d) $\begin{pmatrix} 0{\cdot}5 & 0{\cdot}866 \\ -0{\cdot}866 & 0{\cdot}5 \end{pmatrix}$

4. (0, 0), (−1, 2), (−0·4, 2·8), 53° **5.** (0, 0), (2·4, 3·2), (5, 0); −37° (or 323°)

6. $\mathbf{T} = \begin{pmatrix} 0{\cdot}866 & -0{\cdot}5 \\ 0{\cdot}5 & 0{\cdot}866 \end{pmatrix}$, $\mathbf{S} = \begin{pmatrix} 0{\cdot}5 & -0{\cdot}866 \\ 0{\cdot}866 & 0{\cdot}5 \end{pmatrix}$

(a) $\begin{pmatrix} 0{\cdot}5 & -0{\cdot}866 \\ 0{\cdot}866 & 0{\cdot}5 \end{pmatrix}$ (b) $\begin{pmatrix} 0 & -1 \\ 1 & 0 \end{pmatrix}$ (c) $\begin{pmatrix} -0{\cdot}5 & -0{\cdot}866 \\ 0{\cdot}866 & -0{\cdot}5 \end{pmatrix}$; 120°

$\mathbf{T}^3$ represents a rotation of 90°; $\mathbf{S}^3$ represents a rotation of 180°.

7. $\begin{pmatrix} \cos\theta\cos\phi - \sin\theta\sin\phi & -\cos\theta\sin\phi - \sin\theta\cos\phi \\ \sin\theta\cos\phi + \cos\theta\sin\phi & \cos\theta\cos\phi - \sin\theta\sin\phi \end{pmatrix}$

(a) $\cos\theta\cos\phi - \sin\theta\sin\phi$ (b) $\sin\theta\cos\phi + \cos\theta\sin\phi$

Exercise 8.8 page 154

3. $\begin{pmatrix} 1 & 0 \\ 2 & 1 \end{pmatrix}$, $\begin{pmatrix} 1 & 0 \\ -3 & 1 \end{pmatrix}$, $\begin{pmatrix} 1 & 0 \\ 1 & 1 \end{pmatrix}$ **4.** (a) y-axis (b) $\begin{pmatrix} 1 & 0 \\ \frac{3}{2} & 1 \end{pmatrix}$

5. (0, 0), (1, 0), ($1\frac{1}{2}$, 1), ($\frac{1}{2}$, 1), $\begin{pmatrix} 1 & -\frac{1}{2} \\ 0 & 1 \end{pmatrix}$ **6.** inverse matrices are $\begin{pmatrix} 1 & -3 \\ 0 & 1 \end{pmatrix}$, $\begin{pmatrix} 1 & 0 \\ 2 & 1 \end{pmatrix}$

7. (a) (1, −3), (1, −1), (2, −4) (b) (1, −2), (1, 0), (2, −2)
(c) (1, 0), (5, 2), (6, 2) (d) (1, −3), (5, −13), (6, −16)

(b) yes $\begin{pmatrix} 1 & 0 \\ -2 & 1 \end{pmatrix}$, (d) no

9. $y = -x$ **10.** $y = -x$ **11.** $y = -x$ **12.** (0, 0), (2·5, 5); $y = 2x$

Exercise 8.9a page 156

2. (a) (−2, 0) (b) 90° (c) $\begin{pmatrix} -2 \\ 2 \end{pmatrix}$ (d) $\begin{pmatrix} 0 & -1 \\ 1 & 0 \end{pmatrix}\begin{pmatrix} x \\ y \end{pmatrix} + \begin{pmatrix} -2 \\ 2 \end{pmatrix}$

3. (a) (0, 3) (b) (2, 4), (4, 4), (4, 2), (2, 2)

(c) $\begin{pmatrix} 0 \\ -3 \end{pmatrix}$ (d) $\mathbf{M} = \begin{pmatrix} 2 & 0 \\ 0 & 2 \end{pmatrix}$, $\mathbf{V} = \begin{pmatrix} 0 \\ -3 \end{pmatrix}$

4. (a) $\mathbf{M} = \begin{pmatrix} 0 & 0 \\ 0 & 0 \end{pmatrix}$, $\mathbf{v} = \begin{pmatrix} 2 \\ -1 \end{pmatrix}$ (b) $\mathbf{M} = \begin{pmatrix} -1 & 0 \\ 0 & -1 \end{pmatrix}$, $\mathbf{v} = \begin{pmatrix} 0 \\ 0 \end{pmatrix}$

(c) $\mathbf{M} = \begin{pmatrix} 0 & 1 \\ 1 & 0 \end{pmatrix}$, $\mathbf{v} = \begin{pmatrix} -1 \\ 1 \end{pmatrix}$ (d) $\mathbf{M} = \begin{pmatrix} -2 & 0 \\ 0 & -2 \end{pmatrix}$, $\mathbf{v} = \begin{pmatrix} 12 \\ 9 \end{pmatrix}$

Exercise 8.9b page 157

1. (a) 9 (b) 1 (c) 1 (d) 3
2. (a) 16 (b) 4 (c) 5 (d) 0
3. area unchanged, shape inverted
5. $\begin{pmatrix} 7 & 3 \\ 2 & 5 \end{pmatrix}$, area = 29
6. (0, 0), (5, 2), (4, 6), (−1, 4), area = 22
7. $\begin{pmatrix} 0 & -1 \\ 1 & 2 \end{pmatrix}$
8. $a = 4, a = 5$

Part 9

Exercise 9.1 page 159

1. (a) 1500 (b) 450 (c) $33\frac{1}{3}\%$
2. width 5 cm, height 1·86 cm
3. (a) 130 000 000 km^2
   (b) Africa 33°, Antarctic 39°, Asia 122°, Australasia 22°, Europe 28°, N. America 66°, S. America 50°.

Exercise 9.3 page 163

1. mode 4, median 4
2. mode 100·2 mm, median 100·2 mm
3. Arithmetic: mode 14, 15, 16, 17 (equal values); median 14.
   English: mode 14; median 14
4. mode 3; median 3
5. 3·1
6. 61·5
7. (a) 2 children (b) 3 children
8. mode 121 cm; median 122 cm
9. 166 cm
10. (a) 40–45 mm (b) 42·2 mm

Exercise 9.4 page 165

1. (a) 836 (b) 106 158·6 (c) 8·612
2. (a) 18 040 (b) 0·00 36 (c) $1{\cdot}035 \times 10^{-2}$
3. 7·61
4. 3·1
5. 1·52
6. (a) 12 years 5·7 months (b) 12 years 8·7 months
7. 104 mm
8. 104 mm
9. 51·6
10. 51·6

Exercise 9.5 page 167

1. 6007, 500, 380
2. 32·5 mm
3. percentiles: 9 days, 17 days, 27 days; semi-interquartile range: 9 days
4. 3, 2, 4
5. 161 cm, 171 cm, 17%
6. 33 mm, 49 mm
7. 12·5 years
8. semi-interquartile range ≈ 9, $k \approx 7{\cdot}5$

Exercise 9.6 page 169

1. (a) $\frac{1}{4}$ (b) $\frac{3}{4}$ (c) $\frac{1}{52}$
2. (a) $\frac{1}{52}$ (b) $\frac{1}{13}$ (c) $\frac{12}{13}$
3. (a) $\frac{3}{51}$ (b) $\frac{12}{51}$ (c) $\frac{13}{51}$
4. (a) $\frac{48}{51}$ (b) $\frac{39}{51}$ (c) $\frac{13}{51}$
5. (a) $\frac{1}{4}$ (b) 0 (c) 0 (d) 1
6. (a) 1 (b) $\frac{1}{3}$ (c) $\frac{1}{13}$ (d) $\frac{1}{39}$
7. (a) $\frac{5}{26}$ (b) $\frac{21}{26}$ (c) $\frac{1}{26}$; $\frac{1}{5}$
8. (a) $\frac{3}{8}$ (b) $\frac{3}{7}$ (c) $\frac{1}{2}$
9. (a) $\frac{1}{9}$ (b) $\frac{1}{9}$ (c) $\frac{5}{18}$
10. (a) $\frac{21}{36}$ (b) $\frac{1}{4}$

Exercise 9.7 page 170

1. (a) yes (b) yes (c) no
2. (a) yes (b) yes (c) yes
3. $\frac{1}{78}$
4. $\frac{1}{16}$
5. (a) $\frac{1}{16}$ (b) $\frac{1}{16}$ (c) $\frac{1}{169}$ (d) $\frac{1}{52}$ (e) $\frac{1}{52}$ (f) $\frac{3}{169}$
6. (a) $\frac{1}{36}$ (b) $\frac{1}{12}$ (c) $\frac{1}{4}$ (d) $\frac{5}{18}$ (e) $\frac{5}{18}$
7. (a) $\frac{9}{25}$ (b) $\frac{12}{25}$ (c) $\frac{6}{25}$ (d) $\frac{4}{25}$
8. (a) $\frac{9}{25}$ (b) $\frac{4}{25}$
9. (a) $\frac{1}{156}$ (b) $\frac{1}{16}$ (c) $\frac{1}{624}$ (d) $\frac{3}{52}$
10. (a) $\frac{1}{2}$ (c) $\frac{1}{3}$ (d) $\frac{1}{6}$
11. $\frac{3}{250}$, $\frac{36}{125}$
12. $\frac{1}{63}$, $\frac{55}{252}$

Exercise 9.8 page 173

1. (a) $\frac{3}{8}$ (b) $\frac{1}{4}$ (c) $\frac{5}{16}$
2. $\frac{1}{3}$
3. 8 out of 20 results, $p = \frac{2}{5}$
4. $\frac{1}{6}$
6. $\frac{1}{26}$
7. (a) $\frac{1}{4}$ (b) $\frac{21}{50}$ (c) $\frac{1}{10}$
8. (a) $\frac{3}{13}$ (b) $\frac{9}{169}$ (c) $\frac{100}{169}$
9. $\frac{2}{9}$
10. $\frac{4}{15}$

Exercise 9.9 page 175

1. (a) $\frac{1}{5}$ (b) $\frac{11}{15}$
2. (a) $\frac{1}{12}$ (b) $\frac{7}{12}$
3. (a) $\frac{1}{169}$ (b) $\frac{25}{169}$
4. $\frac{7}{16}$
5. (a) $\frac{1}{4}$ (b) $\frac{6}{13}$ (c) $\frac{6}{52}$ (d) $\frac{31}{52}$
6. (a) $\frac{1}{13}$ (b) $\frac{1}{52}$ (c) $\frac{4}{13}$
7. (a) $\frac{1}{36}$ (b) $\frac{11}{36}$
8. (a) $\frac{1}{9}$ (b) $\frac{5}{9}$
9. (a) $\frac{15}{28}$ (b) $\frac{3}{7}$ (c) $\frac{1}{14}$
10. $\frac{11}{20}$

Part 10

Exercise 10.1a page 176

1. (a) 5 (b) 1 (c) 3 (d) 3 (e) 5 (f) 3
2. (a) 3 (b) 5 (c) 5 (d) 3 (e) 4
3. (a) 4 (b) 1 (c) 2 (d) 2 (e) 4 (f) 3 (g) 3 (h) 4
4. (a) 1 (b) 3 (c) 2 (d) 0
5. (a) 2 (b) 4 (c) 4 (d) 3 (e) 0
6. (a) 6 (b) 1 (c) 3 (d) 2 (e) 4 (f) 5
7. (a) 6 (b) 1 (c) 3 (d) 1 (e) 3 (f) 3
8. (a) 3 (b) 6 (c) 3 (d) 3 (e) 5

9. (5 + 6) in Mod 7, (5 + 8) in Mod 9 and (10 + 6) in Mod 12
10. (5 × 2) in Mod 7 and (3 × 7) in Mod 9
11. (a) 0 (b) 0 (c) 7 (d) 1 **12.** (a) 3 (b) 3 (c) 0 (d) 2
13. (a) 4 (b) 3 (c) 1 (d) 2, 5, or 8

Exercise 10.1b page 177

1. (a) 4 (b) 4 (c) 4 (d) 3 (e) 2 or 5
2. (a) 2 (b) 4 (c) 4 (d) 1 (e) 2 or 3
3. (a) 1 (b) 5 (c) 0, 2, 3, and 4
4. (a) 1 (b) 1 (c) 2 (d) 3 (e) 2 (f) 1
5. (a) 1 (b) 1, 7, 3, 9 (c) 9
6. (a) 6 (b) 8, 4, 6, 2 (c) 4
7. (a) 4 (b) 5 (c) 3 or 7 (d) 0, 2, 4, or 6 (e) 2 or 6 (f) 1, 3, 5, or 7
8. (a) 2, 3 ; 3, 4 (b) 2, 4; 4, 6; 4, 4 (c) 2, 5; 4, 5; 6, 5; 8, 5
(d) 2, 6; 4, 6; 6, 6; 8,6; 10,6; 4, 3; 8, 3; 9, 4; 9, 8
(e) 2, 12; 4, 12; 6, 12; 8, 12; 10, 12; 12, 12; 14, 12; 16, 12; 18, 12; 20, 12; 22, 12; 3, 8; 6, 8; 9, 8; 12, 8; 15, 8; 18, 8; 21, 8; 4, 6; 16, 6; 20, 6; 16, 9; 16, 3; 15, 16; 16, 18; 16, 21; 18, 4; 18, 20; 20, 6
(f) 2, 30; 3, 20; 3, 40; 4, 15; 4, 30; 4, 45; 5, 12; 5, 24; 5, 36; 5, 48; 6, 10; 6, 20; 6, 30; 6, 40; 6, 50; 8, 15; 8, 30; 8, 45; 9, 20; 9, 40; 10, 12; 10, 18; 10, 24; 10, 30; 10, 36: 10, 42; 10, 48; 10, 54; 12, 15; 12, 20; 12, 25; 12, 30; 12, 35; 12, 40; 12, 45; 12, 50; 14, 30; 15, 16; 15, 20; 15, 24; 15, 28; 15, 32; 15, 36; 15, 40; 15, 44; 15, 48; 15, 52; 15, 56; 16, 30; 16, 45; 18, 20; 18, 30; 18, 40; 18, 50; 20, 21; 20, 24; 20, 27; 20, 30; 20, 33; 20, 36; 20, 39; 20, 42; 20, 45; 20, 48; 20, 51; 20, 54; 20, 57; 21, 40; 22, 30, 24, 25; 24, 30; 24, 35; 24, 40; 24, 45; 24, 50; 24, 55; 25, 36; 25, 48; 26, 30; 27, 40; 28, 30; 28, 45; 30, 30; 30, 32; 30, 34; 30, 36; 30, 38; 30, 40; 30, 42; 30, 44; 30, 46; 30, 48; 30, 50; 30, 52; 30, 54; 30, 56; 30, 58; 32, 45; 33, 40; 35, 36; 35, 48; 36, 40; 36, 45; 36, 50; 36, 55; 39, 40; 40, 42; 40, 45; 40, 48; 40, 51; 40, 54; 40, 57; 42, 50; 44, 45; 45, 48; 45, 52; 45, 56; 48, 50; 48, 55; 50, 54
9. 21, 33, 45, 57, 69, 81, 93
10. 12, 32, 52, 72, 92
11. 6
12. (a) 6 (b) 7 (c) No solutions

Exercise 10.2 page 178

1. (a) the first five odd numbers (b) the first five square numbers
(c) factors of 16 or powers of 2 (d) the first four multiples of 10
(e) the first five letters of the alphabet (f) the five vowels
(g) the first five prime numbers (h) the first five consonants
2. (a) the first five multiples of 3 (b) the whole numbers from 13 to 16 inclusive
(c) the last five letters of the alphabet (d) the capital letters with two lines of symmetry
(e) the letters of the alphabet whose positions are the first five odd numbers
(f) the odd numbers from 21 to 29 inclusive
(g) the first five cubes (h) the multiples of 4 from 80 to 96 inclusive
3. (a) 7, 14, 21, 28 (b) Clubs, Diamonds, Hearts, Spades (c) 1, 2, 3, 4, 6, 12
(d) France, Belgium, West Germany, Denmark, Holland, Italy, UK, Luxemburg, Irish Republic

4. (a) 4, 8, 12, 16, 20 (b) n, o, p, q (c) 2, 3, 5, 7, 11, 13
(d) January, March, May, July, August, October, December
5. (a) $\in$ (b) $\subset$ (c) $\notin$ (d) $\supset$ (e) $\cap$
6. (a) $\subset$ (b) $\notin$ (c) $\cup$ (d) $\not\subset$
7. (a) True (b) False (c) True (d) False
8. (a) False (b) True (c) False (d) False
9. (a) {c, o} (b) {1, 4, 16, 64, 256} (c) $\{1, 3, 6, 10\} \cap \{4, 6, 8, 10\} = \{6, 10\}$
(d) $\{j, l, n, p\} \cup \{a, p\} = \{a, j, l, n, p\}$
10. (a) ϕ (b) {1, 3, 5, 7, 9} (c) $\{2, 6, 8, 9\} \cap \{8, 9, 10\} = \{8, 9\}$
(d) $\{a, m, r, z\} \cup \{a, *, y\} = \{a, m, r, y, z\}$ * could be m, r, or z
11. (a) {10, 20} (b) {20, 30} (c) {5, 20} (d) {5, 10, 15, 20, 25, 30}
(e) {1, 2, 3, 5, 10, 20, 30} (f) {1, 2, 3, 5, 10, 15, 20, 25, 30}
12. (a) {12} (b) {4, 12} (c) {3, 6, 12} (d) {3, 4, 6, 8, 9, 12, 15, 16, 20}
(e) {1, 2, 3, 4, 6, 8, 12, 16, 20} (f) {1, 2, 3, 4, 6, 9, 12, 15}
13. A = {the first five multiples of 5}; B = {the first three multiples of 10}
C = {the factors of 60 excluding 4, 6, 10, 12, 15 and 60};
P = {the first five multiples of 3}; Q = {the first five multiples of 4};
R = {the factors of 12}
14. (a) {20} (b) {1, 2, 3, 5, 10, 15, 20, 25, 30} (c) {1, 2, 3, 5, 10, 20, 30}
15. (a) {12} (b) {1, 2, 3, 4, 6, 8, 9, 12, 15, 16, 20} (c) {3, 4, 6, 12}
16. (a) {1, 3, 5, 7, 9, 10} (b) {2, 4, 5, 7, 8, 9} (c) {6}
(d) {1, 2, 3, 4, 5, 7, 8, 9, 10} (e) {1, 2, 3, 4, 6, 8, 10} (f) {5, 7, 9}
17. (a) {4, 16, 36, 64} (b) {9, 25, 36, 49, 81} (c) {1}
(d) {4, 9, 16, 25, 36, 49, 64, 81} (e) {1, 4, 9, 16, 25, 49, 64, 81}
(f) {36}
18. (a) 4 (b) 7 (c) 8 (d) 0
19. (a) 4 (b) 1 (c) 2 (d) 8 (e) 7
20. (a) 6 (b) 9 (c) 4 (d) 2 (e) 2 (f) 0 (g) 13 (h) 11
(i) 10 (j) 0
21. (a) 13 (b) 8 (c) 7 (d) 12
22. (a) 7 (b) 3 (c) 6 (d) 16
23. (a) False (b) True (c) True, but strictly the odd numbers from 1 to 20
(d) True (e) False (f) False
24. (a) {1, 9, 25} (b) 3 (c) {1, 3, 5, 9, 15, 25}
(d) {7, 11, 13, 17, 19, 21, 23, 27, 29} (e) 6
(f) {1, 3, 7, 9, 11, 13, 17, 19, 21, 23, 27, 29} (g) 12
(h) {1, 5, 9, 15, 25} (i) 3

Exercise 10.3a page 180

1. (a) {1} (b) {1, 2, 3, 4, 6, 8, 10} (c) {2, 4, 5, 7, 8, 9}
(d) {3, 5, 6, 7, 9, 10} (e) {2, 3, 4, 5, 6, 7, 8, 9, 10} (f) {5, 7, 9}
2. (a) ϕ (b) $\mathcal{E}$ (c) Q (d) P (e) $\mathcal{E}$ (f) ϕ
3. (a) {15, 21} (b) {2, 6, 7, 8, 11, 15, 21} (c) {4, 6, 8, 9, 16}
(d) {2, 4, 7, 9, 11, 16} (e) {2, 7, 11} (f) {6, 8}
(g) {2, 4, 6, 7, 8, 9, 11, 16} (h) {4, 9, 16} (i) {4, 9, 16}
4. (a) {b, e} (b) {a, b, c, d, e, f, g, i} (c) {d, g, h, i, j, k}
(d) {a, c, f, h, j, k} (e) {d, g, i} (f) {a, b, c, e, f, h, j, k}
(g) {a, c, f, d, g, i, h, j, k} (h) {h, j, k} (i) {a, c, f, d, g, i, h, j, k}
5. (a) R' (b) $X \cap Y$ (c) $M \cap L'$ (d) $(P \cup Q)'$

6. (a) { 2, 6, 10, 14, 18} (b) { 4, 6, 10, 12, 16, 18} (c) { 6, 10, 18}
(d) { 2, 4, 8, 12, 14, 16, 20}; { 2, 4, 8, 12, 14, 16, 20}$'$ = {6, 10, 18}

7. (a) { b, d, f, i, k, l} (b) { a, c, e, g, i, k} (c) { a, b, c, d, e, f, g, i, k, l}
(d) { j, h}; { j, h}$'$ = {a, b, c, d, e, f, g, i, k, l}

9. (a) ϕ (b) & (c) A (d) & (e) ϕ (f) ϕ (g) A (h) ϕ
(i) & (j) &

10. (a) A (b) P (c) $X \cup Y$ (d) $R \cap S$

11. (a) $A \cap B$ (b) $P' \cap Q$ (c) & (d) $R \cup S$

12. (a) False (b) True (c) True (d) False

Exercise 10.3b page 181

1. (a) { a, i, j} (b) { f, g, i, j} (c) { a, b, c, e, f, g, i, j}
(d) { i, j} (e) { a } (f) ϕ

2. (a) { 6, 12} (b) { 1, 3, 6, 9, 10, 12} (c) { 6 }
(d) { 1, 2, 3, 4, 6, 8, 9, 10, 12} (e) { 12} (f) { 9 }

3. (a) { 3, 4, 5} (b) { 3, 4, 5, 6, 7, 8, 9, 10} (c) { 6, 7, 8, 9, 11, 12}
(d) { 6, 11, 12} (e) { 1, 2, 3, 6, 9, 10, 11, 12} (f) { 4, 5}
(g) { 1, 2, 3, 4, 5, 6, 7, 8, 9, 10} (h) { 3 } (i) { 1, 2}
(j) { 1, 2, 3, 4, 5, 6, 9, 10, 11, 12} (k) { 11, 12} (l) { 4, 5, 6, 7, 8, 9, 11, 12}

4. (a) True (b) True (c) False (d) True

Exercise 10.4a page 182

1. 3 **2.** (a) 8 (b) 5 2

4. 2 take English and Goegraphy only, 1 takes English and History only.

5. 7

6. 5 people require all three currencies; 119 people will require at least one currency.

Exercise 10.4b page 183

1. 3 **2.** 8 **3.** 3; 8 **4.** 8 **5.** 7

Exercise 10.5a page 184

1. (a) { 0, 1, 2, 3, 4} (b) { 4, 5, 6, 7, 8, 9, 10 } (c) { 0, 1, 2, 3} (d) { 7, 8, 9, 10}

2. (a) { 0, 1, 2, 3, 4, 5, 6, 7, 8, 9, 10} (b) { 15, 16, 17, 18, 19, 20}
(c) { 0, 1, 2, 3, 4, 5, 6} (d) { 8, 9, 10, 11, 12, 13, 14, 15, 16, 17, 18, 19, 20}

3. (a) $\{x : x > 6\}$ (b) $\{x : x < 22\}$ (c) $\{x : x > 7\}$ (d) $\{x : x < 8\}$
(e) $\{x : x < 5\}$

4. (a) $\{x : x < 5\}$ (b) $\{x : x > 4\}$ (c) $\{x : x > 4\}$ (d) $\{x : x > 7\}$
(e) $\{x : x < 9\}$

5. (a) $\{x : 5 \leqslant x < 9\}$ (b) $\{x : 2 < x < 12\}$ (c) $\{x : 6 \leqslant x \leqslant 14\}$ (d) $\{x : 2 \leqslant x \leqslant 19\}$

6. (a) $A = \{x : 3 \leqslant x \leqslant 6\}$, $B = \{x : 1 < x \leqslant 4\}$, $C = \{x : 4 < x < 9\}$
and $D = \{x : 2 \leqslant x < 6\}$
(b) $A = \{x : -1 < x < 3\}$, $B = \{x : -2 \leqslant x \leqslant 5\}$, $C = \{x : -3 < x \leqslant 2\}$
and $D = \{x : 1 \leqslant x < 7\}$

7. (a) $\{x : 3 \leqslant x \leqslant 4\}$; $\{x : -1 < x < 3\}$ (b) ϕ; $\{x : -2 \leqslant x \leqslant 2\}$
(c) $\{x : 4 < x < 6\}$; $\{x : 1 \leqslant x \leqslant 2\}$ (d) $\{x : 1 < x \leqslant 6\}$; $\{x : -2 \leqslant x \leqslant 5\}$
(e) $\{x : 1 < x < 9\}$; $\{x : -3 < x \leqslant 5\}$ (f) $\{x : 2 \leqslant x < 9\}$; $\{x : -3 < x < 7\}$
(g) $\{x : 6 \leqslant x < 9\}$; $\{x : -3 < x < 1\}$
8. (a) $\{x : 6 < x < 12\}$ (b) $\{x : x < 10\}$ (c) $\{x : 6 < x < 13\}$
(d) $\{x : x \leqslant 9\}$ (e) ϕ (f) $\mathscr{E}$ (g) $\{x : x \geqslant 3\}$
(h) $\{x : 0 < x < 5\}$

Exercise 10.5b page 185

1. $\{x : x < 2\} \cup \{x : x > 3\}$
2. $\{x : 1 < x < 4\}$
3. $\{x : -2 < x < 3\}$
4. $\{x : x < -4\} \cup \{x : x > -3\}$
5. (a) $x > 2$ (b) $x > 5$ (c) $x > 5$ (d) $x < 2$ (e) $x < 5$
(f) $x < 2$; $\{x : x < 2\} \cup \{x : x > 5\}$
6. (a) $x > 3$ (b) $x < 4$ (c) $3 < x < 4$ (d) $x < 3$ (e) $x > 4$
(f) ϕ; $\{x : 3 < x < 4\}$
7. $\{x : x < -3\} \cup \{x : x > 4\}$
8. $\{x : -5 < x < -4\}$
9. $x > 5$; $x < 3$; $\{x : x < 3\} \cup \{x : x > 5\}$
10. $-4 < x < 1$; no values; $\{x : -4 < x < 1\}$
11. $\{x : x < -2\} \cup \{x : x > 4\}$
12. $\{x : -3 < x < 3\}$
13. (a) $\{x : x < 1\} \cup \{x : x > 6\}$ (b) $\{x : -2 < x < 3\}$
(c) $\{x : -5 < x < 4\}$ (d) $\{x : -2 < x < -1\}$
(e) $\{x : x < 2\} \cup \{x : x > 3\}$ (f) $\{x : x < \frac{1}{2}\} \cup \{x : x > 4\}$
(g) $\{x : -3 < x < \frac{1}{4}\}$ (h) $\{x : -\frac{1}{2} < x < \frac{1}{5}\}$

Exercise 10.6a

1. (a) $4 \to 8, 7 \to 11, 6 \to 10$; *is four less than*
(b) $15 \to 45, 7 \to 21, 21 \to 63$; *is a third of*
(c) $20 \to 30, 30 \to 45$; *is two-thirds of*
2. (a) *is a factor of* (b) *is three more than* (c) *is a half of*
3. (a) $\{(4, 1), (5, 2), (6, 3), (7, 4), (8, 5)\}$
(b) $\{(1, 4), (3, 12), (4, 16), (12, 48), (16, 64)\}$
(c) $\{(1, 1), (2, 4), (3, 9), (4, 16), (5, 25)\}$
(d) $\{(\frac{1}{2}, \frac{2}{4}), (\frac{2}{4}, \frac{1}{2}), (\frac{1}{3}, \frac{2}{6}), (\frac{2}{6}, \frac{1}{3}), (\frac{1}{4}, \frac{2}{8}), (\frac{2}{8}, \frac{1}{4}), (\frac{1}{2}, \frac{3}{6}), (\frac{3}{6}, \frac{1}{2}), (\frac{1}{3}, \frac{3}{9}), (\frac{3}{9}, \frac{1}{3}), (\frac{2}{4}, \frac{3}{6}), (\frac{3}{6}, \frac{2}{4}), (\frac{2}{6}, \frac{3}{9}), (\frac{3}{9}, \frac{2}{6}),$
$(\frac{1}{2}, \frac{1}{2}), (\frac{1}{3}, \frac{1}{3}), (\frac{1}{4}, \frac{1}{4}), (\frac{2}{4}, \frac{2}{4}), (\frac{2}{6}, \frac{2}{6}), (\frac{3}{6}, \frac{3}{6}), (\frac{3}{9}, \frac{3}{9}), (\frac{2}{8}, \frac{2}{8}), (\frac{9}{12}, \frac{9}{12})\}$

Exercise 10.6b page 187

1. (a) $1 \twoheadrightarrow 5$; *is four less than* (b) $3 \twoheadrightarrow 9$; *is one-third of*
(c) $2 \twoheadrightarrow 11$; *is one-fifth of one less than*
2. (a) *has as its initial letter*; (M–1) (b) *is the number of days in*; (1–M)
(c) *is called a*; (1–1) (d) *is a factor of*; (M–M)
3. (a) (1–M); $\{(1, 1), (4, 2), (4, -2), (9, 3), (9, -3), (16, 4), (25, 5)\}$
(b) (M–M); $\{(8, 2), (8, 4), (10, 2), (10, 5), (12, 2), (12, 3), (12, 4), (12, 6), (16, 2) (16, 4), (18, 2), (18, 3), (18, 6)\}$
(c) (1–1); $\{(1, 10), (2, 100), (3, 1000), (4, 10000)\}$
(d) (M–1); {(Ann, A), (June, J), (Jack, J), (Jill, J), (Andrew, A)}

Exercise 10.6c page 187

1. 2(a), 2(c), 3(c) and 3(d)
2. (a), (c), (d), and (f) are functions; (b) is (1–M); (e) is (1–M)

Exercise 10.7a page 188

1. (a) (p, 1), (p, 2), (q, 1), (q, 2), (r, 1), (r, 2)
   (b) (x, 1), (x, 2), (x, 3), (y, 1), (y, 2), (y, 3)
   (c) (p, x), (p, y), (q, x), (q, y)
2. (a) (1, p), (2, p), (1, q), (2, q), (1, r), (2, r)
   (b) (1, x), (2, x), (3, x), (1, y), (2, y), (3, y)
   (c) (x, p), (y, p), (x, q), (y, q)
3. (a) (1, p), (2, p), (3, p), (4, p), (1, q), (2, q), (3, q), (4, q)
   (b) (p, 1), (p, 2), (q, 1), (q, 2), (r, 1), (r, 2), (s, 1), (s, 2)
   (c) (x, 1), (x, 2), (x, 3), (y, 1), (y, 2), (y, 3), (z, 1), (z, 2), (z, 3)
   (d) (a, 5), (b, 5), (c, 5), (d, 5)
4. (a) (p, 1), (p, 2), (p, 3), (p, 4), (q, 1), (q, 2), (q, 3), (q, 4)
   (b) (1, p), (2, p), (1, q), (2, q), (1, r), (2, r), (1, s), (2, s)
   (c) (1, x), (2, x), (3, x), (1, y), (2, y), (3, y), (1, z), (2, z), (3, z)
   (d) (5, a), (5, b), (5, c), (5, d)
5. (a) 8 (b) 9 (c) 4 (d) 3
6. (b) (1, 1), (1, 2), (1, 3), (2, 1), (2, 2), (2, 3), (3, 1), (3, 2), (3, 3)

Exercise 10.7b page 189

1. (a) (a, 1), (a, 2), (a, 3), (a, 4), (b, 1), (b, 2), (b, 3), (b, 4), (c, 1), (c, 2), (c, 3), (c, 4)
   (b) (b, 1), (b, 2), (c, 1), (c, 2)
   The ordered pairs in $A \times B$ also appear in $P \times Q$.
2. (a) $P \cap Q = \{q, r, s\}$ (b) $S \cap T = \{t, u\}$; $\{(q, t), (q, u), (r, t), (r, u), (s, t), (s, u)\}$
4. $A \times A = \{(1, 1), (1, 2), (1, 3), (1, 4), (1, 5), (2, 1), (2, 2), (2, 3), (2, 4), (2, 5), (3, 1), (3, 2), (3, 3), (3, 4), (3, 5), (4, 1), (4, 2), (4, 3), (4, 4), (4, 5), (5, 1), (5, 2), (5, 3), (5, 4), (5, 5)\}$
   The ordered pairs represent the 25 points where the grid lines cross. $B \times B$ represents nine of the 25 points of $A \times A$.
5. (a) $P \cup Q = \{a, b, c, d, e\}$ (b) $R \cup S = \{1, 2, 3, 4, 5\}$;
   $(P \cup Q) \times (R \cup S)$ is a set of 25 ordered pairs with one of a, b, c, d or e as first element and one of 1, 2, 3, 4 or 5 as second element.
6. $(P \cup Q) \times (R \cup S) \neq (P \times R) \cup (Q \times S)$; the first includes (1, d), the second does not.
7. (a) There are 36 possible results when throwing two different dice, given by $\{1, 2, 3, 4, 5, 6\} \times \{1, 2, 3, 4, 5, 6\}$
   (b) There are 9 possible mixed pairs chosen from 3 boys and 3 girls, given by $\{b_1, b_2, b_3\} \times \{g_1, g_2, g_3\}$
   (c) There are 16 possible matches using one player from each team, given by $\{t_1, t_2, t_3, t_4\} \times \{s_1, s_2, s_3, s_4\}$
8. (a) $A \times B = \{(x, y) : 1 \leqslant x \leqslant 6, 2 \leqslant y \leqslant 5\}$; i.e. a rectangle
   (b) $C \times D = \{(x, y) : 3 \leqslant x \leqslant 4, 1 \leqslant y \leqslant 6\}$; i.e. a rectangle
   (c) $(A \times B) \cap (C \times D) = \{(x, y) : 3 \leqslant x \leqslant 4, 2 \leqslant y \leqslant 5\}$; i.e. the overlapping rectangle

Exercise 10.8a page 190

1. (a) 6 (b) 15 (c) 30 **2.** (a) 1 (b) 13 (c) 29
3. (a) {2, 4, 6, 8} (b) {3, 4, 5, 6} (c) {2, 7, 12, 17} (d) {9, 8, 7, 6}
4. (a) {4, 12, 28, 32} (b) {8, 14, 26, 29} (c) {1, 17, 49, 57} (d) {14, 10, 2, 0}
5. (a) {−15, −5, 10, 20} (b) {−6, −4, −1, 1} (c) {−7, −1, 8, 14}
(d) {10, 8, 5, 3}
6. (a) $\{\frac{1}{2}, 1, \frac{3}{2}, 2\}$ (b) $\{2, 2\frac{1}{2}, 3, 3\frac{1}{2}\}$ (c) $\{1, \frac{1}{2}, \frac{1}{3}, \frac{1}{4}\}$ (d) {1, 4, 9, 16}
7. (a) 1. {1, 2, 3}; 2. {1, 2, 3, 4, 5, 6}; 3. {2, 4, 6};
(b) 1. {3, 5, 8}; 2. {4, 5, 6, 7, 10}; 3. {5, 7, 10};
(c) 1. {1, 2, 3, 4}; 2. {1, 4, 7, 9, 12, 16}; 3. {1, 4, 9, 16};
8. (a) $x \to 2x$ (b) $x \to x + 2$ (c) $x \to x^2$
10. (a) 1. {3, 4, 5, 6, 7}; 2. {4, 5, 6, 7};
(b) 1. {0, 1, 2, 3, 4}; 2. {1, 2, 3};
(c) 1. {0, 1, 2, 4}; 2. {0, 1, 4}.
12. (a) {4, 5, 6} (b) {−11, −2, 10} (c) {85, 64, 7}

Exercise 10.8b page 191

1. (a) {−3, −1, 0, 2, 5, 6} (b) {−3, 0, 5} (c) 0 (d) −3 (e) −3
(g) 0 (h) 5
2. (a) 1. 10 2. 45 3. −4 4. −25
(b) 1. 27 2. 47 3. 19 4. 7
(c) 1. −9 2. −34 3. 1 4. 16
(d) 1. −2 2. −17 3. 4 4. 13
3. (a) 1. 6 2. 9 3. 21 4. $5\frac{1}{4}$
(b) 1. 0 2. 2 3. 12 4. $-\frac{1}{4}$
(c) 1. 0 2. −7 3. −63 4. $\frac{7}{8}$
(d) 1. 2 2. $2\frac{1}{2}$ 3. $4\frac{1}{4}$ 4. $2\frac{1}{2}$
5. (a) {4, 7, 8, 10, 12}; {7, 8, 10, 12}; 7
(b) {14, 33, 77, 113, 117}; {14, 77, 113}; 14
(c) {26, 16, −4, −14}; {26, 16, −4, −14}; −4
7. (a) 39 (b) −23 (c) −4 (d) 20
8. (a) 1. 29 2. 125 3. −31 4. 5
(b) 1. 2 2. −38 3. 27 4. 12
(c) 1. 2 2. 90 3. 12 4. 0
9. (a) −2 (b) $-2\frac{1}{2}$ (c) −1 (d) 0
10. (a) 5 (b) 13 (c) 6
11. 7
12. (a) 5; 25 (b) 20; 23

Exercise 10.9a page 192

1. (a) $x \to 3(x + 2)$ (b) $x \to 5x - 3$ (c) $x \to 7(x - 4)$
2. (a) $x \to x^2 + 4$ (b) $x \to (x + 4)^2$ (c) $x \to \dfrac{1}{(x + 1)}$
3. (a) $x \to 5(x - 3)$ (b) $x \to 2x + 5$ (c) $x \to (x - 2)^2$ (d) $x \to \dfrac{1}{x + 3}$
4. (a) {−25, −20, −15, −10, −5} (b) {1, 3, 5, 7, 9}
(c) {16, 9, 4, 1, 0} (d) $\{1, \frac{1}{2}, \frac{1}{3}, \frac{1}{4}, \frac{1}{5}\}$

5. (a) $x \to 2(3x+4)+1 = 6x+9$ (b) $x \to 3(2x-7)+5 = 6x-16$
(c) $x \to 4(5x+1)-3 = 20x+1$ (d) $x \to (4x-1)^2+2 = 16x^2-8x+3$
6. (a) $\{9, 15, 21, 27, 33, 39\}$ (b) $\{-16, -10, -4, 2, 8, 14\}$
(c) $\{1, 21, 41, 61, 81, 101\}$ (d) $\{3, 11, 51, 123, 227, 363\}$
7. $x \to 2(3x+1)+5 = 6x+7$; (d)
8. $x \to 3(2x+5)+1 = 6x+16$; Yes, (c)
9. $x \to 3(5x-2)-1 = 15x-7$; (c)
10. $x \to 5(3x-1)-2 = 15x-7$; Yes, (c)
11. $x \to 3(4x+3)+2 = 12x+11$; $\{23, 35, 47, 59, 71,\}$
$x \to 4(3x+2)+3 = 12x+11$; $\{23, 35, 47, 59, 71,\}$; Yes.

Exercise 10.9b page 193

1. (a) 7 (b) 49 (c) 49 (d) 99 (e) 225 (f) 19;
$gf(x) = (2x+1)^2$; $fg(x) = 2x^2+1$
2. (a) 8 (b) 64 (c) 64 (d) 318 (e) 1444 (f) 18;
$gf(x) = (5x-2)^2$; $fg(x) = 5x^2-2$
3. $gf: x \to 2(3x-7)+5 = 6x-9$; $fg: x \to 3(2x+5)-7 = 6x+8$;
(a) 3 (b) 20 (c) -27 (d) -10
4. $gf: x \to 3(2x-7)+5 = 6x-16$; $fg: x \to 2(3x+5)-7 = 6x+3$;
(a) 8 (b) 9 (c) -16 (d) -9
5. (a) $(3x-1)^2$ (b) $3x^2-1$ (c) x^4 (d) $3(3x-1)-1 = 9x-4$
6. (a) $(4x+1)^2$ (b) $4x^2+1$ (c) x^4 (d) $4(4x+1)+1 = 16x+5$
7. (a) $x \to (2x-1)^3$ (b) $x \to 2x^3-1$ (c) $x \to x^6$ (d) $x \to 2(2x-1)-1 = 4x-3$
8. (a) $x \to (1-x)^3$ (b) $x \to 1-x^3$ (c) $x \to x^6$ (d) $x \to 1-(1-x) = x$
9. (a) gf (b) fg (c) gg (d) ff
10. (a) hk (b) kh (c) kkk (d) hkh
11. (a), (b), and (d)

Exercise 10.10a page 194

1. (a) $x \to 3x+2$, $x \to \frac{1}{3}(x-2)$ (b) $x \to 4x-3$, $x \to \frac{1}{4}(x+3)$ (c) $x \to 5(x+3)$, $x \to \frac{1}{5}x-3$
2. (a) $x \to 4(x-5)$, $x \to \frac{1}{4}x+5$ (b) $x \to \frac{1}{3}(x+2)$, $x \to 3x-2$ (c) $x \to \frac{1}{4}x-3$, $x \to 4(x+3)$
3. (a) $x \to \frac{1}{4}(x-7)$ (b) $x \to \frac{1}{3}(x+5)$ (c) $x \to \frac{1}{5}x-2$ (d) $x \to \frac{1}{7}x+5$
4. (a) $x \to \frac{1}{6}(x+1)$ (b) $x \to 2x-3$ (c) $x \to \sqrt{x-3}$ (d) $x \to \sqrt{x}-1$
5. (a) $x \to \frac{1}{3}(x+7)$ (b) $x \to \frac{1}{2}x-5$ (c) $x \to 3(x-\frac{2}{3})$ (d) $x \to 3x+4$
6. (a) $x \to \frac{1}{4}(x-1)$ (b) 9 (c) 2 (d) $\frac{1}{4}$
7. (a) $x \to \frac{1}{2}x-5$ (b) 12 (c) 1 (d) $-4\frac{1}{2}$
8. (a) $x \to 2x+3$ (b) 0 (c) 3 (d) 9
9. (a) $x \to \sqrt{x+4}$ (b) 21 (c) ± 5 (d) ± 3
10. (a) $x \to \frac{1}{2}[\frac{1}{3}(x-4)-1]$ (b) 25 (c) 3 (d) $-\frac{1}{2}$
11. 5 12. 3 13. (a) 7 (b) 7 (c) -1 (d) $3\frac{1}{2}$
14. (a) 7 (b) 11 (c) -1 (d) -3
15. (a) 4 (b) 7 (c) -2 16. (a) 3 (b) 11 (c) -4
17. (a) $\frac{1}{2}$ (b) -2 (c) 62
18. $f^{-1}: x \to \frac{1}{2}[\frac{1}{7}(2x-5)+3]$; $x = 4$; $f^{-1}(6) = 2$

Exercise 10.10b page 195

1. (a) $(x+2)^2-16$ (b) $(x+3)^2-16$ (c) $(x+4)^2-36$ (d) $(x-2)^2-9$
2. (a) $(x+2)^2-1$ (b) $(x-5)^2-4$ (c) $(x-6)^2-25$ (d) $(x+8)^2$
3. (a) $2, -6$ (b) $1, -7$ (c) $2, -10$ (d) $3, -7$ (e) $3, -9$ (f) $1, -9$
4. (a) $5, -1$ (b) $10, -6$ (c) $1, 3$
5. (a) $-1, -3$ (b) $3, 7$ (c) $1, 11$ (d) $1, -5$ (e) $1, 9$ (f) $14, -2$
6. (a) -8 (b) $-3, -13$ (c) $1, -17$

Exercise 10.10c page 195

1. (a) $x \to 3x+2$ (b) $x \to \frac{1}{3}x$ (c) $x \to x-2$ (d) $x \to \frac{1}{3}(x-2)$
   $fg: x \to 3(x+2)$; both are $x \to \frac{1}{3}x - 2$
2. (a) $x \to 4(x-3)$ (b) $x \to \frac{1}{4}x$ (c) $x \to x+3$ (d) $x \to \frac{1}{4}x+3$
   $gf: x \to 4x-3$; both are $x \to \frac{1}{4}(x+3)$
3. (a) $x \to 4(3x+2)$ (b) $x \to \frac{1}{3}(\frac{1}{4}x-2)$ (c) $x \to 12x+2$ (d) $x \to \frac{1}{12}(x-2)$
4. (a) $x \to 2x+4$ (b) $x \to \frac{1}{2}(x-4)$ (c) $x \to 2(x+5)-1$
   (d) $x \to \frac{1}{2}(x+1)-5$
5. (a) 13 (b) 15 (c) 3
6. (a) 19 (b) 6 (c) 4
7. (a) 3 (b) 4

Exercise 10.11a page 196

1. (a) 18 (b) 22 (c) 20 (d) 56 (e) 32; No; No.
2. (a) 72 (b) 72 (c) 36 (d) 432 (e) 432;
   $a \square b = 3ab = b \square a$; $(a \square b) \square c = 9abc = a \square (b \square c)$.
3. (a) 13 (b) 13 (c) 25 (d) 185 (e) 629; Yes; No.
   $p * q = p^2 + q^2 = q * p$; $(2 * 3) * 4 \neq 2 * (3 * 4)$.
4. (a) 36 (b) 36 (c) 144 (d) 20736 (e) 82944; Yes; No;
   $p \square q = p^2q^2 = q \square p$; $(2 \square 3) \square 4 \neq 2 \square (3 \square 4)$.
5. (a) $3 * 4 = 3, 4 * 3 = 4$ (b) $(x * y) * z = x, x * (y * z) = x$
6. (a) $5 \circ 12 = 13 = 12 \circ 5$; $a \circ b = \sqrt{a^2+b^2} = b \circ a$.
   (b) $(3 \circ 4) \circ 12 = 13 = 3 \circ (4 \circ 12)$;
   Yes, $(a \circ b) \circ c = \sqrt{(\sqrt{a^2+b^2})^2 + c^2} = \sqrt{a^2+b^2+c^2} = a \circ (b \circ c)$
7. (a) $5 \circ 12 = \dfrac{1}{5+12} = 12 \circ 5$; $a \circ b = \dfrac{1}{(a+b)} = b \circ a$
   (b) $(3 \circ 4) \circ 12 = \dfrac{1}{\dfrac{1}{3+4}+12}$, $3 \circ (4 \circ 12) = \dfrac{1}{3+\dfrac{1}{4+12}}$
   No; $(a \circ b) \circ c = \dfrac{1}{\dfrac{1}{a+b}+c}$, $a \circ (b \circ c) = \dfrac{1}{a+\dfrac{1}{b+c}}$; these are not equal.
8. (a) $2 * 3 = 11 = 3 * 2$; $p * q = pq + p + q = q * p$.
   (b) $(2 * 3) * 4 = 59 = 2 * (3 * 4)$
   Yes, $(p * q) * r = pqr + pr + qr + pq + p + q + r = p * (q * r)$

Exercise 10.11b page 197

1. (a) 2 (b) 48 (c) 24 (d) 12 (e) 12

   Yes, $(x * y) \circ z = 2\left(\dfrac{x-y}{2}\right)z = xz - yz = (x \circ z) * (y \circ z)$
2. (a) 36 (b) 14 (c) 10 (d) 72 (e) 72

   Yes, $p \circ (q * r) = \dfrac{p \times 3(q+r)}{2} = \dfrac{3}{2}(pq + pr) = (p \circ q) * (p \circ r)$
3. (a) No (b) No
4. (a) Yes (b) No
5. (a) 12 (b) 60 (c) 105 (d) 180 (e) 166. No, (d) and (e) are not equal.
   $(4 \square 7) * 5 = 180, (4 * 5) \square (7 * 5) = 166$. No, since these are not equal.
6. (a) Yes (b) Yes
7. (a) 5 (b) 5 (c) 0 (d) 0; Yes.
   (e) 2 (f) 2; Yes
   (g) 6 (h) 6; Yes.
   $5 * (6 \oplus 3) = 5 * 2 = 3;\ (5 * 6) \oplus (5 * 3) = 2 \oplus 1 = 3.$

Exercise 10.12a page 198

1. (a) $0{\cdot}333\dot{3}$ (b) $0{\cdot}166\dot{6}$ (c) $0{\cdot}111\dot{1}$ (d) $0{\cdot}\dot{1}4285\dot{7}$ (e) $0{\cdot}1818\dot{1}\dot{8}$
2. (a) $\frac{4}{9}$ (b) $\frac{4}{11}$ (c) $\frac{4}{7}$
3. (a) $\frac{1}{2}(\frac{2}{5} + \frac{3}{7})$ (b) $0{\cdot}3$ (c) $\frac{1}{2}(0{\cdot}197 + \frac{3}{16})$ or $0{\cdot}19$
4. (a) $\sqrt{2}$ (b) π (c) $\frac{1}{2}(\sqrt{2} + 2)$ (d) $\frac{1}{2}(\pi + 4)$ (e) $\frac{1}{2}(\sqrt{2} + \pi)$

Exercise 10.12b page 198

1. $7 - 4 = 3, 4 - 7 = {}^-3;\ 8 \div 4 = 2, 4 \div 8 = \frac{1}{2}$
2. (a) True (b) False (c) False (d) True
3. $(7-4) - 1 = 2, 7 - (4-1) = 4;\ (8 \div 4) \div 2 = 1, 8 \div (4 \div 2) = 4$
4. (a) True (b) False (c) False

Exercise 10.12c page 199

1. (a) True (b) True (c) False (d) True (e) False (f) False
2. (a) $3 \times (7-2) = 15 = (3 \times 7) - (3 \times 2)$; (b) $(8-3) \times 4 = 20 = (8 \times 4) - (3 \times 4)$
3. (a) 2 (b) $1{\cdot}4 + 0{\cdot}6 = 2$ (c) 6 (d) $9 - 3 = 6$;
   $(35 + 25) \div 5 = 12 = (35 \div 5) + (25 \div 5)$
4. $(15 + 6) \div 3 = 7 = (15 \div 3) + (6 \div 3)$;
   $4 \div (2 + 8) = 0{\cdot}4$ but $(4 \div 2) + (4 \div 8) = 2{\cdot}5$

Exercise 10.12d page 199

1. (a) ${}^-7$ (b) 5 (c) ${}^-\frac{1}{3}$ (d) ${}^-\frac{2}{5}$ (e) $\frac{1}{8}$ (f) $\frac{5}{7}$
2. (a) $\frac{1}{4}$ (b) 5 (c) $\frac{7}{2}$ (d) ${}^-\frac{1}{3}$ (e) ${}^-6$ (f) ${}^-\frac{8}{3}$
3. (a) commutative, associative, inverse, identity
   (b) distributive, associative, inverse, identity
   (c) commutative, associative, inverse, identity

Exercise 10.13a page 200

1. (a) 0 (b) 0, 0; 1, 3; 2, 2; 3, 1
2. (a) 1 (b) 1, 1; 2, 3; 3, 2; 4, 4
3. (a) $\mathbf{R}_{360}$ (b) $\mathbf{R}_{90}, \mathbf{R}_{270}$; $\mathbf{R}_{180}, \mathbf{R}_{180}$; $\mathbf{R}_{270}, \mathbf{R}_{90}$; $\mathbf{R}_{360}, \mathbf{R}_{360}$
4. (a) $\begin{pmatrix} 1 & 0 \\ 0 & 1 \end{pmatrix}$

   (b) $\begin{pmatrix} 1 & 0 \\ 0 & 1 \end{pmatrix}, \begin{pmatrix} 1 & 0 \\ 0 & 1 \end{pmatrix}$; $\begin{pmatrix} -1 & 0 \\ 0 & -1 \end{pmatrix}, \begin{pmatrix} -1 & 0 \\ 0 & -1 \end{pmatrix}$; $\begin{pmatrix} 1 & 0 \\ 0 & -1 \end{pmatrix}, \begin{pmatrix} 1 & 0 \\ 0 & -1 \end{pmatrix}$; $\begin{pmatrix} -1 & 0 \\ 0 & 1 \end{pmatrix}, \begin{pmatrix} -1 & 0 \\ 0 & 1 \end{pmatrix}$
5. Only (b) and (e) represent groups.
   In (a) E appears twice in the bottom row and right-hand column.
   In (c) 3 is not a member of the original set.
   In (d) 0 is not a member of the original set.
6. (a) 1; b (b) 1, 1 and $-1, -1$; a, c and b, b and c, a

Exercise 10.13b page 201

1. (a) r (b) p, p : q, s : r, r : s, q : t, u : u, t (c) u (d) s
2. (a) q (b) p (c) u (d) all values
3. (a), (b) and (c) represent groups
   In (a) a is the identity. The inverses of a, b, c, and d are a, b, c, and d respectively.
   In (b) d is the identity. The inverses of a, b, c, and d are c, b, a, and d respectively.
   In (c) d is the identity. The inverses of a, b, c, and d are a, c, b, and d respectively.
   In (d) there is no identity.
4. (a) d; a; d; b (b) d; b; a; b (c) d; b; a; c (d) a; b; a; d
5. (a) d; a; a; a (b) b; a; b; b (c) a, b, c or d; none; b or c; b
   (d) a; c; d; d or c depending on which pair in b $*$ x $*$ c is considered first.
6. $\begin{pmatrix} 3 & -5 \\ -1 & 2 \end{pmatrix}, \begin{pmatrix} \frac{7}{3} & -\frac{8}{3} \\ -\frac{4}{3} & \frac{5}{3} \end{pmatrix}$ (a) $\begin{pmatrix} -2 & -2 \\ 1 & 1 \end{pmatrix}$ (b) $\begin{pmatrix} -\frac{8}{3} & \frac{7}{3} \\ \frac{5}{3} & -\frac{4}{3} \end{pmatrix}$
7. $\begin{pmatrix} 2 & -5 \\ -3 & 8 \end{pmatrix}, \begin{pmatrix} 1 & -2 \\ -\frac{1}{2} & \frac{3}{2} \end{pmatrix}$ (a) $\begin{pmatrix} 1 & -2 \\ -1 & 4 \end{pmatrix}$ (b) $\begin{pmatrix} \frac{3}{2} & -\frac{1}{2} \\ -2 & 1 \end{pmatrix}$

Part 11

Exercise 11.1a page 202

1. $a \to 7a, b \to 7b$, so $\dfrac{7b - 7a}{b - a} = 7$
2. (a) 1 (b) 5 (c) 7 (d) 9 (e) 4
3. (a) 3 (b) 9 (c) 15 (d) 21 (e) 12
   Where the intervals are the same, each result is three times the corresponding result in question 2.
4. $3(a + b)$; $6a$
5. (a) 1 (b) 7 (c) 19 (d) 9 (e) $a^2 + ab + b^2 = \dfrac{(b^3 - a^3)}{(b - a)}$
6. (a) 2 (b) 14 (c) 38 (d) 18 (e) $2(a^2 + ab + b^2)$
   Each result is twice the corresponding result in question 5.

7. (a) $-\frac{1}{2}$ (b) $-\frac{1}{6}$ (c) $-\frac{1}{12}$ (d) $-\frac{1}{4}$ (e) $-\frac{1}{14}$
8. (a) $-\frac{5}{2}$ (b) $-\frac{5}{6}$ (c) $-\frac{5}{12}$ (d) $-\frac{5}{4}$ (e) $-\frac{5}{14}$
   Each result is five times the corresponding result in question 7.
9. $-\dfrac{1}{ab}$; $-\dfrac{1}{a^2}$
10. (a) 5 (b) 7 (c) 9 (d) $(a+b)+4$; $2a+4$
11. (a) 3 (b) $(a+b)$ (c) $2(a+b)+1$ (d) $-\dfrac{(a+b)}{a^2b^2}$
12. (a) 3; $2a$; $4a+1$; $-\dfrac{2}{a^3}$ (b) 3; 20; 41; $-0{\cdot}002$

Exercise 11.1b page 203

1. 7; 7
2. 13; 13
3. $(a+b)$; $2a$; (a) $2a$ (b) 10
4. $-\dfrac{3}{(a \times b)}$; $-\dfrac{3}{a^2}$; (a) $-\dfrac{3}{a^2}$ (b) $-\dfrac{3}{25}$
5. $-\dfrac{1}{(p \times q)}$; $-\dfrac{1}{p^2}$
6. (a) $(p+q)-\dfrac{1}{pq}$; $2p-\dfrac{1}{p^2}$ (b) $1-\dfrac{p+q}{p^2q^2}$; $1-\dfrac{2}{p^3}$

Exercise 11.2a page 204

1. (a) $3x^2$ (b) $4x^3$ (c) $10x^9$ (d) $-7x^{-8}$ (e) $-2x^{-3}$
2. (a) $5x^4$ (b) $-4x^{-5}$ (c) $\frac{4}{3}x^{\frac{1}{3}}$ (d) $-\frac{3}{2}x^{-\frac{5}{2}}$ (e) $\frac{1}{2}x^{-\frac{1}{2}}$
3. (a) $7x^6$ (b) $-5x^{-6}$ (c) $-\frac{1}{3}y^{-\frac{4}{3}}$ (d) $\frac{2}{3}z^{-\frac{1}{3}}$
4. (a) x^9+c (b) $\dfrac{x^{10}}{10}+c$ (c) $x^{\frac{5}{2}}+c$ (d) $-\dfrac{x^{-9}}{9}+c$ (e) $-3x^{-\frac{1}{3}}+c$
5. (a) $2t+3t^2$ (b) $5t^4-10$ (c) $-3t^{-4}-4t^{-5}$ (d) $3t^{\frac{1}{2}}-\frac{2}{3}t^{-\frac{5}{3}}$
6. (a) $10r^4+24r^7$ (b) $-12r^{-5}-42r^{-7}$ (c) $-\frac{5}{2}r^{-\frac{7}{2}}-\frac{1}{4}r^{-\frac{5}{4}}$
7. (a) x^4+x^6 (b) $t^{-\frac{1}{3}}+t^{-\frac{3}{2}}$ (c) $\frac{1}{2}x^{-\frac{1}{2}}+\frac{1}{3}x^{-\frac{2}{3}}$
8. (a) 2π (b) $2\pi r$ (c) $4\pi r^2$ (d) $2\pi rh$ (e) $4\pi r+\pi h$
9. 6 10. 13 11. 23 12. $60x^2-60x^{-6}$

Exercise 11.2b page 205

1. (a) $5x^4-2x$; $20x^3-2$ (b) $-7x^{-8}+1$; $56x^{-9}$
   (c) $\frac{5}{2}x^{\frac{3}{2}}-\frac{1}{2}x^{-\frac{3}{2}}$; $\frac{15}{4}x^{\frac{1}{2}}+\frac{3}{4}x^{-\frac{5}{2}}$ (d) $\frac{4}{3}x^{\frac{1}{3}}+\frac{3}{4}x^{-\frac{7}{4}}$; $\frac{4}{9}x^{-\frac{2}{3}}-\frac{21}{16}x^{-\frac{11}{4}}$
2. (a) $12t^3+2t^{-2}$; $36t^2-4t^{-3}$ (b) $3t^2-14t$; $6t-14$
   (c) $2t-5$; 2 (d) $\frac{1}{2}t^{-\frac{1}{2}}+16$; $-\frac{1}{4}t^{-\frac{3}{2}}$
3. (a) $4x^3+3x^{-4}, 256+\frac{3}{256}$; $12x^2-12x^{-5}, 192-\frac{3}{256}$
   (b) $\frac{1}{2}x^{-\frac{1}{2}}-\frac{1}{2}x^{-\frac{3}{2}}, \frac{1}{4}-\frac{1}{16}$; $-\frac{1}{4}x^{-\frac{3}{2}}+\frac{3}{4}x^{-\frac{5}{2}}, -\frac{1}{32}+\frac{3}{128}$

(c) $4x^3 + 2x, 256 + 8;\ 12x^2 + 2, 192 + 2$

(d) $4x^3 - 1 - \frac{1}{x^2}, 256 - 1 - \frac{1}{16};\ 12x^2 + \frac{2}{x^3}, 192 + \frac{1}{32}$

4. $-3, 3$; the curve is symmetrical about $x = 1\frac{1}{2}$
5. $4, 0, -4;\ y = 4(x+2), y = 4, y = -4(x-2)$; the curve is parallel to the x-axis when $x = 0$
6. $\frac{dy}{dx} = \frac{3}{2}x^{\frac{1}{2}} + \frac{1}{2}x^{-\frac{1}{2}} : \frac{d^2y}{dx^2} = \frac{3}{4}x^{-\frac{1}{2}} - \frac{1}{4}x^{-\frac{3}{2}}$

$$2x[\tfrac{3}{4}x^{-\frac{1}{2}} - \tfrac{1}{4}x^{-\frac{3}{2}}] - [\tfrac{3}{2}x^{\frac{1}{2}} + \tfrac{1}{2}x^{-\frac{1}{2}}] = -\frac{1}{x^{\frac{1}{2}}};$$

so the square of the left-hand side is $\frac{1}{x}$.

Exercise 11.3a page 206

1. (a) 4 (b) 2 (c) $\frac{1}{2}$
2. 21
3. (a) 4 (b) 14 (c) 9
4. (a) 60 km/h (b) 48 km/h (c) 151·2 km/h
5. (a) 44 km/h^2 (b) 540 km/h^2
6. (a) 12π cm^2/cm (b) $2{\cdot}7\pi$ cm^2/cm (c) 4π cm^2/cm
7. (a) 52 cm^3/cm (b) 4·33 cm^3/cm (c) 2·77 cm^3/cm
8. 84π cm^3/cm; 504π cm^3/s

Exercise 11.3b page 207

1. (a) 6 (b) 28 (c) 27 (d) $\frac{8}{9}$
2. (a) 256 (b) 40 (c) $-\frac{1}{256}$ (d) $3\frac{1}{4}$
3. (a) 3 (b) -4 (c) -11 (d) 20
4. (a) 6π (b) 10π (c) $2{\cdot}4\pi$
5. (a) 16π (b) 64π (c) 400π
6. (a) 32 (b) 36 (c) 24; speed is greatest at $t = 2$ for $s = 9t^2 - 3$
7. (a) 17 (b) 6 (c) -1; acceleration is smallest at $t = 9$ for $v = 81t^{-1} + 1$
8. 73; 95
9. 290; 261 : 320

Exercise 11.4a page 208

1. (a) 70 m/s, 50 m/s, 30 m/s, 10 m/s, -10 m/s, -30 m/s, -50 m/s, -70 m/s,
   (b) 60 m/s, 40 m/s, 20 m/s, 0 m/s, -20 m/s, -40 m/s, -60 m/s
   The average speed decreases to 0 m/s and then increases negatively.
2. 60 m/s, 40 m/s, 20 m/s, 0 m/s, -20 m/s, -40 m/s, -60 m/s, -80 m/s
   The particle is moving in the opposite direction.
3. (a) $t = 4$ (b) $t = 0$ and $t = 8$
4. (a) 7 m/s, 1 m/s, 1 m/s, 7 m/s (b) 3 m/s, 0 m/s, 3 m/s, 12 m/s (c) $t = 2$
5. (a) 5 m (b) 5 m/s (c) 4 m/s, 6 m/s
6. (a) 54 m, 3 s (b) 27 m/s, 23 m/s

Exercise 11.4b page 209

1. (a) 20 m/s, 0 m/s^2 (b) -12 m/s, -48 m/s^2
2. (a) 0 m/s^2, 24 m/s^2 (b) 6 m (c) 6 m/s
3. 0 m/s^2 or 36 m/s^2; when $t = 0$ or $t = 2$ (a) positive for $t > 2$ (b) negative for $0 < t < 2$

4. $-\frac{3}{10}$ m/s^2; 6 m/s, $-\frac{1}{2}$m/s^2 (a) none (b) for all values of $t > 0$
5. (a) 0·4 m/s, 6·8 m/s, 13·2 m/s, 19·6 m/s, 26 m/s, 32·4 m/s, 38·8 m/s, 45·2 m/s
   (b) 23 m/s
   (c) The acceleration is constant and equal to 6·4 m/s^2

Exercise 11.5a page 210

1. $x = 2$, minimum; $y = -1$
2. $x = -\frac{3}{2}$; minimum
3. $-5\frac{1}{4}$
4. 4
5. $(2, -16)$, $(-2, 16)$; minimum at $x = 2$, maximum at $x = -2$
6. $(0, 5)$, $(2, 9)$; minimum at $x = 0$, maximum at $x = 2$
7. $x = 1$ or $x = 3$; $x = 3$ gives a minimum value for $f(x)$
8. $x = 1, x = 3$, or $x = 5$
   $x = 1$ gives a minimum value for $g(x)$;
   $x = 3$ gives a maximum value for $g(x)$;
   $x = 5$ gives a minimum value for $g(x)$

Exercise 11.5b page 211

1. $A = x(10 - x)$; greatest area is 25 cm^2
2. $V = x^2(72 - 2x)$; greatest volume is 13 824 cm^3
3. 4·9 metres
4. $t = 1$ or $t = 2$; 5 m, 4 m; 9 m, 0 m
5. $V = x(20 - 2x)^2$; $x = \frac{10}{3}$, 592·59 cm^3
6. $\frac{4}{27}$, 0
7. $V = \pi r(48 - r^2)$; $r = 4$; 128π cm^3, $h = 8$

Exercise 11.6a page 212

1. (a) $\frac{x^4}{4} + c$ (b) $\frac{x^5}{5} + c$ (c) $\frac{x^{11}}{11} + c$ (d) $-\frac{x^{-6}}{6} + c$ (e) $-x^{-1} + c$
2. (a) $\frac{x^6}{6} + c$ (b) $-\frac{x^{-2}}{2} + c$ (c) $\frac{3}{7}x^{\frac{7}{3}} + c$ (d) $-\frac{2}{3}x^{-\frac{3}{2}} + c$ (e) $2x^{\frac{1}{2}} + c$
3. (a) $\frac{x^8}{8} + c$ (b) $-\frac{x^{-3}}{3} + c$ (c) $\frac{3}{5}y^{\frac{5}{3}} + c$ (d) $\frac{3}{2}z^{\frac{2}{3}} + c$
4. (a) x^6 (b) $9x^8$ (c) $x^{-\frac{2}{3}}$ (d) $-\frac{3}{4}x^{-\frac{7}{4}}$
5. (a) $\frac{t^3}{3} + \frac{t^4}{4} + c$ (b) $t^5 - 10t + c$ (c) $-\frac{1}{2}t^{-2} - \frac{1}{3}t^{-3} + c$ (d) $\frac{4}{5}t^{\frac{5}{2}} + 3t^{\frac{1}{3}} + c$
6. (a) $\frac{r^6}{3} + \frac{r^9}{3} + r + c$ (b) $-r^{-3} - \frac{7}{5}r^{-5} - 5r + c$ (c) $-\frac{2}{3}r^{-\frac{3}{2}} + \frac{4}{3}r^{\frac{3}{4}} - 0{\cdot}5r + c$
7. (a) $x^5 + \frac{3}{4}x^8 + c$ (b) $t^{\frac{5}{3}} - t^{\frac{1}{2}} + c$ (c) $\frac{2}{3}x^{\frac{3}{2}} + \frac{3}{4}x^{\frac{4}{3}} + c$
8. (a) $-s^{-1} - \frac{1}{2}s^{-2} - 4s + c$ (b) $\frac{4}{5}t^{\frac{5}{4}} + \frac{5}{6}t^{\frac{6}{5}} - \frac{t^2}{2} + c$
   (c) $\frac{y^{21}}{21} + \frac{y^{-19}}{19} + \frac{20}{21}y^{\frac{21}{20}} + c$

9. $y = x^3 + 3$

10. $y = x^2 + \frac{x^4}{4}$

11. $s = t^3 - \frac{t^2}{2}$, 56 m

12. $\frac{x^6}{6} - \frac{1}{2}x^{-2} + Cx + D$

Exercise 11.6b page 213

1. (a) $\frac{7}{3}$ (b) $\frac{211}{5}$ (c) $\frac{3}{8}$ (d) 20
2. (a) $\frac{3}{4}$ (b) $\frac{3}{10}$ (c) $-\frac{3}{2}$ (d) 2
3. (a) $\frac{29}{6}$ (b) 0 (c) $\frac{23}{3}$
4. (a) $\frac{95}{2}$ (b) 8 (c) $\frac{11}{6}$
5. (a) $\frac{5}{4}$ (b) $\frac{44}{3}$ (c) $\frac{709}{120}$
6. (a) $\frac{5}{6}$ (b) $\frac{16}{15}$ (c) $-\frac{3}{4}$
7. (a) $\frac{23}{6}$ (b) $\frac{52}{3}$ (c) $\frac{9}{2}$
8. (a) $\frac{q^3}{3} - \frac{p^3}{3}$ (b) $\frac{n^4}{4} + \frac{n^2}{2}$ (c) $\frac{2r^5}{5}$

Exercise 11.7a page 214

1. (a) $\frac{16}{3}$ (b) $\frac{20}{3}$ (c) $\frac{4}{3}$
2. (a) $\frac{5}{6}$ (b) $\frac{53}{6}$ (c) $\frac{57}{2}$
3. (a) $\frac{1}{4}$ (b) 20 (c) $\frac{81}{4}$
4. (a) $\frac{9}{4}$ (b) $\frac{55}{4}$ (c) 16; $\int_1^3 (x^3 - x)dx = \int_1^2 (x^3 - x)\,dx + \int_2^3 (x^3 - x)\,dx$
5. $\frac{7}{3}, -\frac{16}{3}$. For the first, the curve is above the x-axis; for the second, the curve is below the x-axis.
6. $A = \int_2^3 (x^3 - 4x)\,dx$, $B = \int_{-2}^0 (x^3 - 4x)\,dx$, $C = \int_0^2 (x^3 - 4x)\,dx$;

   $A = \frac{25}{4}$, $B = 4$, $C = -4$;

   $\int_{-2}^2 (x^3 - 4x)\,dx = B + C = 0.$

Exercise 11.7b page 215

1. 14, $12\frac{2}{3}$; $\frac{4}{3}$
2. $\int_1^2 (x^2 - 2)\,dx = \frac{1}{3}$, $\int_1^2 (3x - 4)\,dx = \frac{1}{2}$; $\frac{1}{6}$
3. $(-1, 2), (3, 10)$; $13\frac{1}{3}, 24, 10\frac{2}{3}$
4. $A = \int_0^1 3x\,dx - \int_0^1 (x^3 - x)\,dx$, $B = \int_1^2 3x\,dx - \int_1^2 (x^3 - x)\,dx$,

   $C = \int_{-2}^0 (x^3 - x)\,dx - \int_{-2}^0 3x\,dx$;

   $A = \frac{7}{4}$, $B = \frac{9}{4}$, $C = 4$; $\int_0^2 (x^3 - x)\,dx = 2.$

Exercise 11.8a page 216

1. 40 m
2. $166\frac{2}{3}$ m, $312\frac{1}{2}$m. 225 m
3. 42 m, 42 m
4. $12\frac{2}{3}$m, $24\frac{5}{6}$m
5. (a) 28 m (b) $15\frac{1}{3}$m (c) 50 m; 7 m/s, $15\frac{1}{3}$ m/s, 10 m/s; the acceleration is greatest when $t = 5$.

Exercise 11.8b page 217

1. (a) $11\frac{1}{6}$ (b) $20\frac{1}{6}$ (c) (i) 3 (ii) -3
2. (a) $4\frac{5}{6}$ (b) $4\frac{1}{2}$ (c) (i) -3 (ii) 1 (d) $\frac{3}{4}$ (e) 13
3. (a) $48\frac{2}{3}$ (b) (i) 13 (ii) -7 (iii) 8
(c) 26, 7·48 (d) 350, 7·48 (e) $-\frac{25}{3}$
4. 65 m/s, 210 m **5.** 20 m/s, $28\frac{2}{3}$ m **6.** 7500 m
7. 7000 m **8.** 20 s, 1000 m

Exercise 11.9a page 218

1. 21, $21\frac{1}{3}$ **2.** $21\frac{1}{4}$ **3.** 200, $202\frac{2}{3}$ **4.** 202
5. 19·96875, actual area = 20·25 **6.** 7·34, actual area = $7\frac{1}{3}$

Exercise 11.9b page 219

1. 22 **2.** (a) 24 (b) 21·5 **3.** 208
4. (a) 224 (b) 204 **5.** 20·8125 **6.** 7·32
7. (a) 11 (b) 21